# PHENOTYPIC INTEGRATION

# PHENOTYPIC INTEGRATION

## Studying the Ecology and Evolution of Complex Phenotypes

---

Edited by

*Massimo Pigliucci*
*Katherine Preston*

2004

Oxford New York
Auckland Bangkok Buenos Aires Cape Town Chennai
Dar es Salaam Delhi Hong Kong Istanbul Karachi Kolkata
Kuala Lumpur Madrid Melbourne Mexico City Mumbai Nairobi
São Paulo Shanghai Taipei Tokyo Toronto



Published by Oxford University Press, Inc.,
198 Madison Avenue, New York, New York 10016

www.oup.com



Library of Congress Cataloging-in-Publication Data
Phenotypic integration: studying the ecology and evolution of complex phenotypes / [edited by] Massimo Pigliucci, Katherine Preston.
p. cm.
Includes bibliographical references and index.
ISBN 0-19-516043-6
I. Phenotype. 2. Evolutionary genetics. I. Pigliucci, Massimo, 1964–
II. Preston, Katherine (Katherine A.)
QH438.5 .P475 2003
576.5′3—dc21 2003002295

9 8 7 6 5 4 3 2 1
Printed in the United States of America
on acid-free paper

# Contents

# Contributors

*David D. Ackerly*, Department of Biological Sciences, 371 Serra Mall, Stanford University, Stanford, CA 94305-5020, USA. [dackerly@stanford.edu] http://www.stanford.edu/group/ackerly

*Rebecca Rogers Ackermann*, Department of Archaeology, Faculty of Science, University of Cape Town, Private Bag, Rondebosch 7701, South Africa. [becky@science.uct.ac.za] www.stillevolving.com

*Cerisse E. Allen*, Division of Biological Sciences, University of Montana, Missoula, MT 59812-4824, USA. [ceallen@selway.umt.edu]

*W. Scott Armbruster*, Department of Biology, Norwegian University of Science and Technology, NO-7491 Trondheim, Norway, and Institute of Arctic Biology, University of Alaska, Fairbanks, AK 99775, USA. [ffwsa@uaf.edu] http://mercury.bio.uaf.edu/~scott_armbruster/

*Alexander V. Badyaev*, Department of Ecology and Evolutionary Biology, University of Arizona, Tucson, AZ 85721-0088, USA. [abadyaev@email.arizona.edu] http://www.auburn.edu/~badyaal

*Mats Björklund*, Department of Animal Ecology, Evolutionary Biology Centre, Uppsala University, Norbyvägen 18D, SE-752 36 Uppsala, Sweden. [mats.bjorklund@ebc.uu.se] http://www.ebc.uu.se/zooeko/MatsB/matshome.shtml

*James M. Cheverud*, Department of Anatomy and Neurobiology, Washington University School of Medicine, 660 S. Euclid Ave., St. Louis, MO 63110, USA. [cheverud@pcg.wustl.edu] http://thalamus.wustl.edu/cheverudlab/

*Gunther J. Eble*, Interdisciplinary Centre for Bioinformatics, University of Leipzig, Kreuzstrasse 7b, D-04103 Leipzig, Germany and Centre National de la Recherche Scientifique, UMR 5561 Biogeosciences (Macroevolution and Biodiversity Dynamics Group), 6 boulevard Gabriel, F-21000 Dijon, France. [eble@santafe.edu, eble@bioinf.uni-leipzig.de]

*W. Anthony Frankino,* Institute of Evolutionary and Ecological Sciences, Leiden University, 2300 RA Leiden, The Netherlands. [frankino@alumni.indiana.edu]

*Russell D. Gray*, Department of Psychology, University of Auckland, Auckland 92019, New Zealand. [rd.gray@auckland.ac.nz] http://www.psych.auckland.ac.nz/Psych/research/Evolution/GrayRes.htm

*Paul E. Griffiths*, Department of History and Philosophy of Science, 1017 Cathedral of Learning, University of Pittsburgh, Pittsburgh, PA 15260, USA. [pauleg@pitt.edu] http://www.pitt.edu/~pauleg

*Thomas F. Hansen*, Department of Biological Science, Florida State University, Conradi Building, Tallahassee, FL 32306-1100, USA. [thomas.hansen@bio.fsu.edu]

*David Houle*, Department of Biological Science, Florida State University, Conradi Building, Tallahassee, FL 32306-1100, USA. [dhoule@bio.fsu.edu] http://bio.fsu.edu/~dhoule/

*Christian Peter Klingenberg*, School of Biological Sciences, 3.614 Stopford Building, University of Manchester, Oxford Road, Manchester M13 9PT, UK. [cpk@man.ac.uk]

*Paula X. Kover*, Department of Ecology and Evolutionary Biology, University of Tennessee, Knoxville, TN 37996, USA. [pkover@utk.edu]

*Juha Merilä*, Ecological Genetics Research Unit, Department of Ecology and Systematics, PO Box 65, FIN-00014 University of Helsinki, Finland. [juha.merila@helsinki.fi]

*Rosa A. Moscarella*, Department of Zoology, Michigan State University, East Lansing, MI 48824-1115, USA. [moscarel@msu.edu]

*Christa P. H. Mulder*, Department of Biology, Norwegian University of Science and Technology, N-7491 Trondheim, Norway. [ffcpm2@uaf.edu]

*Courtney J. Murren*, Department of Biology, University of Maryland, 1210 H.J. Patterson Hall, College Park, MD 20742, USA. [cmurren@wam.umd.edu]

*Christophe Pélabon*, Department of Biology, Norwegian University of Science and Technology NO-7491 Trondheim, Norway. [Christophe.Pelabon@chembio.ntnu.no]

*Massimo Pigliucci*, Departments of Botany, Ecology and Evolutionary Biology, and Philosophy, University of Tennessee, Knoxville, TN 37996-1100, USA. [pigliucci@utk.edu] www.genotype-environment.org.

*Katherine A. Preston*, Department of Biological Sciences, 371 Serra Mall, Stanford University, Stanford CA 94305-5020, USA. [kap1@stanford.edu]

*Rick A. Relyea*, Department of Biological Sciences, University of Pittsburgh, 101 Clapp Hall, Pittsburgh, PA 15260, USA. [relyea@pitt.edu] http://www.pitt.edu/~relyea/

*Derek Roff*, Department of Biology, University of California, Riverside, CA 92521, USA. [Derek.Roff@ucr.edu]

*Carl Schlichting*, Department of Ecology and Evolutionary Biology, University of Connecticut, Storrs, CT 06269-3043, USA. [schlicht@uconnvm.uconn.edu]

*Kurt Schwenk*, Department of Ecology and Evolutionary Biology, University of Connecticut, Storrs, CT 06269-3043, USA. [kurt.schwenk@uconn.edu] http://www.eeb.uconn.edu/faculty/schwenk.htm

*Scott J. Steppan*, Department of Biological Science, Florida State University, Conradi Building, Tallahassee, FL 32306-1100, USA. [steppan@bio.fsu.edu] http://bio.fsu.edu/~steppan

*Günter P. Wagner*, Department of Ecology and Evolutionary Biology, Yale University, 165 Prospect Street, POB 208106, New Haven, CT 06520-8106, USA. [Gunter.Wagner@yale.edu]

*Jason B. Wolf*, Department of Ecology and Evolutionary Biology, University of Tennessee, Knoxville, TN 37996, USA. [jbwolf@utk.edu] www.evolutionarygenetics.org

*Miriam Leah Zelditch*, Museum of Paleontology, University of Michigan, Ann Arbor, MI 48109-1079, USA. [zelditch@umich.edu]

# Introduction

The word "integration" as it refers to complex aspects of an organism's phenotype has been around in ecology and evolutionary biology for quite some time. Olson and Miller (1958) wrote a whole book developing the concept of integration just a few years after the discovery of the structure of DNA, while their botanical colleagues, Clausen and collaborators (e.g., Clausen and Hiesey 1960) had been carrying out a multiple decade-long research program aimed at studying what they referred to as "character coherence." Berg (1960) was among the first to propose specific hypotheses concerning under what ecological circumstances higher or lower degrees of integration among specific sets of characters should be favored, hypotheses that are still used today to guide empirical research (e.g., Armbruster et al. 1999).

Yet, studies of phenotypic integration have been in the background of evolutionary ecological research throughout the second half of the twentieth century, partly because of the unparalleled explosion of molecular techniques that has focused attention elsewhere, partly because long-standing conceptual and analytical problems have remained largely unaddressed for a great part of that time. Among the problems faced by researchers interested in integration were basic obstacles, such as the absence of a coherent and practical definition of "integration" itself, the rather vague conceptual framework in which to fit empirical and theoretical studies of integration, and the formidable challenges posed by multivariate statistical analyses of the relevant quantities.

Things have begun to change during the last decade, and research in this area is now rapidly maturing to the point where it has been possible to assemble this book with the aim of highlighting the major advances and challenges in the field.

Perhaps the beginning of the renewal of studies on integration came from the realization in the mid-1980s (Schlichting 1986, 1989) of its close ties with another long-neglected biological phenomenon that was beginning to experience a renaissance, phenotypic plasticity (Pigliucci 2001). Soon afterwards, the so-called "new morphometrics," an ensemble of sophisticated multivariate statistical techniques for the analysis of shape changes, came of age (Rohlf and Marcus 1993; Marcus et al. 1996), finally making it possible to quantify rigorously what was portrayed intuitively by D'Arcy Thompson (1917) at the beginning of the twentieth century through his famous "grid diagrams" of shape change.

Another important set of advances occurred in the mid-1990s, when Günter Wagner and his collaborators tackled the difficult task of providing a coherent evolutionary framework for studying integration. They first analyzed the role of modularity in the evolution of body plans, distinguishing between the phenomena of integration (increased genetic and functional relationship among traits) and parcellation (i.e., a decrease in integration by decoupling of formerly related traits to form quasi-independent modules) (Wagner 1995; Wagner and Altenberg 1996). They have more recently provided an overarching conceptual treatment of integration as it relates to adaptation and constraints over evolutionary time scales (Wagner and Schwenk 2000), and have explored in detail the very concept of "character" in evolutionary biology (Wagner 2001).

In the decades before these innovations, the field of integration studies as a whole had been held back by the combined lack of proper statistical and analytical tools, and of a clear conceptual framework. We think it is a testimonial to the increased perceived importance of understanding the evolution of complex phenotypes that so many researchers have finally begun to take on the complexities inherent in conducting empirical and theoretical research on integration. Judging from the constant flow of new papers and from the variety of contributions to this book, it seems clear to us that the field is experiencing a healthy rejuvenation, fueled in part by the enthusiasm for the new studies of the evolution of development based on modern molecular genetics, as well as by the resurgence of interest in phenotypic plasticity and organismal-level studies. Far from being a problem, the current fluid stage of thinking about phenotypic integration is a sign of most welcome vitality in a field that had remained largely in the background for almost a century, and is finally coming of age.

There are, of course, plenty of challenges. For example, when it comes to incorporating the conceptual and analytical advances that have characterized the study of phenotypic integration into the more general framework of the neo-Darwinian theory of evolution, there is still no single obvious path. Instead, we have before us numerous alternative routes, some of which will likely turn out to be dead ends. Promising attempts are surely being made (this volume includes several examples), but they are clearly only the initial forays into what promises to be a stimulating intellectual enterprise over the next decade or two. Some of the major challenges facing researchers interested in the expansion of current thinking on phenotypic integration discussed in this book include these:

- Research on the relative advantages of different analytical methods to analyze integration data sets needs to continue, if necessary to yield the development of yet newer methods. It will not be necessary to settle on one "best" approach, in the same way in which many other problems in quantitative biology can be tackled using more than one statistical tool. However, we do need a better understanding of the robustness of our approaches and of the limits they place on our ability to infer causal connections among interesting biological phenomena (Shipley 2000).

- We need new theoretical and empirical tools that allow us to operationalize the insight that natural selection and the genetic-developmental milieu in which it occurs are both necessary players locked into a continuous dialectical relationship, and cannot simply be thought of as dichotomous alternatives ("constraints" versus selection).
- The metaphor of evolution on "adaptive landscapes," which has recently been fundamentally criticized in its classic form for the misleading imagery it evokes when one considers unrealistically low-dimensional landscapes (Gavrilets 1997; Gavrilets et al. 1998), needs to be abandoned or fundamentally revised to be useful when it comes to the high-dimensionality of research problems in phenotypic integration.
- Research on integration (and, we would argue, more generally in evolutionary ecology) will benefit greatly from a more nuanced philosophy of science, one that emphasizes competition among different complex models as opposed to more standard, often simplistic, null hypotheses. Similarly, we need to explore methods of investigation that aim at converging on the best inference from different, partial, angles of attack to a particular problem (consilience of evidence), instead of continuing with the practice of standard falsificationism (Popper 1968) or "strong inference" (Platt 1964) more typical of the "hard" sciences.
- Finally, we need to assess how well the view of phenotypic evolution that emerges from integration studies fits with the standard neo-Darwinian paradigm, or perhaps whether it prompts us to an expansion of such paradigm (Gould 2002) to include new insights from the field of the evolution of development (Oyama et al. 2001).

This is a tall order indeed, and it will require the joint efforts of researchers interested in developmental biology, molecular genetics, biostatistics, evolutionary ecology, and even philosophy of science. The ultimate prize will be a truly satisfactory, and repeatedly invoked (Schlichting and Pigliucci 1998), integration of evolutionary and developmental biology.

This book aims at providing a reference point for researchers interested in the above-mentioned goals of organismal biological research. For the reader's convenience, it is divided into five parts, preceded by a foreword (Chapter 1) by Carl Schlichting, someone who has contributed fundamentally to the contemporary conceptual framework supporting the study of integration. The first part of the book comprises five chapters on various aspects of adaptations and constraints; the second part includes two related chapters on phenotypic plasticity and integration; we then proceed with three contributions discussing various aspects of the molecular and genetic study of phenotypic integration, three chapters on the relationship between macroevolution and integration follow, and we conclude with five chapters ranging from issues of data analysis to the broader conceptual framework within which integration needs to be studied and how it relates to modern evolutionary theory.

Our hope is to have provided significant food for thought to our colleagues and students, some of whom may be inspired to lend their talents to the study of phenotypic evolution in general and to the many challenges posed by phenotypic integration in particular. We could not have achieved all of this without the help of many people, starting of course with the contributors to the book, among the brightest researchers in the field. We would also like to thank our editor, Kirk Jensen, who has been enthusiastic about this idea since its inception, as well as all those colleagues and students at our respective universities who have generously devoted their time to read or discuss with us issues concerning the study of integration. It is our sincere wish that all this intellectual energy will catalyze further efforts by many more people, and that they will enjoy the challenge at least as much as we did.

### *Literature Cited*


Armbruster, W. S., V. S. D. Stilio, J. D. Tuxill, T. C. Flores, and J. L. V. Runk. 1999. Covariance and decoupling of floral and vegetative traits in nine Neotropical plants: a re-evaluation of Berg's correlation-pleiades concept. American Journal of Botany 86:39–55.

Berg, R. L. 1960. The ecological significance of correlation pleiades. Evolution 14:171–180.

Clausen, J., and W. M. Hiesey. 1960. The balance between coherence and variation in evolution. Proceedings of the National Academy of Sciences USA 46:494–506.

Gavrilets, S. 1997. Evolution and speciation on holey adaptive landscapes. Trends in Ecology and Evolution 12:307–312.

Gavrilets, S., H. Li, and M. D. Vose. 1998. Rapid parapatric speciation on holey adaptive landscapes. Proceedings of the Royal Society of London B 265:1483–1489.

Gould, S. J. 2002. The Structure of Evolutionary Theory. Harvard University Press, Cambridge, MA.

Marcus, L. F., M. Corti, A. Loy, G. J. P. Naylor, and D. E. Slice. 1996. Advances in Morphometrics. Plenum Press, New York.

Olson, E. C., and R. L. Miller. 1958. Morphological Integration. University of Chicago Press, Chicago.

Oyama, S., P. E. Griffiths, and R. D. Gray. 2001. Cycles of Contingency: Developmental Systems Theory and Evolution. MIT Press, Cambridge, MA.

Pigliucci, M. 2001. Phenotypic Plasticity: Beyond Nature and Nurture. Johns Hopkins University Press, Baltimore, MD.

Platt, J. R. 1964. Strong inference. Science 146:347–353.

Popper, K. R. 1968. Conjectures and refutations: the growth of scientific knowledge. Harper & Row, New York.

Rohlf, F. J., and L. F. Marcus. 1993. A revolution in morphometrics. Trends in Ecology and Evolution 8:129–132.

Schlichting, C. D. 1986. The evolution of phenotypic plasticity in plants. Annual Review of Ecology and Systematics 17:667–693.

Schlichting, C. D. 1989. Phenotypic integration and environmental change. BioScience 39:460–464.

Schlichting, C. D., and M. Pigliucci. 1998. Phenotypic Evolution: A Reaction Norm Perspective. Sinauer, Sunderland, MA.

Shipley, B. 2000. Cause and Correlation in Biology: A User's Guide to Path Analysis, Structural Equations and Causal Inference. Cambridge University Press, Cambridge.

Thompson, D. 1917. On Growth and Form. Cambridge University Press, Cambridge.

Wagner, G. 2001. The Character Concept in Evolutionary Biology. Academic Press, San Diego.

Wagner, G. P. 1995. Adaptation and the modular design of organisms. Pp. 317–328 in F. Moran, A. Moreno, J. J. Merelo, and P. Chacon, eds. Advances in Artificial Life. Springer, Berlin.

Wagner, G. P., and L. Altenberg. 1996. Complex adaptations and the evolution of evolvability. Evolution 50:967–976.

Wagner, G. P., and K. Schwenk. 2000. Evolutionarily stable configurations: functional integration and the evolution of phenotypic stability. Evolutionary Biology 31:155–217.

# PHENOTYPIC INTEGRATION

# 1

# The Diversity of Complexity

CARL SCHLICHTING

Life is infinitely stranger than anything which the mind of man could invent.
*Sherlock Holmes, A Case of Identity (1891)*

Holmes was of course not referring to biological complexity, but this phrase distills the sense of amazement shared when we are confronted with yet another of life's labyrinthine contrivances to increase the likelihood of survival and reproduction. However, as strange and complicated as life is, it also exhibits order and regularity, and it is the job of evolutionary biologists to discover common underlying patterns and discern how they can be modified to generate diversity. The chapters that comprise this book confront some of the intricacies and dilemmas of investigating the complexity and integration of organisms in the modern world.

The standing diversity of living things is wonderfully varied, from single cells to colonies to massive multicellular beings. Even the simplest of these possesses a fairly elaborate epigenetic system for transforming genotypic instructions into phenotypic form, and ultimately, one of the goals in the study of complexity is to characterize this transformation. There are other questions of interest as well: Are there patterns of increasing complexity during evolutionary time? (McShea 1991, 1996; Valentine et al. 1994; Adami et al. 2000; Alroy 2000; Turney 2000); How are increases in numbers of genes related to the evolution of complexity? (Bird 1995; Holland 1999; Edelman and Gally 2001; Zuckerkandl 2001; Betrán and Long 2002; Gregory 2002; Lespinet et al. 2002); Is increased complexity linked with evolutionary radiation? (Lauder 1981; Schaefer and Lauder 1996; Friel and Wainwright 1999; Sarà 1999); Is there a relationship between complexity and constraint? (Bonner 1988; Wagner 1988; Arnold 1989; Wagner and Altenberg 1996; Duboule and Wilkins 1998; Waxman and Peck 1998).

A panoply of approaches have been tried—disparate even from the beginning, with D'Arcy Thomson's intuitive grids (Thompson 1917) versus Julian Huxley's rigorous mathematics (Huxley 1932). The botanical study of morphological integration dates to at least 1939 in a paper by Davey and Lang (1939). These authors grew 102 plants each of three populations of *Plantago* (two of *P. maritima*, one of *P. serpentina*), and measured 20 traits on each plant. They calculated all pairwise correlations, as well as up to three partial correlations per character (imagine doing all those by hand!). As we might expect, they found two similar populations and one distinctive. However, it was the coastal population of *P. maritima* that stood out. At odds to explain this discrepancy, they suggested that this population may have had lower soil fertility—a harbinger of the focus of later botanical researchers on environmental shifts in correlations. The evolutionary significance of suites of correlated characters at higher taxonomic levels was also debated by botanists (e.g., Sporne 1948, 1954; Corner 1949; Stebbins Jr. 1951), and is embodied in the notion of pollination syndromes.

Major interest in integration per se at the population/species level dates to a triumvirate of groups in the 1950s.[1] University of Chicago zoologists Everett Olson and Robert Miller (1951) advocated the use of correlation analysis to examine the extent to which strongly correlated traits were also functionally related (greatly elaborated in Olson and Miller 1958; see Chernoff and Magwene 1999 for a review). Botanists Jens Clausen and William Hiesey at the Carnegie Institution in Stanford were also avid practitioners of correlation analysis. They investigated the architecture of correlations in intraspecific hybrids to examine the "coherence" of the character complexes typical of ecotypic races (Clausen and Hiesey 1958, 1960; see also Grant 1979). The third investigator was Raisa Berg, a botanist at St. Peter's University in Leningrad. Berg's thesis was two-pronged: correlations should be higher within functional modules than between such modules (e.g., reproductive versus vegetative characters of plants), and selection for integration itself should produce stronger correlations. Both were borne out by her data—the second by observations that correlations among different floral traits were much stronger for animal-pollinated than wind-pollinated species (see review by Murren 2002).

Modern modes of attack range from mostly methodological to largely conceptual, and include[2] (1) geometric morphometrics (landmark and allometric analyses); (2) quantitative genetics; (3) modularity analyses; (4) developmental reaction norms; (5) historical and morphospace investigations; and (6) computational approaches. These approaches are not all entirely distinct from one another; in fact some authors explicitly combine methods from one with concepts from others.

## Geometric Morphometrics

The mathematical description of *allometry* (differences in rates of growth of two features) was elaborated by Huxley (1932; see chap. 4 in Schlichting and Pigliucci 1998 and Gayon 2000). Allometric analyses ranging from simple correlation studies to complicated multiple regressions are still quite common to describe the relationships among traits of organisms in an evolutionary context (e.g., Thompson 1992; Klingenberg 1998; Danforth and Desjardins 1999; Gilchrist et al. 2000; Kidson and Westoby 2000; Collard and O'Higgins 2001; Moczek et al. 2002; Niklas and Enquist 2002), and allometries are also being investigated from molecular and physiological perspectives (e.g., Nijhout and Wheeler 1996; Moen et al. 1999). I shall deal only briefly with areas explicitly covered in this volume. Readers interested in greater depth on some topics are referred to more extensive reviews in Schlichting and Pigliucci (1998) and Chernoff and Magwene (1999).

*Landmark analysis*, conceived by Bookstein (e.g., 1991), is the intellectual descendant of the transformation grids of D'Arcy Thompson. This involves identifying homologous landmarks on the body or skeleton of an organism, and statistically disentangling change in size from that in shape (see chap. 7 in Schlichting and Pigliucci 1998; sections II and III in Chernoff and Magwene 1999). Changes in location of those points can be mapped, and integrated sets of features identified, either across a phylogeny (e.g., Corti and Rohlf 2001; Fadda and Corti 2001; Rüber and Adams 2001), or during ontogeny (Loy et al. 1998; Monteiro et al. 1999; Collard and O'Higgins 2001). In this volume, several contributions deal with allometry or landmark analysis: Preston and Ackerly (Chapter 4), Klingenberg (Chapter 10), and Zelditch and Moscarella (Chapter 13).

## Quantitative Genetics

Despite its genesis as a key component of evolutionary theory, quantitative genetics (QG) was used for a long period mostly as a tool by plant and animal breeders, that is, until Lande and Arnold (Lande 1979, 1982; Lande and Arnold 1983) reminded evolutionary biologists of its utility. QG, via analysis of variance, separates components of phenotypic variation into statistical bins: over there "environmental," over here "additive genetic." Statistical representations of phenotypic and genetic architectures can be produced. QG remains the predominant paradigm for describing/dissecting phenotypic complexity (see Schlichting and Pigliucci 1998, chaps. 1, 7, and 9; Chernoff and Magwene 1999, sections IV and VI). In this volume, numerous chapters employ QG (Ackermann and Cheverud; Armbruster et al.; Badyaev; Hansen and Houle; Klingenberg; Merilä and Björklund; Pigliucci; Relyea; Roff; Steppan; Wolf). Quantitative trait locus (QTL) studies also build on the QG framework (Murren and Kover, Chapter 9, this volume).

## Modularity Analyses

The concept of modularity has by now achieved a central role in studies of complexity and integration (Bonner 1988; Raff 1996; von Dassow and Munro 1999; Schlosser 2002; but see Csete and Doyle 2002). Although many studies of modularity and integration can be traced to the stimuli provided by Olson and

Miller (1958; see also Bonner 1958) and Berg (1960), some modern investigations spring from complexity theory, and others from the concept of the Genotype-to-Phenotype (G→P) map (Raff 1996; Wagner and Altenberg 1996; Mezey et al. 2000; Stadler et al. 2001). Various authors differ in their meaning of the term "modularity" and in the hierarchical level examined (see Raff and Raff 2000; Winther 2001). Although many studies are centered on organismal or morphological modularity (in this volume: Preston and Ackerly; Armbruster et al.; Badyaev; Eble; Hansen and Houle; Klingenberg; Schwenk and Wagner; Wolf), other workers have examined lower levels.

Ancel and Fontana (2000) examined modularity of RNA molecules, and, via simulated evolution, found that phenotypic stability was favored and was associated with modularity of the RNA molecule. As temperatures increased, evolutionarily more successful forms had more regions that could unfold without simultaneous change in the rest of the structure, and the phenotypic state of these modular sequences of nucleotides tended to be maintained even when the sequence was inserted between two random sequences (for a similar study with proteins, see Bornberg-Bauer 2002).

Modularity of sets of genes, proteins, or cells has also received attention (Hartwell et al. 1999; Thieffry and Romero 1999). An increase of studies at the molecular level is being fueled by the explosion of gene expression data (especially from microarrays). These studies investigate modularity (or networks) of genes and proteins, expression pathways, and regulatory control (e.g., von Dassow et al. 2000; Yi et al. 2000; Holter et al. 2001; Keplinger et al. 2001; Abouheif and Wray 2002; Davidson et al. 2002; Guet et al. 2002; Klebes et al. 2002; Rogozin et al. 2002; Ronen et al. 2002; Shen-Orr et al. 2002; Trewavas 2002; Wilkins 2002). As an example, Miki et al. (2001) used microarrays to analyze expression patterns for over 18,000 cDNAs in various embryonic and adult mouse tissues. Tissues with common embryological origins or related functions showed quite similar expression patterns. Enzymes involved in various metabolic activities also showed extensive coregulation during development and in the adult organism. Their detailed analysis of the glycolytic pathway showed that tissue-specific genes tended to cluster together (e.g., muscle, testis, kidney/liver genes), and genes that were closer together in the pathway had higher similarity of expression.

Gene duplication remains a fundamental topic related to modularity and complexity because of its ability to create de novo the raw material, either for new modules or for their regulation (Averof et al. 1996; Carroll 2001; Dermitzakis and Clark 2001; Wagner 2001; Wang and Gu 2001; Cohen et al. 2002; Hogeweg 2002; Mellgren and Johnson 2002; Skaer et al. 2002; Wilkins 2002).

## Developmental Reaction Norms

The phrase *developmental reaction norms* was introduced by Pigliucci et al. (1996) to denote a perspective on the evolution of the phenotype that gives explicit consideration to ontogeny, correlations among traits, and environmental influences as modifiers of selection (for a full explication see Schlichting and Pigliucci

1998). The insights of I. I. Schmalhausen (1949) had a fundamental influence on our thinking, but the aggregation drew from the contributions of dozens of workers.[3] Others have advocated similar approaches (Scharloo 1987, 1989; McKinney and Gittleman 1995; Gedroc et al. 1996; Sarà 1996a, 1996b). In this volume, Relyea (Chapter 8), Pigliucci (Chapter 7), and Wolf (Chapter 17) employ this perspective to different extents.

That selection may operate at various stages during ontogeny has of course been recognized by many (e.g., Buchholz 1922; Clegg and Allard 1973; Atchley and Rutledge 1980; Cane 1993; Moolman et al. 1996; Rhees and Atchley 2000; Apiolaza and Garrick 2001). Various researchers have carried out ontogenetically explicit studies of selection or response to environmental variation (Bernardo 1994; Carrier 1996; Pigliucci et al. 1997; Cam and Monnat 2000; Tammaru et al. 2000; Ostrowski et al. 2002). For example, Hjelm et al. (2001) employed a reaction norm approach to examine the ontogeny of body shape changes in perch. They found that body shape changed naturally during growth, but this progression was diet-dependent. In pelagic enclosures with mainly zooplankton as food, perch with high growth rates had a more fusiform body morphology than slow-growing perch; the opposite pattern is found for littoral zone perch whose diets contained more macroinvertebrates (see also Svanback and Eklov 2002).

Another focus has been on the environment-dependent nature of trait correlations. Clausen and Hiesey (1958) were among the first to observe this (trait correlations varied for clones grown at different altitudes), and to discuss its potential significance (see also Núñez-Farfán and Schlichting 2001). Subsequent studies have often found substantial changes in correlation matrices measured in different environmental conditions (Adams 1967; Antonovics 1976; Huhn and Schuster 1982; Schlichting 1986, 1989b; Mazer and Schick 1991; Biere 1995; Donohue et al. 2000; Etterson and Shaw 2001).

Another view of this issue comes from examination of the correlation of plastic responses across macroenvironments—plasticity integration (Schlichting 1986, 1989a). Schlichting (1989a) proposed that selection would favor developmental systems that reduced plasticity integration, to allow adjustments of the relationships among traits in response to environmental circumstances: regularly encountered conditions could be met with specific, coordinated responses. This view is reinforced by observations that particular patterns of trait covariation appear to be advantageous in certain environments (Lechowicz and Blais 1988; Geber and Dawson 1997; Sandquist and Ehleringer 1998) and that such patterns can be achieved through plasticity (Blais and Lechowicz 1989; Dorn et al. 2000; Merilä et al. 2000). As discussed above, one way to achieve such differential control of adaptive plastic responses is by specialization of different members of gene families (e.g., the phytochromes: Smith 1990; Schlichting and Smith 2002).

Recently, there have been more overt considerations of development itself as a reaction norm, with the view that development proceeds due to a series of changes in the local environment of cells that in turn give rise to changes in gene expression and morphology (Hogeweg 2000a; Newman and Müller 2001; Schlichting and Smith 2002; Schlichting 2003).

## Historical and Morphospace Analyses

Most of these studies use paleontological data to examine trends in the evolution of groups of organisms. One branch of this category deals with whether disparity ("a measure of the range or significance of morphology in a given sample of organisms": Wills et al. 1994) has increased or decreased in evolutionary time. Contributions range from speculative to statistical, many centered around the issue of whether peak disparity occurred in the Cambrian (Gould 1991; Briggs et al. 1992; Foote 1997; Smith and Lieberman 1999; Ciampaglio 2002). Others are interested in whether there are associations between patterns of disparity and evolutionary events or processes (Jernvall et al. 1996; Gatesy and Middleton 1997; Dommergues et al. 2000; Eble 2000; Magniez-Jannin et al. 2000; Middleton and Gatesy 2000).

Hulsey and Wainwright (2002) examine disparity/diversity issues by focusing on the biomechanical feeding apparatus of modern labrid fish (parrotfish and wrasses), with intriguing results. Members of this group employ a remarkable diversity of feeding modes, ranging from mollusc- and coral-crushing to plankton and ecto-parasite feeders, despite sharing the same basic functional feeding unit: an anterior jaw four-bar linkage. There was no relationship of the disparity of morphology with that of kinematics in a comparison of ten genera: morphologically similar[4] species could differ by an order of magnitude in their kinematic properties.

The exploration of morphospace, pioneered by Raup (1961, 1966; Raup and Michelson 1965), has provided interesting perspectives on which morphologies have been tried and which may or may not be possible (see the review by McGhee Jr. 1999). Recent works have forged several new directions: examination of ontogeny (Stone 1998) and developmental mechanisms (Boyce and Knoll 2002); functional morphology (de Visser and Barel 1998; Hulsey and Wainwright 2002); theoretical examination of plant evolution in different environments (Niklas 1999); modeling of changes in disparity (Gavrilets 1999a); and combined phylogenetic-morphospace analyses (Wagner 1997; Hertel and Lehman 1998; Zweers and Berge 1998; O'Keefe 2002).

## Computational Approaches

Complexity theory arrived with a bang with the promise of *NK* models (*N* genes with *K* connections), self-organization, and allusions to "life at the edge of chaos," but to my view has not generated much in the mainstream of evolutionary work. However, there are some recent reviews and applications: network evolution (Dorogovtsev and Mendes 2002), self-organization (Richardson 2001; Roces 2002; Seeley 2002), and nonlinear dynamics (Goodwin 2000; Ranta et al. 2000; Aviles et al. 2002; Fitch et al. 2002; Hutt and Luttge 2002). Some studies of modularity have found that the number of interactions among genes appears to follow a power law distribution (Wagner 2001; Ravasz et al. 2002), a predicted property of self-organizing systems.

There are a variety of simulation, algebraic, and artificial life explorations of the evolution of complex systems. These investigate issues concerning gene duplication and modularity (Calabretta et al. 2000; Rotaru-Varga and Komosinski 2001), differentiation/morphogenesis (Nijhout et al. 1986; Ofria et al. 1999; Furusawa and Kaneko 2000; Hogeweg 2000a, 2000b; Salazar-Ciudad et al. 2000), and limits to complexity (Nehaniv and Rhodes 2000; Lu and Li 2001).

Conceptually similar are the empirical studies or *in vitro* approaches to evolution of proteins (Zhang and Rosenberg 2002) or RNAs (Hanczyc and Dorit 1998; Lehman et al. 2000). The study of RNA evolution holds special interest for its nearly direct Genotype→Phenotype map. An interesting outcome of these investigations is that the topography of the phenotypic landscapes differs strikingly depending on whether the space is organized by the similarities of the morphologies themselves (the way we normally do it), or according to genetic accessibility. When genotypes are connected through a network of single base substitutions (like a gene tree), genetic nearest neighbors may occasionally have markedly distinctive phenotypes, and these same phenotypes may arise in very different parts of the network (e.g., Huynen et al. 1996; Fontana and Schuster 1998a, 1998b; Cupal et al. 2000). As Stadler et al. (2001) point out, "Simulated populations of replicating and mutating RNA sequences under selection exhibit many phenomena known from organismal evolution: neutral drift, punctuated change, plasticity, environmental and genetic canalization, and the emergence of modularity." What remains to be seen is whether general principles derived from studies of RNA systems will also obtain for structures and functions organisms with more complex G→P maps. For example, is genetic accessibility facilitated or inhibited by the genetic architecture of pleiotropy, redundancy, and modularity (Gavrilets 1997, 1999b; Schlichting and Pigliucci 1998, pp. 319–321)?

In sum, there is a rich ferment of investigation of aspects of biological complexity. The syntheses and advances described in this volume will add to that mix as we move toward a more complete understanding of the mechanisms underlying the production of both biological form and diversity.

## Notes

1 See Schlichting and Pigliucci, (1988, chap. 7) for a detailed discussion of these pioneers.
2 Although striving for comprehensiveness, I've probably overlooked whole areas of inquiry. And, since I did not attempt completeness, I'm sure to have missed important citations. *Mea malus!* Cited works are mostly recent, allowing the reader to traverse the literature vicariously. Older citations are classic works, distinctive contributions, or may have been overlooked.
3 Especially S. J. Gould, C. H. Waddington, J. Clausen, R. Lewontin, J. Cheverud, G. Wagner, and R. Lande.
4 Occupying a cube of three-dimensional morphospace constructed from the values of the lengths of the maxilla, lower jaw, and nasal bones.

## Literature Cited

Abouheif, E., and G. A. Wray. 2002. Evolution of the gene network underlying wing polyphenism in ants. Science 297:249–252.

Adami, C., C. Ofria, and T. C. Collier. 2000. Evolution of biological complexity. Proceedings of the National Academy of Sciences USA 97:4463–4468.

Adams, M. W. 1967. Basis of yield component compensation in crop plants with special reference to the field bean, *Phaseolus vulgaris*. Crop Science 7:505–510.

Alroy, J. 2000. Understanding the dynamics of trends within evolving lineages. Paleobiology 26:319–329.

Ancel, L. W., and W. Fontana. 2000. Plasticity, evolvability, and modularity in RNA. Journal of Experimental Zoology (Molecular and Developmental Evolution) 288:242–283.

Antonovics, J. 1976. The nature of limits to natural selection. Annals of the Missouri Botanical Garden 63:224–247.

Apiolaza, L. A., and D. J. Garrick. 2001. Analysis of longitudinal data from progeny tests: some multivariate approaches. Forest Science 47:129–140.

Arnold, S. J. 1989. Group report: How do complex organisms evolve? Pp. 403–433 in D. B. Wake and G. Roth, eds. Complex Organismal Functions: Integration and Evolution in Vertebrates. John Wiley and Sons, Chichester, UK.

Atchley, W. R., and J. J. Rutledge. 1980. Genetic components of size and shape. I. Dynamics of components of phenotypic variability and covariability during ontogeny in the laboratory rat. Evolution 34:1161–1173.

Averof, M., R. Dawes, and D. Ferrier. 1996. Diversification of arthropod Hox genes as a paradigm for the evolution of gene functions. Seminars in Cell and Developmental Biology 7:539–551.

Aviles, L., P. Abbot, and A. D. Cutter. 2002. Population ecology, nonlinear dynamics, and social evolution. I. Associations among nonrelatives. American Naturalist 159:115–127.

Berg, R. L. 1960. The ecological significance of correlation pleiades. Evolution 14:171–180.

Bernardo, J. 1994. Experimental analysis of allocation in two divergent, natural salamander populations. American Naturalist 143:14–38.

Betrán, E., and M. Long. 2002. Expansion of genome coding regions by acquisition of new genes. Genetica 115:65–80.

Biere, A. 1995. Genotypic and plastic variation in plant size: effects on fecundity and allocation patterns in *Lychnis flos-cuculi* along a gradient of natural soil fertility. Journal of Ecology 83:629–642.

Bird, A. P. 1995. Gene number, noise reduction and biological complexity. Trends in Genetics 11:94–100.

Blais, P. A., and M. J. Lechowicz. 1989. Variation among populations of *Xanthium strumarium* (Compositae) from natural and ruderal habitats. American Journal of Botany 76:901–908.

Bonner, J. T. 1958. The Evolution of Development. Cambridge University Press, Cambridge.

Bonner, J. T. 1988. The Evolution of Complexity by Means of Natural Selection. Princeton University Press, Princeton, NJ.

Bookstein, F. L. 1991. Morphometric Tools for Landmark Data: Geometry and Biology. Cambridge University Press, Cambridge.

Bornberg-Bauer, E. 2002. Randomness, structural uniqueness, modularity and neutral evolution in sequence space of model proteins. International Journal of Research in Physical Chemistry and Chemical Physics 216:139–154.

Boyce, C. K., and A. H. Knoll. 2002. Evolution of developmental potential and the multiple independent origins of leaves in Paleozoic vascular plants. Paleobiology 28:70–100.

Briggs, D. E. G., R. A. Fortey, and M. A. Wills. 1992. Morphological disparity in the Cambrian. Science 256:1670–1673.

Buchholz, J. T. 1922. Developmental selection in vascular plants. Botanical Gazette 73:249–286.

Calabretta, R., S. Nolfi, D. Parisi, and G. P. Wagner. 2000. Duplication of modules facilitates the evolution of functional specialization. Artificial Life 6:69–84.

Cam, E., and J. Y. Monnat. 2000. Stratification based on reproductive state reveals contrasting patterns of age-related variation in demographic parameters in the kittiwake. Oikos 90:560–574.

Cane, W. P. 1993. The ontogeny of postcranial integration in the common tern, *Sterna hirundo*. Evolution 47:1138–1151.

Carrier, D. R. 1996. Ontogenetic limits on locomotor performance. Physiological Zoology 69:467–488.

Carroll, S. B. 2001. Chance and necessity: the evolution of morphological complexity and diversity. Nature 409:1102–1109.

Chernoff, B., and P. M. Magwene. 1999. Afterword—Morphological Integration: Forty years later. Pp. 319–353 in E. C. Olson and R. L. Miller. Morphological Integration. University of Chicago Press, Chicago.

Ciampaglio, C. N. 2002. Determining the role that ecological and developmental constraints play in controlling disparity: examples from the crinoid and blastozoan fossil record. Evolution and Development 4:170–188.

Clausen, J., and W. M. Hiesey. 1958. Experimental studies on the nature of species. IV. Genetic structure of ecological races. Publication 615, Carnegie Institution of Washington, Washington, DC.

Clausen, J., and W. M. Hiesey. 1960. The balance between coherence and variation in evolution. Proceedings of the National Academy of Sciences USA 46:494–506.

Clegg, M. T., and R. W. Allard. 1973. Viability versus fecundity selection in the slender wild oat, *Avena barbata* L. Science 181:667–668.

Cohen, B. A., Y. Pilpel, R. D. Mitra, and G. M. Church. 2002. Discrimination between paralogs using microarray analysis: application to the *Yap1p* and *Yap2p* transcriptional networks. Molecular Biology of the Cell 13:1608–1614.

Collard, M., and P. O. O'Higgins. 2001. Ontogeny and homoplasy in the papionin monkey face. Evolution and Development 3:322–331.

Corner, E. J. H. 1949. The Durian theory or the origin of the modern tree. Annals of Botany 13:367–414.

Corti, M., and F. J. Rohlf. 2001. Chromosomal speciation and phenotypic evolution in the house mouse. Biological Journal of the Linnean Society 73:99–112.

Csete, M. E., and J. C. Doyle. 2002. Reverse engineering of biological complexity. Science 295:1664–1669.

Cupal, J., S. Kopp, and P. F. Stadler. 2000. RNA shape space topology. Artificial Life 6:3–23.

Danforth, B. N., and C. A. Desjardins. 1999. Male dimorphism in *Perdita portalis* (Hymenoptera, Andrenidae) has arisen from preexisting allometric patterns. Insectes Sociaux 46:18–28.

Davey, V. M., and J. M. S. Lang. 1939. Experimental taxonomy. III. Correlation of characters within a population. New Phytologist 38:32–61.

Davidson, E. H., J. P. Rast, P. Oliveri, A. Ransick, C. Calestani, C.-H. Yuh, T. Minokawa, G. Amore, V. Hinman, C. Arenas-Mena, O. Otim, C. T. Brown, C. B. Livi, P. Y. Lee, R. Revilla, A. G. Rust, Z. J. Pan, M. J. Schilstra, P. J. C. Clarke, M. I. Arnone, L. Rowen, R. A. Cameron, D. R. McClay, L. Hood, and H. Bolouri. 2002. A genomic regulatory network for development. Science 295:1669–1678.

Dermitzakis, E. T., and A. G. Clark. 2001. Differential selection after duplication in mammalian developmental genes. Molecular Biology and Evolution 18:557–562.

de Visser, J., and C. D. N. Barel. 1998. The expansion apparatus in fish heads, a 3-D kinetic deduction. Netherlands Journal of Zoology 48:361–395.

Dommergues, J. L., P. Neige, and S. Von Boletzky. 2000. Exploration of morphospace using Procrustes analysis in statoliths of cuttlefish and squid (Cephalopoda: Decabrachia)—Evolutionary aspects of form disparity. Veliger 43:265–276.

Donohue, K., E. H. Pyle, D. Messiqua, M. S. Heschel, and J. Schmitt. 2000. Density dependence and population differentiation of genetic architecture in *Impatiens capensis* in natural environments. Evolution 54:1969–1981.

Dorn, L. A., E. H. Pyle, and J. Schmitt. 2000. Plasticity to light cues and resources in *Arabidopsis thaliana*: testing for adaptive value and costs. Evolution 54:1982–1994.

Dorogovtsev, S. N., and J. F. F. Mendes. 2002. Evolution of networks. Advances in Physics 51:1079–1187.

Duboule, D., and A. S. Wilkins. 1998. The evolution of "bricolage." Trends in Genetics 14:54–59.

Eble, G. J. 2000. Contrasting evolutionary flexibility in sister groups: disparity and diversity in Mesozoic atelostomate echinoids. Paleobiology 26:56–79.

Edelman, G. M., and J. A. Gally. 2001. Degeneracy and complexity in biological systems. Proceedings of the National Academy of Sciences USA 98:13763–13768.

Etterson, J. R., and R. G. Shaw. 2001. Constraint to adaptive evolution in response to global warming. Science 295:151–154.

Fadda, C., and M. Corti. 2001. Three-dimensional geometric morphometrics of *Arvicanthis*: implications for systematics and taxonomy. Journal of Zoological Systematics and Evolutionary Research 39:235–245.

Fitch, W. T., J. Neubauer, and H. Herzel. 2002. Calls out of chaos: the adaptive significance of nonlinear phenomena in mammalian vocal production. Animal Behaviour 63:407–418.

Fontana, W., and P. Schuster. 1998a. Continuity in evolution: on the nature of transitions. Science 280:1451–1455.

Fontana, W., and P. Schuster. 1998b. Shaping space: the possible and the attainable in RNA genotype-phenotype mapping. Journal of Theoretical Biology 194:491–516.

Foote, M. 1997. The evolution of morphological diversity. Annual Review of Ecology and Systematics 28:129–152.

Friel, J. P., and P. C. Wainwright. 1999. Evolution of complexity in motor patterns and jaw musculature of tetraodontiform fishes. Journal of Experimental Biology 202:867–880.

Furusawa, C., and K. Kaneko. 2000. Complex organization in multicellularity as a necessity in evolution. Artificial Life 6:265–281.

Gatesy, S. M., and K. M. Middleton. 1997. Bipedalism, flight, and the evolution of theropod locomotor diversity. Journal of Vertebrate Paleontology 17:308–329.

Gavrilets, S. 1997. Evolution and speciation on holey adaptive landscapes. Trends in Ecology and Evolution 12:307–312.

Gavrilets, S. 1999a. Dynamics of clade diversification on the morphological hypercube. Proceedings of the Royal Society of London Series B 266:817–824.

Gavrilets, S. 1999b. A dynamical theory of speciation on holey adaptive landscapes. American Naturalist 154:1–22.

Gayon, J. 2000. History of the concept of allometry. American Zoologist 40:748–758.

Geber, M. A., and T. E. Dawson. 1997. Genetic variation in stomatal and biochemical limitations to photosynthesis in the annual plant, *Polygonum arenastrum*. Oecologia 109:535–546.

Gedroc, J. J., K. D. M. McConnaughay, and J. S. Coleman. 1996. Plasticity in root/shoot partitioning: optimal, ontogenetic, or both? Functional Ecology 10:44–50.

Gilchrist, A. S., R. B. R. Azevedo, L. Partridge, and P. O'Higgins. 2000. Adaptation and constraint in the evolution of *Drosophila melanogaster* wing shape. Evolution and Development 2:114–124.

Goodwin, B. C. 2000. The life of form. Emergent patterns of morphological transformation. Comptes Rendus de l'Académie des Sciences Série Iii: Sciences de la Vie—Life Sciences 323:15–21.

Gould, S. J. 1991. The disparity of the Burgess Shale arthropod fauna and the limits of cladistic analysis—why we must strive to quantify morphospace. Paleobiology 17:411–423.

Grant, V. 1979. Character coherence in natural hybrid populations in plants. Botanical Gazette 140:443–448.

Gregory, T. R. 2002. Genome size and developmental complexity. Genetica 115:131–146.

Guet, C. C., M. B. Elowitz, W. Hsing, and S. Leibler. 2002. Combinatorial synthesis of genetic networks. Science 296:1466–1470.

Hanczyc, M. M., and R. L. Dorit. 1998. Experimental evolution of complexity: *in vitro* emergence of intermolecular ribozyme interactions. RNA 4:268–275.

Hartwell, L. H., J. J. Hopfield, S. Leibler, and A. W. Murray. 1999. From molecular to modular cell biology. Nature 402:C47–C52.

Hertel, F., and N. Lehman. 1998. A randomized nearest-neighbor approach for assessment of character displacement: the vulture guild as a model. Journal of Theoretical Biology 190:51–61.

Hjelm, J., R. Svanback, P. Bystrom, L. Persson, and E. Wahlstrom. 2001. Diet-dependent body morphology and ontogenetic reaction norms in Eurasian perch. Oikos 95:311–323.

Hogeweg, P. 2000a. Evolving mechanisms of morphogenesis: on the interplay between differential adhesion and cell differentiation. Journal of Theoretical Biology 203:317–333.

Hogeweg, P. 2000b. Shapes in the shadow: evolutionary dynamica of morphogenesis. Artificial Life 6:85–101.

Hogeweg, P. 2002. Multilevel evolution: the fate of duplicated genes. Zeitschrift für Physikalische Chemie—International Journal of Research in Physical Chemistry and Chemical Physics 216:77–90.

Holland, P. W. H. 1999. Gene duplication: past, present and future. Seminars in Cell and Developmental Biology 10:541–547.

Holter, N. S., M. Maritan, M. Cieplak, N. V. Fedoroff, and J. R. Banavar. 2001. Dynamic modeling of gene expression data. Proceedings of the National Academy of Sciences USA 98:1693–1698.

Huhn, M., and W. Schuster. 1982. Einige experimentelle Ergebnisseüber den Einfluss unterschiedlicher Konkurrenzsituationen auf die Korrelation zwischen Einzelpflanzmerkmalen (Ertragskomponenten) bei Winterraps. Zeitschrift für Pflanzenzüchtung 89:60–73.

Hulsey, C. D., and P. C. Wainwright. 2002. Projecting mechanics into morphospace: disparity in the feeding system of labrid fishes. Proceedings of the Royal Society of London Series B:Biological Sciences 269:317–326.

Hutt, M. T., and U. Luttge. 2002. Nonlinear dynamics as a tool for modelling in plant physiology. Plant Biology 4.281–297.

Huxley, J. S. 1932. Problems of Relative Growth. MacVeagh, Dial Press, London.

Huynen, M., P. F. Stadler, and W. Fontana. 1996. Smoothness within ruggedness: the role of neutrality in adaptation. Proceedings of the National Academy of Sciences USA 93:397–401.

Jernvall, J., J. P. Hunter, and M. Fortelius. 1996. Molar tooth diversity, disparity, and ecology in Cenozoic ungulate radiations. Science 274:1489–1492.

Keplinger, B. L., X. Guo, J. Quine, Y. Feng, and D. R. Cavener. 2001. Complex organization of promoter and enhancer elements regulate the tissue- and developmental stage-specific expression of the *Drosophila melanogaster Gld* gene. Genetics 157:699–716.

Kidson, R., and M. Westoby. 2000. Seed mass and seedling dimensions in relation to seedling establishment. Oecologia 125:11–17.

Klebes, A., B. Biehs, F. Cifuentes, and T. B. Kornberg. 2002. Expression profiling of *Drosophila* imaginal discs. Genome Biology 3:research0038.1–0038.16.

Klingenberg, C. P. 1998. Heterochrony and allometry: the analysis of evolutionary change in ontogeny. Biological Reviews of the Cambridge Philosophical Society 73:79–123.

Lande, R. 1979. Quantitative genetic analysis of multivariate evolution, applied to brain:-body size allometry. Evolution 33:402–416.

Lande, R. 1982. A quantitative genetic theory of life history evolution. Ecology 63:607–615.

Lande, R., and S. J. Arnold. 1983. The measurement of selection on correlated characters. Evolution 37:1210–1226.

Lauder, G. V. 1981. Form and function: structural analysis in evolutionary morphology. Paleobiology 7:430–442.

Lechowicz, M. J., and P. A. Blais. 1988. Assessing the contributions of multiple interacting traits to plant reproductive success: environmental dependence. Journal of Evolutionary Biology 1:255–273.

Lehman, N., M. D. Donne, M. West, and T. G. Dewey. 2000. The genotypic landscape during in vitro evolution of a catalytic RNA: implications for phenotypic buffering. Journal of Molecular Evolution 50:481–490.

Lespinet, O., Y. I. Wolf, E. V. Koonin, and L. Aravind. 2002. The role of lineage-specific gene family expansion in the evolution of eukaryotes. Genome Research 12:1048–1059.

Loy, A., L. Mariani, M. Bertelletti, and L. Tunesi. 1998. Visualizing allometry: geometric morphometrics in the study of shape changes in the early stages of the two-banded sea bream, *Diplodus vulgaris* (Perciformes, Sparidae). Journal of Morphology 237:137–146.

Lu, X., and Y. D. Li. 2001. Simulation of the evolution of genomic complexity. Biosystems 61:83–94.

Magniez-Jannin, F., B. David, J.-L. Dommergues, Z.-H. Su, T. S. Okada, and S. Osawa. 2000. Analysing disparity by applying combined morphological and molecular approaches to French and Japanese carabid beetles. Biological Journal of the Linnean Society 71:343–358.

Mazer, S. J., and C. T. Schick. 1991. Constancy of population parameters for life-history and floral traits in *Raphanus sativus* L. II. Effects of planting density on phenotype and heritability estimates. Evolution 45:1888–1907.

McGhee Jr., G. R. 1999. Theoretical Morphology: The Concept and Its Applications. Columbia University Press, New York.

McKinney, M. L., and J. L. Gittleman. 1995. Ontogeny and phylogeny: tinkering with covariation in life history, morphology and behaviour. Pp. 21–47 in K. J. McNamara, ed. Evolutionary Change and Heterochrony. John Wiley and Sons, New York.

McShea, D. W. 1991. Complexity and evolution: what everybody knows. Biology and Philosophy 6:303–324.

McShea, D. W. 1996. Metazoan complexity and evolution: is there a trend? Evolution 50:477–492.

Mellgren, E. M., and S. L. Johnson. 2002. The evolution of morphological complexity in zebrafish stripes. Trends in Genetics 18:128–134.

Merilä, J., A. Laurila, M. Pahkala, K. Rasanen, and A. T. Laugen. 2000. Adaptive phenotypic plasticity in timing of metamorphosis in the common frog *Rana temporaria*. Ecoscience 7:18–24.

Mezey, J. G., J. M. Cheverud, and G. P. Wagner. 2000. Is the genotype-phenotype map modular? A statistical approach using mouse quantitative trait locus data. Genetics 156:305–311.

Middleton, K. M., and S. M. Gatesy. 2000. Theropod forelimb design and evolution. Zoological Journal of the Linnean Society 128:149–187.

Miki, R., K. Kadota, H. Bono, Y. Mizuno, Y. Tomaru, P. Carninci, et al. 2001. Delineating developmental and metabolic pathways *in vivo* by expression profiling using the RIKEN set of 18,816 full-length enriched mouse cDNA arrays. Proceedings of the National Academy of Sciences USA 98:2199–2204.

Moczek, A. P., J. Hunt, D. J. Emlen, and L. W. Simmons. 2002. Threshold evolution in exotic populations of a polyphenic beetle. Evolutionary Ecology Research 4:587–601.

Moen, R. A., J. Pastor, and Y. Cohen. 1999. Antler growth and extinction of Irish elk. Evolutionary Ecology Research 1:235–249.

Monteiro, L. R., L. G. Lessa, and A. S. Abe. 1999. Ontogenetic variation in skull shape of *Thrichomys apereoides* (Rodentia: Echimyidae). Journal of Mammalogy 80:102–111.

Moolman, A. C., N. Vanrooyen, and M. W. Vanrooyen. 1996. The effect of drought stress on the morphology of *Anthephora pubescens* Nees. South African Journal of Botany 62:36–40.

Murren, C. J. 2002. Phenotypic integration in plants. Plant Species Biology 17:89–99.

Nehaniv, C. L., and J. L. Rhodes. 2000. The evolution and understanding of hierarchical complexity in biology from an algebraic perspective. Artificial Life 6:45–67.

Newman, S. A., and G. B. Müller. 2001. Epigenetic mechanisms of character origination. Journal of Experimental Zoology (Molecular and Developmental Evolution) 288:304–317.

Nijhout, H. F., and D. E. Wheeler. 1996. Growth models of complex allometries in holometabolous insects. American Naturalist 148:40–56.

Nijhout, H. F., G. A. Wray, C. Kremen, and C. K. Teragawa. 1986. Ontogeny, phylogeny and evolution of form: an algorithmic approach. Systematic Zoology 35:445–457.

Niklas, K. J. 1999. Evolutionary walks through a land plant morphospace. Journal of Experimental Botany 50:39–52.

Niklas, K. J., and B. J. Enquist. 2002. Canonical rules for plant organ biomass partitioning and annual allocation. American Journal of Botany 89:812–819.

Núñez-Farfán, J., and C. D. Schlichting. 2001. Evolution in changing environments: the "synthetic" work of Clausen, Keck and Hiesey. Quarterly Review of Biology 76:433–457.

Ofria, C., C. Adami, T. C. Collier, and G. K. Hsu. 1999. Evolution of differentiated expression patterns in digital organisms. Advances in Artificial Life, Proceedings Lecture Notes in Artificial Intelligence 1674:129–138.

O'Keefe, F. R. 2002. The evolution of plesiosaur and pliosaur morphotypes in the Plesiosauria (Reptilia: Sauropterygia). Paleobiology 28:101–112.

Olson, E. C., and R. L. Miller. 1951. A mathematical model applied to a study of the evolution of species. Evolution 5:325–338.

Olson, E. C., and R. L. Miller. 1958. Morphological Integration. University of Chicago Press, Chicago.

Ostrowski, M. F., P. Jarne, O. Berticat, and P. David. 2002. Ontogenetic reaction norm for binary traits: the timing of phallus development in the snail *Bulinus truncatus*. Heredity 88:342–348.

Pigliucci, M., C. D. Schlichting, C. S. Jones, and K. Schwenk. 1996. Developmental reaction norms: the interactions among allometry, ontogeny and plasticity. Plant Species Biology 11:69–85.

Pigliucci, M., P. J. DiIorio, and C. D. Schlichting. 1997. Phenotypic plasticity of growth trajectories in two species of *Lobelia* in response to nutrient availability. Journal of Ecology 85:265–276.

Raff, E. C., and R. A. Raff. 2000. Dissociability, modularity, evolvability. Evolution and Development 2:235–237.

Raff, R. A. 1996. The Shape of Life: Genes, Development, and the Evolution of Animal Form. University of Chicago Press, Chicago.

Ranta, E., V. Kaitala, S. Alaja, and D. Tesar. 2000. Nonlinear dynamics and the evolution of semelparous and iteroparous reproductive strategies. American Naturalist 155:294–300.

Raup, D. M. 1961. The geometry of coiling in gastropods. Proceedings of the National Academy of Sciences USA 47:602–609.

Raup, D. M. 1966. Geometric analysis of shell coiling: general problems. Journal of Paleontology 40:1178–1190.

Raup, D. M., and A. Michelson. 1965. Theoretical morphology of the coiled shell. Science 147:1294–1295.

Ravasz, E., A. L. Somera, D. A. Mongru, Z. N. Oltvai, and A.-L. Barabási. 2002. Hierarchical organization of modularity in metabolic networks. Science 297:1551–1555.

Rhees, B. K., and W. R. Atchley. 2000. Body weight and tail length divergence in mice selected for rate of development. Journal of Experimental Zoology (Molecular and Developmental Evolution) 288:151–164.

Richardson, R. C. 2001. Complexity, self-organization and selection. Biology and Philosophy 16:655–683.

Roces, F. 2002. Individual complexity and self-organization in foraging by leaf-cutting ants. Biological Bulletin 202:306–313.

Rogozin, I. B., K. S. Makarova, J. Murvai, E. Czabarka, Y. I. Wolf, R. L. Tatusov, L. A. Szekely, and E. V. Koonin. 2002. Connected gene neighborhoods in prokaryotic genomes. Nucleic Acids Research 30:2212–2223.

Ronen, M., R. Rosenberg, B. I. Shraiman, and U. Alon. 2002. Assigning numbers to the arrows: parameterizing a gene regulation network by using accurate expression kinetics. Proceedings of the National Academy of Sciences USA 99:10555–10560.

Rotaru-Varga, A., and M. Komosinski. 2001. Comparison of different genotype encodings for simulated three-dimensional agents. Artificial Life 7:395–418.

Rüber, L., and D. C. Adams. 2001. Evolutionary convergence of body shape and trophic morphology in cichlids from Lake Tanganyika. Journal of Evolutionary Biology 14:325–332.

Salazar-Ciudad, I., J. Garcia-Fernández, and R. V. Solé. 2000. Gene networks capable of pattern formation: from induction to reaction-diffusion. Journal of Theoretical Biology 205:587–603.

Sandquist, D. R., and J. R. Ehleringer. 1998. Intraspecific variation of drought adaptation in brittlebush: leaf pubescence and timing of leaf loss vary with rainfall. Oecologia 113:162–169.

Sarà, M. 1996a. The problem of adaptations: an holistic approach. Rivista de Biologia 82:75–101.

Sarà, M. 1996b. A "sensitive" cell system. Its role in a new evolutionary paradigm. Rivista de Biologia 89:139–156.

Sarà, M. 1999. Innovation and specialization in evolutionary trends. Rivista di Biologia 92:247–272.

Schaefer, S. A., and G. V. Lauder. 1996. Testing historical hypotheses of morphological change: biomechanical decoupling in Loricarioid catfishes. Evolution 50:1661–1675.

Scharloo, W. 1987. Constraints in selection response. Pp. 125–149 in V. Loeschcke, ed. Genetic Constraints on Adaptive Evolution. Springer-Verlag, Berlin.

Scharloo, W. 1989. Developmental and physiological aspects of reaction norms. BioScience 39:465–471.

Schlichting, C. D. 1986. The evolution of phenotypic plasticity in plants. Annual Review of Ecology and Systematics 17:667–693.

Schlichting, C. D. 1989a. Phenotypic integration and environmental change. BioScience 39:460–464.

Schlichting, C. D. 1989b. Phenotypic plasticity in *Phlox*. II. Plasticity of character correlations. Oecologia 78:496–501.

Schlichting, C. D. 2003. The origins of differentiation via phenotypic plasticity. Evolution and Development 5:98–105.

Schlichting, C. D., and M. Pigliucci. 1998. Phenotypic Evolution: A Reaction Norm Perspective. Sinauer Associates, Sunderland, MA.

Schlichting, C. D., and H. Smith. 2002. Phenotypic plasticity: linking molecular mechanisms with evolutionary outcomes. Evolutionary Ecology 16:189–211.

Schlosser, G. 2002. Modularity and the units of evolution. Theory in Biosciences 121:1–80.

Schmalhausen, I. I. 1949. Factors of Evolution. Blakiston, Philadelphia, PA.

Seeley, T. D. 2002. When is self-organization used in biological systems? Biological Bulletin 202:314–318.

Shen-Orr, S. S., R. Milo, S. Mangan, and U. Alon. 2002. Network motifs in the transcriptional regulation network of *Escherichia coli*. Nature Genetics 31:64–68.

Skaer, N., D. Pistillo, J. M. Gibert, P. Lio, C. Wulbeck, and P. Simpson. 2002. Gene duplication at the achaete-scute complex and morphological complexity of the peripheral nervous system in Diptera. Trends in Genetics 18:399–405.

Smith, H. 1990. Signal perception, differential expression within multigene families and the molecular basis of phenotypic plasticity. Plant, Cell and Environment 13:585–594.

Smith, L. H., and B. S. Lieberman. 1999. Disparity and constraint in olenelloid trilobites and the Cambrian radiation. Paleobiology 25:459–470.

Sporne, K. R. 1948. Correlation and classification in dicotyledons. Proceedings of the Linnean Society of London 160:40–47.

Sporne, K. R. 1954. Statistics and the evolution of dicotyledons. Evolution 8:55–64.

Stadler, B. M. R., P. F. Stadler, G. P. Wagner, and W. Fontana. 2001. The topology of the possible: formal spaces underlying patterns of evolutionary change. Journal of Theoretical Biology 213:241–274.

Stebbins Jr., G. L. 1951. Natural selection and the differentiation of angiosperm families. Evolution 5: 299–324.

Stone, J. R. 1998. Ontogenic tracks and evolutionary vestiges in morphospace. Biological Journal of the Linnean Society 64:223–238.

Svanback, R., and P. Eklov. 2002. Effects of habitat and food resources on morphology and ontogenetic growth trajectories in perch. Oecologia 131:61–70.

Tammaru, T., K. Ruohomäki, and M. Montola. 2000. Crowding-induced plasticity in *Epirrita autumnata* (Lepidoptera: Geometridae): weak evidence of specific modifications in reaction norms. Oikos 90:171–181.

Thieffry, D., and D. Romero. 1999. The modularity of biological regulatory networks. BioSystems 50:49–59.

Thompson, D. B. 1992. Consumption rates and the evolution of diet-induced plasticity in the head morphology of *Melanoplus femurrubrum* (Orthoptera, Acrididae). Oecologia 89:204–213.

Thompson, D. W. 1917. On Growth and Form. Cambridge University Press, Cambridge.

Trewavas, A. 2002. Plant cell signal transduction: the emerging phenotype—Foreword. Plant Cell 14:S3–S4.

Turney, P. D. 2000. A simple model of unbounded evolutionary versatility as a largest-scale trend in organismal evolution. Artificial Life 6:109–128.

Valentine, J. W., A. G. Collins, and C. P. Meyer. 1994. Morphological complexity increase in metazoans. Paleobiology 20:131–142.

von Dassow, G., and E. Munro. 1999. Modularity in animal development and evolution: elements of a conceptual framework for EvoDevo. Journal of Experimental Zoology (Molecular and Developmental Evolution) 285:307–325.

von Dassow, G., E. Meir, E. M. Munro, and G. M. Odell. 2000. The segment polarity network is a robust developmental module. Nature 406:188–192.

Wagner, A. 2001. The yeast protein interaction network evolves rapidly and contains few redundant duplicate genes. Molecular Biology and Evolution 18:1283–1292.

Wagner, G. P. 1988. The significance of developmental constraints for phenotypic evolution by natural selection. Pp. 222–229 in G. de Jong, ed. Population Genetics and Evolution. Springer-Verlag, Berlin.

Wagner, G. P., and L. Altenberg. 1996. Perspective: Complex adaptations and the evolution of evolvability. Evolution 50:967–976.

Wagner, P. J. 1997. Patterns of morphologic diversification among the Rostroconchia. Paleobiology 23:115–150.

Wang, Y. F., and X. Gu. 2001. Functional divergence in the caspase gene family and altered functional constraints: statistical analysis and prediction. Genetics 158:1311–1320.

Waxman, D., and J. Peck. 1998. Pleiotropy and the preservation of perfection. Science 279:1210–1213.

Wilkins, A. S. 2002. The Evolution of Developmental Pathways. Sinauer Associates, Sunderland, MA.

Wills, M. A., D. E. G. Briggs, and R. A. Fortey. 1994. Disparity as an evolutionary index: a comparison of Cambrian and Recent arthropods. Paleobiology 20:93–130.

Winther, R. G. 2001. Varieties of modules: kinds, levels, origins, and behaviors. Journal of Experimental Zoology (Molecular and Developmental Evolution) 291:116–129.

Yi, T.-M., Y. Huang, M. I. Simon, and J. Doyle. 2000. Robust perfect adaptation in bacterial chemotaxis through integral feedback control. Proceedings of the National Academy of Sciences USA 97:4649–4653.

Zhang, J., and H. F. Rosenberg. 2002. Complementary advantageous substitutions in the evolution of an antiviral RNase of higher primates. Proceedings of the National Academy of Sciences USA 99:5486–5491.

Zuckerkandl, E. 2001. Intrinsically driven changes in gene interaction complexity. I. Growth of regulatory complexes and increase in number of genes. Journal of Molecular Evolution 53:539–554.

Zweers, G. A., and J. C. V. Berge. 1998. Evolutionary transitions in the trophic system of the wader-waterfowl complex. Netherlands Journal of Zoology 47:25–287.

PART I

# ADAPTATION AND CONSTRAINTS

The relationship between adaptation and constraints lies at the core of many, if not most, persistent questions about the nature of phenotypic evolution. Without an interplay between the two, many subtle and interesting research programs could degenerate into nothing more than parameterization of single-trait optimality models. It is not surprising, then, that essentially all of the chapters in this volume address this complex relationship in some way. Part I highlights those chapters that draw on empirical work and take questions of adaptation and constraint as a central theme.

Although adaptive evolution and constraints on adaptation are often characterized as opposing forces (see Schwenk and Wagner, Chapter 18), the nature of their relationship depends a great deal on the role that limitations on trait variation and covariation play in selection. It will be clear from the following chapters that a variety of factors may underlie trait covariation within or among species. In those cases where the alternative trait combinations are developmentally possible but are nonfunctional or relatively less fit, then selection is expected to eliminate them. Moreover, in such cases selection should also favor developmental or genetic mechanisms that reduce the occurrence of maladaptive trait combinations in the first place (see Merilä and Björklund, Chapter 5). In other words, selection may favor phenotypic integration. Yet, by limiting phenotypic variation to those trait combinations with high fitness in the current environment, integrating mechanisms also limit the range of phenotypic variation on which selection may act if the optimal trait relationship changes. Consequently, limitations on the production of certain trait combinations are properly viewed as constraints on evolution (Schwenk, 1995), even when these same constraints are themselves adaptations. Thus the relationship between adaptation and constraint is not simply one of opposition, but is instead the result of a continuous dynamic interaction.

In their chapter, Armbruster, Pélabon, Hansen, and Mulder address this relationship directly, asking whether genetic integration of floral traits represents constraint or adaptation. In animal-pollinated blossoms, correspondence in the placement of male and female structures both within and among blossoms is under strong selection. Mechanisms that ensure male-female spatial correspondence and reduce individual variation in placement should be adaptive but potentially limit response to selection on these same traits. By examining patterns of trait variation and covariation over a range of scales (genets, populations, species) in three different groups of plants, the authors were able to test alternative hypotheses about the role of constraint and adaptation in maintaining trait correlations.

Similarly, Badyaev uses the apparent paradox of constraint-as-adaptation to resolve another paradox arising from sexual selection. Females favor male sexual ornaments that are good indicators of condition (and thus well integrated with the rest of the organism), yet exaggerated (and therefore evolutionarily decoupled from other organismal functions). In addition, although traits that better indicate condition will cost more, the fittest males will be able to modify their investment costs according to environmental conditions. Badyaev resolves this paradox by reinterpreting observed patterns of sexual trait covariation in terms of phenotypic modularity and integration. He cites numerous examples in which a suite of traits involved in sexual ornamentation is itself composed of relatively independent modules that reflect different aspects of organismal condition across a range of environmental situations.

Both the chapters by Armbruster et al. and by Badyaev describe systems in which the functional relationship between traits depends on outside factors. In the case of floral traits, the probability of donating or receiving pollen depends on the location of pollen transfer sites in other blossoms and on the size and behavior of the pollinators. In the case of sexual selection, choosy females determine male fitness. Very often, however, the adaptive value of a given trait combination depends on how the traits interact within the same organism. Allometric (scaling) relationships between morphological traits, measured either within or among species, are thought to reflect adaptive relationships or fundamental developmental or functional constraints (e.g., Niklas 1994; West et al. 1999). In organisms composed of reiterated structural units (morphological modules), such as most plants and some animals, it is also possible to characterize patterns of trait covariation within a single individual. In their chapter, Preston and Ackerly discuss the interpretation of allometric relationships at this lower level of morphological organization. They argue that in modular organisms, a range of trait combinations is expressed all at once, with each instance located in its own ontogenetic and morphological context. Consequently, an entire allometric relationship is subject to selection within a single individual. Drawing on recent studies from the literature, they show how allometry can be seen as an adaptive strategy generating appropriate trait combinations across a range of developmental conditions.

A common theme underlying all the chapters in Part I is that the status of phenotypic integration as adaptation or constraint depends very much on the scale at which it is examined, be it phylogenetic distance or ecological or developmental time. The point is especially developed by Merilä and Björklund. After reviewing the evidence for integration as adaptation and as constraint, they show that phenotypic response to selection should be sensitive to the frequency and magnitude of change in the selective environment, relative to the strength of the constraint over evolutionary time. Since integration is more likely to limit variation in trait relationships over short time scales, it is also more likely to constrain evolutionary responses in rapidly fluctuating environments than under more stable conditions.

The picture that emerges from these chapters, then, is that integration generally reflects adaptation within a certain environment and may act as a proximal constraint on evolution when conditions change over the short term, but ultimately the constraint can be broken under selection for a different pattern of integration. Yet, as Hansen and Houle discuss in Chapter 6, this picture cannot account for long-term evolutionary stasis. Some traits, such as wing shape in *Drosophila*, have remained unchanged over very long periods of time across diverse ecological conditions. Unsatisfied with the usual explanation of stabilizing selection, Hansen and Houle explore several mechanisms besides limited genetic variation that may restrict trait evolvability even under directional selection. Over longer time scales, the functional architecture (mapping genotype to phenotype) and the amount of new mutational variation that can be generated by the genome (variability) may both contribute to evolutionary stasis. Consistent with an important theme of this section, however, their influence is context-dependent and both may either retard or enhance evolvability, depending on the genetic and evolutionary background.

### *Literature Cited*

Niklas, K. J. 1994. Plant Allometry: The Scaling of Form and Process. University of Chicago Press, Chicago.

Schwenk, K. 1995. A utilitarian approach to evolutionary constraint. Zoology 98:251–262.

West, G. B., J. H. Brown, and B. J. Enquist. 1999. A general model for the structure and allometry of plant vascular systems. Nature 400:664–667.

# 2

# Floral Integration, Modularity, and Accuracy

## *Distinguishing Complex Adaptations from Genetic Constraints*

W. SCOTT ARMBRUSTER
CHRISTOPHE PÉLABON
THOMAS F. HANSEN
CHRISTA P. H. MULDER

The origin of flowers is clearly one of the most important "key innovations" in the history of plant evolution. In addition to representing a number of functional advancements over the reproductive structures of nonflowering plants, flowers are modular organs that usually express limited variation within populations and yet huge variation among species and higher taxa. This variation in floral traits has fascinated evolutionary biologists since at least Darwin's time. Because flowers are highly evolved structures that interact in complex but analyzable ways with pollinators, they make ideal study systems for the evolutionary analysis of morphology, integration, and modularity. In this chapter we address the functional significance of floral integration and modularity, focusing particularly on the role of integration in the evolution of accuracy and precision of pollination, as compared across three different study systems.

Patterns of variation and covariation are interesting in an evolutionary context because they can yield insights into the processes controlling past evolution of traits and probable trajectories of future evolution (Lande and Arnold 1983). Trait variation and covariation reflect the interactions between phenotypic response to the environment, genetic response to natural selection, and the limits to such responses imposed by genetic and developmental constraints (see review in Schlichting and Pigliucci 1998). Olson and Miller (1958) and Berg (1960) were among the first to suggest explicitly that particular patterns of phenotypic variance/covariance may reflect the direct operation of natural selection for such patterns. Berg (1960) proposed that variance/covariance patterns should differ in two important ways between plant species with specialized pollination and those with generalized pollination. Because flowers of species with specialized pollination have to "fit" their pollinators, selection should favor phenotypic integration and decoupling or buffering of floral traits from variation in vegetative traits (i.e., flowers being modular). Covariation is the default situation given the homology and developmental connections between floral and vegetative traits; see Armbruster et al. (1999) for further discussion.

Rather than reflecting the results of adaptive evolution, however, the tight integration of floral parts and partial decoupling of floral from vegetative traits could be by-products of the genetic/developmental architecture of the organism (see Olson and Miller 1958). Traits that are developmentally more closely related to one another tend to be more tightly correlated phenotypically and genetically, probably because of overlapping genetic control and developmental regulation (Olson and Miller 1958; Cheverud 1982, 1984, 1996; Diggle 1992; C. Herrera 2001). For example, we might expect parts of a single floral whorl (e.g., sepals, petals, or stamens) to be tightly integrated (covary more tightly), because they are developmentally interrelated, and parts of different floral whorls to be less integrated (e.g., sepals versus stamens). Similarly, parts of a blossom inflorescence (pseudanthium) would be less integrated than parts of a flower. Following this logic, floral and vegetative parts would tend to covary least of all. Thus, it remains unresolved whether observed patterns of floral integration and modularity largely reflect the operation of natural selection or instead the "constraints" of the genetic and developmental systems that underlie the floral traits. Note that genetic/developmental constraints can also have "positive" effects on evolution (Gould 2002); as we describe below, tight structural integration may create genetic constraints, but at the same time improve floral precision and accuracy (see also Diggle 1992).

Another aspect of the evolution of correlated traits that has been little explored is the issues of scale and phylogeny. Are patterns of floral-trait integration observed within populations conserved throughout each species? Are among-population and among-species patterns of trait covariance similar to those seen within populations? Are patterns at these three levels generated by the same processes (see Armbruster 1991; Armbruster and Schwaegerle 1996)? What is the role of selection versus drift in creating among-population differences in phenotypic integration (see Herrera et al. 2002)? Finally, do the answers to these questions differ dramatically among various higher taxa, such as among genera or families, or among species with different life-history strategies (see Armbruster et al. 1999)?

The goals of this chapter are to (1) examine the evolutionary dynamics that have influenced floral form and function; (2) explore ways to determine whether floral integration is the result of adaptive processes, genetic/developmental constraints, or a combination of these two processes; and (3) assess the role of selection, drift, and constraint in generating among-population and among-species patterns of trait covariance. We examine a variety of approaches that have been taken to describe and analyze the causes of floral integration, but place special emphasis on multilevel analyses. We illustrate the utility of the multilevel approach with data drawn from three of our own study systems.

## Accuracy, Precision, and Bias in Pollination and Evolution

Adaptation is usually conceptualized as the closeness of the population mean to a fitness optimum, and it is rarely appreciated that this does not necessarily imply that the individual organisms in the population are well adapted (Orzack and Sober 1994). Even if organisms are not systematically different from the optimum, poor precision in their developmental ability to realize the optimum may make the typical individual poorly adapted. Orzack and Sober (1994) argued that the study of adaptation must take both the individual and the population into account. For example, while the Lande-Arnold regression approach (Lande and Arnold 1983) may look at deviation of individuals from an "optimum" observed within the population range, it does not, in its nonexperimental manifestation, allow estimation of the deviation of the population mean from a true optimum. Thus, in analogy with statistical measurement, we may distinguish between the accuracy and precision of adaptation. Accuracy is the closeness of a measured quantity to its true value, and precision is the closeness of repeated measurements of the same quantity to each other (Sokal and Rohlf 1995). The accuracy of adaptation may thus be defined as the closeness of individuals to an optimum and precision as the closeness of individuals to each other. The distance of a population mean to its optimum would then be analogous to statistical bias. Thus, if we treat inaccuracy as the mean square distance from the optimum, we get the following relationship:

$$\text{Inaccuracy} = \text{Imprecision} + (\text{Bias})^2$$

where imprecision is the variance in the trait value, and bias is the difference between the mean of observed values and the optimum value.

Because bias has numerous nonstatistical connotations that could cause confusion in an evolutionary context, we shall refer to closeness of the population mean to the selective optimum as "optimality in mean." Natural selection will always work on accuracy, but due to tradeoffs this may not necessarily improve both precision and optimality in mean. In general, constraints that reduce variability may often improve precision at the same time that they make optimality in mean more difficult to achieve.

In plants, flowers, more than vegetative structures, have clear connections to the above concepts of accuracy, precision, and optimality in mean. Indeed, flowers are excellent study systems for the analysis of adaptation and phenotypic integration in general because they have two reproductively important functions that can be analyzed mechanistically: (1) placing pollen on the pollinator (the first step in getting pollen to the stigmas of other flowers, the male function in pollination), and/or (2) picking up pollen from the pollinator with the stigmas (female function in pollination). Bisexual flowers perform both functions, unisexual flowers only one. Accuracy, precision, and bias in both functions can be assessed through analysis of the variance and covariance of floral morphological traits. Interaction with pollinators of particular size and behavior requires precision, while the functional interaction of placing and receiving pollen by different structures in the bisexual flower (stamens and pistils respectively) requires accuracy.

Precision in this context refers to the variance within populations in position of anther contact or stigma contact with pollinators. High male precision means that pollen is placed consistently in the same location on the pollinator. High female precision means that the stigmas contact the pollinator consistently in the same location. Precision in both functions is determined by the consistent morphology and orientation of those floral parts that contact the pollinator and/or influence the pollinator's position and movements on the flower, both across flowers on a plant and across plants in a population. Thus, male and female precision is determined by within-population variance in critical floral traits as generated by genetic variation, environmental variation, and/or developmental instability (see Fenster and Galloway 1997).

Optimality in mean in a pollination context refers to how close the mean position of pollen placement (male optimality in mean) or stigma contact (female optimality in mean) is to the selective optimum. Male optimality in mean refers to the correspondence of the mean position of pollen placement relative to the mean position of stigma contact in the population. (The selective "target" for pollen placement is the site on the pollinator from which most of the population retrieves pollen.) Similarly, female optimality in mean refers to deviation of mean stigma position from the site on pollinators where the pollen from the rest of the population is placed, on average. Joint optimality in mean of male and female pollination function can be assessed at the among-population level by examining the correlation between population means of the interacting male and female traits.

Accuracy refers to the combined effects of precision and bias on departure of the distribution of pollen on pollinators (male accuracy) or positions of stigma contact with pollinators (female accuracy) from the respective selective optima. The male accuracy of a flower is how closely the anthers contact the pollinator in the place that stigmas are expected to contact it, and similarly for female accuracy. Measurement of accuracy is not straightforward, however. We would like to relate individual male positions to expected (mean) female positions, and individual female positions to expected (mean) male positions. However, these statistics are challenging to estimate and work with, and are beyond the scope of the present chapter.

The positions of male and female parts of a flower or blossom are also affected by selection against self-pollination. The selective scenario described in the previous paragraphs works quite well for unisexual flowers or bisexual flowers that express male and female functions sequentially (e.g., protandry or protogyny). However, plants without temporal separation of sexual functions have often evolved spatial separation of the sexes (herkogamy). Hence, in these plants there is selection for departure of anther position from stigma position within a flower, obviously acting in conflict with selection for correspondence of the two positions among flowers. The response to these conflicting selective forces may be difficult to predict, but one well-known outcome is the evolution of male and female parts having reciprocal positions (e.g., heterostyly; Barrett 2002).

Thus, we see that the accuracy and effectiveness of pollination function is affected by patterns of variance and covariance of certain floral traits. Examination of variance statistics within populations allows assessment of the precision of pollination function, and examination of the correlation statistics among means of populations or species allows assessment of optimality in mean for pollination function.

## Modularity and Regulation of Variance and Covariance

It is the modular nature of flowers that allows the evolution of precise fit with pollinators in the face of huge vegetative plasticity expressed by most plant populations (Schlichting 1986; Armbruster et al. 1999; Sultan 2000; Pigliucci and Hayden 2001). Modularity and phenotypic integration of flowers are interrelated concepts because they both concern the "regulation" of variances and covariances of plant traits. Modules have three characteristics (Wagner 1996; Wagner and Altenberg 1996): (1) the parts of a module collectively serve a joint function; (2) the parts of a module are, to some extent, genetically (or epigenetically) integrated; and (3) variation in parts of a module is quasi-independent of variation in other modules.

Floral modularity has important evolutionary consequences beyond plasticity. Flower modules can evolve quasi-independently of other plant parts; for example, floral evolution can be rapid and labile while vegetative traits remain relatively unchanged, or vice versa. One way to assess the evolutionary dependencies of traits within and among modules is to measure the "conditional evolvability" of traits. This estimates the genetic variance available to respond to directional selection while holding constant (as under strong stabilizing selection) other traits genetically correlated with the first (Hansen et al. 2003b; see also Hansen and Houle, Chapter 6, this volume).

Berg (1960) suggested that plants that relied on precise fit between their flowers and one or a few pollinator species would display floral and vegetative traits belonging to separate modules, or "correlation pleiades" in her terminology. While environmentally and genetically induced variation in size of vegetative traits can be adaptive (adaptive plasticity; e.g., sun and shade leaves; see Schlichting 1986; Sultan 2000; but see Winn 1999), such variation in floral traits would disrupt the fit between flowers and their pollinators (Berg 1960). Thus, plants with specialized relationships with a few pollinator species should have flowers exhibiting developmental homeostasis (Fenster and Galloway 1997) and phenotypic independence from variation in vegetative traits. Correlations between floral and vegetative traits are thus predicted to be weak. Most study systems (e.g., Berg 1960; Conner and Sterling 1996; Pigliucci and Hayden 2001), although not all, support this prediction (see Armbruster et al. 1999).

With regard to the genetic-integration requirement of modules, Berg (1960) also considered this issue in flowers and predicted that floral parts should covary more tightly in species with specialized relationships with animal pollinators than in species with more generalized relationships. This is because many floral traits function interactively in the pollination process, and the tighter the relationship with pollinators, the more likely traits are to interact. Observations of strong genetic and phenotypic correlations among floral parts are consistent with this idea (e.g., Berg 1960; Schwaegerle and Levin 1991; Conner and Via 1993; Conner and Sterling 1995), although the interpretation of correlations among floral traits is certainly more complicated than Berg (1960) envisaged (see Armbruster et al. 1999 for further discussion).

## Multilevel Analyses of Phenotypic Integration

Most studies of phenotypic integration have focused on covariation within populations. However, as Armbruster (1991), Armbruster and Schwaegerle (1996), and Herrera et al. (2002) have pointed out, there are advantages in viewing integration from a hierarchical perspective. If multivariate divergence of populations or species follows a trajectory predicted from the pattern of the genetic correlations seen within populations, it suggests that proximal genetic constraints are of major significance in limiting population differentiation (Sokal 1978; Primack 1987; Armbruster 1991; Schluter 1996; Andersson 1997). (Note, however, that genetic correlations can themselves evolve in response to natural selection [Cheverud 1982, 1984, 1996; Zeng 1988; Wagner 1996], so proximal genetic constraints may be ultimately adaptive. For purposes of discussion, however, we shall treat genetic correlations as proximal constraints unless there is evidence that they have evolved through adaptation.) Alternatively, if the course of population divergence cannot be predicted from the genetic correlations exhibited within populations, then we may be able to conclude that evolution in response to natural selection or drift is not constrained by pleiotropy or linkage during population divergence (Armbruster 1991; Armbruster and Schwaegerle 1996). The weakening of trait correlations at higher organizational levels (e.g., species) would indicate selection overcoming the genetic constraint over longer periods of time, or a change in the fitness landscape away from the original axis of integration. Precisely because genetic constraints are likely to be weaker over longer periods of evolutionary divergence (see Zeng 1988), stronger trait correlations at population and species levels than within populations should reflect the shape of the underlying fitness surface governing floral evolution. Thus, studies of the divergence of populations and species in relation to genetic and within-population phenotypic covariances can yield insights into the frequency with which phenotypic integration reflects constraints, and how persistent those constraints are (see, for example, Primack's (1987) comparative analysis of pleiotropic constraints on flower and fruit evolution).

We can recognize at least four distinct levels in the genetic hierarchy: within genetic individuals (genets), among genets within a population, among populations within a species, and among species (see Armbruster 1988, 1991). The covariance of traits within genetic individuals reflects ontogenetic processes and correlated response to developmental instability and environmental variation (see Armbruster 1991). Genetic and morphometric studies generally attempt to minimize the effects of ontogenetic variance and covariance by holding developmental "age" constant. However, complete control of ontogenetic stage is rarely possible. Another way to deal with this source of covariance is to control or estimate it statistically (Armbruster 1991).

The covariance of traits among genetic individuals (genets) within populations reflects the effects of genetic correlations (caused by pleiotropy or linkage disequilibrium) and environmental covariance, plus statistical effects carried up from the lowest level. The covariance of traits among populations within a species may be the result primarily of pleiotropy or linkage, environmental covariance, natural selection, and gene flow, plus statistical contributions from lower levels. In turn, covariance of traits among species may be the result of genetic correlations, environmental covariance, adaptive and stochastic speciation, further differentiation generated by natural selection and drift, plus statistical contributions from lower levels (see Armbruster and Schwaegerle 1996 and Hansen and Martins 1996 for reviews of these issues; see also Arnold and Phillips 1999; Roff 2000; Waldmann and Andersson 2000; Begin and Roff 2001).

Multilevel comparisons can be useful in distinguishing between the contributions of genetic/developmental constraints and adaptation in the course of population and species differentiation. Analysis of within-population variance of critical traits assesses adaptation through the evolution of precision (control of phenotypic variation via reduced genetic variance/covariance and/or increased developmental homeostasis/canalization). Covariation of population and species means can be used to assesses the degree to which traits are correlated as a result of evolutionary divergence of populations and species (see Armbruster 1990, 1991), and hence may give insights into the roles of genetic constraints and adaptation by movement of population means toward respective optima (optimality in means).

One of the first questions one might wish to address at the level of populations or species is the degree to which patterns of phenotypic integration are conserved across related groups (as would be expected if they largely reflect strong proximal constraints or consistent, long-term patterns of selection; see Westoby et al. 1995). Alternatively, if phenotypic integration is neutral or reflects functional relationships and natural selection, it might be extremely labile in its evolution, at least if the selective pressures vary across populations and species. Indeed, Herrera et al. (2002) found significant differences in patterns of floral integration among related populations of *Helleborus*. They interpreted this lability to be the result of genetic drift, which certainly argues against phenotypic integration being the result of strong genetic constraints (or selection). Murren et al. (2002) also found no evidence for a strong phylogenetic signal in patterns of integration (i.e., phylogenetic inertia or genetic "constraint") among related species in the Brassicaceae. Although some phenotypic correlations were conserved across most taxa, overall patterns of phenotypic correlations differed in major ways and idiosyncratically across closely related species.

As discussed above and by Armbruster et al. (1999), Herrera et al. (2002), and Murren et al. (2002), the most productive approach to analyzing modularity and phenotypic integration may be to examine specific form-function relationships and use this information to make predictions about statistical patterns of covariance. In the following sections we address the evolution of floral integration by considering the details of the function and development of a few floral parts involved in the pollination process, and we compare the patterns across three plant groups that differ in degree of developmental and structural integration of reproductive parts.

As discussed above, the positions of anther and stigma contact with the body of pollinators have two properties: precision and optimality in mean position. Both may affect the evolution of floral integration. In turn, there are two components of precision: (1) the absolute size of the pollen load deposited on pollinator ("pollen spot") by a single flower; (2) the among-flower and among-plant variation in position of the pollen placed on the pollinator. These factors together determine the size of the pollen spot on pollinators that have visited more than one plant. We therefore wished to assess the integration of the male and female parts of the flower in terms of optimality in mean and precision of pollination. We compared patterns of covariation, specialization, and pollination ecology across three distantly related groups of plants with different degrees of structural integration of flower parts and different degrees of precision in pollen delivery and pickup: *Dalechampia* (Euphorbiaceae), *Collinsia* and *Tonella* (Scrophulariaceae), and *Stylidium* (Stylidiaceae).

### *Dalechampia* (Euphorbiaceae)

*Dalechampia* have unisexual flowers secondarily clustered together into functionally bisexual pseudanthial inflorescences (blossoms). Pollen deposition and pollen pickup are performed by different flowers in the same blossom. Response to selection for accuracy in pollination may be limited by this architectural feature, which reflects the phylogenetic history of the group (all members of the Euphorbiaceae have unisexual flowers). However, compared to its closest relatives in tribe Plukenetieae and more distant relatives in subfamily Acalyphoideae (Webster 1994; Baldwin and Armbruster, unpublished data; K. Wurdack, pers. comm.), the structural, developmental, and ecological integration of staminate and pistillate flowers has increased greatly, coincident with the origin of the functionally bisexual blossoms that characterize the genus.

Most species of *Dalechampia* have bilaterally arranged blossom parts, with ca. 10 staminate flowers in a subinflorescence positioned above three pistillate flowers (Fig. 2.1). Above the staminate flowers is a specialized gland-like structure that secretes resin (the "resin gland"). The resin is collected by certain bee species that use resin in nest construction. Hence the resin serves as a pollinator reward. The bees position themselves moderately consistently on the blossom, placing their heads at the resin gland as they collect resin with their mandibles. Thus, the landmark to which bees position themselves is this resin gland. However, the resin is produced over a two-dimensional surface of considerable area (ca. 5–50 mm$^2$, depending on the population and species). Hence the position of a bee on the blossom varies among visits and even within visits as the bee repositions itself slightly while collecting resin, and precision in pollen placement is thereby reduced. For example, in the well-studied *D. scandens* population from Tulum, the range in bee positions might be as great as the width of the resin gland (mean = 2.97 mm; range 1.40–4.40 mm).

Another component of precision, the size of the pollen load deposited by a single blossom on the underside of the bee, has not been measured directly. It is determined by the bee's movements as it collects resin and the size of the globular cluster of stamens in an open male flower. The length of pollen-load spot on a bee visiting one blossom would then be, at a minimum, approximately the diameter of one open male flower (e.g., for *D. scandens* from Tulum, mean = 2.80 mm), and over two diameters if the bee moves across the width of the gland while collecting resin. If the bee touches more than one open staminate flower while collecting resin, the size of the pollen spot would be even larger, as

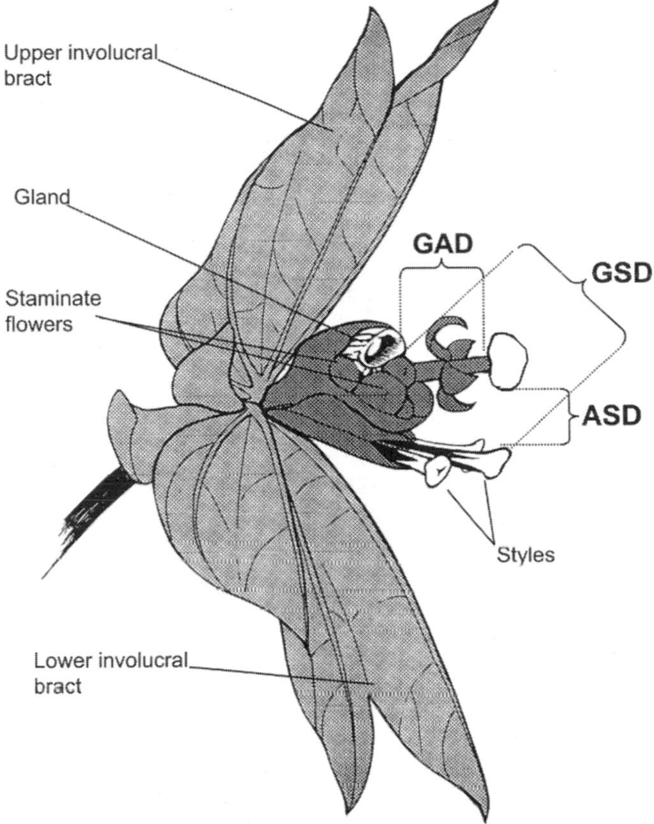


Figure 2.1 Side view of *Dalechampia scandens* blossom. The lower involucral bract is ca. 20 mm in length. Traits analyzed in this chapter are indicated: gland-anther distance (GAD), gland-stigma distance (GSD), and anther-stigma distance (ASD).

additional variance is introduced by within-blossom variation in the position of the staminate flowers.

As bees move among blossoms, the size of the pollen spot grows because of variation in size and shape of blossoms, as well as additional variation in the position of the bee. For example, the among-genet standard deviation (SD) for anther-contact position (gland-anther distance, GAD) in the Tulum *D. scandens* population was 0.54 mm, with a mean coefficient of variance (CV) of 0.145 (range of CVs: 0.112–0.299 across five populations of *D. scandens*). Thus, after visiting a few blossoms, an 11 mm *Euglossa* (the main pollinator in this population of *D. scandens*; Armbruster 1985) is likely to have pollen spread over more than 50% of the underside of its body. The large pollen spot and low precision may, of course, be adaptive if there is low mean optimality (correspondence) in where anthers and stigmas contact pollinating bees.

From the perspective of precision in female function, we must concern ourselves with the variation in position of the bee and the size and variation in position of the stigmas within and among blossoms. In the Tulum population of *D. scandens* the mean stigma diameter is 1.39 mm. Within this same population the among-genet SD of stigma position relative to the resin gland (gland-stigma distance, GSD; Fig. 2.1) is 0.85 mm, with a mean CV of 0.124 (range of CVs: 0.114–0.198 across five populations of *D. scandens*).

Thus, the *Dalechampia* system shows limited structural/developmental integration of male and female floral parts (they are in separate flowers within a single inflorescence). The degree of consistency and precision in contact of sexual parts with pollinators is relatively low compared with other zygomorphic flowers and blossoms (see below).

In light of these structural limitations, it is interesting to examine the statistical integration of the blossom components. First consider the within-population correlation between traits affecting pollen deposition and pickup. The phenotypic correlation in the greenhouse (among genets within the Tulum population of *D. scandens*) between GAD and GSD was only 0.29. This indicates low accuracy and phenotypic integration, due presumably to the many sources of variance in the form of environmental/developmental noise affecting the numerous blossom components that influence the two traits (lengths and angles of peduncles and pedicels, size of floral parts, etc.). Interestingly, when we factor out the environmental noise we find a rather close correspondence between GAD and GSD: the additive genetic correlation was 0.87 (Hansen et al. 2003b). This large difference between the phenotypic and genetic correlations may reflect response to selection for among-genet correspondence of pollen placement and pickup (the genetic correlation), but the limitations of loose architectural construction as reflected in the phenotypic correlation.

Other blossom traits showed weak to strong phenotypic correlations at the within-population level, although the phenotypic correlations were generally weaker than the genetic correlations (Fig. 2.2A, B). Many of the strongest genetic and phenotypic correlations were among traits within developmental modules (Fig. 2.2), suggesting that the genetic and phenotypic integration reflected here is to some extent the result of developmental relationships (hence potentially proximal constraints), and that a constraint imposed by stabilizing selection on one blossom part would limit the evolution of other parts of the same structural/developmental unit (i.e., condition the evolvability of those traits; Hansen et al. 2003b; see Hansen and Houle, this volume, Fig. 6.26).

In our studies of four additional populations of *Dalechampia scandens*, we found that most blossom traits were again intercorrelated, and the correlation patterns were fairly similar across all populations. The overall phenotypic integration ($I$) can be summarized by the variance in the eigenvalues of the phenotypic correlation matrix (Wagner 1984; Cheverud et al. 1989; Herrera et al. 2002). To compare analyses with different sample sizes, it is necessary to correct the index (to $I'$) by subtracting the random expected level of integration: $T - 1/N$, where $T$ is the number of traits and $N$ is the number of objects measured (Wagner 1984). If different numbers of traits are measured, the index should be scaled by the maximum possible value, $T$ (Herrera et al. 2002). The corrected index did not differ significantly among populations (overlapping 95% confidence intervals (CIs), calculated by randomization). Closely related populations tended to have similar patterns of phenotypic integration, and hence there was an apparent phylogenetic

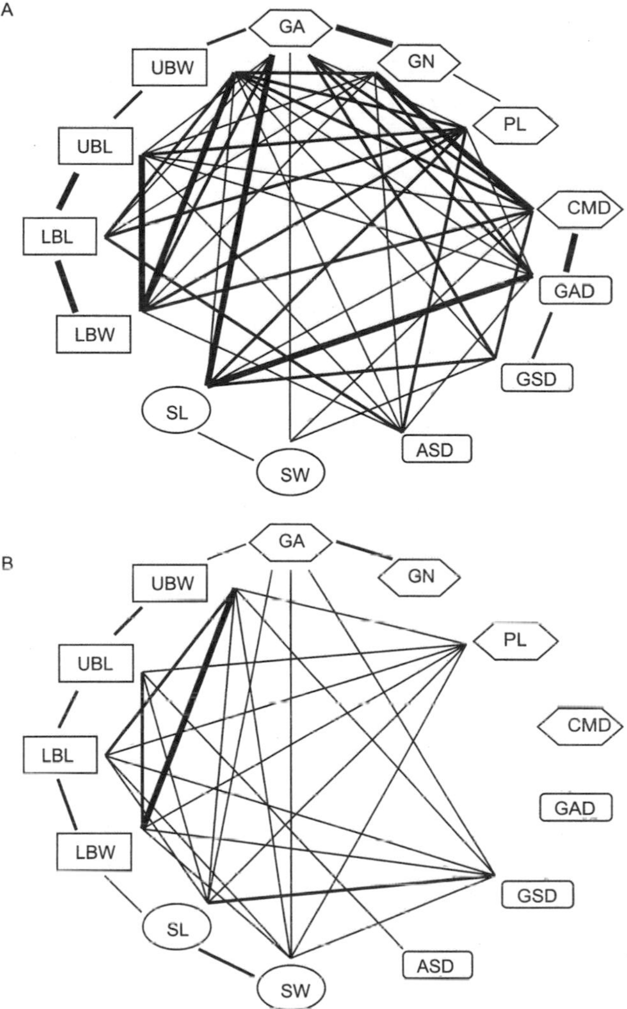


Figure 2.2 Genetic and phenotypic correlations among several blossom traits of plants from the Tulum population of *Dalechampia scandens* raised in the greenhouse in a diallel crossing experiment (see Hansen et al. 2003a). A. Additive genetic correlations ($N = 1046$); integration index = 5.48. B. Phenotypic correlations ($N = 1046$); integration index = 2.90. Integration-index values differ at $p < 0.05$. Only correlations $> 0.5$ are reported on the figures; the thickness of the lines reflects the strength of the absolute correlation in increments of 0.1. Solid lines connecting traits indicate positive correlations, dotted lines, when present, indicate negative correlations. Abbreviations: Traits related to the male inflorescence (in hexagons, following the graphical approach of Murren et al. 2002): GA, gland area; GN, number of bractlets forming the gland; PL, length of the peduncle supporting the male inflorescence; CMD, diameter of the central staminate flower. Traits related to the female inflorescence (in ovals): SW, style width; SL, style length. Traits related to the involucral bracts (in rectangles): UBL, upper bract length; UBW, upper bract width; LBL, lower bract length; LBW, lower bract width. Functional traits (in rounded rectangles): GAD, gland-anther distance; GSD, gland stigma distance; ASD, anther-stigma distance (see Hansen et al. 2003a for additional details on these measurements).

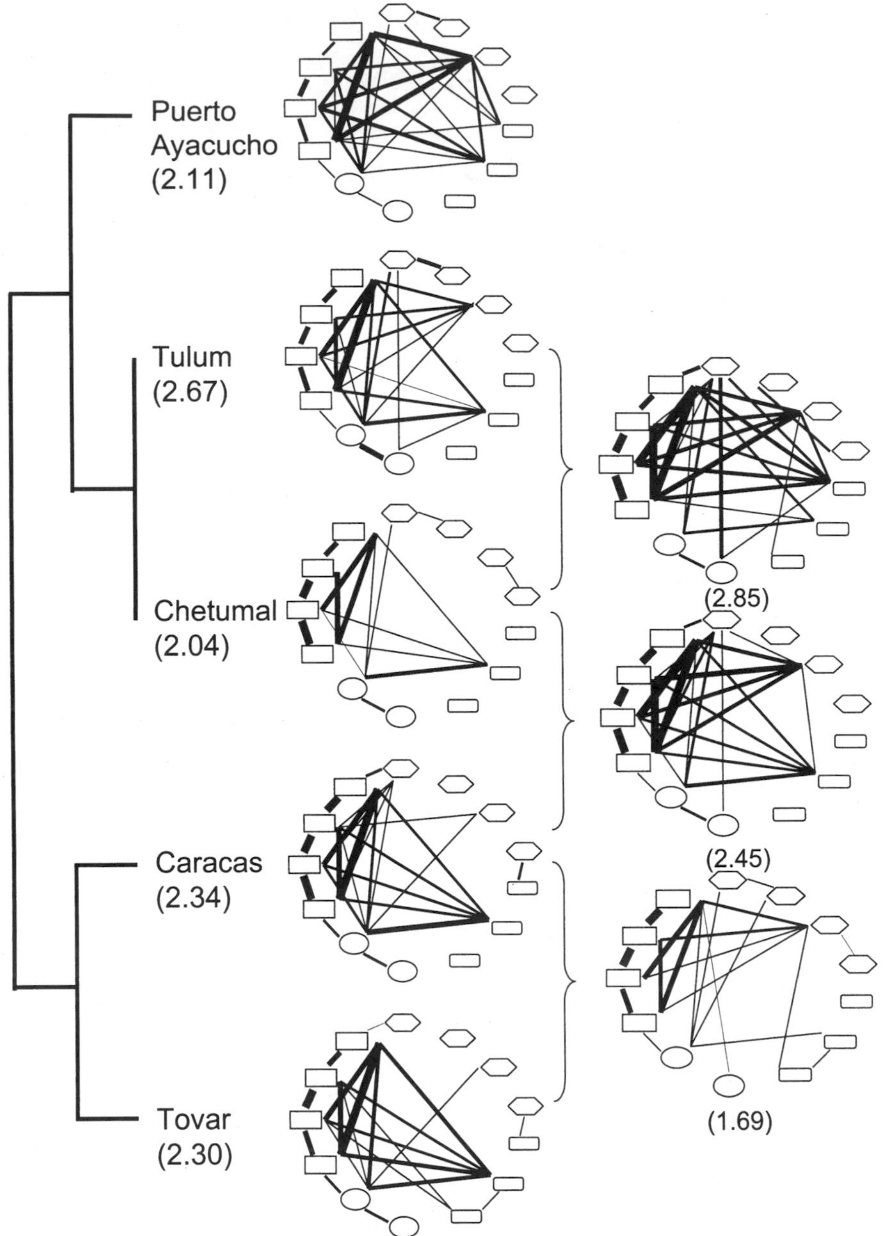


Figure 2.3 Phenotypic correlations in five populations of *D. scandens* and three $F_1$ interpopulation crosses. Phylogeny is derived from ISSR data (Pélabon et al. in press). The data for each population have been recorded from individuals raised together in a single greenhouse (Tulum, $N = 33$; Chetumal, $N = 93$; Caracas, $N = 43$; Tovar, $N = 39$; Puerto Ayacucho, $N = 16$; Chetumal × Tulum, $N = 24$; Caracas × Tovar, $N = 19$; Chetumal × Caracas, $N = 45$). Corrected integration index ($I'$) indicated next to each population name; all have overlapping confidence intervals, estimated by randomization. Traits, positions, and abbreviations as in Fig. 2.2

component to the evolutionary pattern (Fig. 2.3). The matrix correlation between the between-population difference in $I$ and the genetic distance between populations (estimated from ISSR data; Armbruster et al. in prep.) was marginally significant ($r = 0.517$, $p = 0.09$, based on a Mantel test). We also calculated the similarity between the genetic-distance matrix and the similarity matrix of trait-correlation patterns. However, the matrix correlation was only 0.09 ($p > 0.50$, Mantel test). As did Clausen and Heisey (1958) and Murren et al. (2002), we found that interpopulation hybrids ($F_1$'s) tended to show higher levels of phenotypic integration than either parental population (two out of three crosses; Fig. 2.3), although the differences were not significant (overlapping CIs, calculated by randomization). This general pattern (if there is one) may be the result of the increased $F_1$ variance in traits that show some effects of dominance or epistasis, or decreased developmental stability due to outbreeding depression (see Pélabon et al. in press).

We also calculated correlations among the five population means for *Dalechampia scandens* grown in the greenhouse. Interestingly, the correlations tended to be stronger among populations than within populations, although the difference in the values of corrected integration index was not significant (compare Figs. 2.2B and 2.4). Whether this is a general trend requires more work on additional species (see below).

A subset of the blossom traits most directly related to pollination function was measured in the field at three hierarchical levels: among genets within one population of *D. scandens* (Tulum), among 18 populations of *D. scandens*, and among 14 species belonging to the large *Dalechampia* clade containing *D. scandens* (clade

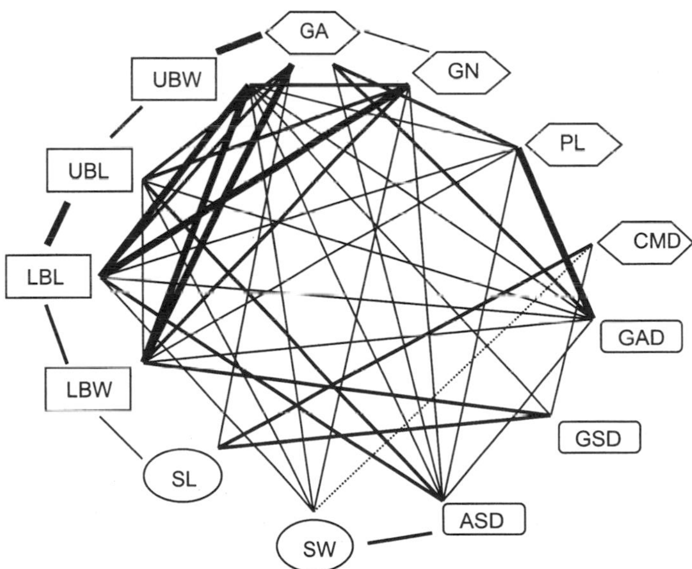


Figure 2.4 Phenotypic correlations between population means of blossom traits, from five populations of *D. scandens* grown in a common greenhouse. Populations were Tulum, Chetumal, Tovar, Caracas, and Puerto Ayacucho. Abbreviations, symbols, and line definition as in Fig. 2.2.

membership determined by analysis of the sequence of the ITS region of the nuclear ribosomal DNA; Armbruster and Baldwin 1998, unpublished data). The corrected integration index ($I'$) was fairly similar across all 18 populations, and there was no statistical evidence of heterogeneity in $I'$ (all CIs overlapped, determined by randomization). The correlations between functionally related traits increased dramatically from within population, to among populations, to among species, although the differences in the values of the corrected integration index were not significant (Table 2.1). Precision and accuracy appeared to be low: GAD and GSD were rather weakly correlated within the population (although apparently stronger than in the greenhouse). The optimalities in means were higher, however: correlations of population means were very strong ($r = 0.92$), as they were at the species level ($r = 0.90$). If this relationship were the result of developmental constraint, we would expect the opposite: a weakening relationship at higher levels because species have had the most time to break down genetic/developmental constraints. Thus the population and species patterns appear to reflect the evolution of optimality in mean anther and stigma position. The strong correlations at population and species levels may indicate the shape of the underlying fitness surface governing the divergence of *Dalechampia* populations and species (apparently a multidimensional "ridge"; see Phillips and Arnold 1989; Armbruster 1990; Armbruster and Schwaegerle 1996). Note that this result is testable through phenotypic-selection studies conducted in the field on traits manipulated (genetically or mechanically) to deviate from the predicted combinations of high fitness.

### *Collinsia* and *Tonella* (Scrophulariaceae, *sensu lato*)

In *Collinsia* and the closely related sister genus, *Tonella*, pollen deposition and pollen pickup are performed by two separate floral whorls of a single flower. This is of course the norm in angiosperms, but represents an "advancement" in structural/developmental integration over that seen in gymnosperms and monoecious or dioecious angiosperms. The portion of the family (Plantaginaceae; formerly Scrophulariaceae), to which *Collinsia* and *Tonella* belong (tribe Cheloneae; Wolfe et al. 2002), shows significant reduction in number of parts and/or connation within each floral whorl, representing further structural/developmental integration over more basal angiosperms. Specifically, pistils have been reduced to two, and they are fully fused (see Armbruster et al. 2002a for general analysis of carpel fusion), and the stamens have been reduced to four. Variation in structural integration is seen within the study group itself. While *Tonella* has open flowers that have four stamens and a style splayed in three dimensions, *Collinsia* has pea-flower-like flowers, with the five corolla lobes arranged as banner (two fused lobes), two wings, and a folded keel (Fig. 2.5). The folded keel encloses the distal two-thirds of the stamens and the style (and the terminal stigma). In addition to keeping out "unwanted" pollen thieves (Armbruster, unpublished data), this arrangement reduces the positional variation within (higher precision) and among (optimality in mean) sexual parts to essentially one dimension, hence structurally improving precision and optimality of mean contact with the pollinator. At the base of the banner the corolla narrows, forming an aperture and

Table 2.1 Patterns of variation and covariation at three hierarchical levels of organization in *Dalechampia*. A. Among genets within one population of one species (Tulum, *D. scandens*; $N = 22$). B. Among populations within one species (*D. scandens*; based on population means; $N = 18$). C. Among resin-reward species within the larger clade to which *D. scandens* belongs (based on species means; $N = 14$).

| | Gland-stigma | Gland-anther | Anther-stigma | Gland length | Gland width | Gland area | Lower bract length |
|---|---|---|---|---|---|---|---|
| A. | | | | | | | |
| Gland-stigma | 0.203 | | | | | | |
| Gland-anther | **0.554**$^{*}$ | *0.213* | | | | **rel.*I*′ = 16.9%**[a] | |
| Anther-stigma | **0.491**$^{*}$ | **0.507**$^{*}$ | *0.248* | | | | |
| Gland length | 0.339 | 0.335 | 0.392 | *0.104* | | | |
| Gland width | 0.031 | 0.009 | 0.160 | **0.585**$^{**}$ | *0.145* | | |
| Gland area | 0.166 | 0.154 | 0.306 | **0.844**$^{***}$ | **0.926**$^{***}$ | *0.217* | |
| Lower-bract length | 0.342 | 0.365 | 0.142 | 0.352 | **0.522**$^{*}$ | **0.473**$^{*}$ | *0.252* |
| B. | | | | | | | |
| Gland-stigma | *0.213* | | | | | | |
| Gland-anther | **0.921**$^{***}$ | *0.185* | | | | **rel.*I*′ = 39.8%**[a] | |
| Anther-stigma | **0.615**$^{**}$ | **0.619**$^{**}$ | *0.416* | | | | |
| Gland length | **0.774**$^{***}$ | **0.708**$^{***}$ | **0.623**$^{***}$ | *0.240* | | | |
| Gland width | **0.645**$^{***}$ | **0.577**$^{*}$ | 0.460 | 0.431 | *0.400* | | |
| Gland area | **0.695**$^{***}$ | **0.622**$^{**}$ | **0.492**$^{*}$ | **0.947**$^{***}$ | **0.984**$^{***}$ | *0.646* | |
| Lower-bract length | **0.863**$^{***}$ | **0.832**$^{***}$ | **0.412**$^{*}$ | **0.687**$^{**}$ | **0.566**$^{*}$ | **0.641**$^{**}$ | *0.169* |
| C. | | | | | | | |
| Gland-stigma | *0.510* | | | | | | |
| Gland-anther | **0.897**$^{***}$ | *0.500* | | | | **rel.*I*′ = 48.8%**[a] | |
| Anther-stigma | 0.458 | **0.693**$^{**}$ | *0.480* | | | | |
| Gland length | **0.610**$^{*}$ | **0.737**$^{**}$ | **0.691**$^{**}$ | *0.320* | | | |
| Gland width | **0.744**$^{**}$ | **0.841**$^{***}$ | **0.722**$^{**}$ | **0.880**$^{***}$ | *0.430* | | |
| Gland area | **0.744**$^{**}$ | **0.828**$^{***}$ | **0.716**$^{**}$ | **0.939**$^{***}$ | **0.980**$^{***}$ | *0.720* | |
| Lower-bract length | **0.721**$^{**}$ | **0.860**$^{***}$ | **0.772**$^{**}$ | **0.857**$^{***}$ | **0.860**$^{***}$ | **0.911**$^{***}$ | *0.500* |

*Notes*: Coefficients of variation for the appropriate level are given in italics on the diagonal. The relativized corrected integration index (rel. $I'$) is given above the diagonal. Shared superscripts indicate no significant difference in rel. $I'$. Traits in row 1 connected by solid lines are structurally related, and those connected by dashed lines are expected to covary by selection for accuracy in pollen placement on pollinators. Significant $r$ values are shown in bold. * $p < 0.05$, ** $p < 0.01$, *** $p < 0.001$.

tube, through which pollinating bees can insert their proboscides but not their heads. Thus, the banner base and aperture form a fixed "landmark" that orients the bee consistently relative to the sexual parts of the flower. This increased mechanical integration of floral parts increases the potential for precision and mean optimality in pollination.

The statistical data for within-population variation and correlations bear out the greater accuracy and precision in *Collinsia*, but not *Tonella*, compared to

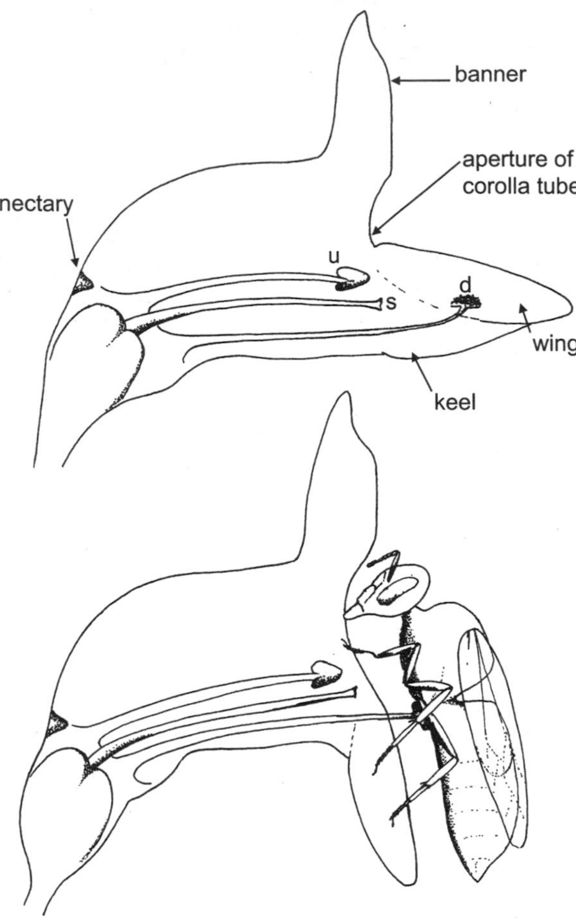


Figure 2.5 Schematic side view of a flower of *Collinsia torreyi*. The corolla is 8 mm long. Top: Details of the "internal" floral parts. Abbreviations: u = undehisced stamen, d = dehisced stamen, s = stigma (tip of the style). Bottom: Floral shape when keel is depressed by pollinator. This flower is in a semi-male phase, placing pollen on the pollinator but the stigma not yet contacting it. The style elongates later in flower development (see Armbruster et al. 2002b).

*Dalechampia*. The absolute size of pollen-load spot on a bee visiting a single flower can be determined from the size of the pollen zone within the keel tip. This is the region where pollen is released as the four anthers dehisce one at a time over the life of the flower. The size of this zone within the keel is determined by the size of the dehiscing anthers and the "ability" of the flower to position the anthers in the same location during the course of late floral development (Armbruster et al. 2002b). We found that the pollen zone ranged from 60% (1.58 mm) of the lower corolla-lobe length in *Tonella floribunda* down to 13% (1.42 mm) of the keel length in *C. heterophylla* (Sierra Nevada). These percentages are roughly indicative of the portion of the underside of the bee being dusted with pollen, since the typical bee attracted to a species is about as long as the keel (Armbruster, unpublished data). It appears, then, that pollen placement is more precise and hence potentially more accurate in *Collinsia* than in *Tonella* (and *Dalechampia*). Indeed, the within-population phenotypic correlation (among genets within a population of *C. sparsiflora arvensis* from Lake County, California) between style length and stamen length in late floral development (stage 4) was 0.90. This indicates surprisingly high phenotypic integration and precision within the population, considering the several sources of variance, including different developmental schedules of the two structures (see Armbruster et al. 2002b).

It seems reasonable to interpret the derived floral shape in *Collinsia* as at least partly an adaptation for increased accuracy of anther and stigma contact with pollinating bees. The divergence of *Collinsia* from the *Tonella* floral plan may be the result of selection for more precise pollen placement on and pickup from the pollinating bees or, alternatively, selection for tighter correspondence of the positions of dehisced anthers and receptive stigmas as a mechanism promoting autonomous self-pollination. Which scenario is more likely depends on the order of evolution of outcrossing and selfing in the genus, which is not yet resolved (see Armbruster et al. 2002b). The greater structural integration of male and female parts, resulting from the keel-like nature of the lower corolla lobe, represents an adaptation for greater accuracy in pollen placement/pickup and a preaptation for greater selfing ability, or vice versa. In either case, response to selection for accurate pollen placement and pickup in *Tonella* and *Collinsia* should be less limited by architectural "looseness" and associated developmental noise than in *Dalechampia*. This is perhaps reflected in the lower proportion of the underside of the body bearing pollen in *Collinsia* pollinators (ca. 13 40%). *Tonella* has poor precision compared to *Collinsia* and is a species-poor genus, despite being of the same age as *Collinsia*. This supports the idea of the keel as a key innovation allowing pollination accuracy (and efficiency) and clade success in *Collinsia*.

The only strong phenotypic correlations seen within *Collinsia* populations, besides the style-stamen correlation discussed above, were between measured structures belonging to the same whorl (e.g., the lengths of four stamens were generally highly correlated; not shown) or between measurements that are structurally and hence genetically correlated (Table 2.2). For example, the two components of the corolla were correlated with the total corolla length. The corrected integration index ($I'$) was fairly similar across all eight populations, and there was no statistical evidence of heterogeneity in $I'$ (all confidence intervals overlapped, determined by randomization).

Among populations of *C. sparsiflora*, the correlation of average stamen and style lengths was very high ($r = 0.988$; Table 2.2), indicating high optimality in mean anther and stigma contact with pollinators. All other floral traits were also highly correlated and the overall integration index was much higher than within populations. This indicates that population divergence of floral traits has proceeded along a scaling trajectory. The general scarcity of strong correlations within populations suggests that this scaling is the result of the interplay of selective pressures governing population divergence, although genetic constraints cannot be ruled out without genetic data. Species-level analyses indicated a similar high level of correlations and high overall integration index. Thus, the same "adaptive-scaling" rule seems to hold at the species level.

It appears that evolutionary divergence of *Collinsia* populations and species results in scaling of size of all flower parts, much as we saw in *Dalechampia*. There is very little change in shape, and the integration metric is thus higher at the population and species levels than it was among genets within a population

Table 2.2 Patterns of variation and covariation at three hierarchical levels of organization in *Collinsia* and *Tonella*. A. Among genets within one population of one species ($N = 7$). B. Among populations within one species (*C. sparsiflora*, based on population means; $N = 8$). C. Among species of *Collinsia* and *Tonella* (based on species means; $N = 22$).

| | Corolla length | Tube length | Keel length | Style length | Stamen length | Anther length |
|---|---|---|---|---|---|---|
| A. | | | | | | |
| Corolla length | *0.052* | | | | | |
| Tube length | **0.829*** | *0.114* | | | **rel.*I*′ = 12.9%**[a] | |
| Keel length | **0.928**** | 0.562 | *0.100* | | | |
| Style length | −0.326 | −0.411 | −0.210 | *0.140* | | |
| Stamen length | −0.137 | −0.409 | 0.070 | **0.900**** | *0.075* | |
| Anther length | 0.226 | −0.271 | 0.515 | 0.068 | 0.382 | *0.017* |
| B. | | | | | | |
| Corolla length | *0.311* | | | | | |
| Tube length | **0.875**** | *0.228* | | | **rel.*I*′ = 70.2%**[a] | |
| Keel length | **0.983***** | **0.772*** | *0.389* | | | |
| Style length | **0.974***** | **0.777*** | **0.986***** | *0.502* | | |
| Stamen length | **0.973***** | **0.808*** | **0.973***** | **0.988***** | *0.454* | |
| Anther length | **0.891**** | **0.780*** | **0.876**** | **0.869**** | **0.894**** | *0.366* |
| C. | | | | | | |
| Corolla length | *0.453* | | | | | |
| Tube length | **0.896***** | *0.445* | | | **rel.*I*′ = 68.4%**[a] | |
| Keel length | **0.976***** | **0.779***** | **0.531** | | | |
| Style length | **0.915***** | **0.766***** | **0.927***** | *0.575* | | |
| Stamen length | **0.921***** | **0.679***** | **0.970***** | **0.926***** | *0.577* | |
| Anther length | **0.833***** | **0.759***** | **0.807***** | **0.786***** | **0.724***** | *0.439* |

*Notes*: Coefficients of variation for the appropriate level are given in italics on the diagonal. The relativized corrected integration index (rel. $I'$) is given above the diagonal. Shared superscripts indicate no significant difference in rel. $I'$. All measurements were taken on flowers in the final stage of floral development (d-4). Traits in row 1 connected by solid lines are structurally related, and those connected by dashed lines are expected to covary by selection for "accuracy" in pollen placement on pollinators. Significant $r$-values are shown in bold. $^{*}p < 0.05$, $^{**}p < 0.01$, $^{***}p < 0.00$.

(Table 2.2), although the differences were not significant (all CIs overlapped, determined by randomization).

### *Stylidium* (Stylideaceae)

From an insect's perspective, *Stylidium* flowers resemble those of *Collinsia* in having a landing platform (the "lower" pair of petals) and a showy, banner-

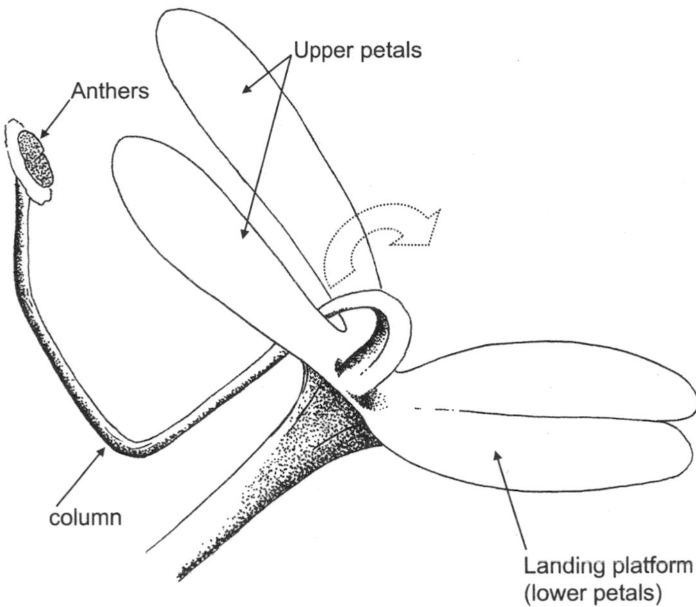


Figure 2.6 A flower of *Stylidium bicolor*. The dotted arrow represents the movement of the column when it snaps forward in response to being triggered by a pollinator. The column is 13.2 mm long.

like pair of "upper" petals (Fig. 2.6). The nectar is also borne in a tube at the base of the banner, although the tube is much narrower in *Stylidium*. Pollen deposition and pollen pickup are both performed by the motile "column," a structure derived from the complete fusion of staminal and pistillate tissues. This intimate structural arrangement of staminal and pistillate tissues is characteristic of the entire family (Stylidiaceae, *sensu stricto;* e.g., Cronquist 1981), and may represent the ultimate solution to selection for integration of male and female function (see Diggle 1992). The anthers are borne near the tip of the column during the first phase of flower maturation. When a pollinator lands on the flower platform and probes the floral tube for nectar, the column snaps forward and places a small spot of pollen on the dorsum, side, or underside of the insect, depending on the species of *Stylidium* (Armbruster et al. 1994). In the second phase of floral maturation, the empty anther sacs curl to either side of the column, and the stigma protrudes from the same location where the anthers had been. The column tip continues to rap the pollinators in the same location as in the first phase.

Also as in *Collinsia*, but unlike in *Dalechampia*, there is a fixed landmark at the base of the upper petals, against which the insect visitor places its head (or proboscis) when obtaining nectar. When the insect is properly positioned and probing for nectar, it touches a sensitive point at the base of the column with its head, triggering the spring-like column movement. Thus the insect acts as a pollinator only when it is precisely positioned on the flower, which, in combination with the small area bearing pollen at the tip of the column, the perfect repetition of column movement in both male and female phases, and the exact replacement of the anthers with the stigma in the female phase, results in a level of precision and accuracy of pollen placement and pickup seen in very few other plants. Indeed, there is nearly perfect correlation, expressed within populations, among populations within a species, and among species, between the position of pollen placement and the position of stigma contact with the pollinator (data not presented).

The anthers are relatively small and there is very little variation in the position of the pollinator on the flower when it triggers the column. Hence the pollen-load spot on a pollinating bee or bee-fly is quite small, about 1 $mm^2$ after visiting one or more flowers in a population (Armbruster et al. 1994), or about 10–15% of the length of the pollinator's body.

The within-population variance in the position of pollen placement on an artificial "pollinator" was quite small: an SD of 0.34 mm (population no. 27, *S. brunonianum, sensu lato*). This represents a CV of 0.082 (range across 12 populations of *S. brunonianum*: 0.058–0.179). The coefficient of variation in stigma contact with the pollinator is presumably similar. This indicates relatively high precision in pollen placement and pickup.

The phenotypic correlation between the location of anther contact and the location of stigma contact was not measured per se. However, the structural integration of the stamens and style into one structure, with anther and stigma in virtually the same position, guarantees the nearly perfect correlation between the two. The corrected integration index ($I'$) was fairly similar across all twelve populations, and there was no statistical evidence of heterogeneity in $I'$ (CIs overlapping, estimated by randomization).

The patterns of variance and covariance within populations, among populations, and among species revealed some interesting patterns quite different from what we saw in the other two genera (Table 2.3). Most within-population phenotypic correlations were weak, and among-population correlations between traits were all very high, as we saw in *Dalechampia* and *Collinsia*. Thus, divergence among populations within species is a scaling function, with shape constant and the value of the integration index becoming very high. However, divergence of species followed a different pattern. Unlike in *Dalechampia* and *Collinsia*, among-species correlations of measured traits were small. Thus, divergence of species does not follow a scaling function, but reflects instead important changes in shape. The integration index at the species level was accordingly very low (Table 2.3).

## Comparisons of Study Systems

These three groups of plants show several similarities and several differences in their floral organization and patterns of integration. The pollination units (blossoms) in *Dalechampia* are collections of independent flowers (pseudanthia), whereas the pollination units in *Collinsia* and *Stylidium* are single flowers. Following the idea that structural and developmental organization of floral parts determines most of the covariation within populations and constrains evolutionary divergence of populations and species (e.g., Stebbins 1974, Schluter

Table 2.3 Patterns of variation and covariation at three hierarchical levels of organization in *Stylidium*. A. Among genets within one population (no. 27) of one species (no. 9; $N = 8$). B. Among populations within one species (no. 9, based on population means; $N = 12$). C. Among species of *Stylidium* (based on species means; $N = 30$).

| | Tube length | Column length | Platform width | Platform length |
|---|---|---|---|---|
| A. | | | | |
| Tube length | *0.094* | | | |
| Column length | 0.644 | *0.076* | | **rel.$I'$ = 26.9%**[a] |
| Platform width | 0.240 | 0.483 | *0.173* | |
| Platform length | 0.398 | **0.808*** | **0.818*** | *0.129* |
| B. | | | | |
| Tube length | *0.159* | | | |
| Column length | **0.834***** | *0.276* | | **rel.$I'$ = 70.0%**[b] |
| Platform width | **0.906***** | **0.907***** | *0.177* | |
| Platform length | **0.899** | **0.839***** | **0.855***** | *0.145* |
| C. | | | | |
| Tube length | *0.389* | | | |
| Column length | 0.281 | *0.0467* | | **rel.$I'$ = 18.8%**[ab] |
| Platform width | **0.532**** | 0.200 | *0.287* | |
| Platform length | **0.434*** | 0.314 | **0.763***** | *0.489* |

*Notes*: Coefficients of variation for the appropriate level are given in italics on the diagonal. The relative integration index (rel. $I'$ is given above the diagonal. Shared superscripts indicate no significant difference in rel. $I'$. Traits in row 1 connected by solid lines are structurally related. Significant $r$-values are shown in bold. $^{*}p < 0.05$, $^{**}p < 0.01$, $^{***}p < 0.001$.

1996), we would predict the fewest correlations within and among populations and species of *Dalechampia*, and much stronger correlations in *Collinsia* and *Stylidium*. In all three systems the degree of phenotypic integration within populations was relatively low (Tables 2.1–2.3). Traits that were obviously structurally or mensurally related (hence strongly genetically correlated) showed large phenotypic correlations. Traits affecting the accurateness of pollen placement and pickup also tended to covary within populations in all three groups. Other traits tended to be weakly correlated or uncorrelated. In contrast to our prediction, *Collinsia* and *Stylidium* flowers were not noticeably more integrated within populations than were *Dalechampia* blossoms.

At the level of populations, all three study systems exhibited dramatic rises in numbers of correlations and hence integration index (cf. Tables 2.1–2.3). Nearly all traits in all three study groups were strongly intercorrelated at the among-population level. This indicates that population divergence is either highly constrained by genetic/developmental relationships, or is governed by an adaptive surface that follows a multivariate axis scaling with size. The lack of strong phenotypic correlations within populations does not support the genetic/developmental constraint hypothesis and instead suggests selection is responsible. However, low within-population, phenotypic integration of blossoms could be an artifact of high environmental variance, high developmental noise, and/or limited trait variation within populations. (Covariance cannot exist without variance, so the concept of integration as a correlation becomes meaningless without variation within or among populations.) Indeed, quantitative-genetic analyses suggest that genetic correlations may, in fact, constrain short-term evolution in *D. scandens* (Hansen et al. 2003a, 2003b). Consistent with having structurally more integrated flowers (compared to *Dalechampia* blossoms), *Collinsia* and *Stylidium* showed more constrained evolutionary divergence of populations than *Dalechampia* (cf. Tables 2.1–2.3).

The three study groups differed sharply in patterns of covariation at the level of species. While *Dalechampia* and *Collinsia* showed strong correlations at this level, species divergence appearing much as an extension of population divergence, *Stylidium* showed a qualitatively different evolutionary pattern at the species level. In *Stylidium*, correlations between flower parts were much weaker at the among-species level than at the among-population level. However, according to the structural-integration hypothesis, species of *Dalechampia*, not *Stylidium*, should show the least constrained differentiation. The unconstrained species differentiation in *Stylidium* can be explained by one or a combination of two processes: (1) Genetic/developmental constraints that limited population divergence to an axis of size scaling in all three study groups have broken down at the species level (presumably as a result of the longer duration or different nature of species divergence). (2) There has been a change in the shape of the adaptive surface governing the evolution at the species level. As discussed below, we suspect that both of these processes are involved in generating the observed patterns.

## Discussion and Conclusions

A review of the literature suggests that flowers are largely modular, with floral variation somewhat decoupled from variation in vegetative traits (Berg 1960; Fenster 1991; Andersson 1994; Conner and Sterling 1996; Cresswell 1998, 2000; Armbruster et al. 1999; Wolfe and Krstolic 1999; J. Herrera 2001). Flowers are also largely integrated phenotypically: size-related floral traits tend to covary both within and among populations and related species (Armbruster 1991; Diggle 1992; Conner and Via 1993; Conner and Sterling 1995; Mitchell-Olds 1996; Andersson 1997; Armbruster et al. 1999; Waitt and Levin 1993; Herrera et al. 2002; Hansen et al. 2003b). The specific sources of this covariation in each case are difficult to determine, however, because they could include genetic, physiological, developmental, and/or functional (selective) relationships among traits. Clues to the causes of phenotypic integration come from examination of patterns of variance and covariance of traits within populations and, especially valuable, from comparisons of patterns of phenotypic integration across populations and species. There is particular value in using an understanding of probable functional and developmental relationships between traits to make explicit, testable predictions about which phenotypic traits should covary for adaptive or developmental reasons (see Gould and Lewontin 1979; Herrera et al. 2002). With respect to understanding the function and evolution of flowers, it seems particularly useful to assess the influence of various forms of integration on the precision, bias, and accuracy of pollen placement on, and capture from, pollinators. This perspective provides a valuable conceptual framework for mechanistic and adaptive analyses of floral form, variation, and integration.

## Does Floral Precision Influence Macroevolutionary Patterns?

The data we have presented here indicate that precision of pollen placement and pickup range from relatively low in *Dalechampia*, to intermediate in *Collinsia*, to extremely high in *Stylidium*. The correspondence of degree of precision with the degree of structural and presumably developmental integration across these three groups suggests that selection for precision may often promote the evolution of floral integration (see also Diggle 1992).

Variation in degree of precision, and hence in one component of accuracy, may also affect how populations adapt to co-occurrence with closely related species. The presence of co-occurring congeners imposes selection against sending pollen to heterospecific stigmas or receiving heterospecific pollen, which may lead to specialization in pollination ecology (Waser 1983; Armbruster 1985, 1986; Armbruster et al. 1994; Johnson and Steiner 2000). This specialization can occur along one or more of three "axes." If accuracy and precision in pollen placement are high, populations can specialize by placing pollen in different locations on the same individual pollinators. But this will not work in species with low precision, and specialization must take the form of utilizing different pollinator species or blooming at different times of the day or season. In the case of *Dalechampia* and, to some extent, *Collinsia*, specialization and coexistence of co-flowering congeners is manifested though utilizing pollinator species of different body size (e.g., Armbruster 1985, 1986, 1988). This is effected by selection for blossoms/flowers of different sizes but similar shapes (scaling). In contrast, *Stylidium*, with its extreme precision in pollen placement and pickup, can specialize in response to sympatry with congeners by placing pollen in a new, "unused" location on the same insects, that is, evolving a new flower shape rather than a new size. Genetic/developmental constraints may limit this off-axis evolution within a species, but we see it fully expressed in among-species divergence (Table 2.3; Armbruster et al. 1994).

The differences in hierarchical patterns of variance and covariance across study groups may be explained by differences in the degree of "sexual integration" that flowers/blossoms achieve within populations. Accuracy of pollen placement and pollen pickup, arguably the most important adaptive consequences of floral integration, may thereby have a major influence over the course of floral macroevolution. Low precision and/or optimality in mean, as a result of limited integration of male and female parts (and other developmental and structural limitations such as low developmental homeostasis and/or lack of a narrowly defined landmark that positions the pollinator), may limit macroevolution to a trajectory of allometric scaling. High precision may allow floral macroevolution to respond to selective challenges by exploring new regions of phenotypic space far away from the allometric trajectory. Under certain circumstances we may also see the evolution of integration being driven by selection for precision, and hence greater integration evolving, for example, in populations that share pollinators with more taxa.

In contrast, the evolution of precision may be limited by departure of the mean phenotype from optimality in mean. For example, the low precision in *Dalechampia* could reflect selection for spreading pollen widely enough over pollinating bees to overcome a mismatch in where anthers and stigmas contact these insects.

## Distinguishing Among Causes of Genetic Integration

One of the most difficult issues to resolve in the study of floral integration is the extent to which genetic correlations should be viewed as proximal constraints versus adaptations promoting adaptive covariance of functionally interacting traits (cf. Armbruster 1990, 1991; Conner and Via 1993; Conner and Sterling 1995; Armbruster and Schwaegerle 1996). Our task, in general, is to try to estimate the relative contribution of adaptation and constraint to phenotypic and genetic correlations. Any one genetic correlation, however, is likely to be either a by-product of the organism's genetic architecture or the result of selection for a specific phenotypic correlation. If one is interested in studying adaptation, it is not particularly instructive to assume, in the absence of independent evidence, that a particular genetic correlation is of adaptive origin (cf. Gould and Lewontin 1979; Cheverud 1984, 1996; Conner and Via 1993). Nor is it valid, of course, to assume that a particular genetic correlation is purely a by-product of genetic constraints (see Eberhard and Gutierrez 1991). The tools available for distinguishing between these alternatives for any given pair of traits, unfortunately, remain inadequate to the task, and the particular alternative hypothesis employed must depend on the kind of questions one wishes to address (see Armbruster 2002). A promising methodology is, however, to generate explicit model-based predictions that are then tested with comparative analyses of correlation structure across multiple populations and species.

*Acknowledgments* We thank numerous assistants for help in the field or greenhouse, Matthew Carlson for Fig. 2.1, and Massimo Pigliucci, Katherine Preston, and Alice Winn for constructive comments on previous versions of this contribution. Support for the research was provided by the US National Science Foundation (grants DEB-906607, DEB-9020265, DEB-9318640, DEB-9708333 to W.S.A.), the Norwegian Research Council (grants to W.S.A. and T.F.H.), and the Norwegian Academy of Sciences (Nansen Fund grant to W.S.A. and C.P.).

### *Literature Cited*

Andersson, S. 1994. Floral stability, pollination efficiency, and experimental manipulation of the corolla phenotype in *Nemophila menziesii* (Hydrophyllaceae). American Journal of Botany 81:1397–1402.

Andersson, S. 1997. Genetic constraints on phenotypic evolution in *Nigella* (Ranunculaceae). Biological Journal of the Linnean Society 62:519–532.

Armbruster, W. S. 1985. Patterns of character divergence and the evolution of reproductive ecotypes of *Dalechampia scandens* (Euphorbiaceae). Evolution 39:733–752.

Armbruster, W. S. 1986. Reproductive interactions between sympatric *Dalechampia* species: are natural assemblages "random" or organized? Ecology 67:522–533.

Armbruster, W. S. 1988. Multilevel comparative analysis of morphology, function, and evolution of *Dalechampia* blossoms. Ecology 69:1746–1761.

Armbruster, W. S. 1990. Estimating and testing the shapes of adaptive surfaces: the morphology and pollination of *Dalechampia* blossoms. American Naturalist 135:14–31.

Armbruster, W. S. 1991. Multilevel analyses of morphometric data from natural plant populations: insights into ontogenetic, genetic, and selective correlations in *Dalechampia scandens*. Evolution 45:1229–1244.

Armbruster, W. S. 2002. Can indirect selection and genetic context contribute to trait diversification? A transition-probability study of blossom-color evolution in two genera. Journal of Evolutionary Biology 15:468–486.

Armbruster, W. S., and B. G. Baldwin. 1998. Switch from specialized to generalized pollination. Nature 394:632.

Armbruster, W. S., and K. E. Schwaegerle. 1996. Causes of covariation of phenotypic traits among populations. Journal of Evolutionary Biology 9:261–276.

Armbruster, W. S., M. E. Edwards, and E. M. Debevec. 1994. Character displacement generates assemblage structure of Western Australian triggerplants (*Stylidium*). Ecology 75:315–329.

Armbruster, W. S., V. S. Di Stilio, J. D. Tuxill, T. C. Flores, and J. L. Velasquez Runk. 1999. Covariance and decoupling of floral and vegetative traits in nine neotropical plants: A reevaluation of Berg's correlation-pleiades concept. American Journal of Botany 86:39–55.

Armbruster, W. S., E. M. Debevec, and M. F. Willson. 2002a. The evolution of syncarpy in angiosperms: theoretical and phylogenetic analyses of the effects of carpel fusion on offspring quantity and quality. Journal of Evolutionary Biology 15:657–672.

Armbruster, W. S., C. P. H. Mulder, B. G. Baldwin, S. Kalisz, B. Wessa, and H. Nute. 2002b. Comparative analysis of floral development and mating-system evolution in tribe Collinsieae (Scrophulariaceae, s.l.). American Journal of Botany 89:37–49.

Arnold, S. J., and P. C. Phillips. 1999. Hierarchical comparison of genetic variance-covariance matrices. II. Coastal-inland divergence in the garter snake, *Thamnophis elegans*. Evolution 53:1516–1527.

Barrett, S. C. H. 2002. Sexual interference of the floral kind. Heredity 88:154–159.

Begin, M., and D. A. Roff. 2001. An analysis of G matrix variation in two closely related cricket species, *Gryllus firmus* and *G. pennsylvanicus*. Journal of Evolutionary Biology 14:1–13.

Berg, R. L. 1960. The ecological significance of correlation pleiades. Evolution 14:171–180.

Cheverud. J. M. 1982. Phenotypic, genetic, and environmental integration in the cranium. Evolution 36: 499–516.

Cheverud, J. M. 1984. Quantitative genetics and developmental constraints on evolution by selection. Journal of Theoretical Biology 110:155–171.

Cheverud, J. M. 1996. Developmental integration and the evolution of pleiotropy. American Zoologist 36:44–50.

Cheverud, J. M., G. P. Wagner, and M. M. Dow. 1989. Methods for the comparative analysis of variation patterns. Systematic Zoology 38:201–213.

Clausen, J. and W. M. Heisey. 1958. Experimental studies on the nature of plant species. IV. Genetic structure of ecological races. Publication 615, Carnegie Institution of Washington, Washington, DC.

Conner, J. K., and A. Sterling. 1995. Testing hypotheses of functional relationships: a comparative survey of correlation patterns among floral traits in five insect-pollinated plants. American Journal of Botany 82:1399–1406.

Conner, J. K., and A. Sterling. 1996. Selection for independence of floral and vegetative traits: evidence from correlation patterns in five species. Canadian Journal of Botany 74:642–644.

Conner, J. K., and S. Via. 1993. Patterns of phenotypic and genetic correlation among morphological and life-history traits in wild radish, *Raphanus raphanastrum*. Evolution 47:704–711.

Cresswell, J. E. 1998. Stabilizing selection and the structural variability of flowers within species. Annals of Botany, London 81:463–473.

Cresswell, J. E. 2000. Manipulation of female architecture in flowers reveals a narrow optimum for pollen deposition. Ecology 81:3244–3249.

Cronquist. A. 1981. An Integrated System of Classification of Flowering Plants. Columbia University Press, New York.

Diggle, P. K. 1992. Development and the evolution of plant reproductive characters. Pp. 326–355 in R. Wyatt (ed.), Ecology and Evolution of Plant Reproduction. Chapman & Hall, New York.

Eberhard, W. G. and Gutierrez, E. E. 1991. Male dimorphisms in beetles and earwigs and the question of developmental constraints. Evolution 45:18–28.

Fenster, C. B. 1991. Selection on floral morphology by hummingbirds. Biotropica 23: 98–101.

Fenster, C. B., and L. F. Galloway. 1997. Developmental homeostasis and floral form: evolutionary consequences and genetic basis. International Journal of Plant Science 158:S121-S130.

Gould, S. J. 2002. The Structure of Evolutionary Theory. Harvard University Press, Cambridge, MA.

Gould, S.J. and Lewontin, R.C. 1979. The spandrels of San Marco and the Panglossian paradigm: a critique of the adaptationist programme. Proceedings of the Royal Society of London, Series B. 205:581–598.

Hansen, T. F., and E. P. Martins. 1996. Translating between microevolutionary process and macroevolutionary patterns: the correlation structure of interspecific data. Evolution 50:1404–1417.

Hansen, T. F., C. Pélabon, W. S. Armbruster, and M. Carlson. 2003a. Evolvability and genetic constraint in *Dalechampia* blossoms: components of variance and measures of evolvability. Journal of Evolutionary Biology 16:754–757.

Hansen, T. F., W. S. Armbruster, M. Carlson, and C. Pélabon. 2003b. Evolvability and genetic constraint in *Dalechampia* blossoms: genetic correlations and conditional evolvability. Journal of Experimental Zoology Molecular and Developmental Evolution 296B:23–39.

Herrera, C. M. 2001. Deconstructing a floral phenotype: do pollinators select for corolla integration in *Lavandula latifolia*? Journal of Evolutionary Biology 14:574–584.

Herrera, C. M., X. Cerda, M. B. Garcia, J. Guitian, M. Medrano, P. J. Rey, and A. M. Sanchez-Lafuente. 2002. Floral integration, phenotypic covariance structure and pollinator variation in bumblebee-pollinated *Helleborus foetidus*. Journal of Evolutionary Biology 15:108–121.

Herrera, J. 2001. The variability of organs differentially involved in pollination, and correlations of traits in Genisteae (Leguminosae: Papilionoideae). Annals of Botany 88:1027–1037.

Johnson, S. D., and K. E. Steiner. 2000. Generalization versus specialization in plant pollination systems. Trends in Ecology and Evolution 15:140–143.

Lande, R., and S. J. Arnold. 1983. The measurement of selection on correlated characters. Evolution 37:1210–1226.

Mitchell-Olds, T. 1996. Genetic constraints on life-history evolution: quantitative-trait loci influencing growth and flowering in *Arabidopsis thaliana*. Evolution 50:140–145.

Murren, C. J., N. Pendleton, and M. Pigliucci. 2002. Evolution of phenotypic integration in *Brassica* (Brassicaceae). American Journal of Botany 89:655–663.

Olson, E. C., and R. L. Miller. 1958. Morphological Integration. University of Chicago Press, Chicago.

Orzack, S. H., and E. Sober. 1994. Optimality models and the test of adaptationism. American Naturalist 143:361–380.

Pélabon, C., M. L. Carlson, T. F. Hansen, N. G. Yoccoz, and W. S. Armbruster. In press. Consequences of inter-population crosses on developmental stability and canalization of floral traits in *Dalechampia scandens* (Euphorbiaceae). Journal of Evolutionary Biology.

Phillips, P. C., and S. J. Arnold. 1989. Visualizing multivariate selection. Evolution 43:1209–1222.

Pigliucci, M., and K. Hayden. 2001. Phenotypic plasticity is the major determinant of changes in phenotypic integration in *Arabidopsis*. New Phytologist 152:419–430.

Primack, R. B. 1987. Relationships among flowers, fruits, and seeds. Annual Review of Ecology and Systematics 18:409–40.

Roff, D. 2000. The evolution of the G matrix: selection or drift? Heredity 84:135–142.

Schlichting, C. D. 1986. The evolution of phenotypic plasticity in plants. Annual Review of Ecology and Systematics 17:667–693.

Schlichting, C. D., and M. Pigliucci. 1998. Phenotypic Evolution: A Reaction Norm Perspective. Sinauer, Sunderland, MA.

Schluter, D. 1996. Adaptive radiation along genetic lines of least resistance. Evolution 50:1766–1774.

Schwaegerle, K. E., and D. A. Levin. 1991. Quantitative genetics of fitness traits in a wild population of *Phlox*. Evolution 45:169–177.

Sokal, R. R. 1978. Population differentiation: something different or more of the same? Pp. 215–239 in P. F. Brussard (ed.), Ecological Genetics: The interface. Springer-Verlag, New York.

Sokal, R. R., and F. J. Rohlf. 1995. Biometry, 3rd ed. Freeman, New York.

Stebbins, G. L. 1974. Flowering Plants: Evolution Above the Species Level. Belknap Press, Cambridge, MA.

Sultan, S. E. 2000. Phenotypic plasticity for plant development, function and life history. Trends in Plant Sciences 5:537–542.

Wagner, G. P. 1984. On the eigenvalue distribution of genetic and phenotypic dispersion matrices: evidence for a nonrandom organization of quantitative character variation. Journal of Mathematical Biology 21:77–95.

Wagner, G. P. 1996. Homologues, natural kinds, and the evolution of modularity. American Zoologist 36:36–43.

Wagner, G. P., and L. Altenberg. 1996. Complex adaptations and the evolution of evolvability. Evolution 50:967–976.

Waitt D. E., and D. A. Levin. 1993. Phenotypic integration and plastic correlations in *Phlox drummondii* (Polemoniaceae). American Journal of Botany 80:1224–1233.

Waldmann, P., and S. Andersson. 2000. Comparison of genetic (co)variance matrices within and between *Scabiosa canescens* and *S. columbaria*. Journal of Evolutionary Biology 13:826–835.

Waser, N. M. 1983. Competition for pollination and floral character differences among sympatric plant species: a review of evidence. Pp. 277–293 in C. E. Jones and R. J. Little (eds.), Handbook of Experimental Pollination Ecology. Van Nostrand Reinhold, New York.

Webster, G. L. 1994. Synopsis of the genera and suprageneric taxa of Euphorbiaceae. Annals of the Missouri Botanical Garden 81:33–144.

Westoby, M., M. R. Leishman, and J. M. Lord. 1995. On misinterpreting the "phylogenetic correction". Journal of Ecology 83:531–534.

Winn, A. A. 1999. Is seasonal variation in leaf traits adaptive for the annual plant *Diceranda linearifolia*? Journal of Evolutionary Biology 12:306–313.

Wolfe, A. D., S. L. Datwyler, and C. P. Randle. 2002. A phylogenetic and biogeographic analysis of the Cheloneae (Scrophulariaceae) based on ITS and matK sequence data. Systematic Botany 27:138–148.

Wolfe, L. M., and J. L. Krstolic. 1999. Floral symmetry and its influence on variance in flower size. American Naturalist 154:484–488.

Zeng, Z.-B. 1988. Long-term correlated response, interpopulation covariation, and interspecific allometry. Evolution 42:363–374.

# 3

# Integration and Modularity in the Evolution of Sexual Ornaments

## *An Overlooked Perspective*

ALEXANDER V. BADYAEV

### Paradox of an Ideal Sexual Ornament: Exaggerated and Flexible Yet Honest

The expression of sexual ornaments often reflects male health and overall physiological condition, and females mating with the most ornamented males are assumed to produce the best-adapted offspring and receive the most benefits from such males (Andersson 1994). However, there are a number of unresolved issues in the evolution of sexual traits that reflect males' condition.

On one hand, sexual traits, such as deer antlers, beetle horns, or elongated bird tails, are under strong directional selection for greater expression, and this selection favors reduced integration (e.g., favors modified allometric relationships) between sexual traits and the rest of the organism (Eberhard 1985; Emlen and Nijhout 2000). Indeed, most sexual traits are "stand-alone" structures (e.g., deer antlers are far less integrated and more variable than frontal bones of which they are a part). On the other hand, sexual traits are expected to indicate the physiological condition and health of an individual, such that the expression of sexual traits represents a complex summary of many organismal processes (Wedekind 1992; Johnstone 1995). However, to be such a summary, sexual traits should be highly integrated into many organismal functions. This represents a paradox where sexual traits are expected to be both less integrated for greater expression, and more integrated to better indicate physiological quality.

Similarly, to be a reliable reflection of organismal processes, sexual ornaments (or, more precisely, the pathways that lead to development of sexual ornaments) are expected to be well integrated in the ontogeny of an organism. To be an indicator of health, these traits need to be costly to an organism (Zahavi 1975; Andersson 1982; Grafen 1990), yet there is an advantage to modify investment of resources into the development of sexual ornamentation depending on life history and context of breeding (Höglund and Sheldon 1998; Kokko 1998; Badyaev and Qvarnström 2002). Facultative investment into the production of sexual ornamentation should favor relative independence of developmental pathways of non-sexual traits and sexual ornaments, which will make the latter less reliable indicators of overall quality. This represents another paradox where costly indicators of individual condition can evolve only if it is possible for an organism to accomplish and "survive" their ontogeny. Yet developmental mechanisms enabling this survival will make sexual ornaments less integrated with organismal functions and thus less reliable indicators of them.

The concepts of morphological integration and modularity have provided valuable insights into the evolution of complex biological structures (Olson and Miller 1958; Schlichting and Pigliucci 1998; Wagner 2001). Whereas it is clear from the above discussion that these concepts are central to the ontogeny, function, and evolution of sexual ornamentation, they are mostly overlooked. Here I discuss how the concepts of integration and modularity can facilitate our understanding of unresolved issues in the evolution of sexual traits.

## Designing Sexual Traits: Relative Importance of Internal and External Selection

### Why Sexual Traits?

Sexual traits are unique in that the environment in which they function is mostly external to the organism, and the selection pressures that affect their evolution are mostly due to phenotypes of other individuals. For example, some sexual ornaments function as signals affecting the behavior of the opposite sex, in which case the sensory characteristics of the opposite sex exert selection pressure on the sexual ornament design and function (Guilford and Dawkins 1991; Endler 1992; Rowe 1999). Other sexual traits function to facilitate mate choice, copulation, or gamete transfer mechanistically, as in the case of animals' genitalia or plants' flower displays. In this case the morphology of the other sex (Eberhard 1985; Arnqvist and Rowe 2002; Dixson and Anderson 2002), or that of pollinators (Creswell 1998; Giurfa et al. 1999), exerts selection pressures on the design of the sexual trait (see Armbruster et al., Chapter 2, this volume).

Organismal traits are subject to two general kinds of selection pressures: internal and external. Internal selection is selection for the internal cohesiveness of an organism during development or function (Whyte 1965; Schlichting and Pigliucci 1998; Wagner and Schwenk 2000). Such selection typically is not sensitive to the external environment, but it is an outcome of the internal selection that determines which phenotype will experience external, or environmental, selection (Fusco 2001; Arthur 2002). Whereas the morphology of sexual ornaments might be affected mostly by external selection (e.g., by other individuals), as exemplified by low genetic correlations of sexual ornaments with other morphological structures (e.g., Preziosi and Roff 1998, see below), external selection for greater condition-dependence of sexual ornaments acts, indirectly, on the developmental aspects of the sexual ornament (Fig. 3.1). Thus, whereas the proximate target of external selection is the elaboration of sexual ornaments, the ultimate target is the underlying relationship between the expression of the ornament and condition of the organism, that is, the developmental integration of the ornament.

## Relative Importance of External and Internal Selections

While often not stated explicitly, the relative importance of internal and external processes is central to current debate on the evolution of sexual ornamentation. Some authors argue that it is the internal selection (i.e., a trait's developmental processes) that determines which morphological traits are most suitable for elaboration by sexual selection. This is because the development of a trait determines its integration into organismal functions and, thus, its condition-dependence.

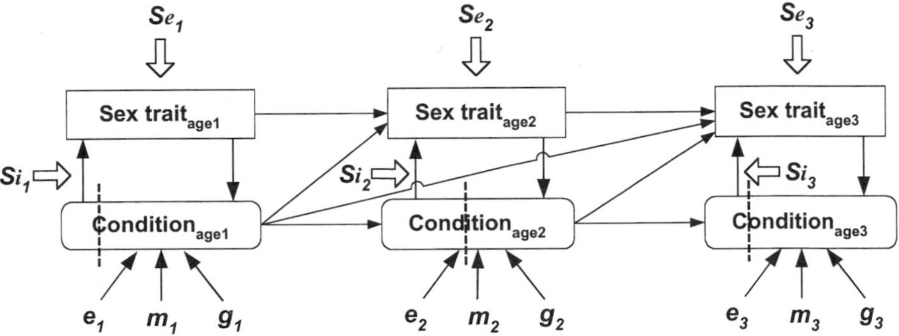


Figure 3.1 Conceptual model of the ontogeny of a sexual trait. At each age $n$, organismal condition depends on environmental ($e_n$), maternal ($m_n$), and direct genetic effects ($g_n$). At each age, the allocation of overall condition to the production of a sexual trait (indicated by vertical dotted line and an arrow pointing up) is governed by internal developmental processes *Si* (internal selection). At each age, the condition-dependent sexual trait (or its precursor) is a subject of external selection (*Se*). Age-specific *Se* acts, indirectly, on the aspects of internal development of a sexual trait (or its precursor), i.e., on *Si* as well as on the overall condition (arrow pointing down), because of the organism-wide costs of expression of the sexual trait. Depending on the similarity in the age-specific effects of *e*, *m*, and *g*, organismal condition at each age can influence organismal condition at subsequent ages. Depending on the duration of development and its complexity, the sexual trait (or its precursor) at each age can affect sexual traits at subsequent ages. Thus, at each age, the sexual trait is affected by (1) direct effects of age-specific condition, (2) direct effects of the sexual trait (or its precursor) at preceding ages, (3) indirect effect of organismal condition at preceding ages on current condition and on allocation of condition to the sexual trait, and (4) indirect effect of the sexual trait (or its precursor) at preceding ages on preceding and current condition as well as on preceding and current allocations to sexual traits. At each age, limiting the number of these effects or decreasing their strength (i.e., by decreasing the overlap in the effects of *e*, *m*, and *g* among ages) will decrease the condition-dependence of sexual ornamentation.

Therefore, some morphological traits may be developmentally predisposed to be targeted by sexual selection because of their existing dependence on condition or high phenotypic variability (Endler 1992; Schluter and Price 1993). Other authors suggest that greater integration into organismal functions can be accomplished easily when sexual selection favors greater trait elaboration, and a prior developmental predisposition to condition-dependence is not required (Grafen 1990; Price et al. 1993). Moreover, elaboration of sexual traits is often triggered by a sensory bias of the receiver and different traits are susceptible to this bias in different environments regardless of their initial developmental integration and condition-dependence (Endler 1992; Schluter and Price 1993).

For example, some researchers have argued that developmental properties make carotenoid-based coloration a more reliable indicator of individual condition in animals, and thus a more frequent target of sexual selection, than melanin- or non-pigment-based coloration (which have both fewer environmental components in their production and stronger developmental integration among these components (Gray 1996; Hill 1999; Badyaev and Hill 2000; McGraw and Hill 2000; Badyaev and Young 2003). Other researchers disagree strongly (Jawor and Breitwisch 2003) and there are many empirical examples supporting each point of view. There are sexually selected carotenoid indicators of condition (Kodric-Brown 1989; review in Hill 1999), melanin-based indicators of condition (Veiga and Puetra 1996; Griffith et al. 1999; Fitze and Richner 2002), non-pigment-based coloration indicators of condition (Keyser and Hill 2000), as well as examples of the lack of strong condition-dependence in and corresponding lack of sexual selection on each of these sources of ornament coloration (Seehausen et al. 1999; Dale 2000; McGraw and Hill 2000; Pryke et al. 2001).

On the other hand, developmental patterns in sexual ornaments can themselves be a product of external selection. For example, some authors suggest that external sexual selection for symmetry in bilateral sexual ornaments favors the evolution of developmental pathways that minimize developmental instability in sexual traits caused by their great exaggeration (Møller 1992a, 1992b; Thornhill 1992; Swaddle and Cuthill 1994a, 1994b; Badyaev et al. 1998; Møller and Thornhill 1998; Morris 1998). Other authors argue that asymmetry is a by-product of selection on ornament size and not itself a target of sexual selection (Evans 1993; Tomkins and Simmons 1996; Hunt and Simmons 1997; David et al. 1998; Cuervo and Møller 1999b; Breuker and Brakefield 2002).

Similarly, two prevalent explanations for the maintenance of genetic variance in sexual traits focus on the relative importance of internal and external processes for accomplishing an ornament's condition-dependence. Pomiankowski and Møller (1995) suggested that the continuous elaboration of a sexual ornament and increasing benefits of such elaboration (i.e., due to external selection) favor the accumulation of developmental modifiers that, by limiting developmental integration, facilitate production of an ever larger ornament. By contrast, Rowe and Houle (1996) suggested that selection for greater elaboration of a sexual ornament results in an increase in the number of condition-dependent (and not specific to an ornament) inputs into ever more expensive production of an ornament, thus increasing integration of the ornament with the rest of the organism. Because greater elaboration of sexual ornamentation requires both weaker developmental integration and stronger external selection, it seems that neither of these viewpoints can, by itself, fully account for the evolution of exaggerated and condition-dependent sexual ornaments. But, importantly, both theories suggest that external selection acts on aspects of developmental integration, either limiting it—by evolution of ornament-specific developmental modifiers (Pomiankowski and Møller 1995)—or strengthening it, by evolution of greater pleiotropy of ornament development (Rowe and Houle 1996; Kotiaho et al. 2001).

Here I suggest that both viewpoints are correct and that the argument over the primacy of internal development versus external selection can be resolved if one considers a likely sequence in the evolution of sexual ornaments. This is because the coevolution of male strategies to reduce condition-dependence in sexual ornaments and female strategies to restore the condition-dependence results in distinct temporal patterns of developmental integration of sexual ornaments into organismal functions (i.e., its condition-dependence).

## Process of Organismal Integration in Sexual Ornaments

### Directional Selection on Sexual Ornaments

Sexual ornaments are expected to be under directional sexual selection favoring their further exaggeration (Andersson 1994). The reliability of exaggerated ornaments as indicators of individual condition is reinforced by viability costs of the expression of large ornaments (Zahavi 1975), by costs of ornament production (e.g., Arnqvist 1994), or by the processes that are proximately unrelated to either production or expression, but reflect general viability (Hamilton and Zuk 1982; Fölstad and Karter 1992, reviewed in Iwasa et al. 1991).

From an ontogenetic perspective, the greater expression of a sexual ornament in adults reflects an organism's ability to successfully accomplish the ornament's expensive development (Fig. 3.1). Two general processes produce ornament elaboration during development. First, ornament elaboration is accomplished by the increased and more efficient allocation of resources and condition so that a progressively smaller increase in condition is amplified into a progressively larger sexual ornament. Second, ornament elaboration is enabled by a decrease in the integration (i.e., by the evolution of flexible allometric relationships) between sexual ornaments and the rest of an organism. This might be accomplished by weakening the pathways that affect the ornament development. Both of these processes might ultimately reduce the developmental costs of sexual ornamentation for males and will lead to the evolution of female strategies to restore these costs.

### Males: Evolution of Cost-Reducing Strategies in the Development of Sexual Ornaments

Consistent directional selection for greater exaggeration of costly sexual ornaments favors the evolution of cost-reducing strategies in their development (Fig. 3.2). These strategies can be divided into two general groups. First, selection might favor the evolution of ornament-specific developmental pathways that are

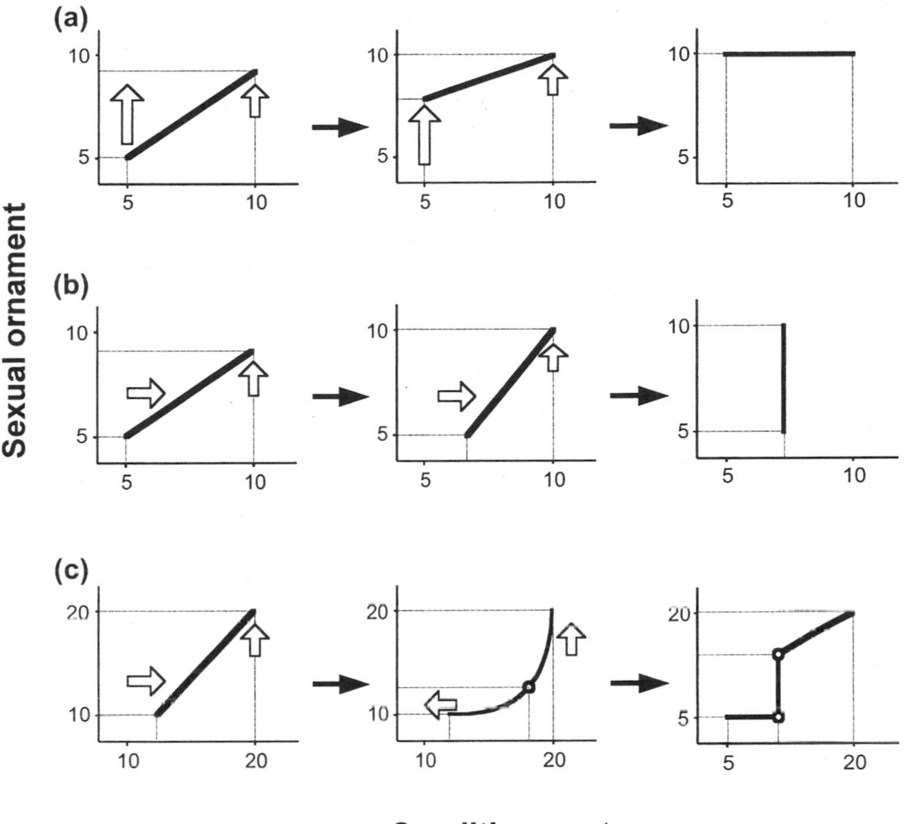


Figure 3.2 Hypothetical scenarios of evolution of cost-reducing strategies in the expression of sexual ornament. (a) Selection (arrows) for cheaper expression of more exaggerated ornament is greater on the ornaments with smaller costs and favors an increase in the amount of precursor of the ornament (i.e., increase in the intercept) leading to the loss of the condition-dependent expression and to the maximum expression of sexual ornament. (b) Selection for higher condition-dependence and greater exaggeration leads to an increase in the rate and to a decrease in duration of growth of the sexual ornament ultimately resulting in the loss of condition-dependence in ornament expression. (c) Same as in (b), but disproportionally greater costs of ornament expression at intermediate condition lead to selection against expressing ornament at low condition and selection for greater expression of ornament at high condition, favoring facultative trait expression regulated by a condition-dependent developmental switch.

less dependent on organismal condition, that is, favor greater modularity of a sexual ornament's *development*. Second, selection might favor the evolution of facultative and context-dependent expression of condition-dependent sexual ornaments, that is, favor greater modularity in a sexual ornament's *expression*.

The initial elaboration of a condition-dependent ornament under directional selection can be accomplished by an increase in the amount of its developmental precursor or a more efficient developmental pathway (Figs. 3.2a and 3.3a), or by

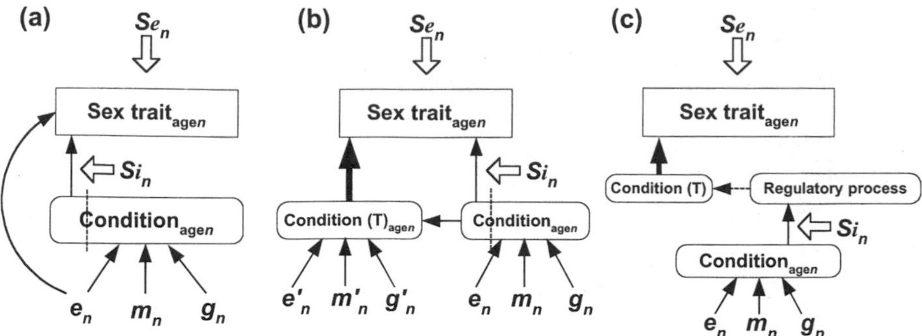


Figure 3.3 Conceptual model of cost-reducing strategies in sexual ornament expression corresponding to Fig. 3.2. (a) Increase in the amount of precursor of sexual ornament (Fig. 3.2a) is enabled by the evolution of a direct path from environment (*e*) to sexual trait expression. (b) Increase in rate and decrease in duration of sexual trait development (Fig. 3.2b) is enabled by the evolution of an ornament-specific developmental pathway (T) in addition to the pathway from general condition to the ornament. (c) Facultative expression of exaggerated sexual ornament (Fig. 3.2c) is enabled by the evolution of a condition-sensitive regulatory switch activating the ornament-specific developmental pathway (T) without direct transfer of resources.

the faster growth of a sexual ornament relative to an increase in condition (Figs. 3.2b and 3.3b). In both cases, selection for cost reduction acts to modify the link between organismal condition and the production of sexual ornament in order to enable greater expression of sexual ornamentation without a corresponding change in cost.

For example, because animals cannot synthesize carotenoids, the presence of carotenoid-based pigmentation in integuments can indicate individual foraging ability (Endler 1983). Thus, females that prefer males with greater expression of carotenoid-based colors will select better-quality mates (Hill 1999). However, once female preference for carotenoid-based coloration is established, there are a number of ways for males to express greater carotenoid ornamentation at lower costs (Hill 1994; Badyaev and Duckworth 2003). For example, males can preferentially forage on carotenoid-rich foods or types of carotenoids that can be deposited directly (i.e., without energetically expensive metabolism) (Fig. 3.3a). Males can alter the expression of carotenoid ornaments by concentrating carotenoid deposition (e.g., limiting the ornamental patch size to increase the intensity of color: Hill 1993), by altering integument structures to increase the display of carotenoids already absorbed (Brush and Seifried 1968; Olson 1970), or by displaying carotenoids for a longer time after consumption (Kodric-Brown 1998). Some bird species, such as flamingoes (*Phoenicopterus* spp.), orioles (*Icterus* spp.), and scarlet ibises (*Endocimus ruber*), apparently have evolved highly specialized pathways for carotenoid metabolism (Figs. 3.2b and 3.3b) that enable them to efficiently extract and deposit carotenoids from food with minimal cost to the organism (e.g., Fox et al. 1969; Mulvihill et al. 1992; Fitze and Richner 2002).

However, female choice of an ornament does not seem to persist until complete loss of its condition-dependence (i.e., the last stages in Fig. 3.2a,b); that is, condition-dependent sexual traits rarely form multiple ornaments (Møller and Pomiankowski 1993b; Iwasa and Pomiankowski 1994; Prum 1997; Badyaev and Hill 2003; Badyaev et al. 2002). This suggests that females either switch to a more informative trait or modify their preference to restore the condition-dependence of the ornament.

Selection for greater elaboration of sexual ornaments can favor their environment- and context-dependent expression and, accordingly, in many taxa the most elaborate sexual ornaments are expressed facultatively (Figs. 3.2c and 3.3c) (Emlen and Nijhout 2000). Facultative expression might be enabled by the evolution of temporal modularity in developmental pathways of an ornament (Figs. 3.1 and 3.3c), and there are many advantages to such expression. First, it enables greater capitalization on environmental condition in production of sexual ornaments when such condition improves (i.e., by lessening the link between condition at $age_{n-1}$ and at $age_n$ in Fig. 3.1). A costly and exaggerated ornament might be expressed only when food is abundant, when predators are rare, or when the benefits from the trait display are the highest. Emlen and Nijhout (2000) showed that the evolution of threshold expression of sexual ornamentation is favored when the high costs of exaggerated ornaments provide increasingly lower reproductive benefits to individuals of intermediate quality and intermediate ornament elaboration (Zahavi 1975).

Evolution of temporal modularity in an ornament's developmental pathways also enables age-specific expression of sexual ornaments, which is beneficial when older individuals are able to channel more resources into production of larger traits (Hansen and Price 1995; Kokko 1997; Badyaev and Qvarnström 2002). For example, male zebra finches (*Taeniopygia guttata*) that were subjected to nutritional stress during growth were nevertheless able, when adults, to develop sexual ornamentation indistinguishable from that of control birds (Birkhead et al. 1999). Moreover, temporal modularity in ornament production can enable the sex-limited expression of an ornament for which the developmental pathways are shared between the sexes (Badyaev 2002). Thus, an exaggerated sexual ornament of a male may indicate the elaboration of a physiological process that is beneficial to both male and female offspring of this male (Kodric-Brown and Brown 1984). For example, the decoupling of carotenoid consumption, which is present in both sexes, from carotenoid deposition, which occurs only in the male's integument, might be accomplished by temporal modularity of ornament development. This enables the evolution of female preference for foraging characteristics that are important for both sexes because of immunological and other health benefits of carotenoid consumption.

Proximately, the facultative expression of sexual ornaments might be enabled by the decoupling of organismal condition from the ornament-specific developmental pathways (Figs. 3.2c and 3.3c; Badyaev and Duckworth 2003). Eventually, the interaction between an organism and an ornament-specific developmental pathway might occur without the transfer of resources and thus might not represent a continuous tradeoff (Fig. 3.3c). Instead, the interaction between an organism and ornament development can be mediated by threshold-like regulatory mechanisms (e.g., hormones; see Emlen and Nijhout 2000 for examples in insects) in which development of an ornament is triggered by the release of resource-level sensitive hormones, but without the actual material transfer. Once this mechanism is in place, differences in costs and benefits of ornament expression between environments can lead to rapid evolution of threshold sensitive controls (such as the degree and timing of sensitivity to hormones) without corresponding changes in the rules of allocation of resources to sexual traits (Moczek and Nijhout 2002). In contrast, in species where thresholds are governed by condition, environmental induction of trait development modifies actual allocation of resources to a sexual trait and comes at the expense of overall organismal condition (Radwan et al. 2002). In addition, hormone-sensitive thresholds can enable sex-limited and age-limited expression of an indicator of a process that is shared between the sexes or across ages (see above, Badyaev 2002). However, the evolution of complete modularity in facultatively expressed sexual ornaments might be rare because sexual ornaments often require prolonged development and female choice will favor greater temporal integration of sexual ornamentation.

### Females: Evolution of Cost-Restoring Strategies in the Development of Sexual Ornaments

Females' mate choice should favor maintenance of condition-dependence in male sexual ornaments (Fig. 3.4). First, female choice might favor the evolution of amplifiers of quality within sexual ornaments. Second, females might base their choice on within-ornament traits that are necessary for ornament maintenance and production, but not directly related to the established pathways of ornament elaboration. Finally, females might base their preference on complex sexual ornaments or ornaments that require prolonged development, because these ornaments better summarize individual condition and are less likely to be produced by developmental pathways that are independent of males' condition.

Within-ornament amplifiers of condition-dependence in sexual ornamentation can increase the precision of females' discrimination among males' ornaments (Fig. 3.4). For example, tail markings and pigment-free spots are more common in birds with longer tails (Hasson 1989; Fitzpatrick 1998). Such markings make tail feathers more susceptible to abrasion and to damage by parasites and thus can reveal individual quality in species in which females prefer males with longer tails, but where individual variation in the condition-dependence of tail length itself might be reduced (Fitzpatrick 1998, 1999). Amplifiers of condition may take the form of displays, such as in the dark-eyed juncos (*Junco hyemalis*) where the expression of condition-dependent melanin spots is reinforced by dynamic plumage displays of males (J. A. Hill et al. 1999). Similarly, in birdsongs, narrow-frequency bandwith notes are strongly amplified by transmission through dense vegetation, and this environment-enhanced transmission increases the efficiency of male song in attracting a female (Slabbekoorn et al. 2002).

In the song of some birds, different elements are more difficult to produce than others. For example, the duration of pauses between the syllables and the time a male is able to maintain a maximum sound amplitude during rapid frequency modulations is limited by the costs of song production (e.g., Lambrechts 1996;

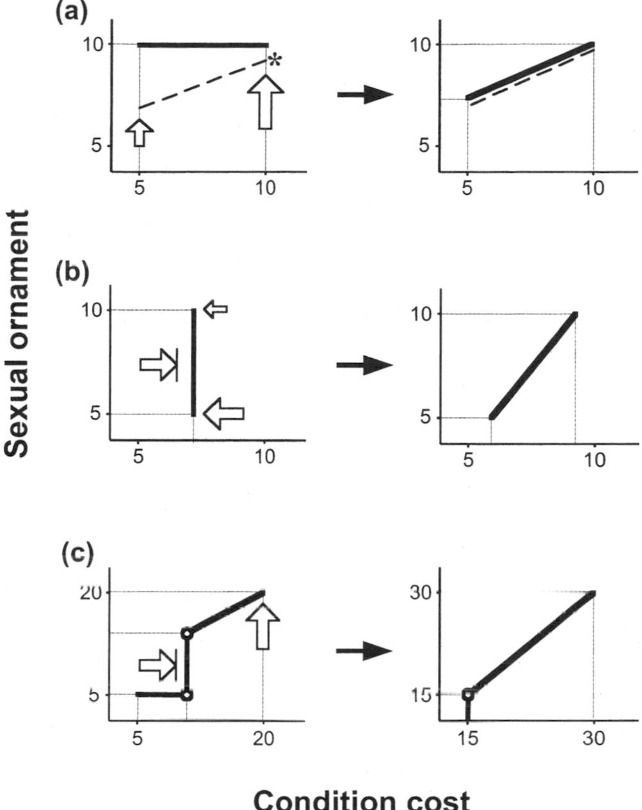


Figure 3.4 Hypothetical scenarios of evolution of strategies to restore condition-dependence in sexual ornaments (corresponding to Fig. 3.2). (a) Female preference (vertical arrows) for the within-ornament features, such as morphological amplifiers or behavioral displays—(shown by dashed line and an asterisk), exaggeration of which is not related to the established pathways of ornament elaboration, but that are necessary for ornament expression (Fig. 3.2a). Female preference for greater exaggerated ornaments of individuals in higher condition leads to a restoration of condition-dependence of an ornament. (b) Female preference for greater condition-dependence of an ornament (right-pointing arrow) in combination with a greater selection for a decrease in condition-dependence of smaller ornaments (due to their reduced benefits and higher costs; larger lower arrow), restores the condition-dependence of an ornament (Fig. 3.2b). (c) As in (b), but selection for greater condition-dependence of the threshold itself in combination with selection for greater exaggeration increases the slope between the ornament and the condition at the large values of ornamentation (Fig. 3.2c).

Podos 1997) and delivery (e.g., Richards and Wiley 1980; Badyaev and Leaf 1997). A recent study of dusky warblers (*Phylloscopus fuscatus*) showed that females have stronger preference for a male's ability to produce these particularly costly elements (e.g., maintenance of high sound amplitude) than for other song structures (Forstmeier et al. 2002; see also Vallet et al. 1998).

Similarly, fluctuating asymmetry (random developmental deviations from perfect symmetry in bilateral structures) in sexual ornaments can amplify the condition-dependence of development and production of exaggerated sexual ornaments. Female preference can favor greater integration of ornament development with organismal functions by targeting asymmetry itself (Fig. 3.4). For example, accomplishing symmetry of floral sexual displays in plants often requires organism-wide (i.e., not trait-specific) integration (Conner and Sterling 1996; Creswell 1998). Continuous female preference for an ornament that is both exaggerated and integrated into organismal functions may ultimately produce a negative relationship between fluctuating asymmetry of an ornament and ornament elaboration (Møller and Pomiankowski 1993a). However, this relationship is only expected at the advanced stages of ornament elaboration. For example, in wild turkeys (*Meleagris gallopavo*), the relationship between the asymmetry of right and left tarsal spurs and mean spur length was distinct between young and older males (Badyaev et al. 1998). The relationship was negative only in older males that attained near-maximum spur length and experienced strong sexual selection on spur length and asymmetry. The asymmetry was not related to spur size in young males that did not participate in mating and had shorter spurs.

Sexual ornaments are often expressed only late in ontogeny, but may require prolonged development. Long-growing and long-lasting ornaments thus better reflect an individual's average condition than ornaments the development and expression of which is shorter and less integrated. For example, although the ability to produce complex songs in birds is not expressed until adult stages, it is dependent on the precise development of brain structures and neural pathways necessary for song learning (Nowicki et al. 1998). An individual's ability to buffer nutritional stress during the development of song-related brain nuclei might be reflected in the ability to produce complex songs at adult stages. Thus, a female preference for more complex male songs might translate into a preference for individuals that are in higher condition and health over the life span (Nowicki et al. 2000). Similarly, the expression of a sexual ornament, a wattle, in adult ring-necked pheasants (*Phasianus colchicus*) reflected nutritional condition early in life. Males that were raised under nutritional stress showed lesser development of wattle at subsequent ages (Ohlsson et al. 2002). In wild turkeys, tarsal spurs that grow continuously throughout life were a better indicator of an individual long-term condition and viability than feather beards that grow mostly during older ages (Badyaev et al. 1998). An interesting example of displays that have a relatively short expression but prolonged development is the lifelong advancement to the center of leks in the black grouse (*Tetrao tetrix*) (Kokko et al. 1998, 1999). Proximity to center of the lek is costly to achieve and maintain, and males at the center of leks are preferred by females.

Female preference favors male ornaments that are closely linked to individual condition, but because male condition and the benefits that females receive are specific to the environment, different aspects of male condition are important to different females in different environments (Wedekind 1994; Qvarnström 2001; Badyaev and Qvarnström 2002). Moreover, a consistent female preference for higher elaboration of condition-dependent ornaments is accompanied by the evolution of male strategies to reduce the cost of such elaboration (Fig. 3.2). A combination of female preference for different aspects of male condition and male strategies to weaken developmental integration in sexual ornamentation, should favor the evolution of composite sexual ornaments whose components are linked to different organismal processes and can reliably reflect condition across a wide range of environments. Here I outline one hypothetical scenario of how a combination of male and female strategies can produce a sexual ornament that is both integrated into organismal functions and sufficiently flexible to allow for greater elaboration. There are three main stages (Fig. 3.5).

### Stage 1: Selection for Greater Exaggeration of Sexual Ornaments

Initial female selection of a male's sexual ornament favors traits with greater detectability and high phenotypic variation (Guilford and Dawkins 1991; Endler 1992; Schluter and Price 1993). Thus, traits that have weaker developmental integration with other traits and traits with greater environmental components in their development or expression (such as displays dependent on ambient light or diet-dependent pigmentation) may be predisposed to be targeted and elaborated by sexual selection. Once a trait is targeted, female preference for both cheaper and more efficient ways to discriminate among potential mates and for displays with stronger condition-dependence should favor larger expression and greater individual variation in the male's sexual ornament (Fig. 3.5). Initially, selection for greater elaboration of sexual display favors its greater condition-dependence, strengthening and expanding the existing links between ornament and condition (Rowe and Houle 1996). Eventually, greater integration with organismal functions increases the phenotypic variance in the sexual ornament and thus limits the effectiveness of external selection to accomplish progressively stronger condition-dependence (Price et al. 1993; see below).

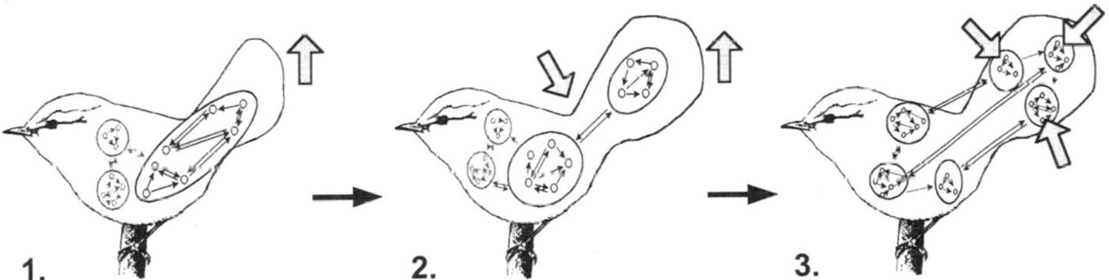


Figure 3.5 Conceptual illustration of the evolution of modularity in sexual ornaments. Small circles and thin arrows within drawings indicate individual physiological processes and their interactions. Large ellipses indicate modules of these processes. Large arrows illustrate external selection. Stage 1: Selection favors greater exaggeration of sexual traits, leading to decrease in their integration with organismal functions. Stage 2: Selection favors both further exaggeration of a sexual ornament (that is increasingly produced by ornament-specific pathways) and greater integration of ornament-specific pathways into organismal functions. Stage 3: Different components of the sexual ornament reflect different organismal processes and are selected in different environments producing a composite and dynamic sexual ornament. See text for discussion.

Selection for greater ornament exaggeration is selection against the developmental pathways that limit variation in sexual ornamentation (Møller and Pomiankowski 1993a; Pomiankowski and Møller 1995). Thus, the disruption of developmental integration of sexual ornaments leads not only to greater variation in ornament size but also to an increase in developmental abnormalities associated with ornament production. For example, the initial increase in the exaggeration of a sexual ornament during the transition from a monogamous to a polygynous mating system in birds was accompanied by an increase in both variation in ornament size and in fluctuating asymmetry of the ornaments (Cuervo and Møller 1999a, 2001). Similarly, in several insect species, directional selection on the size of a sexual ornament was accompanied by an increase in its fluctuating asymmetry, indicating disruption of developmental pathways of ornament production (Hunt and Simmons 1997; Civetta and Singh 1998; David et al. 1998; Breuker and Brakefield 2002). Importantly, at this stage, an increase in fluctuating asymmetry is ornament-specific, that is, developmental mistakes arise from a disruption of ornament production, not from the costs of ornament elaboration to the organism. For example, in birds ornamental feathers typically have higher phenotypic variation and greater fluctuating asymmetry than non-ornamental feathers (Alatalo et al. 1988; Møller and Hoglund 1991; Pomiankowski and Møller 1995; Cuervo and Møller 2001). Moreover, there is often no correlation between developmental instability of ornamental and non-ornamental feathers, suggesting that developmental mistakes are trait-specific and not due to organism-wide costs (Cuervo and Møller 1999b).

Overall, at this stage, selection for greater exaggeration of sexual ornamentation results in weakening the integration of the ornament with the rest of the organism (Fig. 3.5).

### Stage 2: Selection for Greater Condition-Dependence of Exaggerated Sexual Ornaments

As a by-product of weakening developmental integration that accompanies selection for greater exaggeration of a sexual ornament, the ornament becomes progressively less informative about processes other than those exclusively involved in its production (Fig. 3.5). Cost-reducing strategies of ornament elaboration facilitate the formation of ornament-specific developmental pathways and further decrease the dependence between ornament expression and organismal condition. An important consequence of weaker condition-dependence is a decrease in phenotypic variation in ornament expression within a population (Price et al. 1993; Rowe and Houle 1996; Fig. 3.5).

A set of female strategies to counterbalance a decrease in phenotypic variance and condition-dependence of male ornaments favors both greater expression of the ornament and its greater condition-dependence (Figs. 3.4 and 3.5). This is selection strengthening the link between ornament expression and the rest of the traits of an organism. At this stage, however, the developmental pathways of exaggerated sexual ornaments have acquired some independence from the organism's condition (Fig. 3.5). Thus, selection for their greater condition-dependence is likely to capitalize on the general costs that well-elaborated ornaments now impose on the entire organism (i.e., organism-wide costs of trait elaboration; Fig. 3.4b).

For example, great elongation of tail feathers or greater asymmetry in tail feathers leads to an organism-wide compensation for both the size of sexual ornaments and their asymmetry (Evans 1993; Evans et al. 1994). Male barn swallows (*Hirunda rustica*) with greatly elongated but asymmetrical tail feathers undergo compensatory muscle development, providing an example of organism-wide compensation for a by-product of great elaboration in sexual ornaments (Møller and Swaddle 1997, p.181). Similarly, individual plants of higher physiological quality had greater developmental stability in flower structures and produced more symmetrical flowers (Møller 1996). In horned beetles, the diversification of sexually selected horns is facilitated by within-species costs that the most exaggerated expression of horns imposes on other traits of the developing organism (Emlen 2001).

Consequently, condition-dependence in the developmental pathways of a sexual ornament can be acquired or lost as a result of selection. For example, populations of the collared flycatcher (*Ficedula albicollis*) differ in condition-dependence of the forehead patch size—a sexually selected plumage ornament—despite the similarity in additive genetic variance and phenotypic variance of this sexual trait across populations (Qvarnström 1999; Hegyi et al. 2002).

Overall, at this stage, selection for stronger condition-dependence of a sexual ornament favors an increase in integration of the sexual ornament into organismal functions (Fig. 3.5). The condition-dependence favored at this stage is largely due to general viability costs imposed by an ornament and only to a small degree due to costs of specific ornament-producing pathways.

### Stage 3: Selection for Complex Sexual Ornaments: Greater Expression and Process-Dependence

As expression of a sexual ornament becomes progressively more integrated into organismal functions, some components of a sexual ornament may reflect specific organismal processes better than others (Wedekind 1992; Fig. 3.5). Selection by female choice is expected to be stronger on components of a sexual ornament that are more relevant locally, that is, better indicate male performance under local environmental conditions (Wedekind 1994; Zuk and Johnsen 1998; Badyaev et al. 2001; Calkins and Burley 2003). Also, the benefits of preference for some features of the male phenotype may differ between females; that is, benefits can be context- and individual-specific (Jang and Greenfield 2000; Qvarnström 2001; reviewed in Badyaev and Qvarnström 2002). Furthermore, selection across variable environments will favor different aspects of ornament elaboration (Brooks and Couldridge 1999; Day 2000), ultimately resulting in lower integration among ornament components and greater integration of components of an ornament with organismal processes that most strongly affect their production (Figs. 3.5 and 3.6).

Stability of composite and condition-dependent sexual ornaments might be maintained by multiple inputs of resources and energy (Johnstone 1995) and by more efficient recognition of a composite ornament by a female (Brooks and

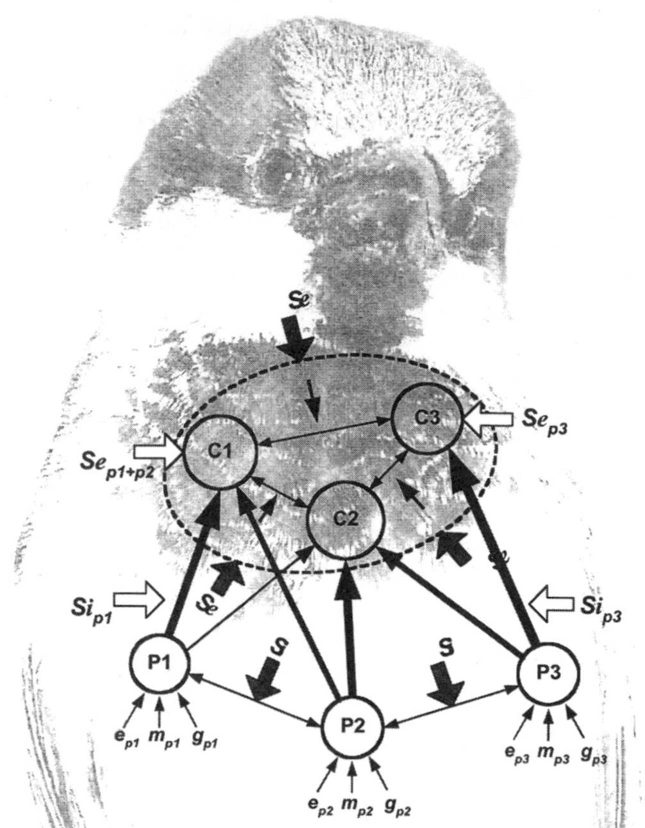


Figure 3.6 Conceptual illustration of the development and maintenance of a composite sexual ornament. Each of the components of the ornament (C1, C2, and C3; e.g., hue, area, and pattern) are mainly affected by one physiological process (P1, P2, and P3). Some organismal processes affect more than one ornamental component (e.g., P2 affects C1 and C2), and some ornamental components are affected by more than one organismal process (e.g., C1 is affected by P1 and P2). External selection acting on individual components of an ornament, proximately acts on the internal processes that produce them (e.g., *Se* on C1 acts on P1 and P2 and on the allocation of P1 and P2 to the production of C1, i.e., *on* $Si_{p1}$ and $Si_{p2}$). External selection (*Se*) acting on the entire ornament, proximately acts on the developmental and functional interactions (double-headed arrows) of ornamental components. Internal selection on the coordination of organismal performance (*Si*) acts on the interactions among physiological processes. Composite sexual ornament is produced and maintained by a combination of developmental and functional integration of its components.

Couldridge 1999; Rowe 1999; Fig. 3.6). For example, carotenoid-based ornamentation of house finches requires a coordination of multiple processes associated with carotenoid consumption, digestion, transportation, and deposition (Fig. 3.7). Each of these processes has different costs in different environments: environmental variation in carotenoid-rich food affects consumption, gut parasite infestation affects digestion, and availability of oils in food affects deposition of lipid-soluble

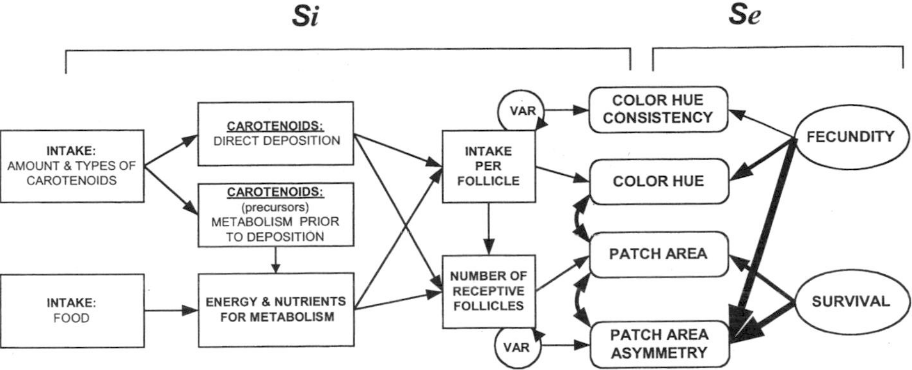


Figure 3.7 Internal (*Si*) and external (*Se*) selection on the components of the carotenoid-based sexual ornament in the house finch. Carotenoid pigments can be either ingested directly or metabolized from suitable carotenoid precursors. Direct use of dietary pigments moves carotenoids along the top of the figure and avoids most dependence of body condition, whereas use of carotenoids metabolized from dietary precursors leads to greater dependence on the energy and health state of the individual. Thus, condition-dependence of individual components of carotenoid ornamentation varies across environments. Thickness of arrows illustrates observed intensity of fecundity and survival selections on individual components of an ornament in the Montana population. Different episodes of external selection act on different components of ornament producing variable patterns of functional integration across environments. Double-headed arrows show observed patterns of developmental integration between ornamental components. Note that, for example, whereas hue and area are developmentally integrated, this integration is not maintained by external selection (modified from Badyaev et al. 2001).

carotenoids into plumage (Brush and Power 1976; Hill and Montgomerie 1994; Thompson et al. 1996). However, despite environmental variation in the degree of condition-dependence of each of these processes (Hill et al. 2002) and corresponding variation in female choice of different components in different environments, production of carotenoid-based sexual ornaments requires some degree of integration among these pathways, thus maintaining some developmental interdependence among the ornament components across environments (Badyaev et al. 2001; Fig. 3.7).

The composite nature of sexual ornaments can be maintained by the requirements of sexual display expression and function, that is, its functional integration (Fig. 3.6). Both ornament-specific and general pathways may be needed for sexual ornament expression and function and this can maintain its complex structure. For example, complex vocalizations of male treefrogs (Hylidae) consist of a combination of highly variable "dynamic" features, production of which is costly, and "static" features that are required to accomplish species-specific patterns of vocalizations or needed to produce "dynamic" features of a song (Gerhardt 1991; Murphy and Gerhardt 2000; see review in Wiley 2000). Greater exaggeration of antler size in cervids might favor the development of additional processes that maintain antler integration (such as within-antler branching patterns or coordination of antler weight on each side of the body). Similarly, greater developmental stability of complex feather ornaments in birds may be maintained by compensatory within-trait integration of growth (Aparicio and Bonal 2002).

Overall, a combination of (1) distinct patterns of selection on individual components of a condition-dependent sexual ornament, (2) partially independent development of these components, and (3) stabilizing selection on the whole sexual ornament, should favor the evolution of composite sexual ornaments whose components reliably reflect the condition of an entire organism across a wide array of environments (Figs. 3.5 and 3.6).

## Meeting the Expectations: Empirical Patterns in Sexual Ornaments

The expected outcome of the evolutionary processes outlined above is that sexual ornaments should have (1) weaker developmental integration with the rest of the organism to enable greater exaggeration at lower cost; (2) greater functional integration and modularity to enable faster and more precise response to changes in external selection; and (3) weaker genetic integration (coinheritance) with the rest of the organismal traits and weaker evolutionary integration between components of an ornament. Taken together, these patterns should produce sexual ornaments that are highly sensitive to environmental variation, variable within and between populations, and have high rates of evolutionary diversification compared to nonsexual traits. Here I review empirical support for these expectations.

### Weaker Developmental Integration

Weaker developmental integration of sexual traits with the rest of an organism and the associated greater sensitivity to environmental variation is well documented. For example, in many species of carnivores and primates, canine teeth play an important role in sexual displays. These teeth are often under directional sexual selection for greater length, and sexual dimorphism in canines is common (Kay et al. 1988; Manning and Chamberlain 1993; Badyaev 1998). A defining feature of canines is the lack of occlusion—a correspondence in tooth cusp patterns and position between upper and lower jaw—which is present in other types of dentition in these species and is necessary for food processing. The lack of constraints imposed by the needs of occlusion during ontogeny is a powerful force behind canine development, evolution, and extensive diversification among mammals. Weaker integration of canines with the rest of the dentition leads to their highly variable growth rates and patterns both among closely related species and even between the sexes (Schwartz and Dean 2001). In grizzly bears (*Ursus arctos horribilis*) and in lowland gorillas (*Gorilla gorilla*), less developmentally integrated canines are under directional selection for increase in males and are sexually dimorphic, whereas more functionally and developmentally integrated premolar and molar teeth are under stabilizing selection and are not different between the sexes (Manning and Chamberlain 1993, 1994; Badyaev 1998). Following strong environmental stress, size variation, developmental abnormalities, and fluctuating asymmetry strongly increased in weakly integrated canine dentition in both species, whereas the variation and developmental stability of well-integrated premolar and molar dentition was only weakly affected (Manning and Chamberlain 1993, 1994; Badyaev 1998). Similarly, in peccaries (Tyassuidae), the weakening of developmental integration of several skull traits with neighboring structures led to rapid elaboration and greater sexual dimorphism in these structures as well as to their greater diversification among species (Wright 1993; see also Badyaev and Foresman 2000).

Numerous studies have documented that weak developmental control of elaboration of sexual traits leads to their greater environmental sensitivity (reviewed in Møller and Swaddle 1997). For example, administering antihelminthic medicine to infected reindeer (*Rangifer tarandus*) had a greater effect on size and asymmetry of less developmentally integrated antlers compared to more integrated skeletal structures (Fölstad et al. 1996). Size and asymmetry of antlers in fallow deer (*Dama dama*) were more responsive to changes in environmental condition than more developmentally integrated skeletal traits (Putman et al. 2000). Similarly, stress had a stronger effect on size and asymmetry of weakly integrated and sexually selected traits in horned beetles (*Onthophagus taurus*) and stalk-eyed flies (*Cyrtodiopsis dalmanni*) compared to nonsexual traits (Hunt and Simmons 1997; David et al. 1998).

In many insect groups, highly variable allometric relationships between exaggerated sexual ornaments and other body traits are enabled by the close dependence of these allometric relationships on environmental variation (Emlen and Nijhout 2000). Consequently, lower developmental integration of exaggerated sexual ornaments with the organism may be responsible for their faster evolution and greater diversification in a number of insect groups (Simmons and Tomkins 1996; Emlen 2000; Baker and Wilkinson 2001; Moczek and Nijhout 2002).

The effect of weak developmental integration on variability and diversification in sexual ornaments is clearly illustrated by contrasting developmental properties of carotenoid-based and melanin-based ornaments of animals, especially of birds and fishes. Because more precursors of carotenoid-based pigmentation have to be acquired from the environment, carotenoid coloration is more dependent on environmental variation than melanin-based ornamental coloration (Endler 1983; Hill 1999; Badyaev and Hill 2000). Environments differ in amount and quality of carotenoid precursors (Grether et al. 1999; Hill et al. 2002), resulting in environmental variation in the amount of metabolism that the consumed carotenoids have to undergo before deposition (Fig. 3.7). Moreover, there are few constraints on carotenoid deposition: the presence of lipid-soluble carotenoids in integument is enabled by the presence of oil or fat there (Brush 1978). Because of the flexible pattern of deposition, carotenoid-based ornaments rarely form clearly delineated patterns of coloration, instead mostly producing patches of variable color with amorphous borders.

In contrast, melanin-based ornamentations have strong developmental and genetic integration of production and deposition. Melanins are synthesized during amino acid metabolism, specifically by a breakdown of tyrosine under the enzyme typosinase, and most of the precursors required for this synthesis have limited environmental or diet-dependence (Fox 1976), although detailed studies are rare (e.g., Murphy and King 1987). Melanin deposition in integuments depends on the duration of melanocyte activity, on the interaction of the melanocyte with the epidermis, and, in some taxa, on concerted interactions among different melanocytes, and these processes are under precise developmental and genetic control (see reviews in Price and Pavelka 1996; Seehausen et al. 1999; Badyaev and Hill 2000; Jawor and Breitwisch 2003). Consequently, melanin-based pigmentation commonly forms complex coloration patterns such as regular spots, stripes, streaks, and clearly delineated caps and bars. Comparative studies suggest that a strong, and invariant with the degree of elaboration (Badyaev and Young 2003), developmental integration of melanin-based pigmentation has an important effect on historical patterns of its elaboration (Price and Pavelka 1996). The effect of differences in developmental integration of carotenoid and melanin ornaments on their variability and diversification is seen in cardueline finches—a group of birds that possess both kinds of ornamentation. In these birds, carotenoid ornaments have greater evolutionary lability and diversification (Gray 1996; Badyaev and Hill 2000), higher variation among environments both within and among species (Hill 1993; Badyaev et al. 2002), and greater response to stress both within and among species (Badyaev and Ghalambor 1998; McGraw and Hill 2000), compared to melanin-based ornaments.

This, however, does not mean that melanin-based ornaments are poor indicators of condition. Instead, their greater integration into multiple organismal functions make them, in a sense, better indicators of the organismal condition. It is exactly because of the multiple controls of their development that melanin-based sexual ornaments rarely achieve a level of variability and environmental sensitivity comparable to that of carotenoid-based ornamentation: stronger developmental integration of melanin coloration constraints its elaboration. Additional mechanisms, such as behavioral displays or development stability, often amplify the condition-dependence of melanin ornaments, and these amplifiers are often under selection for further elaboration (Møller et al. 1996; Fitzpatrick 1998; Hill et al. 1999).

### Greater Functional Integration

Weaker developmental integration of sexual ornaments with the rest of an organism, which enables their greater elaboration, is accompanied by their greater functional integration, which results in their greater and faster diversification.

Copulatory structures in many animal groups provide examples of composite sexual traits in which components of different developmental origins are under external selection that favors their greater functional integration (Eberhard et al. 1998; Arnqvist and Rowe 2002; Kopp and True 2002). For example, abdominal lobes in male sepsid flies (Sepsidae) are used to stimulate females during copulation (Eberhard 2001). Strong female preference for exaggeration of these traits in combination with a diverse developmental origin of their components and their close functional integration, results in both exaggeration and diversification of abdominal lobes in sepsid flies (Eberhard 2001; see also Kopp and True 2002 for similar results in *Drosophila*). Similarly, studies of primates revealed that external genital anatomy, especially penile morphology, is extremely diverse and differs markedly among closely related species and subspecies (e.g., Anderson 2000). For example, distal structures, such as keratinized spines and plates on genitalia, play an important role in female choice in these species (Dixson 1987), and the complexity and evolutionary diversification of distal penile structures is greater in species with a higher intensity of sexual selection (Dixson and Anderson 2002). Overall, weak developmental integration with other organismal traits, a modular structure, and close functional integration of genital components favored by external selection, enables these structures to achieve both rapid exaggeration and precision in their morphological evolution (reviewed in Eberhard 1985; Dixson and Anderson 2002).

In flowering plants that depend on pollinators, selection favors the evolution of structures that enable both a greater attraction of pollinators (which requires greater exaggeration) and a better mechanistic fit between pollinators and flower structures (which requires greater functional integration between flower parts) (Berg 1960; see also Armbruster et al., Chapter 2, this volume). Consequently, stabilizing selection on flower structure is often consistent with the patterns of functional integration favored by pollinator morphology (Giurfa et al. 1999; see Eberhard et al. 1998 for a similar result in copulatory structures). Stronger functional integration within structures of flowers, but low integration between flowers and the rest of a plant, may enable faster and more precise change in flower morphology in order to track the morphology of pollinators (Berg 1960; Conner and Sterling 1996). When selection pressures imposed by pollinators are not consistent (e.g., due to their variable morphologies), selection may favor weaker functional integration among flower parts to enable greater variability (Armbruster et al. 1999).

In the cockroach *Nauphoeta cinerea*, males produce a sex pheromone that consists of many chemical components (Moore 1997). Whereas each of the individual components is under directional selection by different functions and has a different developmental origin (Moore and Moore 1999; Moore et al. 2001), the entire composition of the pheromone is under stabilizing selection (Moore 1997). Similarly, a remarkable structural similarity in complex courtship song among fourteen species of oropendolas despite distinct selection on its components might be accomplished by the need for functional integration of song production (Price and Lanyon 2002). Overall, functional integration of a sexual trait, despite its diverse developmental origins, maintains its stability and composite nature (Gerhardt 1991; Wedekind 1994; Moore 1997; Moore and Moore 1999; Seehausen et al. 1999; Murphy and Gerhardt 2000; Badyaev et al. 2001; Fig. 3.7).

### Weaker Evolutionary and Genetic Integration

Weaker developmental integration of condition-dependent components of sexual ornaments may account for their weak genetic and evolutionary integration. Price (2002) found that the patterns of co-inheritance of sexual displays in avian hybrids show a historical hierarchy of evolutionary lability that roughly corresponds to the genetic and developmental integration in sexual displays.

Ornaments that were affected by fewer genetic factors or developmental pathways (such as color variants) showed higher evolutionary lability (see also Price 1996; Badyaev and Hill 2003; Price and Bontrager 2001). Traits with more integrated development (such as feather structures) showed intermediate lability and diversification, whereas the most complex, long-developing, and pleiotropic sexual displays (such as flight displays) showed the strongest evolutionary integration and stability and persisted throughout multiple hybridization events. Seehausen et al. (1999) analyzed evolution of color patterns in relation to their developmental and functional properties in more than 700 cichid species. They found that a distinct developmental origin and weak developmental integration of components of sexual coloration resulted in their high evolutionary lability. Interestingly, conserved patterns of coloration were maintained across species by the similarity in selection for greater functional integration of color components (e.g., selection for better environmental matching of color patterns) and not by developmental integration. In cardueline finches, the components of sexual ornamentation that showed the highest within-species variability, also varied the most among species along an ecological gradient (Badyaev 1997). In male oropendolas, complex structural features of courtship song were retained across multiple speciation events, whereas song components that were less integrated with the rest of the song and were more susceptible to environmental variation showed the highest evolutionary lability (Price and Lanyon 2002; see also Slabbekoorn and Smith 2002). Similarly, Wiens (2000) found that complex behavioral displays were more constant, whereas display morphology, such as coloration, varied extensively among species of lizards (see Irwin 1996 for review).

Several studies documented weaker genetic integration of sexual traits with the rest of the organism—a pattern consistent with their greater functional integration despite distinct developmental origins. Saetre et al. (2002) documented reduced recombination rates among loci determining components of sexual plumage ornamentation (plumage color, collar area, and forehead patch height) in two species of *Ficedula* flycatchers compared to nonsexually selected plumage traits. Sex linkage, high genetic integration of the sexual ornament's multiple components, and weak genetic integration with other organismal traits contributed to reduced within-species polymorphism and high segregation between flycatcher species (Saetre et al. 2002; see also Ranz et al. 2003 and review in Reinhold 1998). Similarly, Preziosi and Roff (1998) found weak genetic integration between copulatory structures of the water strider (*Aquarius remigis*) and the rest of the organismal traits. These patterns are in close correspondence with the rapid evolution of these traits across taxa (Arnqvist and Rowe 2002).

## Ultimate Integration of Sexual Ornaments: "Mating Phenotype"

Investment into sexual ornaments and investment in mate choice are parts of reproductive allocation of an entire organism and thus are subject to both life-history tradeoffs and the effects of the environment in which breeding occurs. Context- and environment-dependence in the expression of sexual ornaments (which is enabled by their developmental modularity), and in the preference for these ornaments, favor the evolution of reproductive strategies that maximize an organism's reproductive success under variable conditions of breeding (Gross 1996; Höglund and Sheldon 1998; Badyaev and Qvarnström 2002).

Facultative expression of sexual ornamentation (Fig. 3.2) facilitates the evolution of alternative condition-dependent reproductive tactics (Gross 1996; Emlen and Nijhout 2000), and the decision to adopt a particular suite of reproductive behaviors might be phenotypically indicated by the elaboration of a sexual trait. For example in the horned beetle, males employ different reproductive tactics depending on the size of their horns. Large males use their larger horns in direct competition with other males over access to females, whereas small males avoid direct competition with other males by sneaking into female nesting tunnels—a behavioral tactic facilitated by their smaller horns (Emlen 1997). In the mite *Sancassania berlesei*, expression of a most beneficial morph of sexual ornament for a given colony density is accomplished by sensitivity of sexual trait development to a pheromone whose concentration depends on colony density (Radwan et al. 2002).

In the house finch, males with different elaboration of condition-dependent sexual ornamentation adopt a distinct set of parental behaviors (Badyaev and Hill 2002; Duckworth et al. 2003). What is indicated by the elaboration of sexual ornamentation differs between finch populations, but in each one, females base their choice of male parental behaviors on the elaboration of the male's sexual ornamentation. Thus, female choice favors hormonally mediated "functional integration" of male sexual behaviors and male sexual ornamentation. Similarly, composite sexual ornaments can enable more efficient assortative mating (Wedekind 1992; Tregenza and Wedell 2000), thus influencing further elaboration of sexual ornaments. Therefore, to fully understand the evolution of sexual displays, we need to study the entire mating phenotype which includes the functional integration of morphology and behavior for both sexes.

In conclusion, two major, conceptually inspired, approaches to the study of morphological evolution—relative importance of internal and external processes and the evolution of morphological integration—are mostly overlooked in studies of sexual ornamentation. Yet these concepts are central to understanding the evolution of sexual ornamentation and the mechanisms of sexual selection. Moreover, because investment into sexual ornaments is a part of the reproductive strategy of the entire organism, the evolution of sexual displays should be considered in the context of the performance of the entire organism and organism's interactions. Recent synthesis of the sexual selection theory explicitly recognizes the evolutionary continuum of the mechanisms by which sexual selection operates and, on a population level, firmly places sexual selection in the framework of life-history evolution (Gross 1996; Höglund and Sheldon 1998; Badyaev and Qvarnström 2002; Kokko et al. 2002). However, the consequences of the continuum in the mechanisms of sexual selection for the evolution of sexual displays and ornamentation are not well understood. The perspective outlined here, with specific focus on the evolution of development and morphological integration, may provide a useful framework for understanding the evolution of sexual ornamentation.

*Acknowledgments* I thank R. Duckworth, M. Pigliucci, K. Preston, E. Snell-Rood, and R. Young for discussion and comments, and the National Science Foundation (DEB-0075388, DEB-0077804, and IBN-0218313) for funding this work.

## Literature Cited

Alatalo, R. V., J. Hoglund, and A. Lundberg. 1988. Patterns of variation in tail ornament size in birds. Biological Journal of the Linnean Society 34:363–374.

Anderson, M. J. 2000. Penile morphology and classification of bushbabies (subfamily Galagoninae). International Journal of Primatology 21:815–830.

Andersson, M. 1982. Sexual seleciton, natural selection and quality advertisement. Biological Journal of the Linnean Society 17:375–393.

Andersson, M. 1994. Sexual Selection. Princeton University Press, Princeton, NJ.

Aparicio, J. M., and R. Bonal. 2002. Why do some traits show higher fluctuating asymmetry than others? A test of hypotheses with tail feathers of birds. Heredity 89:139–144.

Armbruster, W. S., V. S. Di Stilio, J. D. Tuxill, T. S. Flores, and J. L. V. Runk. 1999. Covariance and decoupling of floral and vegetative traits in nine Neotropical plants: a re-evaluation of Berg's correlation-pleiades concepts. American Journal of Botany 86:39–55.

Arnqvist, G. 1994. The cost of male secondary sexual traits: developmental constraints during ontogeny in a sexually dimorphic water strider. American Naturalist 144:119–132.

Arnqvist, G., and L. Rowe. 2002. Antagonistic coevolution between the sexes in a group of insects. Nature 415:787–789.

Arthur, W. 2002. The emerging conceptual framework of evolutionary developmental biology. Nature 415:757–764.

Badyaev, A. V. 1997. Altitudinal variation in sexual dimorphism: a new pattern and alternative hypotheses. Behavioral Ecology 8:675–690.

Badyaev, A. V. 1998. Environmental stress and developmental stability in dentition of the Yellowstone grizzly bears. Behavioral Ecology 9:339–344.

Badyaev, A. V. 2002. Growing apart: an ontogenetic perspective on the evolution of sexual size dimorphism. Trends in Ecology and Evolution 17:369–378.

Badyaev, A. V., and R. A. Duckworth. 2003. Context-dependent sexual advertisement: plasticity in development of sexual ornamentation throughout the lifetime of a passerine bird. Journal of Evolutionary Biology 16:1065–1076.

Badyaev, A. V., and K. R. Foresman. 2000. Extreme environmental change and evolution: stress-induced morphological variation is strongly concordant with patterns of evolutionary divergence in shrew mandibles. Proceedings of the Royal Society of London Series B: Biological Sciences 267:371–377.

Badyaev, A. V., and C. K. Ghalambor. 1998. Does a trade-off exist between sexual ornamentation and ecological plasticity? Sexual dichromatism and occupied elevational range in finches. Oikos 82:319–324.

Badyaev, A. V., and G. E. Hill. 2000. Evolution of sexual dichromatism: contribution of carotenoid- versus melanin-based coloration. Biological Journal of the Linnean Society 69:153–172.

Badyaev, A. V., and G. E. Hill. 2002. Parental care as a conditional strategy: distinct reproductive tactics associated with elaboration of plumage ornamentation in the house finch. Behavioral Ecology 13:591–597.

Badyaev, A. V., and G. E. Hill. 2003. Avian sexual dichromatism in relation to history and current selection. Annual Reviews of Ecology and Systematics 34:27–49.

Badyaev, A. V., and E. S. Leaf. 1997. Habitat associations of song characteristics in *Phylloscopus* and *Hippolais* warblers. Auk 114:40–46.

Badyaev, A. V., and A. Qvarnström. 2002. Putting sexual traits into the context of an organism: a life-history perspective in studies of sexual selection. Auk 119:301–310.

Badyaev, A. V., and R. L. Young. 2003. Complexity and integration in sexual ornamentation: an example with carotenoid and melanin plumage pigmentation. Evolution and Development. In press.

Badyaev, A. V., W. J. Etges, J. D. Faust, and T. E. Martin. 1998. Fitness correlates of spur length and spur asymmetry in male wild turkeys. Journal of Animal Ecology 67:845–852.

Badyaev, A. V., G. E. Hill, P. O. Dunn, and J. C. Glen. 2001. "Plumage color" as a composite trait: developmental and functional integration of sexual ornamentation. American Naturalist 158:221–235.

Badyaev, A. V., G. E. Hill, and B. V. Weckworth. 2002. Species divergence in sexually selected traits: increase in song elaboration is related to decrease in plumage ornamentation in finches. Evolution 56:412–419.

Baker, R. H., and G. S. Wilkinson. 2001. Phylogenetic analysis of sexual dimorphism and eye-span allometry in stalk-eyed flies (Diopsidae). Evolution 55:1373–1385.

Berg, R. L. 1960. The ecological significance of correlation pleiades. Evolution 14:171–180.

Birkhead, T. R., F. Fletcher, and E. J. Pellatt. 1999. Nestling diet, secondary sexual traits and fitness in the zebra finch. Proceedings of the Royal Society of London Series B: Biological Sciences 266:385–390.

Breuker, C. J., and P. M. Brakefield. 2002. Female choice depends on size but not symmetry of dorsal eyespots in the butterfly *Bicyclus anynana*. Proceedings of the Royal Society of London Series B: Biological Sciences 269:1233–1239.

Brooks, R., and V. Couldridge. 1999. Multiple sexual ornaments coevolve with multiple mating preferences. American Naturalist 154:37–45.

Brush, A. H. 1978. Avian pigmentation. Pp. 141–161 in A. H. Brush, ed. Chemical Zoology. Academic Press, New York.

Brush, A. H., and D. M. Power. 1976. House finch pigmentation: carotenoid metabolism and the effect of diet. Auk 93:725–739.

Brush, A. H., and H. Seifried. 1968. Pigmentation and feather structure in genetic variants of the Gouldian finch, *Poephila gouldiae*. Auk 85:416–430.

Calkins, J. D., and N. Burley. 2003. Mate choice for multiple ornaments in the California quail, *Callipepla californica*. Animal Behaviour 65:69–81.

Civetta, A., and R. S. Singh. 1998. Sex and speciation: genetic architecture and evolutionary potential of sexual vs nonsexual traits in the sibling species of the *Drosophila melanogaster* complex. Evolution 52:1080–1092.

Conner, J. K., and A. Sterling. 1996. Selection for independence of floral and vegetative traits: evidence from correlation patterns in five species. Canadian Journal of Botany 74:642–644.

Creswell, J. E. 1998. Stabilizing selection and the structural variability of flowers within species. Journal of Botany 81:463–473.

Cuervo, J. J., and A. P. Møller. 1999a. Ecology and evolution of extravagant feather ornaments. Journal of Evolutionary Biology 12:986–998.

Cuervo, J. J., and A. P. Møller. 1999b. Phenotypic variation and fluctuating asymmetry in sexually dimorphic feather ornaments in relation to sex and mating system. Biological Journal of the Linnean Society 68:505–529.

Cuervo, J. J., and A. P. Møller. 2001. Components of phenotypic variation in avian ornamental and non-ornamental feathers. Evolutionary Ecology 15:53–72.

Dale, J. 2000. Ornamental plumage does not signal male quality in red-billed queleas. Proceedings of the Royal Society of London Series B: Biological Sciences 267:2143–2149.

David, P., A. Hingle, D. Greig, A. Rutherford, A. Pomiankowski, and K. Fowler. 1998. Male sexual ornament size but not asymmetry reflects condition in stalk-eyed flies. Proceedings of Royal Society of London Series B: Biological Sciences 265:2211–2216.

Day, T. 2000. Sexual selection and the evolution of costly female preferences: spatial effects. Evolution 54:715–730.

Dixson, A., and M. Anderson. 2002. Sexual selection and the comparative anatomy of reproduction in monkeys, apes, and human beings. Annual Review of Sex Research 12:121–144.

Dixson, A. F. 1987. Observations on the evolution of genitalia and copulatory behaviour in male primates. Journal of Zoology 213:423–443.

Duckworth, R. A., A. V. Badyaev, and A. F. Parlow. 2003. Males with more elaborated sexual ornamentation avoid costly parental care in a passerine bird. Behavioral Ecology and Sociobiology. In press.

Eberhard, W. G. 1985. Sexual Selection and Animal Genitalia. Harvard University Press, Cambridge, MA.

Eberhard, W. G. 2001. Multiple origins of a major novelty: moveable abdominal lobes in male sepsid flies (Diptera: Sepidae), and the question of developmental constraints. Evolution and Development 3:206–222.

Eberhard, W. G., B. A. Huber, R. L. Rodriguez, R. D. Briceno, I. Salas, and V. Rodriguez. 1998. One size fits all? Relationship between the size and degree of variation in genitalia and other body parts in twenty species of insects and spiders. Evolution 52:415–431.

Emlen, D. J. 1997. Alternative reproductive tacts and male-dimorphism in the horned beetle *Onthophagus acuminatus* (Coleoptera: Scarabaeidae). Behavioral Ecology and Sociobiology 41:335–341.

Emlen, D. J. 2000. Integrating development with evolution: a case study with beetle horns. BioScience 50:403–418.

Emlen, D. J. 2001. Costs and the diversification of exaggerated animal structures. Science 291:1534–1536.

Emlen, D. J., and H. F. Nijhout. 2000. The development and evolution of exaggerated morphologies in insects. Annual Reviews in Entomology 45:661–708.

Endler, J. A. 1983. Natural and sexual selection on color patterns in poeciliid fishes. Environmental Biology of Fishes 9:173–190.

Endler, J. A. 1992. Signals, signal conditions, and the direction of evolution. American Naturalist 139:S125-S153.

Evans, M. R. 1993. Fluctuating asymmetry and long tails: the mechanical effects of asymmetry may act to enforce honest advertisement. Proceedings of the Royal Society of London Series B: Biological Sciences 253:205–209.

Evans, M. R., T. L. F. Martins, and M. P. Haley. 1994. The asymmetrical cost of tail elongation in red-billed streamertails. Proceedings of the Royal Society of London Series B: Biological Sciences 256:97–103.

Fitze, P. S., and H. Richner. 2002. Differential effect of a parasite on ornamental structures based on melanins and carotenoids. Proceedings of Royal Society of London Series B: Biological Sciences 13:401–407.

Fitzpatrick, S. 1998. Birds' tails as signaling devices: markings, shape, length, and feather quality. American Naturalist 151:157–173.

Fitzpatrick, S. 1999. Tail length in birds in relation to tail shape, general flight ecology and sexual selection. Journal of Evolutionary Biology 12:49–60.

Fölstad, I., and A. J. Karter. 1992. Parasites, bright males, and the immunocompetence handicap. American Naturalist 139:603–622.

Fölstad, I., P. Arneberg, and A. J. Karter. 1996. Parasites and antler asymmetry. Oecologia 105:556–558.

Forstmeier, W., B. Kempenaers, A. Meyer, and B. Leisler. 2002. A novel song parameter correlates with extra-pair paternity and reflects male longevity. Proceedings of the Royal Society of London Series B: Biological Sciences 269:1479–1485.

Fox, D. L. 1976. Avian Biochromes and Structural Colors. University of California Press, Berkeley.

Fox, D. L., V. E. Smith, and A. A. Wolfson. 1969. Carotenoid selectivity in blood and feathers of lesser (Africa) Chilean and Greater (European) flamingos. Comparative Biochemical Physiology 23:225–232.

Fusco, G. 2001. How many processes are responsible for phenotypic evolution? Evolution and Development 3:279–286.

Gerhardt, H. C. 1991. Female mate choice in treefrogs: static and dynamic acoustic criteria. Animal Behaviour 42:615–635.

Giurfa, M., A. Dafni, and P. R. Neal. 1999. Floral symmetry and its role in plant-pollinator systems. International Journal of Plant Sciences 160:S41–S50.

Grafen, A. 1990. Biological signals as handicaps. Journal of Theoretical Biology 144:517–546.

Gray, D. A. 1996. Carotenoids and sexual dichromatism in North American passerine birds. American Naturalist 148:453–480.

Grether, G. F., J. Hudon, and D. F. Millie. 1999. Carotenoid limitation of sexual coloration along an environmental gradient in guppies. Proceedings of the Royal Society of London Series B: Biological Sciences 266:1317–1322.

Griffith, S. C., I. P. F. Owens, and T. Burke. 1999. Environmental determination of a sexually selected trait. Nature 400:358–360.

Gross, M. R. 1996. Alternative reproductive strategies and tactics: diversity within sexes. Trends in Ecology and Evolution 11:92–98.

Guilford, T., and M. S. Dawkins. 1991. Receiver psychology and the evolution of animal signals. Animal Behaviour 42:1–14.

Hamilton, W. D., and M. Zuk. 1982. Heritable true fitness and bright birds: a role for parasites? Science 218:384–387.

Hansen, T. F., and D. K. Price. 1995. Good genes and old age: do old mates provide superior genes? Journal of Evolutionary Biology 8:759–778.

Hasson, O. 1989. Amplifiers and the handicap principle in sexual selection: a different emphasis. Proceedings of the Royal Society of London Series B: Biological Sciences 235:383–406.

Hegyi, G., J. Torok, and L. Toth. 2002. Qualitative population divergence in proximate determination of a sexually selected trait in the collared flycatcher. Journal of Evolutionary Biology 15:710–719.

Hill, G. E. 1993. Geographic variation in the carotenoid plumage pigmentation of male house finches (*Carpodacus mexicanus*). Biological Journal of the Linnean Society 49:63–89.

Hill, G. E. 1994. Trait elaboration via adaptive mate choice: sexual conflict in the evolution of signals of male quality. Ethology, Ecology, and Evolution 6:351–370.

Hill, G. E. 1999. Mate choice, male quality, and carotenoid-based plumage coloration: a review. Proceedings of the XXII International Ornithological Congress, Durban, 1654–1668.

Hill, G. E., and R. Montgomerie. 1994. Plumage colour signals nutritional condition in the house finch. Proceedings of the Royal Society of London Series B: Biological Sciences 258:47–52.

Hill, J. A., D. A. Enstrom, E. D. Ketterson, V. J. Nolan, and C. Ziegenfus. 1999. Mate choice based on static versus dynamic secondary sexual traits in the dark-eyed junco. Behavioral Ecology 10:91–96.

Hill, G. E., C. Y. Inouye, and R. Montgomerie. 2002. Dietary carotenoids predict plumage coloration in wild house finches. Proceedings of the Royal Society of London Series B: Biological Sciences 269:1119–1124.

Höglund, J., and B. C. Sheldon. 1998. The cost of reproduction and sexual selection. Oikos 83:478–483.

Hunt, J., and L. W. Simmons. 1997. Patterns of fluctuating asymmetry in beetle horns: an experimental examination of the honest signalling hypothesis. Behavioral Ecology and Sociobiology 41:109–114.

Irwin, R. E. 1996. The phylogenetic content of avian courtship display and song evolution. Pp. 234–252 in E. P. Martins, ed. Phylogenies and the Comparative Method in Animal Behavior. Oxford University Press, New York.

Iwasa, Y., and A. Pomiankowski. 1994. The evolution of mate preferences for multiple sexual ornaments. Evolution 48:853–867.

Iwasa, Y., A. Pomiankowski, and S. Nee. 1991. The evolution of costly mate preferences. II. The handicap principle. Evolution 45:1431–1442.

Jang, Y., and M. D. Greenfield. 2000. Quantitative genetics of female choice in an ultrasonic pyramid moth, *Achroia grisella*: variation and evolvability or preference along multiple dimensions of the male advertisement signals. Heredity 84:73–80.

Jawor, J. M., and R. Breitwisch. 2003. Melanin ornaments, honesty, and sexual selection. Auk 120:249–265.

Johnstone, R. A. 1995. Honest advertisement of multiple qualities using multiple signals. Journal of Theoretical Biology 177:87–94.

Kay, R. F., J. M. Plavcan, K. E. Glander, and P. C. Wright. 1988. Sexual selection and canine dimorphism in New World monkeys. American Journal of Physical Anthropology 77:385–397.

Keyser, A. J., and G. E. Hill. 2000. Structurally based plumage coloration is an honest signal of quality in male Blue Grosbeaks. Behavioral Ecology 11:202–209.

Kodric-Brown, A. 1989. Dietary carotenoids and male mating success in the guppy: an environmental component to female choice. American Naturalist 124:309–323.

Kodric-Brown, A. 1998. Sexual dichromatism and temporary color changes in the reproduction of fishes. American Zoologist 38:70–81.

Kodric-Brown, A., and J. H. Brown. 1984. Truth in advertising: the kinds of traits favored by sexual selection. American Naturalist 124:309–323.

Kokko, H. 1997. Evolutionarily stable strategies of age-dependent sexual advertisement. Behavioral Ecology and Sociobiology 41:99–107.

Kokko, H. 1998. Should advertising parental care be honest? Proceedings of the Royal Society of London Series B: Biological Sciences 265:1871–1878.

Kokko, H., J. Lindstrom, R. V. Alatalo, and P. T. Rintamaki. 1998. Queuing for territory position in the lekking black grouse (*Tetrao tetrix*). Behavioral Ecology 9:376–383.

Kokko, H., P. T. Rintamaki, R. V. Alatalo, J. Hoglund, E. Karvonen, and A. Lundberg. 1999. Female choice selects for lifetime lekking performance in black grouse males. Proceedings of the Royal Society of London Series B: Biological Sciences 266:2109–2115.

Kokko, H., R. Brooks, J. M. McNamara, and A. I. Houston. 2002. The sexual selection continuum. Proceedings of the Royal Society of London Series B: Biological Sciences 269:1331–1340.

Kopp, A., and J. R. True. 2002. Evolution of male sexual characters in the Oriental *Drosophila melanogaster* species group. Evolution and Development 4:278–291.

Kotiaho, J. S., L. W. Simmons, and J. L. Tomkins. 2001. Towards a resolution of the lek paradox. Nature 410:684–686.

Lambrechts, M. M. 1996. Organization of birdsong and constraints on performance. Pp. 305–321 in D. E. Kroodsma and E. H. Miller, eds. Ecology and Evolution of Acoustic Communication in Birds. Cornell University Press, Ithaca, NY.

Manning, J. T., and A. T. Chamberlain. 1993. Fluctuating asymmetry, sexual selection and canine teeth in primates. Proceedings of the Royal Society of London Series B: Biological Sciences 251:83–87.

Manning, J. T., and A. T. Chamberlain. 1994. Fluctuating asymmetry in gorilla canines: a sensitive indicator of environmental stress. Proceedings of the Royal Society of London Series B: Biological Sciences 255:189–193.

McGraw, K. J., and G. E. Hill. 2000. Differential effects of endoparasitism on the expression of carotenoid- and melanin-based ornamental coloration. Proceedings of the Royal Society of London Series B: Biological Sciences 267:1525–1531.

Moczek, A. P., and H. F. Nijhout. 2002. Developmental mechanisms of threshold evolution in a polyphenic beetle. Evolution and Development 44:252–264.

Møller, A. P. 1992a. Female swallows show preference for symmetrical male sexual ornaments. Nature 357:238–240.

Møller, A. P. 1992b. Patterns of fluctuating asymmetry in weapons: evidence for reliable signaling of quality in beetle horns and bird spurs. Proceedings of the Royal Society of London Series B: Biological Sciences 248:199–206.

Møller, A. P. 1996. Developmental stability of flowers, embryo abortion, and developmental selection in plants. Proceedings of the Royal Society of London Series B: Biological Sciences 1996:53–56.

Møller, A. P., and J. Hoglund. 1991. Patterns of fluctuating asymmetry in avian feather ornaments: implications for models of sexual selection. Proceedings of the Royal Society of London Series B: Biological Sciences 245:1–5.

Møller, A. P., and A. Pomiankowski. 1993a. Fluctuating asymmetry and sexual selection. Genetica 89:267–279.

Møller, A. P., and A. Pomiankowski. 1993b. Why have birds got multiple ornaments? Behavioral Ecology and Sociobiology 32:167–176.

Møller, A. P., and J. P. Swaddle. 1997. Asymmetry, Developmental Stability, and Evolution. Oxford University Press, Oxford.

Møller, A. P., and R. Thornhill. 1998. Bilateral symmetry and sexual selection: a meta-analysis. American Naturalist 151:174–192.

Møller, A. P., R. T. Kimball, and J. Erritzoe. 1996. Sexual ornamentation, condition, and immune defense in the house sparrow *Passer domesticus*. Behavioral Ecology and Sociobiology 39:317–322.

Moore, A. J. 1997. The evolution of social signals: morphological, functional, and genetic integration of the sex pheromone in *Nauphoeta cinerea*. Evolution 51:1920–1928.

Moore, A. J., and P. J. Moore. 1999. Balancing sexual selection through opposing mate choice and male competition. Proceedings of the Royal Society of London Series B: Biological Sciences 266:711–716.

Moore, A. J., P. A. Gowaty, W. G. Wallin, and P. J. Moore. 2001. Sexual conflict and the evolution of female mate choice and male social dominance. Proceedings of the Royal Society of London Series B: Biological Sciences 268:517–523.

Morris, M. R. 1998. Female preference of trait asymmetry in addition to trait size in swordtail fish. Proceedings of the Royal Society of London Series B: Biological Sciences 265:907–911.

Mulvihill, R. S., K. C. Parkes, R. C. Leberman, and D. S. Wood. 1992. Evidence supporting a dietary basis of orange-tipped rectrices in the cedar waxwing. Journal of Field Ornithology 63:212–216.

Murphy, C. G., and H. C. Gerhardt. 2000. Mating preference functions of individual female barking treefrogs, *Hyla gratiosa*, for two properties of male advertisement calls. Evolution 54:660–669.

Murphy, M. E., and J. R. King. 1987. Dietary discrimination by molting white-crowed sparrows given diets differing only in sulfur amino acid concentration. Physiological Zoology 60:279–289.

Nowicki, S., S. Peters, and J. Podos. 1998. Song learning, early nutrition and sexual selection in songbirds. American Zoologist 38:179–190.

Nowicki, S., D. Hasselquist, S. Bensch, and S. Peters. 2000. Nestling growth and song repertoire size in great reed warblers: evidence for song learning as an indicator mechanism in mate choice. Proceedings of the Royal Society of London Series B: Biological Sciences 267:2419–2424.

Ohlsson, T., H. G. Smith, L. Raberg, and D. Hasselquist. 2002. Pheasant sexual ornaments reflect nutrtional conditions during early growth. Proceedings of the Royal Society of London Series B: Biological Sciences 269:21–27.

Olson, E. C., and R. L. Miller. 1958. Morphological Integration. University of Chicago Press, Chicago.

Olson, S. L. 1970. Specializations of some carotenoid-bearing feathers. Condor 72:424–430.

Podos, J. 1997. A performance constraint on the evolution of trilled vocalizations in a songbird family (Passeriformes: Emberizidae). Evolution 51:537–551.

Pomiankowski, A., and A. P. Møller. 1995. A resolution of the lek paradox. Proceedings of the Royal Society of London Series B: Biological Sciences 260:21–29.

Preziosi, R. F., and D. A. Roff. 1998. Evidence of genetic isolation between sexually monomorphic and sexually dimorphic traits in the water strider *Aquarius remigis*. Heredity 81:92–99.

Price, J. J., and S. M. Lanyon. 2002. Reconstructing the evolution of complex bird song in the oropendolas. Evolution 56:1517–1529.

Price, T. 1996. An association of habitat with color dimorphism in finches. Auk 113:256–257.

Price, T., and A. Bontrager. 2001. Evolutionary genetics: the evolution of plumage patterns. Current Biology 11:R405–408.

Price, T., and M. Pavelka. 1996. Evolution of a colour pattern: history, development, and selection. Journal of Evolutionary Biology 9:451–470.

Price, T., D. Schluter, and N. E. Heckman. 1993. Sexual selection when the female directly benefits. Biological Journal of the Linnean Society 48:187–211.

Price, T. D. 2002. Domesticated birds as a model for the genetics of speciation. Genetica 116:311–327.

Prum, R. 1997. Phylogenetic tests of alternative intersexual selection mechanisms: trait macroevolution in a polygynous clade (Aves: Pipridae). American Naturalist 149: 668–692.

Pryke, S. R., S. Andesson, and M. J. Lawes. 2001. Sexual selection on multiple handicaps in the red-collared widowbird: female choice of tail length but not carotenoid display. Evolution 55:1452–1463.

Putman, R. J., M. S. Sullivan, and J. Langbein. 2000. Fluctuating asymmetry in antlers of fallow deer (*Dama dama*): the relative roles of environmental stress and sexual selection. Biological Journal of the Linnean Society 70:27–36.

Qvarnström, A. 1999. Genotype-by-environment interactions in the determination of the size of a secondary sexual character in the collared flycatcher (*Ficedua albicollis*). Evolution 53:1564–1572.

Qvarnström, A. 2001. Context-dependent genetic benefits from mate choice. Trends in Ecology and Evolution 16:5–7.

Radwan, J., J. Unrug, and J. L. Tomkins. 2002. Status-dependence and morphological trade-offs in the expression of a sexually selected character in the mite, *Sancassania berlesei*. Journal of Evolutionary Biology 15:744–752.

Ranz, J. M., C. I. Castillo-Davis, C. D. Meiklejohn, and D. L. Hartl. 2003. Sex-dependent gene expression and evolution of the Drosophila transcriptome. Science 300:1742–1745.

Reinhold, K. 1998. Sex linkage among genes controlling sexually selected traits. Behavioral Ecology and Sociobiology 44:1–7.

Richards, D. G., and R. H. Wiley. 1980. Reverberations and amplitude fluctuations in the propagation of sound in a forest: implications for animal communication. American Naturalist 115:381–399.

Rowe, C. 1999. Receiver psychology and the evolution of multicomponent signals. Animal Behaviour 58:921–931.

Rowe, L., and D. Houle. 1996. The lek paradox and the capture of genetic variance by condition-dependent traits. Proceedings of the Royal Society of London Series B: Biological Sciences 263:1415–1421.

Saetre, G.-P., T. Borge, K. Lindroos, J. Haavie, B. C. Sheldon, C. Primmer, and A.-C. Syvanen. 2002. Sex chromosome evolution and speciation in *Ficedula* flycatchers. Proceedings of the Royal Society of London Series B: Biological Sciences 270:53–59.

Schlichting, C. D., and M. Pigliucci. 1998. Phenotypic Evolution: A Reaction Norm Perspective. Sinauer Associates, Sunderland, MA.

Schluter, D., and T. Price. 1993. Honesty, perception and population divergence in sexually selected traits. Proceedings of the Royal Society of London Series B: Biological Sciences 253:117–122.

Schwartz, G. T., and C. Dean. 2001. Ontogeny of canine dimorphism in exant hominoids. American Journal of Physical Anthropology 115:269–283.

Seehausen, O., P. J. Mayhew, and J. J. M. Van Alphen. 1999. Evolution of colour patterns in East African cichlid fish. Journal of Evolutionary Biology 12:514–534.

Simmons, L. W., and J. L. Tomkins. 1996. Sexual selection and the allometry of earwig forceps. Evolutionary Ecology 10:97–104.

Slabbekoorn, H., and T. B. Smith. 2002. Bird song, ecology and speciation. Philosophical Transactions of the Royal Society of London Series B: Biological Sciences 357:493–503.

Slabbekoorn, H., J. Ellers, and T. B. Smith. 2002. Birdsong and sound transmission: the benefits of reverberations. Condor 104:564–573.

Swaddle, J. P., and I. C. Cuthill. 1994a. Female zebra finches prefer males with symmetric chest plumage. Proceedings of the Royal Society of London Series B: Biological Sciences 258:267–271.

Swaddle, J. P., and I. C. Cuthill. 1994b. Preference for symmetric males by female zebra finches. Nature 367:165–166.

Thompson, C. W., N. Hillgarth, M. Leu, and H. E. McClure. 1996. High parasite load in house finches (*Carpodacus mexicanus*) is correlated with reduced expression of a sexually selected trait. American Naturalist 149:270–294.

Thornhill, R. 1992. Female preference for the pheromone of males with low fluctuating asymmetry in the Japanese scorpionfly (*Panorpa japonica*: Mecoptera). Behavioral Ecology 3:277–283.

Tomkins, J. L., and L. W. Simmons. 1996. Dimorphisms and fluctuating asymmetry in the forceps of male earwigs. Journal of Evolutionary Biology 9:753–770.

Tregenza, T., and N. Wedell. 2000. Genetic compatibility, mate choice and patterns of parentage. Molecular Ecology 9:1013–1027.

Vallet, E., I. R. Beme, and M. Kreutzer. 1998. Two-note syllables in canary songs elicit high levels of sexual displays. Animal Behaviour 55:291–297.

Veiga, J. P., and M. Puetra. 1996. Nutritional constraints determine the expression of a sexual trait in the house sparrow, *Passer domesticus*. Proceedings of the Royal Society of London Series B: Biological Sciences 263:229–234.

Wagner, G. P. 2001. The Character Concept in Evolutionary Biology. Academic Press, San Diego.

Wagner, G. P., and K. Schwenk. 2000. Evolutionary stable configurations: functional integration and the evolution of phenotypic stability. Evolutionary Biology 31:155–217.

Wedekind, C. 1992. Detailed information about parasites revealed by sexual ornamentation. Proceedings of the Royal Society of London Series B: Biological Sciences 247:169–174.

Wedekind, C. 1994. Mate choice and maternal selection for specific parasite resistance before, during and after fertilization. Proceedings of the Royal Society of London Series B: Biological Sciences 346:303–311.

Whyte, L. L. 1965. Internal Factors in Evolution. George Braziller, New York.

Wiens, J. J. 2000. Decoupled evolution of display morphology and display behavior in phrynosomatid lizards. Biological Journal of the Linnean Society 70:597–612.

Wiley, R. H. 2000. A new sense of complexities of bird songs. Auk 117:861–868.

Wright, D. B. 1993. Evolution of sexually dimorphic characters in peccaries (Mammalia, Tyassuidae). Paleobiology 19:52–70.

Zahavi, A. 1975. Mate selection—a selection for a handicap. Journal of Theoretical Biology 53:205–214.

Zuk, M., and T. S. Johnsen. 1998. Seasonal changes in the relationship between ornamentation and immune response in red jungle fowl. Proceedings of the Royal Society of London Series B: Biological Sciences 265:1631–1635.

# 4

# The Evolution of Allometry in Modular Organisms

KATHERINE A. PRESTON
DAVID D. ACKERLY

The variation in organismal form that arises from size-dependent relationships among parts is a fundamental aspect of development and evolutionary change. Allometric analysis was one of the earliest morphometric tools developed to reveal ontogenetic or evolutionary changes in shape as a consequence of changes in organ or body size (Gayon, 2000). As allometric analysis has become more sophisticated, it has continued to provide insights into patterns and mechanisms of multivariate evolution (e.g., Niklas 1994; West et al. 1999). Although studies of allometry have been particularly prevalent in the animal literature, recent work has demonstrated the power of allometric analysis for interpreting morphological variation in plants as well (e.g., Armbruster 1991; Weiner and Thomas 1992; Jones 1993; Mazer and Wheelwright 1993; Niklas 1994; Ackerly and Donoghue 1998; Niklas and Enquist 2002).

A striking feature of plant populations is that individuals range widely in size, often as a result of asymmetric competition for light (Weiner et al. 2001) or patchy distribution of other resources. The ultimate basis for this dramatic variability lies in the way plants are constructed. Plants, and some animals, grow through the repeated addition of similar morphological subunits, that is, metamers and modules. This mode of development contributes to size variation by creating the potential for both indeterminate and exponential growth. Size variation, in turn, affects the expression of many other traits (reproductive allocation, resource capture, biomechanical stability, etc.), and plant functional morphologists have used allometry to examine these evolutionary and ecological consequences (Armbruster 1991; Jones 1993; Weiner and Thomas 1992; Mazer and Wheelwright 1993; Niklas 1994; Ackerly and Donoghue 1998; Niklas and Enquist 2002).

Modularity presents special challenges and opportunities for allometric analysis. In addition to generating size variation, modular construction also complicates the interpretation of allometry by blurring the meaning of apparently straightforward concepts such as size, age, and developmental stage. In this chapter we discuss the expression of allometry in modular organisms, its biological interpretation, and its consequences for multivariate trait evolution. We also ask what can be learned from modular organisms about the evolution of allometrically related traits in general.

## Character Variation and Covariation at Several Levels of Biological Organization

Much of the complexity in biological systems arises from their tendency to be organized into levels. Individuals come from genetic families and exist in populations that belong to communities. Populations constitute species, and species ultimately are grouped into nested series of increasingly inclusive clades. Modular organisms have additional levels of organization because of their clear within-individual substructure. In this section we connect the levels of *inter*individual variation (the usual subject of allometry) with several kinds of *intra*individual variation resulting from modularity. After a brief introduction to the traditional categories of allometry, we explore the sources of within-individual variation and the complex part-whole relationships of modular systems.

### Levels of Allometric Analysis

Organismal traits covary at several levels of biological organization, and at each level it is possible to characterize the form of the relationship allometrically. Interspecific allometry (*sensu* Gould 2002), or "evolutionary allometry" (*sensu* Cheverud 1982; Klingenberg 1998), is based on species means and is thought to reflect adaptive relationships between traits, or fundamental developmental or functional constraints (e.g., Niklas 1994; West et al. 1999). Intraspecific allometry is based on individuals measured either at a common developmental stage ("static," *sensu* Cheverud 1982; Klingenberg 1996) or at different points in development ("ontogenetic"). Several thoughtful reviews have addressed the relationships among evolutionary, static, and ontogenetic allometry (Cheverud 1982; Armbruster, 1991; Klingenberg and Zimmermann 1992; Klingenberg 1998). Here we focus primarily on different ways of measuring allometry within a species.

The raw material for any allometric analysis is variation among individuals, or among modules, in traits of interest. Interpretation of allometry requires careful consideration of the factors contributing to this variation. Four important sources are usually recognized: (1) ontogenetic variation due to differences in age and/or developmental rates; (2) environmental variation due to differences in conditions influencing growth; (3) genetic variation in heritable factors that influence the traits under consideration; and (4) developmental noise or experimental error unexplained by other sources. In experimental settings it may be possible to distinguish and control the relative contributions of each of these factors, but in field populations they are both difficult to measure and intertwined in a complex fashion. For example, ontogenetic variation may arise from the chronological age of an individual or module or its developmental stage (see below). However, age and stage are decoupled (and thus separable) only in individuals growing at different rates, which may in turn be traced to environmental or genetic differences. Moreover, in a temporally varying environment, two individuals of different age will have experienced different environments at corresponding stages in their growth. It is also well known that genetic and environmental variation may be confounded in the field when genotypes occupy different environmental conditions (Falconer 1989). The biological significance of static allometric relationships among individuals or modules depends on these many underlying sources of variation, which we shall revisit below in the context of various experimental studies.

While interspecific allometry reflects the outcome of evolutionary divergence, intraspecific allometry reflects adaptive evolution and developmental constraints operating within populations; as such it has provided insight into patterns of phenotypic integration and multivariate evolution (Olson and Miller 1958; Klingenberg and Zimmermann 1992; Jones 1993, 1995; Huber and Stuefer 1997; Ackerly and Donoghue 1998; Wright et al. 2001). For example, intraspecific allometry itself may be treated as an evolving character, and the slopes and intercepts of intraspecific relationships compared among species to test for evolutionary shifts in functional relationships between traits (Whitehead et al. 1984; Ackerly and Donoghue 1998; Klingenberg 1998; Wright et al. 2001; Preston and Ackerly 2003). More importantly, ontogenetic allometry describes the joint developmental trajectory of correlated traits, and static intraspecific allometry reflects their phenotypic architecture (analogous to genetic architecture). To the extent that these relationships reflect pleiotropic effects of genes acting during development, intraspecific allometry may also be paralleled by genetic correlations between the traits at the population level. Therefore, both types of intraspecific allometric analysis may help predict the way selection will act on correlated traits (Olson and Miller 1958; Schluter 1996; Klingenberg 1998).

This use of intraspecific allometry is complicated, however, because the trait relationships it captures are frequently not fixed for species or genotypes. The nature of covariation between characters may vary during development (Olson and Miller 1958; Coleman et al. 1994; Mazer and Delesalle 1996; Pigliucci et al. 1996; Acosta et al. 1997; Bonser and Aarssen 2001; López et al. 2001), and it can be phenotypically plastic (e.g., Lechowicz and Blais 1988; Weiner and Thomas 1992; Weiner and Fishman 1994; Müller et al. 2000; Bonser and Aarssen 2001). In many such cases, developmental stage and environment alter not merely the strength of the correlation between traits, but even their functional relationship, as measured by shifts in allometric slope or intercept. Thus the joint response to selection by a pair of related traits cannot be predicted reliably from their phenotypic covariation measured at any one time under one set of circumstances (Pigliucci et al. 1996).

The conceptual relationship between ontogenetic and static allometry is also complicated. Armbruster (1991) and Klingenberg and Zimmermann (1992) have argued that a similarity between them indicates that developmental processes contribute to trait covariation at the population level. To the extent that this is true, the developmental component itself may also be plastic. For example, in several different annual plant species, Weiner and Thomas (1992) and Weiner and Fishman (1994) found that ontogenetic and static allometry did coincide for uncrowded populations, but that they diverged under crowded conditions.

Nearly all organisms potentially exhibit ontogenetic and environmental variation in allometry, but some of the mechanisms behind this variation are unique to modular organisms. For example, modular growth permits serial adjustments to the phenotype through the addition of new parts during development. In the following sections we consider in detail the role that modularity plays in generating variation in allometrically related traits. Although we draw exclusively on examples from the plant literature, many of the issues we raise will be relevant to a variety of modular organisms, from bryozoans to eusocial insects.

## Modularity and Trait Expression In Plants

### Types of Modularity

Two related but distinct conceptions of modularity are prominent in morphology and evolutionary biology. The concept we use in this chapter is not the one applied in most work on multivariate trait evolution (cf. other chapters in this volume), so it is important to clarify their differences and draw comparisons between them. In developmental genetics, modules refer to largely autonomous, developmentally and functionally integrated units (Raff 1996; Wagner 1996; Carroll 2001; Chapters 2, 3, and 10 in this volume). In this context, modules usually correspond to developmentally independent nonhomologous parts (e.g., heads and limbs; Magwene 2001) or homologous metameric units that have become highly differentiated (e.g., arthropod limb types; Carroll 2001). There is little overlap between the sets of interacting genes that control each module, thus modularity may be selectively advantageous because it preserves functional coherence of interdependent parts while allowing evolutionary flexibility of the entire organism (Wagner 1996; Wagner and Altenberg 1996; Magwene 2001).

Here we apply another concept of modularity, which originated in comparative morphology and gained prominence within the field of plant population ecology (for historical overview see Harper and White 1974; White 1979). Under this usage, modular growth occurs through the reiteration of the same few types of structure, such as branches and flowers in plants and zooids in bryozoans. In contrast to developmental genetic modules, the modular subunits comprising an organism are similar to each other (serially homologous; Stevens 1984) and loosely analogous to individuals in a population. The concept of the plant as "metapopulation" (*sensu* White 1979) appeared as early as Theophrastus and continued through Goethe and Darwin (Harper and White 1974; White 1979). More recently, the analogy has proved especially fruitful for studies of plant population ecology (White 1979, 1984; Watson and Casper 1984; Sarukhán et al. 1985; Watson and Lu 1999), growth (Bazzaz and Harper 1977; Maillette 1985; McGraw and Garbutt 1990; Colasanti and Hunt 1997), resource allocation (Geber et al. 1997; Acosta et al. 1997; López et al. 2001), and phenotypic plasticity (Diggle 1994; Watson et al. 1995; Bonser and Aarssen 1996; Huber et al. 1999; Bonser and Aarssen 2001). Still, there are limits to the metapopulation metaphor. As most of these studies recognize, subunits may be modeled as individuals, but they are not entirely independent in physiology or development (White 1979; Watson and Casper 1984).

Under both concepts, an evolutionary advantage of modularity is the flexibility it affords to complex organisms. Developmental genetic modularity allows separate parts, organs, or structures to respond more or less independently to selection (Wagner 1996; Magwene 2001), and this decoupling may explain some major evolutionary trends within multicellular lineages (Carroll 2001). Presumably, developmentally autonomous parts could also express separate adaptive plastic responses, providing some morphological adaptability to individuals (Berg 1959; Armbruster 1991). The second kind of modularity also provides flexibility since modular growth allows individual plasticity in size, shape, and resource allocation. Examples of adaptive modular plasticity include the shade avoidance response (Schmitt et al. 1995; Dudley and Schmitt 1996), regrowth following herbivory (Haukioja 1991; Rosenthal and Welter 1995), foraging for resources in a heterogeneous environment (Slade and Hutchings 1987; de Kroon and Hutchings 1995; Huber 1996), shifting root-to-shoot ratio (see Müller et al. 2000), and partitioning fixed resources among structures in different combinations of size and number (Williams 1986). In eusocial animal colonies, modular plasticity can be seen in caste-specific nest defense behaviors in response to intruders (e.g., Duffy et al. 2002).

The most significant difference between the two concepts concerns their assumptions about the relations among modules. In the first, modules are genetically independent, but they interact in the proper functioning of a physiologically integrated organism. They are generally not autonomous with regard to resource uptake and use, and they do not have the potential for truly independent survival. By contrast, the second version of modularity often confers physiological substructure (Watson and Casper 1984) and in some cases the capacity for full physical independence, especially in clonal plants (e.g., Cook 1983; Pitelka and Ashmun 1985); but there is little if any genetic autonomy between modules. Although gene expression patterns may vary, the same set of genes controls development in each iteration of similar structures: there is not a separate set of genes for each successive leaf or branch produced. Of course, dissimilar modules, such as roots, flowers, and vegetative branches, may be genetically decoupled from each other. In that case, different *types* of morphological module also qualify as distinct modules in the developmental genetics sense (see Armbruster et al. Chapter 2, this volume). Even so, phenotypic correlations between the size of an inflorescence and the stems and leaves that support it have been found both within and among species (Corner 1949; Bond and Midgley 1988; Le Maître and Midgley 1991; Ackerly and Donoghue 1998), which suggests that development in vegetative and reproductive modules is not entirely independent.

## Modularity in Plants

In the specific case of plant shoots, we consider modules to be the products of apical meristems, for example, branches, cones, or flowers (following Prévost 1978; White 1979, 1984). Vegetative modules produce new meristems that may give rise to additional vegetative or reproductive modules, creating a nested modular hierarchy (Fig. 4.1). Usually, modules are themselves composed of repeated units, or metamers. For example, a typical vegetative metamer consists of a node and an internode, a leaf (or leaf homolog), and an axillary meristem (White 1979, 1984).

Each level of morphological organization potentially has its own demographic properties, and the interplay between modular and whole-plant demography has important consequences for trait expression at different levels. Some of these morphological, physiological, or life-history traits are expressed by subunits,

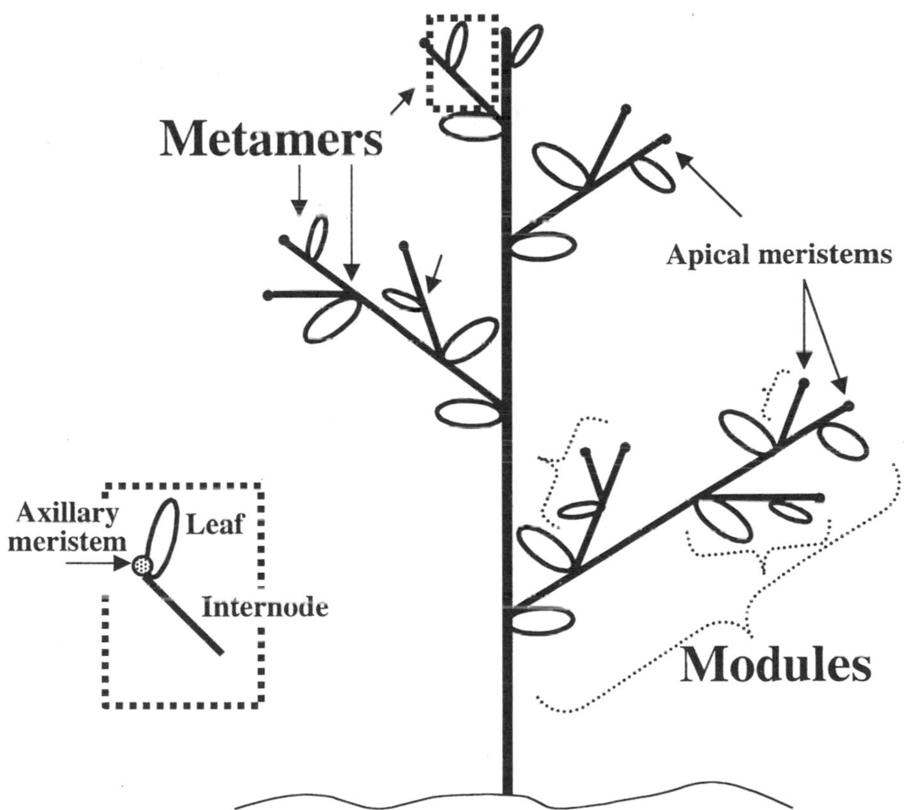


Figure 4.1 Modular construction in plants. We define a module as the product of an apical meristem (following Prévost 1978). New modules arise from existing ones through the differentiation of axillary meristems. Modules are indicated here by curly braces. Metamers are serially homologous repeated units along an axis and are usually subunits of modules. A vegetative metamer (shown in inset) consists of a leaf, the segment of stem subtending it, and its axillary meristem.

and some apply only to the entire organism. For example, modules and metamers originate, mature, senesce, and die, often independently of the whole plant. In addition, individual parts each have their own size (mass and extension), shape, and developmental potential, which may change over ontogeny. Likewise, whole plants age, become reproductively mature, senesce, and die, and they also have size, shape, and architecture arising from growth rules that regulate metamer development in space and time. Although both subunits and whole individuals partition resources between vegetative growth and reproduction, the concept of reproductive effort typically applies to whole organisms. Fitness also is generally assessed at the level of the whole plant because it is a property of the genetic individual. The state of the individual plant provides a developmental context for the parts (age, access to resources), and the properties of the parts influence whole-plant characters (e.g., architecture, reproduction) (Diggle 1994; Colasanti and Hunt 1997; Preston 1999; Jones 2001; Jones and Watson 2001). These whole-part relationships are the source of some key conceptual issues addressed in the next section.

## Allometry in Modular Organisms

Two aspects of modular construction have important implications for both the measurement and interpretation of allometry. First, modular organisms consist of multiple subunits, and traits expressed independently by each subunit can be measured repeatedly on a single individual. Examples include branch length, stem diameter, leaf number and size, flower number, etc. Metameric structures, such as leaves, provide yet another level of replication. As a result, allometric relationships among these traits emerge at the level of a single individual (Fig. 4.2). We refer to this intraindividual trait relationship as "modular allometry."

Second, modules develop sequentially as new growth arises from older growth. Modular construction thereby imposes both temporal and spatial structure on the population of subunits that comprise an individual, and also onto the traits expressed within each module (or metamer). Consequently, modular allometry measured in a single genetic individual may be based on a population of units just as phenotypically and demographically heterogeneous as the population of individuals underlying an intraspecific static or ontogenetic allometry. The modular allometric relationship will reflect several interacting sources of size variation, including age and developmental stage, resource status, and environment.

### Ontogenetic Variation

One of the main sources of size variation in modular allometry is age. We distinguish three main units of developmental age and ontogeny in plants, which may or may not be distinct depending on the life history and phenology of the organism (Table 4.1). The most inclusive unit is *whole-plant ontogeny*, which covers

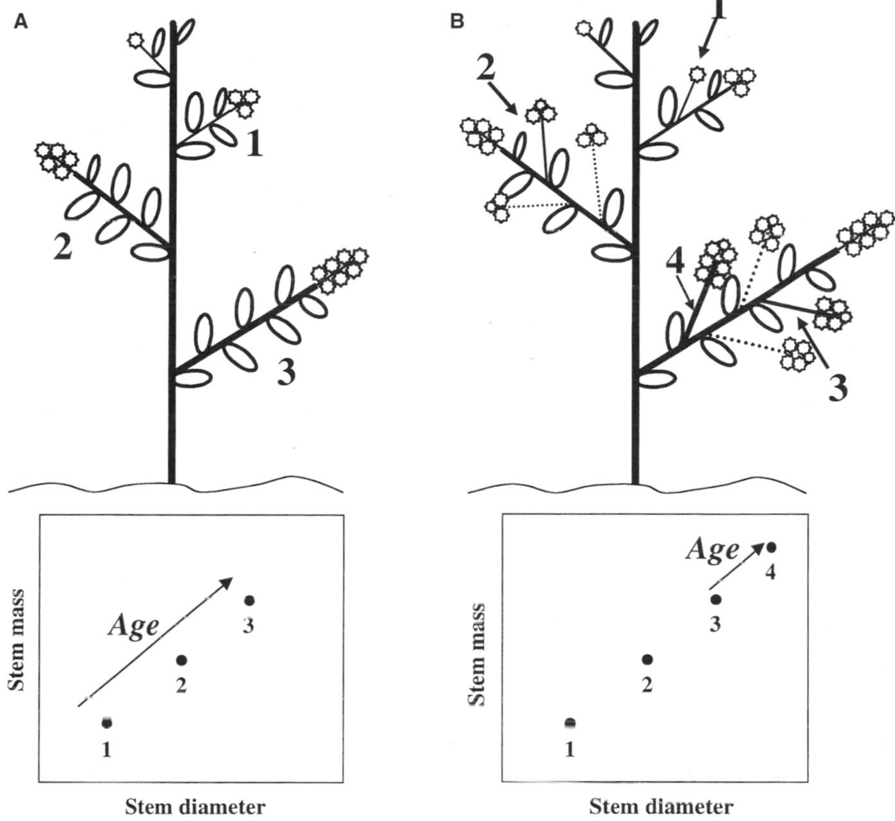


Figure 4.2 Modular allometry. A. Relationship between stem diameter at the base of a branch and total mass of that branch, where size variation is due to branch age. B. Diameter-mass allometry where size variation is due to both architectural position and age. Shoots 1, 2, and 3 are all the same age, as measured by plastochron index (*sensu* Erickson and Michelini 1957), but shoot 4 is older than the others.

germination to plant death. Whole-plant ontogeny may then be divided into a series of distinct reproductive or developmental cycles each with its own *rhythmic ontogeny* (following Hallé et al. 1978). "Rhythmic" ontogeny encompasses any regular sequence of developmental events, so it is not necessarily defined by a particular season or time period, although rhythmic cycles may be adapted to a predictable season length or cued by the environment. Examples include an interval of growth between periods of dormancy, a flowering phase in tropical plants (which may occur several times a year, or once every several years), or the maturation of a cohort of pine cones or acorns (often lasting more than a year). Finally, *modular ontogeny* applies to the development of each module and metamer, beginning with its differentiation and extending over its lifetime. In most woody perennials, these three categories are distinct; however, in annuals, whole-plant lifetime ontogeny, rhythmic ontogeny, and in some cases modular ontogeny,

Table 4.1 Levels at which developmental age may be measured in plants, and their manifestation according to life history, and phenology and architecture, expressed as both time periods and structural units.

| | Developmental level | | | |
|---|---|---|---|---|
| | Whole plant | Rhythmic | Modular | Metameric |
| Time period: | *lifetime* | *varies, and is defined by development* | *seasonal or subseasonal* | *seasonal or subseasonal* |
| Structural unit: | The product of a seed or propagule: also some long-lived ramets (e.g., *Populus*) | Nested set of modules from a meristem activated to start a developmental cycle | Morphological subset of the shoot produced from an apical meristem | Leaf or leaf homolog (e.g., bracts, spines, petals); axillary meristem |
| Woody perennials | * | * | * | * |
| Herbaceous perennials with indeterminate aerial shoots | * | | | * |
| Herbaceous perennials with determinate aerial structures (e.g., *Podophyllum, Dentaria, Arisaema, Sanguinaria*) | * | | | |
| Annuals | | | * | * |

Asterisks indicate distinct levels; double lines join levels that coincide for a given group of plants.

coincide; and in many perennial woodland herbs, a season's rhythmic growth is reduced to the differentiation of one module or even a single metamer (e.g., *Podophyllum peltatum*; Jones and Watson 2001).

Over each of these developmental time periods, allometric relationships among traits may change. For example, size-dependent reproductive allocation (reproductive effort) generally increases with whole-plant age in annuals (King and Roughgarden 1982) and some long-lived perennials (e.g., Piñero et al. 1982; López et al. 2001). Within a season, the total leaf area on a shoot declines relative to the size of the stem supporting it (K. Preston, unpublished data). At the module level, several studies have shown increased allocation to pollen relative to ovules within an inflorescence over development (Diggle 1994, 1995; Ashman and Hitchens 2000).

## Modular Demography

Significantly, ontogenetic changes in allometry at relatively higher morphological levels emerge from demographic processes operating at lower levels (e.g., Colasanti and Hunt 1997). For instance, reproductive allocation at the level of the whole plant depends on the number, type, and characteristics of its modules (Bonser and Aarssen 2001; López et al. 2001). The composition of the subunit population at any one time depends in turn on modular demography (birth rate and life span), identity (vegetative or reproductive), and development (e.g., size, gender expression). Similar lower-level processes affect allometry at the module level; for example, total shoot leaf area depends on the number and size of individual leaves, and gender expression within an inflorescence is a consequence of metamer (stamen and carpel) development (Lloyd 1980). In a simulation of modular plant growth, Colasanti and Hunt (1997) found that whole-plant characteristics as diverse as root-to-shoot ratio, foraging behavior, and even self-thinning under crowded conditions could be generated from purely module-based dynamics, independent of the status of the whole plant.

Nevertheless, some empirical studies suggest that whole-plant allometry is not simply "bottom-up," with module demography and modular traits generating allometric relationships; there are also "top-down" effects, where module development responds to the condition of the whole plant. Bonser and Aarssen (2003) showed that in four different annual herbs, greater overall plant size and resource status influenced module demography, thereby increasing the proportion of meristems devoted to reproduction and branching. In a Mediterranean shrub, reproductive effort rose with plant age through a change in module demography (i.e., the proportion of reproductive modules), with no effect on within-module characteristics (López et al. 2001). Whole-plant ontogeny also has an effect on stem-leaf allometry in several myrmecophytic tropical tree species that develop stem cavities housing ants (caulinary domatia). Brouat and McKey (2001) found that stem-leaf allometry in young plants is consistent with the high biomechanical demands that cavities impose on smaller stems. As stem and leaf size increase with plant age, hydraulic demands exceed structural requirements, and the slope of the stem-leaf allometry shifts to reflect shoot water relations.

## Allometry in Plants

The morphological hierarchy described above grows out of a series of developmental events that unfold over time. At any given moment, the phenotype of a single plant part is influenced by its temporal location within several developmental phases at different hierarchical levels. In addition to expressing its own developmental stage, the part is located within the ontogenetic progression of a larger module. The final size and shape of a leaf, for example, often varies with node position along an axis (Jones 1999; Kaplan 2001). Its phenotype will also depend on when it was initiated during a seasonal or rhythmic cycle. Preformed leaves overwintering in bud have been found to differ morphologically, anatomically, and physiologically from leaves initiated and matured later in the season (Roy et al. 1986). In a Mediterranean grassland shrub (*Baccharis pilularis*), life span is shortest for those leaves produced early in summer, just before the onset of the driest period (K. Preston, unpublished data; see also Abul-Fatih and Bazzaz 1980). Leaf shape in *Viola septemloba* depends on the interaction of a seasonal cue (photoperiod) with an inherent developmental shift (Winn 1996). Finally, the age of the parent plant may influence not only the identity of an individual part (e.g., vegetative versus reproductive; López et al. 2001), but also its shape (as in juvenile versus adult leaves), and the way resources are partitioned within it (e.g., size-number allocation in fruits; Acosta et al. 1997). Thus each plant part develops within a complex ontogenetic context (Diggle 1995; Watson et al. 1995); moreover, at any one time a plant is likely to be composed of many such individual parts, initiated at different points during modular, seasonal, and whole-plant development.

To see how this kind of interplay among morphological and developmental levels affects allometry, it is useful to return to the analogy of the plant as a population of subunits. Modular allometry, which describes covariation between traits expressed by multiple parts of a single plant, can then be likened to static or ontogenetic allometry of multiple individuals in a population. Static allometry for a population of unitary organisms is based on individuals measured at the same developmental stage, typically at reproductive maturity (Cheverud 1982; Klingenberg and Zimmermann 1992). The challenge presented by modularity is that there are several ways to define developmental stage, and one must decide which of these should be constant among the parts measured. For example, if all branches on a shrub are flowering, they are at the same seasonal reproductive stage; however, branches are likely to have been initiated at different times and so will vary in the length of their vegetative phase before the onset of flowering. Therefore, a sample of flowering branches will include branches of different chronological age but at the same stage in seasonal ontogeny. The length of the vegetative period can also be measured on two different scales. If shoots initiate growth at different times and also grow at different rates, they may have the same number of nodes (plastochron index; Erickson and Michelini 1957) but differ in chronological age (Ritterbusch 1990). For similar reasons, modular systems do not lend themselves to the method commonly used to estimate ontogenetic allometry in unitary organisms, in which individuals ranging in age are measured at a single sampling time to construct a composite ontogeny (but see Klingenberg 1996). In modular organisms, a collection of variously aged shoots measured at one time will reflect a variety of developmental responses to environmental conditions present during different segments of the season (rhythmic ontogeny).

Despite these challenges, modular allometry has been used in a number of studies to uncover the mechanisms underlying variation in complex morphological traits. Below we illustrate some interpretive approaches using examples from studies of both modular and whole-plant intraspecific allometry.

## Whole-Plant Intraspecific Allometry

Of the various types of plant allometry discussed above, intraspecific allometry of whole-plant traits is the most comparable to intraspecific allometry in unitary organisms. It differs, however, in that modular construction and indeterminate growth lead to the high degree of plasticity and intraspecific variation in trait relationships that are so characteristic of plants, especially in relation to whole-plant size. Questions about whether plasticity in trait relationships represents an adaptive functional response thus become particularly relevant to understanding the ecology and evolution of modular organisms. The three studies discussed below explore the connection between intraspecific allometry of whole-plant traits and plasticity in the trait relationships.

### Root-Shoot Biomass Partitioning

A frequently cited example of adaptive phenotypic plasticity is environmentally dependent variation in resource allocation to different functions (Bloom et al. 1985; Sultan and Bazzaz 1993; Coleman et al. 1994; Coleman and McConnaughay 1995; Grace 1997). Optimal partitioning models predict that organisms should increase relative allocation to structures (or activities) used to acquire the most limiting resource (Bloom et al. 1985; Sultan, 1996; Grace 1997). To test the generality of this prediction, Müller, Schmid, and Weiner (2000) measured plasticity in biomass allocation to roots, stems, and leaves in response to two nutrient levels in 27 herbaceous plant species. In accordance with predictions, they found that the root-to-stem and root-to-leaf ratios were lower and the stem-to-leaf ratios were greater in the high-nutrient treatment compared to the low-nutrient treatment in nearly all of the species that showed any response. In addition to comparing mean ratios between treatments, Müller and colleagues also examined intraspecific allometric relationships among the three variables at each nutrient level. Interestingly, they found that for more than half of the species, allometric relationships did not differ between treatments, even when the ratios between variables did differ. In those cases, biomass ratios varied with plant size because the allometric slopes departed from isometry; slopes of root biomass versus leaf and stem biomass were less than 1 (negative allometry), and the slope of stem versus leaf biomass were greater than 1 (positive allometry) (Fig. 4.3). Therefore, when plants grown under the high-nutrient conditions were substantially larger than those grown with low nutrients, they occupied a different region of the species' allometric trajectory, which translated into a difference between treatments in the mean ratio (see also Coleman et al. 1994; McConnaughay and Coleman 1998, 1999).

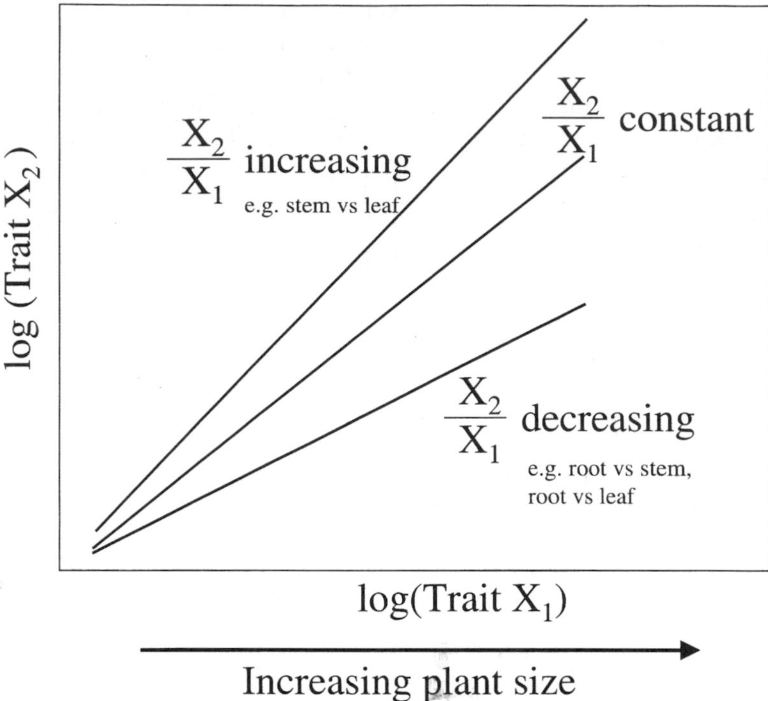


Figure 4.3 Trait relationships characterized allometrically and as ratios. When the slope of the allometric relationship between log-transformed variables is isometric (= 1), then the ratio between the untransformed variables is constant. When the slope is greater than 1 (positive allometry), the ratio increases with size; when the slope is less than 1 (negative allometry), the ratio declines with size. Müller et al. (2000) found a positive allometric relationship between stem and leaf biomass, and consequently an increasing stem-to-leaf ratio with overall plant size. Conversely, they found that root-to-stem and root-to-leaf ratios declined with overall plant size due to their negative allometric relationship.

These results are intriguing because they suggest two very different interpretations. One holds that the treatment differences in biomass ratio were an allometric consequence of differences in plant size, and therefore not a result of adaptive plasticity in biomass allocation (Coleman and McConnaughay 1995; McConnaughay and Coleman 1999). An alternative interpretation focuses on the adaptiveness of allometry itself, emphasizing that optimal biomass allocation varies with plant size in a predictable way, so that the best allometric relationship is one that yields the optimal ratio across a range of plant sizes. From this standpoint, allometry is itself an evolved strategy to maintain the functionally appropriate biomass partitioning as plant size varies over development or in response to environmental variation (cf. Chapter 7 by Pigliucci in this volume). Müller and colleagues see their results as evidence for such an allometric strategy in most of the species they examined. We return to the evolutionary potential of allometric strategies below.

### Foraging in Clonal Plants

Another classic example of apparently optimal allocation is foraging in clonal plants. Because of their spreading habit and adventitious roots, clonal plants can explore horizontal space more efficiently than upright plants can, and they are able to concentrate growth in high-resource patches to the benefit of the entire plant (e.g. Cook 1983; Slade and Hutchings 1987; de Kroon and Hutchings 1995, and references therein). However, the patterns of ramet placement often interpreted as foraging could also arise from nonadaptive growth responses to resource availability by individual modules (ramets), rather than from a whole-plant growth strategy. To distinguish these possibilities, Huber and Stuefer (1997) compared growth rate and branch production by individual stolons of *Potentilla reptans* under three different light treatments: full light (control), neutral shade, and canopy shade (reduced red:far-red). Compared to the controls, both of the shade treatments reduced final biomass, growth rate of the primary axis, and the rate of new branch initiation. Slower branch production resulted in a much less branched architecture at final harvest for the shaded plants relative to the controls. Thus, when plants were compared at the same chronological age, those grown under higher light had allocated resources to new branches, allowing more intensive use of the habitat, whereas shaded plants had adopted a more linear growth form, allowing plants to escape poor local conditions. Nevertheless, Huber and Stuefer's results tell a very different story when plants are compared at the same developmental stage (plastochron index). Under shaded conditions, primary ramets (arising from nodes along the main axis) were chronologically much older than unshaded ramets, but they did not differ in developmental stage at the time they produced their first branch. The authors speculate that if shaded plants had been allowed to reach the same whole-plant developmental age as the unshaded plants at the time they were harvested, there would have been no treatment effect on branching intensity. They conclude that the response of clonal morphology to shading should not be interpreted as an adaptive plastic change in growth form, but rather as a simple artifact of comparing plants at different points along their ontogenetic trajectory.

### Adaptive Interpretations of Plasticity and Allometry

Both the Müller et al. study (2000) and the Huber and Stuefer study (1997) question common assumptions about adaptive plasticity in biomass allocation and use allometric analysis to reveal the developmental mechanisms producing an apparently appropriate phenotype. In both studies, plants adopted the phenotype predicted for their treatment environment under an optimal allocation model, but treatment differences disappeared when plants were compared at the same size or developmental age rather than the same chronological age. Despite these similarities, the authors of the two studies draw contrasting conclusions: Müller and colleagues argue that for most species root-shoot allometry represents an evolved strategy for plasticity, whereas Huber and Stuefer see allocation patterns as simply conforming to a "null" model and reflecting the effects of resource limitation on the expression of a fixed developmental program.

Interestingly, each study's conclusions can be justified. As with any complex character, the question of whether a fixed ontogenetic allometry acts as a constraint on phenotypic expression (as in Huber and Stuefer) or the means to an adaptive response (as in Müller et al.) must be decided in context and on the basis of other evidence. For example, adaptive plasticity must *ceteris paribus* involve the expression of an appropriate phenotype for a given environment (Sultan and Bazzaz 1993). In the first study, root-to-leaf and root-to-stem ratios generally declined with overall plant size, as predicted by "economic" models of allocation for optimal resource acquisition (e.g., Bloom et al. 1985). Moreover, the observation that numerous distantly related species have similar root-shoot allometry suggests either a widespread fundamental constraint on biomass allocation, or convergence on an adaptive strategy. In the second example, plants appeared to adopt a linear clonal morphology consistent with optimal foraging, yet Huber and Stuefer's developmental analysis showed this response to be only temporary. Eventually, the shaded plants would literally grow out of their supposedly adapted phase, while the resource environment remained unchanged.

A second kind of evidence for the adaptive nature of allometry depends on its role in the overall phenotype. Because the root-shoot allometry of a species is a direct manifestation of individual root-to-shoot ratios, it is plausible that an allometric relationship could evolve entirely under selection acting on biomass allocation in individual plants as a function of their size. By contrast, allometry in a branching shoot is likely to be a purely geometric consequence of a complex developmental program, whose evolution has been shaped by numerous biomechanical, physiological, and ecological forces. Such a developmental program is integral to whole-plant functioning, and its features carry constraints as a consequence of this integration. Therefore, the potential for optimal allocation in a particular environment will be limited by the larger phenotypic context. The contrast between these two studies demonstrates that judgments about the role of allometry in adaptive plasticity depend on the relative weights of at least these explanatory considerations: the apparent "fit" between phenotype and environment (including changes in fit over time), and the potential for an allometric strategy to evolve, given the role of allometry in phenotypic integration.

## Allometry of Meristem Allocation

A third example of whole-plant allometry combines features of the two studies described above by including both ontogenetic and static allometry measured across a resource gradient. Bonser and Aarssen (2001) used nutrient treatments to generate size variation among individuals from five genotypes of *Arabidopsis thaliana* and examined patterns of meristem allocation as a function of plant size weekly throughout development. At each census, they characterized allometric relationships among three meristem fates (inactivity, $I$; vegetative growth, $G$; and reproduction, $R$) in individuals of the same chronological age. Each of these relationships could in turn be followed across censuses to generate an ontogeny of allometries (Fig. 4.4). In other words, their data show the effect of size on allocation at each time point, as well as the allometric effect of achieving a given size at different times. For example, reproductive effort ($R$ versus $G + I$) increases

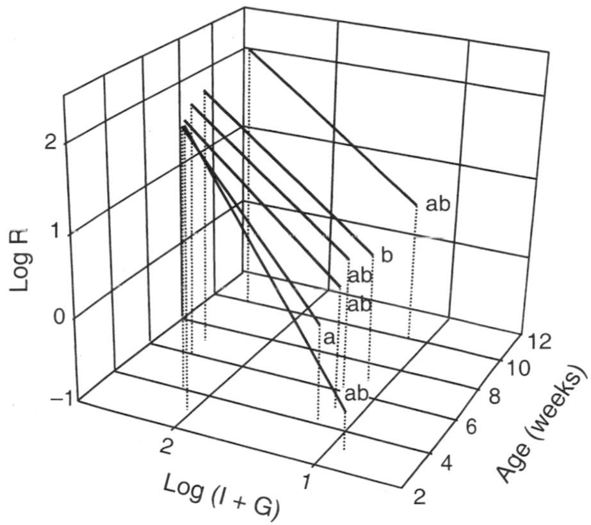


Figure 4.4 Allometry throughout development for one genotype (Kil) of *Arabidopsis thaliana*. Lines represent Model II regression of log $R$ (number of reproductive meristems) versus log $(I + G)$ (number of inactive and vegetative growth meristems). Regressions labeled with different letters differ significantly in slope. Reproduced with permission from Bonser and Aarssen (2001).

more sharply with plant size early in development than late; and for a given size, reproductive effort increases with plant age (cf. Pigliucci et al. 1996). This multidimensional data set allowed the authors to use an ontogenetic series of static allometries to demonstrate the ways in which meristem identity depends on the combination of plant size and plant age.

As discussed by Bonser and Aarssen, allometry is neither a mechanism for adaptive plasticity nor a constraint on it. Rather, allometry measured at a common age is a manifestation of plasticity, where steeper slopes indicate a greater response to nutrient availability in allocation patterns (assuming that most size variation was due to nutrient treatment). Like Müller et al. (2000), Bonser and Aarssen see allometry in allocation as an evolved strategy, but the two versions differ somewhat. Whereas the earlier study found that a fixed allometric relationship generated plasticity in ratios, Bonser and Aarssen found that within-genotype allometry varied with plant age, and this temporal dimension was an essential axis of plant allocation strategy. Interestingly, they found very little variation in allocation patterns among the five genotypes measured. Results from a similar study that included three additional, architecturally diverse, species suggest that these patterns can be generalized beyond *Arabidopsis thaliana* at least to other annuals (Bonser and Aarssen 2003).

## Modular Allometry

As discussed above, modular organisms are composed of reiterated morphological subunits, and traits expressed in each module can be used to construct an intraindividual modular allometry. Here we describe two studies that use modular allometry to examine evolution in functionally related traits.

### Stem-Leaf Allometry and a Novel Stem Function

Leaf performance depends crucially on the hydraulic and biomechanical support provided by stems, and this relationship is reflected in the widely observed positive correlation between leaf area and stem size, both within and among species (Corner 1949; White 1983, Bond and Midgley 1988; Ackerly and Donoghue 1998; Brouat et al. 1998). Brouat and McKey (2001) examined the evolutionary changes in stem-leaf allometry that accompany the integration of a new function into the usual suite of stem functions. In three lineages of tropical trees, they characterized stem-leaf allometry for myrmecophytic species that develop enlarged hollow stems specialized for housing ants, and compared them with stem-leaf allometries found in the species' close nonmyrmecophytic relatives. In the myrmecophytic species, leaf and stem size both increase with whole-plant age, while cavity size remains relatively constant; thus the role of hollowed stems in overall stem function was predicted to change as plants age. The authors therefore based their functional interpretation on whole-plant ontogenetic allometry. Although each sample was taken from a different tree, they minimized the potentially confounding effects of module age and developmental stage (discussed above) by measuring only the terminal leaf on a stem and the internode subtending it.

Whereas nonmyrmecophytic species showed a nearly constant proportionality between leaf area and stem cross-sectional area, the stems of myrmecophytic species were relatively large early in development and did not increase in proportion to leaf size. The resulting shallow allometry can be explained by the age-dependent effect of stem cavities on the relative importance of biomechanical and hydraulic support. The structural integrity of a hollow stem depends on the ratio between the thickness of the ring of wood and the radius of the whole twig (ring and cavity), and on the mechanical properties of the wood. The authors propose that when plants are young, the amount of wood required to supply water to a small leaf is biomechanically adequate for solid stems but not for hollow ones, which must allocate relatively more to stem tissue. As leaf size increases with plant age, the stem tissue required to meet hydraulic demands also meets structural requirements. As predicted under this scenario, the species with hollowed stems showed age-dependent stem-leaf size relationships but a nearly constant ratio of wood thickness to stem size.

Brouat and McKey conclude that in these species, stem-leaf allometry should be interpreted as a line of functional equivalence across a range of plant sizes. Their study also shows that the form of stem-leaf allometry has evolved repeatedly in three independent clades, leading to convergence in slope when a new role is integrated into the usual suite of stem functions. Thus, they argue, the phenotypic correlation between stem and leaf size in solid stems does not represent a constraint on the evolution of this relationship in hollow stems; rather, in both solid and hollow stems, allometry can be seen as the by-product of selection on stem function.

### Within-Module Trait Correlations

A second example of modular allometry explores the tension between mechanisms that may underlie correlations between modular traits. Mazer and Delesalle (1996) examined phenotypic and genetic correlations among floral parts in a highly selfing annual species (*Spergularia marina*) in order to disentangle the various forces that generate trait covariation within flowers. Strictly speaking, sepals, petals, stamens, and carpels are metamers composing a floral module; consequently, these floral parts may show interdependent developmental responses to shared external and internal environmental conditions. Moreover, floral traits have sometimes been found to be more highly integrated than vegetative traits of the same species (Berg 1959; Armbruster et al. 1999). Although strong within-flower trait correlations should be less common in self-pollinating than in animal-pollinated species (Berg 1959), Mazer and Delesalle argue that stabilizing selection should favor equal numbers of male and female gametes within self-pollinating flowers (a special case of isometry). These conditions lead to conflicting predictions. For example, strong stabilizing selection to maintain optimal gender allocation (pollen: ovule = 1) should generate positive genetic correlations between anther number and ovule number. Alternatively, resource limitation within a flower could lead to a tradeoff in allocation to male and female gametes. Mazer and Delesalle found no correlation between family means in the number of male and the number of female floral parts (controlling for flower size), which suggests that stamen and ovule number per flower have evolved independently among these genotypes. They speculate that the evolutionary decoupling of these traits has resulted from the tension between selection favoring isometry in gender allocation and a negative genetic correlation due to tradeoffs in resource allocation.

## Selection on Allometry in Modular Organisms

The studies described above provide evidence that allometry should not always be viewed simply as evidence for a constraint on evolution, but that allometry itself evolves, and that the form of the relationships among traits may be shaped by selection (see Chapters 2, 7, and 18 in this volume). Evolutionary changes in allometry reflect shifts in two or more traits with respect to each other, so they represent a form of multivariate phenotypic evolution. Understanding the evolution of allometry thus requires consideration of the full scope of evolutionary processes involved in multivariate evolution: genetic and phenotypic variation and covariation, plasticity, and selection (Lande and Arnold 1983; Wagner and Altenberg 1996; Schlichting and Pigliucci 1998). Evolution of modular allometry presents yet another factor for consideration, namely the role of selection at different hierarchical levels.

Allometric relationships in unitary organisms are observed among individuals (static) or within individuals over time (ontogenetic). In both cases, the traits involved exhibit a single value in each individual at a particular moment in time (Fig. 4.5A,B). In evolutionary terms, selection acts on variation in the traits and the trait combinations exhibited by individuals, and allometric relationships emerge at the population level. If there is plasticity in the traits or their relationship to one other, then the dynamics of selection depend on patterns of environmental heterogeneity, and the genetics of the underlying reaction norms (Schlichting and Pigliucci 1998). The resulting allometric relationships within populations will depend on the effects of genetic and environmental variation (and their interaction) among individuals.

At a higher level of organization, interpopulational or interspecific allometry may arise as populations and species, respectively, evolve to different trait means (Fig. 4.5A). The important point about variation at this level is that the allometry itself is not under selection. If the optimal relationship of two traits ($X$, $Y$) is $Y = bX^a$, then within each population selection will lead to shifts in $X$ and/or Y, moving toward the joint optimum described by this equation. This would appear as an allometric relationship at a higher level of analysis considering multiple populations or species. Quantitative analysis of the $(X,Y)$ relationship might reveal the underlying values of coefficients $a$ and $b$, describing the optimum value that was favored by selection in each population; yet these values are not properties of the evolving genotypes or populations that are directly under selection. The evolutionary allometry is thus a consequence of selection within populations and species, not a property directly targeted by selection.

The evolution of modular allometry differs from this standard scenario in a way that has significant consequences for the efficacy of selection in shaping allometric relationships. In indeterminate modular organisms, many phenotypic traits contributing to organismal function are expressed at the modular level (e.g., stem diameter, leaf area, etc.). These traits are expressed simultaneously in multiple modules, and as a consequence the allometric relationship itself becomes a feature of the organismal phenotype, in a way not generally possible in unitary organisms (Fig. 4.5C). As a simple example, consider the allometry of basal stem diameter and total stem mass in the branches of a woody plant. Biomechanical theory predicts an allometric relationship between these traits to maintain a safety factor against failure, where the coefficient of the safety factor depends on the role of static versus dynamic loading of the stem (Holbrook and Putz 1989; Niklas 2000). The diameter-mass relationship across a broad size range can be evaluated on a single tree, based on the size and diameter of each branch (e.g., Niklas 1999). Ontogenetic relationships also would emerge during the growth of each branch and of the central trunk itself. The allometric relationship is thus a phenotypic property of the individual, and selection can operate directly on allometry at the level of individuals within populations. Assuming there is an optimal relationship favored by selection, there would be costs to diverging from the optimum at any point in the allometry. Conversely, the success of a genotype with the optimal relationship would result in selection on the entire allometry within one generation.

To understand how modular allometries may evolve as traits of the individual, it is important to consider the multiple interacting sources of modular variability that generate allometric relationships. When the allometry is a property of an individual organism, the integration of the individual phenotype demands that all of these underlying sources of variation be considered together. As noted above, in the course of indeterminate growth modules may be initiated at different times,

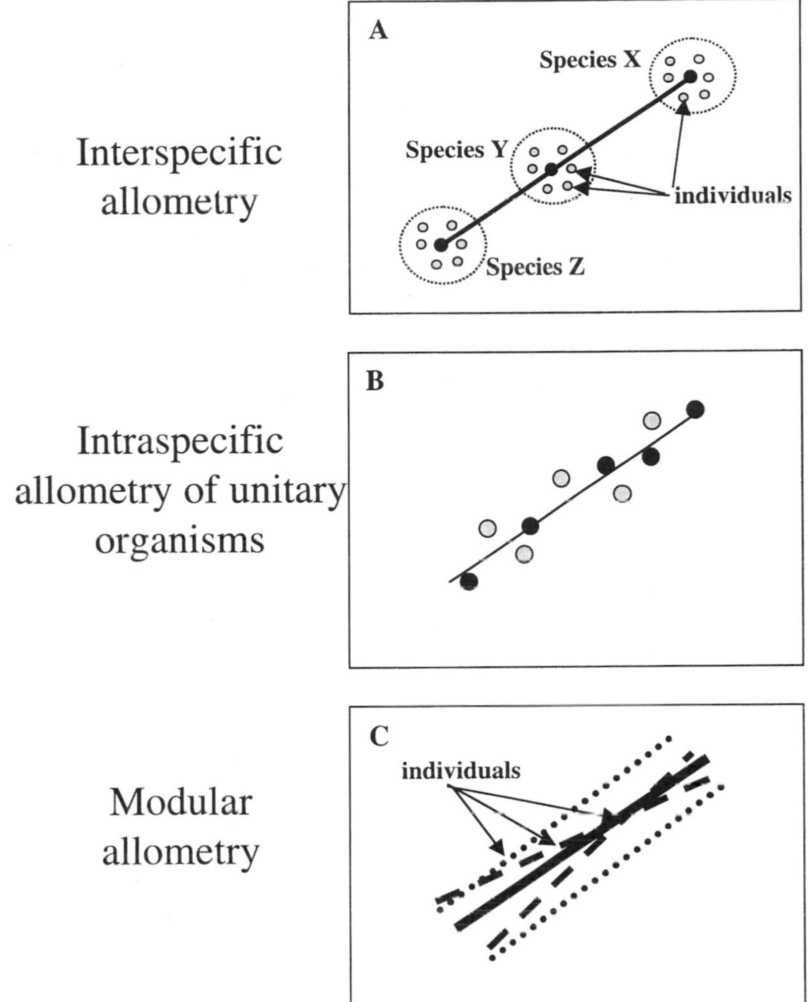


Figure 4.5 Schematic diagram of the role of selection shaping allometric relationships at the interspecific, intraspecific, and intraindividual (modular) levels. Individuals with relatively high fitness are indicated with black symbols, and less fit individuals are depicted in lighter grey. A. Across species, stabilizing selection favoring species means that fall along a bivariate axis leads to interspecific allometry. Within each species, the joint mean for the two traits is favored, but the allometric relationship per se is not under selection. B. Similarly, within populations of unitary organisms, selection favoring genotypes that lie close to the optimal bivariate trait values leads to intraspecific allometry. C. For modular organisms, allometric relationships may be expressed by the individual, based on trait variation among modules (e.g., leaves, zooids). The allometric relationship is thus an individual trait subject to direct selection within populations. See additional discussion in text.

resulting in a range of sizes within the individual that correspond to an ontogenetic sequence (Fig. 4.2a). We have also noted, however, that trait relationships are shaped by both internal and external environmental conditions throughout development. Returning to our biomechanical example, wood production is stimulated by biomechanical demands, a response seen in localized wood growth in areas of compression stress ("reaction wood"; Gartner 1991), and in the greater inherent stability of plants exposed to wind (e.g., Holbrook and Putz 1989). Presumably, then, the typical ontogenetic trajectory of a growing branch is influenced by both external and internal biomechanical forces normally generated during growth (see Niklas 1999). Module development is also sensitive to ambient light levels and the supply of water, nutrients, and plant growth substances at each growing point. These factors depend on the dynamic interaction between plant architecture and the internal and external environment (which are essentially equivalent from the point of view of a module). For example, self-shading produces a vertical gradient of light from upper to lower canopy, affecting leaf physiology and subsequent shoot growth.

In these situations, traits covary across modules as a function of their joint plastic response to fine-grained environmental heterogeneity (Lloyd 1984). We therefore argue that allometric relationships are a special case of the more general potential for selection on reaction norms as individual traits (Winn 1996, 1999). The expression of variable phenotypes within the individual is itself an adaptive strategy (Lloyd 1984; Diggle 1994; Winn 1996). Allometry in modular organisms not only reflects functional and developmental relations among traits, as it does in unitary organisms, but it is also one of the direct targets of selection. As a result, allometric strategies of the sort suggested by Müller and colleagues (2000) are of special interest for empirical study.

## Conclusions

The significance of modular allometry for phenotypic evolution rests on two related questions: whether a given allometric relationship is adaptive, and if so, how it responds to selection. In this chapter, we have drawn on several recent studies presenting examples of apparently adaptive allometric relationships that maintain organismal function across a range of environments (Müller et al. 2000; Bonser and Aarssen 2001) and throughout development (Bonser and Aarssen 2001; Brouat and McKey 2001). We contrasted these studies with one in which a nonadaptive allometric relationship appears to be the consequence of a fixed developmental program (Huber and Stuefer 1997). Together, these results demonstrate that the adaptive value of allometry should be judged not only on its particular "fit" to the environment, but also on its larger phenotypic context. We have also argued that modular allometry itself can become a target of selection, thereby gaining standing as a biologically significant property of the integrated phenotype, in addition to serving as a measure of integration.

The potential for allometry to respond to selection depends on the sources of character variation and covariation, and the extent to which the allometrically related traits are integrated with the rest of the organism. For whole-organism intraspecific allometry, the most common sources of trait variation (genotype, microhabitat) are not fundamentally related to phenotypic integration. By contrast, many of the sources of variation contributing to within-individual modular allometry arise directly from the relationship between the allometric traits and the rest of the organism. As we have discussed in detail, modular traits are expressed within a spatial and temporal context that grows out of the dynamic interplay between body plan (architecture) and phenology. In addition, trait expression in each module is more or less contingent on prior developmental decisions (Diggle 1994, 1995; Watson et al. 1995). The integration of modular allometry with other whole-plant traits and processes (architecture, phenology) should limit the range of allometric relationships among individuals available for selection (see Badyaev, Chapter 3, this volume, for discussion of allometry and integration in a unitary organism). However, selection may act very efficiently on modular allometry, because the allometric relationship is an individual trait and a wide range of allometries may be exposed to selection within populations.

*Acknowledgments* We thank Stephen Bonser for sharing a manuscript in press, and Lanier Anderson, Stephen Bonser, and Massimo Pigliucci for helpful comments on the manuscript. Research contributing to the ideas in this chapter was conducted with support to K.A.P. from a National Science Foundation Postdoctoral Research Fellowship.

### *Literature Cited*

Abul-Fatih, H. A. and F. A. Bazzaz. 1980. The biology of *Ambrosia trifida* L. IV. Demography of plants and leaves. New Phytologist 84:107–111.

Ackerly, D. D. and M. J. Donoghue. 1998. Leaf size, sapling allometry, and Corner's rules: a phylogenetic study of correlated evolution in maples (*Acer*). American Naturalist 152:767–791.

Acosta, F. J., J. A. Delgado, J. M. Serrano, and F. López. 1997. Functional features and ontogenetic changes in reproductive allocation and partitioning strategies of plant modules. Plant Ecology 132:71–76.

Armbruster, W. S. 1991. Multilevel analysis of morphometric data from natural plant populations: insights into ontogenetic, genetic, and selective correlations in *Dalechampia scandens*. Evolution 45:1229–1244.

Armbruster, W. S., V. S. Di Stilio, J. D. Tuxill, T. C. Flores, and J. L. Velasquez Runk. 1999. Covariance and decoupling of floral and vegetative traits in nine Neotropical plants: a re-evaluation of Berg's correlation-pleiades concept. American Journal of Botany 86:39–55.

Ashman, T.-L. and M. S. Hitchens. 2000. Dissecting the causes of variation in intra-inflorescence allocation in a sexually polymorphic species, *Fragaria virginiana* (Rosaceae). American Journal of Botany 87:197–204.

Bazzaz, F. and J. L. Harper. 1977. Demographic analysis of the growth of *Linum usitatissimum*. New Phytologist 78:193–208.

Berg, R. L. 1959. Ecological significance of correlation pleiades. Evolution 14:171–180.

Bloom, A. J., F. S. Chapin, et al. 1985. Resource limitation in plants: an economic analogy. Annual Review of Ecology and Systematics 16:363–392.

Bond, W. J. and J. Midgley. 1988. Allometry and sexual differences in leaf size. American Naturalist 131:901–910.

Bonser, S. P. and L. W. Aarssen. 1996. Meristem allocation: a new classification theory for adaptive strategies in herbaceous plants. Oikos 77:347–352.

Bonser, S. P. and L. W. Aarssen. 2001. Allometry and plasticity of meristem allocation throughout development in *Arabidopsis thaliana*. Journal of Ecology 89:72–79.

Bonser, S. P. and L. W. Aarssen. 2003. Allometry and development in herbaceous plants: functional responses of meristem allocation to light and nutrient availability. American Journal of Botany 90:404–412.

Brouat, C. and D. McKey. 2001. Leaf-stem allometry, hollow stems, and the evolution of caulinary domatia in myrmecophytes. New Phytologist 151:391–406.

Brouat, C., M. Gibernau, L. Ansellem, and D. McKey. 1998. Corner's rules revisited: ontogenetic interspecific patterns in leaf-stem allometry. New Phytologist 139:459–470.

Carroll, S. B. 2001. Chance and necessity: the evolution of morphological complexity and diversity. Nature 409:1102–1109.

Cheverud, J. M. 1982. Relationships among ontogenetic static and evolutionary allometry. American Journal of Physical Anthropology 59:139–150.

Colasanti, R. L. and R. Hunt. 1997. Resource dynamics and plant growth: a self-assembling model for individuals, populations and communities. Functional Ecology 11:133–145.

Coleman, J. S. and K. D. M. McConnaughay. 1995. A non-functional interpretation of a classical optimal-partitioning example. Functional Ecology 9:951–954.

Coleman, J. S., K. D. M. McConnaughay, and D. D. Ackerly. 1994. Interpreting phenotypic variation in plants. Trends in Ecology and Evolution 9:187–191.

Cook, R. E. 1983. Clonal plant populations. American Scientist 71:244–253.

Corner, E. J. H. 1949. The Durian theory or the origin of the modern tree. Annals of Botany 13:367–414.

de Kroon, H. and M. J. Hutchings. 1995. Morphological plasticity in clonal plants: the foraging concept reconsidered. Journal of Ecology 83:143–152.

Diggle, P. K. 1994. The expression of andromonoecy in *Solanum hirtum* (Solanaceae): phenotypic plasticity and ontogenetic contingency. American Journal of Botany 81:1354–1365.

Diggle, P. K. 1995. Architectural effects and the interpretation of patterns of fruit and seed development. Annual Review of Ecology and Systematics 26:531–552.

Dudley, S. A. and J. Schmitt. 1996. Testing the adaptive plasticity hypothesis: density dependent selection on manipulated stem length in *Impatiens capensis*. American Naturalist 147:445–465.

Duffy, J. E., C. L. Morrison, and K. S. Macdonald. 2002. Colony defense and behavioral differentiation in the eusocial shrimp *Synalpheus regalis*. Behavioral Ecology and Sociobiology 51:488–495.

Erickson, R. O., and F. J. Michelini. 1957. The plastochron index. American Journal of Botany 44:297–305.

Falconer, D. S. 1989. Introduction to Quantitative Genetics. Longman Scientific and Technical, New York.

Gartner, B. L. 1991. Stem hydraulic properties of vines vs. shrubs of western poison oak, *Toxicodendron diversilobum*. Oecologia 87:180–189.

Gayon, J. 2000. History of the concept of allometry. American Zoologist 40:748–758.

Geber, M. A., M. A. Watson, and H. de Kroon. 1997. Organ preformation, development, and resource allocation in perennials. Pp. 113–141 in F. A. Bazzaz and J. Grace (eds.), Plant Resource Allocation. Academic Press, San Diego.

Gould, S. J. 2002. The Structure of Evolutionary Theory. Belknap Press, Cambridge, MA.

Grace, J. 1997. Toward models of resource allocation by plants. Pp. 279–291 in F. A. Bazzaz and J. Grace (eds.), Plant Resource Allocation. Academic Press, San Diego.

Hallé, F., R. A. A. Oldeman, and P. Tomlinson. 1978. Tropical Trees and Forests: An Architectural Analysis. Springer-Verlag, New York.

Harper, J. L. and J. White. 1974. The demography of plants. Annual Review of Ecology and Systematics 5:419–463.

Haukioja, E. 1991. The influence of grazing on the evolution morphology and physiology of plants as modular organisms. Philosophical Transactions of the Royal Society of London Series B: Biological Sciences 333:241–248.

Holbrook, N. M. and F. E. Putz. 1989. Influence of neighbors on tree form: effects of lateral shade and prevention of sway on the allometry of *Liquidambar styraciflua* (sweet gum). American Journal of Botany 76:1740–1749.

Huber, H. 1996. Plasticity of internodes and petioles in prostrate and erect *Potentilla* species. Functional Ecology 10:401–409.

Huber, H. and J. F. Stuefer. 1997. Shade-induced changes in the branching pattern of a stoloniferous herb: functional response or allometric effect? Oecologia 110:478–486.

Huber, H., S. Lukács, and M. A. Watson. 1999. Spatial structure of stoloniferous herbs: an interplay between structural blue-print ontogeny and phenotypic plasticity. Plant Ecology 141:107–115.

Jones, C. S. 1993. Heterochrony and heteroblasic leaf development in two subspecies of *Cucurbita argyrosperma* (Cucurbitaceae). American Journal of Botany 80:778–795.

Jones, C. S. 1995. Does shade prolong juvenile development? A morphological analysis of leaf shape changes in *Cucurbita argyrosperma* subsp. *sororia* (Cucurbitaceae). American Journal of Botany 82:346–359.

Jones, C. S. 1999. An essay on juvenility, phase change, and heteroblasty in seed plants. International Journal of Plant Science 160:S105–S111.

Jones, C. S. 2001. The functional correlates of heteroblastic variation in leaves: changes in form and ecophysiology with whole-plant ontogeny. Boletín de la Sociedad Argentina de Botánica 36:171–184.

Jones, C. S. and M. A. Watson. 2001. Heteroblasty and preformation in mayapple *Podophyllum peltatum* (Berberidaceae): developmental flexibility and morphological constraint. American Journal of Botany 88:1340–1358.

Kaplan, D. R. 2001. The science of plant morphology: definition, history, and role in modern biology. American Journal of Botany 88:1711–1741.

King, D. and J. Roughgarden. 1982. Graded allocation between vegetative and reproductive growth for annual plants in growing seasons of random length. Theoretical Population Biology 22:1–16.

Klingenberg, C. P. 1996. Individual variation of ontogenies: a longitudinal study of growth and timing. Evolution 50:2412–2428.

Klingenberg, C. P. 1998. Heterochrony and allometry: the analysis of evolutionary change in ontogeny. Biological Reviews 73:79–123.

Klingenberg, C. P. and M. Zimmermann. 1992. Static, ontogenetic, and evolutionary allometry: a multivariate comparison in nine species of water striders. American Naturalist 140:601–620.

Lande, R. and S. J. Arnold 1983. The measurement of selection on correlated characters. Evolution 37:1210–1226.

Lechowicz, M. J. and P. A. Blais 1988. Assessing the contributions of multiple interacting traits to plant reproductive success: environmental dependence. Journal of Evolutionary Biology 1:255–274.

Le Maître, D. C. and J. J. Midgley. 1991. Allometric relationships between leaf and inflorescence mass in the genus *Protea* (Proteaceae): an analysis of the exceptions to the rule. Functional Ecology 5:476–484.

Lloyd, D. G. 1980. Sexual strategies in plants: 1. An hypothesis of serial adjustment of maternal investment during one reproductive session. New Phytologist 86:69–80.

Lloyd, D. G. 1984. Variation strategies of plants in heterogeneous environments. Biological Journal of the Linnean Society 21:357–385.

López, F., S. Fungairino, P. de las Heras, J. Serrano, and F. Acosta. 2001. Age changes in the vegetative vs. reproductive allocation by module demographic strategies in a perennial plant. Plant Ecology 157:13–21.

Magwene, P. M. 2001. New tools for studying integration and modularity. Evolution 55:1734–1745.

Maillette, L. 1985. Modular demography and growth patterns of two annual weeds (*Chenopodium album* L. and *Spergula arvensis* L.) in relation to flowering. Pp. 17–31 in J. White (ed.), Studies on Plant Demography: A Festschrift for John L. Harper. Academic Press, London.

Mazer, S. J. and V. A. Delesalle. 1996. Covariation among floral traits in *Spergularia marina* (Caryophyllaceae): geographic and temporal variation in phenotypic and among-family correlations. Journal of Evolutionary Biology 9:993–1015.

Mazer, S. J. and N. T. Wheelwright. 1993. Fruit size and shape: allometry at different taxonomic levels in bird-dispersed plants. Evolutionary Ecology 7:556–575.

McConnaughay, K. D. M., and J. S. Coleman. 1998. Can plants track changes in nutrient availability via changes in biomass partitioning? Plant and Soil 202:201–209.

McConnaughay, K. D. M., and J. S. Coleman. 1999. Biomass allocation in plants: ontogeny or optimality? A test along three resource gradients. Ecology 80:2581–2593.

McGraw, J. and K. Garbutt. 1990. The analysis of plant growth in ecological and evolutionary studies. Trends in Ecology and Evolution 5:251–254.

Müller, I., B. Schmid, and J. Weiner. 2000. The effect of nutrient availability on biomass allocation patterns in 27 species of herbaceous plants. Perspectives in Plant Ecology, Evolution and Systematics 3:115–127.

Niklas, K. J. 1994. Plant Allometry: The Scaling of Form and Process. University of Chicago Press, Chicago.

Niklas, K. J. 1999. Changes in the factor of safety within the superstructure of a dicot tree. American Journal of Botany 86:688–696.

Niklas, K. J. 2000. Wind-induced stresses in cherry trees: evidence against the hypothesis of constant stress levels. Trees 14:230–237.

Niklas, K. J. and B. J. Enquist. 2002. On the vegetative biomass partitioning of seed plant leaves, stems and roots. American Naturalist 159:482–497.

Olson, E. C. and R. L. Miller. 1958. Morphological Integration. University of Chicago Press, Chicago.

Pigliucci, M., C. D. Schlichting, C. S. Jones, and K. Schwenk. 1996. Developmental reaction norms: the interactions among allometry, ontogeny, and plasticity. Plant Species Biology 11:69–85.

Piñero, D., J. Sarukhán, et al. 1982. The costs of reproduction in a tropical palm, *Astrocaryum mexicanum*. Journal of Ecology 70:473–481.

Pitelka, L. F. and J. W. Ashmun. 1985. Physiology and integration of ramets in clonal plants. Pp. 339–435 in J. B. C. Jackson, L. W. Buss, and R. E. Cook (eds.), Population Biology and Evolution of Clonal Organisms. Yale University Press, New Haven, CT.

Preston, K. A. 1999. Can plasticity compensate for architectural constraints on reproduction? Patterns of seed production and carbohydrate translocation in *Perilla frutescens*. Journal of Ecology 87:697–712.

Preston, K. A. and D. D. Ackerly. 2003. Hydraulic architecture and the evolution of shoot allometry in contrasting climates. American Journal of Botany 90:1502–1512.

Prévost, M. F. 1978. Modular construction and its distribution in tropical woody plants. Pp. 223–231 in P. B. Tomlinson and M. H. Zimmermann (eds.), Tropical Trees as Living Systems: The Proceedings of the Fourth Cabot Symposium. Cambridge University Press, New York.

Raff, R. A. 1996. The Shape of Life: Genes, Development and the Evolution of Animal Form. University of Chicago Press, Chicago.

Ritterbusch, A. 1990. The measure of biological age in plant modular systems. Acta Biotheoretica 38:113–124.

Rosenthal, J. P. and S. C. Welter. 1995. Tolerance to herbivory by a stemboring caterpillar in architecturally distinct maizes and wild relatives. Oecologia 102:146–155.

Roy, J., B. Thiebaut, and M. A. Watson. 1986. Physiological and anatomical consequences of morphogenetic polymorphism: leaf response to light intensity in young beech trees

(*Fagus sylvatica* L.). Naturalia Monspeliensia—Colloque International sur l'Arbre, 431–449.

Sarukhán, J., D. Piñero, and M. Martínez-Ramos. 1985. Plant demography: a community-level interpretation. Pp. 17–31 in J White (ed.), Studies on Plant Demography: A Festschrift for John L. Harper. Academic Press, London.

Schlichting, C. D. and M. Pigliucci. 1998. Phenotypic Evolution: A Reaction Norm Perspective. Sinauer Associates, Sunderland, MA.

Schluter, D. 1996. Adaptive radiation along genetic lines of least resistance. Evolution 50:1766–1774.

Schmitt, J., A. C. McCormac, and H. Smith. 1995. A test of the adaptive plasticity hypothesis using transgenic and mutant plants disabled in phytochrome-mediated responses to neighbors. American Naturalist 146:937–953.

Slade, A. J. and M. J. Hutchings. 1987. The effects of light intensity on foraging in the clonal herb *Glechoma hederacea*. Journal of Ecology 75:639–650.

Stevens, P. F. 1984. Homology and phylogeny: morphology and systematics. Systematic Botany 9:395–409.

Sultan, S. E. 1996. Phenotypic plasticity for offspring traits in *Polygonum persicaria*. Ecology 77:1791–1807.

Sultan, S. E. and F. A. Bazzaz. 1993. Phenotypic plasticity in *Polygonum persicaria*. I. Diversity and uniformity in genotypic norms of reaction to light. Evolution 47:1009–1031.

Wagner, G. P. 1996. Homologues, natural kinds and the evolution of modularity. American Zoologist 36:36–43.

Wagner, G. P. and L. Altenberg. 1996. Complex adaptations and the evolution of evolvability. Evolution 50:967–976.

Watson, M. A. and B. B. Casper. 1984. Morphogenetic constraints on patterns of carbon distribution in plants. Annual Review of Ecology and Systematics 15:233–258.

Watson, M. A. and Y. Lu. 1999. Timing of shoot senescence and demographic expression in the clonal perennial *Podophyllum peltatum* (Berberidaceae). Oikos 86:67–78.

Watson, M. A., M. A. Geber, and C. S. Jones. 1995. Ontogenetic contingency and the expression of plant plasticity. Trends in Ecology and Evoltuion 10:474–475.

Weiner, J. and L. Fishman. 1994. Competition and allometry in *Kochia scoparia*. Annals of Botany 73:263–271.

Weiner, J. and S. C. Thomas. 1992. Competition and allometry in three species of annual plants. Ecology 73:648–656.

Weiner, J., P. Stoll, H. Muller-Landau, and A. Jasentuliyana. 2001. The effects of density, spatial pattern and competitive symmetry on size variation in simulated plant populations. American Naturalist 158:438–450.

West, G. B., J. H. Brown, and B. J. Enquist. 1999. A general model for the structure and allometry of plant vascular systems. Nature 400:664–667.

White, J. 1979. The plant as metapopulation. Annual Review of Ecology and Systematics 10:109–145.

White, P. S. 1983. Corner's rules in eastern deciduous trees: allometry and its implications for the adaptive architecture of trees. Bulletin of the Torrey Botanic Club 110:203–212.

White, J. 1984. Plant metamerism. Pp. 15–47 in R. Dirzo and J. Sarukhán (eds.), Perspectives on Plant Population Ecology. Sinauer Associates, Sunderland, MA.

Whitehead, D., W. R. N. Edwards, and P. G. Jarvis. 1984. Conducting sapwood area, foliage area, and permeability in mature trees of *Picea sitchensis* and *Pinus contorta*. Canadian Journal of Forest Research 14:940–947.

Williams, G. C. 1986. Retrospect on modular organisms. Philosophical Transactions of the Royal Society of London Series B: Biological Sciences 13:245–250.

Winn, A. A. 1996. The contributions of programmed developmental change and phenotypic plasticity to within-individual variation in leaf traits in *Dicerandra linearifolia*. Journal of Evolutionary Biology 9:737–752.

Winn, A. A. 1999. The functional significance and fitness consequences of heterophylly. International Journal of Plant Science 160:S113–S121.

Wright, I. J., P. B. Reich, and M. Westoby. 2001. Strategy shifts in leaf physiology, structure and nutrient content between species of high- and low-rainfall and high- and low-nutrient habitats. Functional Ecology 15:423–434.

# 5

# Phenotypic Integration as a Constraint and Adaptation

JUHA MERILÄ
MATS BJÖRKLUND

> An angel whose muscles developed no more power weight for weight than those of an eagle or a pigeon would require a breast for about four feet to house the muscles engaged in working its wings, while to economize in weight, its legs would have to be reduced to mere stilts.
>
> *J. B. S. Haldane (1927)*

Despite the enormous diversity of living forms, it is quite clear that not all imaginable life forms—such as angels—have ever come into existence. There are two broad explanations for this. First, according to what can be called the genetic/epigenetic constraint view (e.g., Maynard Smith et al. 1985), organisms are to some degree victims of their own history: solutions available for a given organism or clade are constrained by its evolutionary history, as for instance dictated by its genetic architecture and/or ancestral developmental pathways. Second, according to what can be termed the functionalist/adaptationist view (Gould and Lewontin 1979), almost any solution to a problem posed by a selective environment is possible, insofar as it is not opposed by the laws of physics or by natural selection. Hence, organisms like angels might not exist either because of constraints of development, or because they are not fit for selective reasons. These two world views, although frequently polarized, are of course not mutually exclusive (Maynard Smith et al. 1985; Schlichting and Pigliucci 1998). According to a more balanced view, organisms are capable of adapting to most challenges posed by their environment given sufficient time and genetic variability, but the possible solutions will always be constrained to some degree by their history. Consequently, given that both constraints and selection can create similar macroevolutionary patterns, it is generally speaking difficult to distinguish between the two as causes of phenotypic similarity among related taxa (Björklund and Merilä 1993; Armbruster and Schwaegerle 1996; Schluter 1996; Arnold et al. 2001).

Phenotypic integration (Olson and Miller 1958) refers to a pattern and magnitude of covariation among a set of traits. In the case of morphometric traits, a common pattern of trait covariation among closely related taxa is that they vary often almost entirely in terms of overall size, but little in shape (e.g., Cheverud 1989; Voss et al. 1990; Björklund 1991, 1996a; Björklund and Merilä 1993), a pattern often manifested already in early ontogeny (Björklund, 1993, 1994a, 1996b, 1997). This sort of conservatism in the pattern of trait covariation can be understood either in terms of bias in the processes controlling the evolution of biological systems (i.e., shared developmental constraints; Maynard Smith et al. 1985), or in terms of shared adaptive landscape among closely related taxa. Differentiating between these alternative explanations, or at least understanding the relative importance of these processes as explanations for patterns of integration within and among different taxa, is one of the major challenges for contemporary evolutionary biology (Schlichting and Pigliucci 1998; Arnold et al. 2001; Steppan et al. 2002). Although considerable theoretical, methodological, and conceptual progress has been made toward understanding these issues during the past decades (see reviews in Maynard Smith et al. 1985; Arnold 1992; Cheverud 1996; Wagner et al. 1997; Schlichting and Pigliucci 1998; Schluter 2000; Arnold et al. 2001; Steppan et al. 2002), it is perhaps fair to say that the extent to which evolutionary patterns are dictated by constraints versus natural selection is still not very well understood.

In this chapter we shall focus on the concept of phenotypic integration and its dual role in dictating the directions and pace of evolutionary transformations. We shall start with a brief theoretical treatment explaining how integration is thought to constrain evolutionary transformations, and what is known about the importance of these constraints in the light of empirical studies conducted so far. After looking at integration as a constraint, we shall turn toward its more positive connotation, and address the question of how integration is thought to evolve as an adaptation, and whether it can actually facilitate further adaptation. We conclude our treatment with a brief discussion about the evolution of phenotypic integration patterns (modularity), drawing specific attention to concurrence between theory and data, as well as to a possibility of studying the evolution of modularity with a neural network approach. Throughout the treatment we pay special attention to technical and interpretational problems, as well as to the identification of issues that might benefit from further studies.

## Phenotypic Integration as Constraint

### Theory in Brief

Phenotypic integration, as reflected in the patterns of genetic variation and covariation of characters, can exert a strong influence on the rate and direction of evolution (e.g., Lande 1976; Arnold 1992; Björklund 1996a; Schluter 1996). Genetic constraints can exert their effects in two ways: either through low levels of genetic variation, or through genetic tradeoffs among components of fitness. In quantitative genetics, these constraints are summarized in the genetic variance-covariance matrix (**G**), which is a symmetric matrix with the genetic variance for each trait on the diagonal, and the genetic covariances in off-diagonal cells (Lynch and Walsh 1998). This matrix is the central element of Lande's (1979) general model of evolutionary change:

$$\Delta\mathbf{z} = \mathbf{G}\beta \tag{5.1}$$

where **z** is the vector of trait means and $\beta$ is the vector of selection gradients for each trait. The important aspect of this equation is that the evolution of a given trait is not only a function of the additive genetic variance and selection on that trait, but also of the genetic variance of other traits with which it covaries genetically. By simple numerical calculations, it is straightforward to demonstrate that for a given $\beta$, the response vector $\Delta\mathbf{z}$ can result in directions quite different from the direction of $\beta$, that is, the shortest distance to the nearest adaptive peak (e.g., Björklund 1996a). The reason for this is that when the genetic covariances increase in magnitude, the response to selection is determined chiefly by the direction of the eigenvector associated with the largest eigenvalue ($\mathbf{g}_{max}$) rather than the direction of the selection gradient vector (Björklund 1996a; Schluter 1996). This is illustrated in Fig. 5.1a, where the evolutionary trajectory toward the adaptive peak is initially biased toward the line of $\mathbf{g}_{max}$. In this example, the adaptive peak is eventually reached, but this is not self-evident: if the genetic covariances are sufficiently strong (i.e., little variance perpendicular to $\mathbf{g}_{max}$), then the adaptive peak may never be reached (Kirkpatrick and Lofsvold 1992). Generally speaking, this can be seen from the distribution of eigenvalues of **G** (Wagner 1984). The total genetic variance is simply the sum of the variances of each trait. If covariances are high, then one eigenvalue is much larger than the remaining ones. This in turn means that one or a few eigenvalues will become very

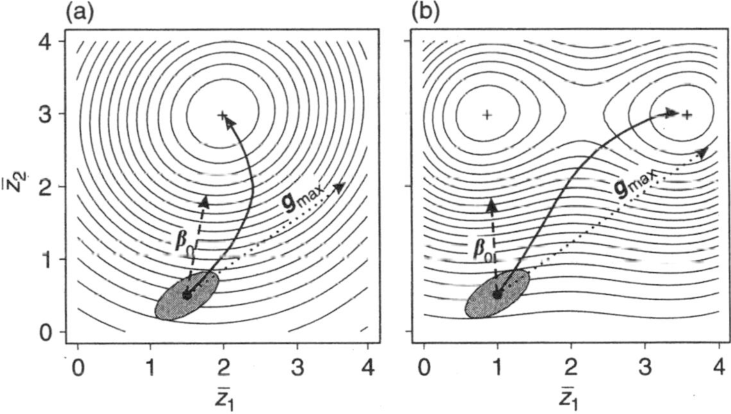


Figure 5.1 Effects of genetic covariance on direction of evolution in two traits ($z_1$ and $z_2$). Depicted are initial population means (filled circles) and their paths over subsequent generations (solid line) over an adaptive landscape when there are (a) one and (b) two adaptive peaks. Ellipses surround 95% of the breeding values in each population, and the dotted line is the direction of maximum genetic variance in the population ($\mathbf{g}_{max}$). Contours represent increments of mean fitness in adaptive landscape where adaptive peaks are indicated with crosses. $\boldsymbol{\beta}_0$ is the section gradient at the start showing the direction of steepest ascent in mean fitness. Adapted from Schluter (2000) with permission from Oxford University Press.

small (by simple subtraction) or even zero. Thus, the corresponding eigenvectors represent directions where there is no or very little genetic variance, and hence are "forbidden" evolutionary trajectories (Kirkpatrick and Lofsvold 1992), in other words, constraints in the strict sense. Another example of possible bias in the evolutionary trajectory caused by **G** is shown in Fig. 5.1b. Here, genetic covariances steer the population away from the closest adaptive peak toward a more distant one that lies closer to the line of $\mathbf{g}_{max}$. Note also that although selection in this case favors a slight decrease in trait $z_1$, it increases as a consequence of genetic covariance with $z_2$ (Fig. 5.1b). Hence, these examples highlight the fact that even if the constraints posed by genetic covariances can be transitory, their effects can endure if genetic variation of certain dimensions is entirely lacking, or if the adaptive landscape has several optima (Lande 1979; Arnold 1992; Price et al. 1993).

## Empirical Evidence

Turning to the empirical evidence for integration as a constraint, there appears to be ample additive genetic variance for most traits, at least in wild animal populations (Mousseau and Roff 1987; Houle 1992), and hence, response to natural selection is unlikely to be prevented or slowed down by lack of additive genetic variance. However, the situation may be different in the case of plants, where studies of natural populations have revealed typically lower levels of variability (see Mitchell-Olds 1996 and references therein). Constraints imposed by genetic covariances among traits can be severe as illustrated by analyses of **G**-matrices, which typically show shortage of genetic variance in one or more morphological dimensions (Kirkpatrick and Lofsvold 1992; Gomulkiewicz and Kirkpatrick 1992; Björklund 1996b; Schluter 2000, p. 221). For instance, Björklund (1996a) analyzed patterns of integration in 27 published **G**- matrices of morphological characters, and found that the median standardized variance of eigenvalues (a measure of the extent of integration expressed in percent of maximum possible) was 41%. This corresponds to an average genetic correlation of approximately 0.65. In most (21/27) matrices examined, the first eigenvector ($\mathbf{g}_{max}$) was an isometric size vector, suggesting that most evolutionary change (assuming random selection vectors) in the short term will be in terms of size, rather than shape. By creating random selection vectors, Björklund (1996a) further found that in 70% of the simulated cases ($n = 500$), selection created only changes in size. Hence, he concluded that most evolutionary change in the wild—at least in the short term—is predicted to be biased toward size changes as dictated by $\mathbf{g}_{max}$.

The idea that evolutionary transformations in the wild are frequently biased toward $\mathbf{g}_{max}$ as dictated by the structure of **G** was explicitly tested by Schluter (1996; see also Schluter 1994, 2000). His approach represents perhaps the strongest empirical test of the role of **G** in biasing directions of evolutionary transformations to date, and makes three predictions about the long-term rates and direction of evolutionary divergence under natural selection. First, for populations and species that have been recently derived from the common ancestor, the direction of this divergence should be generally biased toward $\mathbf{g}_{max}$. The magnitude of this bias can be measured as the angle ($\theta$) between $\mathbf{g}_{max}$ and the line connecting the "ancestral" and "derived" populations (Fig. 5.2; see Schluter 1996 for details). Second, $\theta$ is predicted to increase with time since divergence, as shown by the progressive departure of the population mean from $\mathbf{g}_{max}$ as illustrated in Fig. 5.2. Third, the rate of divergence should be inversely related to deviation from $\mathbf{g}_{max}$. This is because there is relatively little genetic variation into other (see above) directions than $\mathbf{g}_{max}$, reducing the expected response to selection for a given intensity of selection (Schluter 1996). Schluter (1996) tested these predictions using data from recently diverged populations of three-spined sticklebacks (*Gasterosteus aculeatus*) and published data from four other species of vertebrates. As predicted, it was found that the direction of divergence was biased toward the $\mathbf{g}_{max}$ more than would be expected by chance (Fig. 5.3a). Likewise, conforming to the second prediction, there was a tendency for deviations from $\mathbf{g}_{max}$ to decay with increasing time as measured by Nei's genetic distance between the pairs species (Fig. 5.3a). Finally, the rate of divergence was inversely related to the deviation from $\mathbf{g}_{max}$ (Fig. 5.3b). Taken together, these results suggest that **G** can constrain adaptive differentiation in quantitative traits for considerable time periods, in this particular case, millions of years.

The evidence for profound genetic constraints in adaptive evolution from Schluter's (1996) study is subject to a couple of assumptions. First, his approach assumes that the location of the adaptive peak (cf. Fig. 5.2) is random in respect to orientation of the $\mathbf{g}_{max}$ vector (Schluter 1996). One argument for why this might not be so is that the **G**-matrices are themselves subjects of natural selection (Cheverud 1984; Wilkinson et al. 1990; Steppan et al. 2002), and the principal axes of **G** ($\mathbf{g}_{max}$) might be aligned with the principal axes of adaptive landscapes ($\boldsymbol{\omega}_{max}$; Arnold et al. 2001). We return to this issue below. Second, Phillips et al. (2001) pointed out that a finding that divergence has occurred mainly along $\mathbf{g}_{max}$ is also entirely consistent with the interpretation that the observed divergence has nothing to do with selection, as it is also predicted by models of genetic drift (cf. Lande

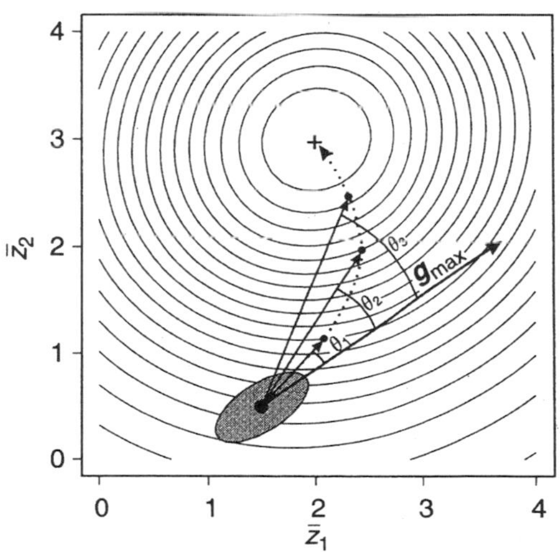


Figure 5.2 A schematic illustration of the expected decay in bias toward $\mathbf{g}_{max}$ with time. Contours are increments of mean fitness on an adaptive landscape with a maximum at +. The ellipse represents the distribution of breeding values in the ancestral population. The dotted line indicates the evolutionary trajectory in the second population derived from the first. $\theta$ is the angle between $\mathbf{g}_{max}$ and the line connecting the mean of the derived population (three filled dots) from the mean of the ancestral population. Adapted from Schluter (2000) with permission from Oxford University Press.

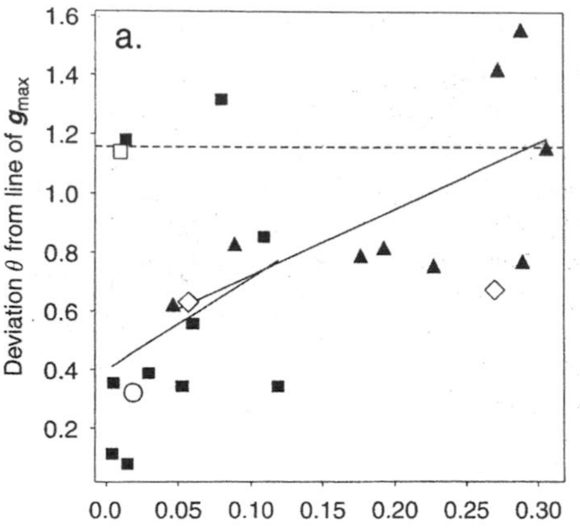


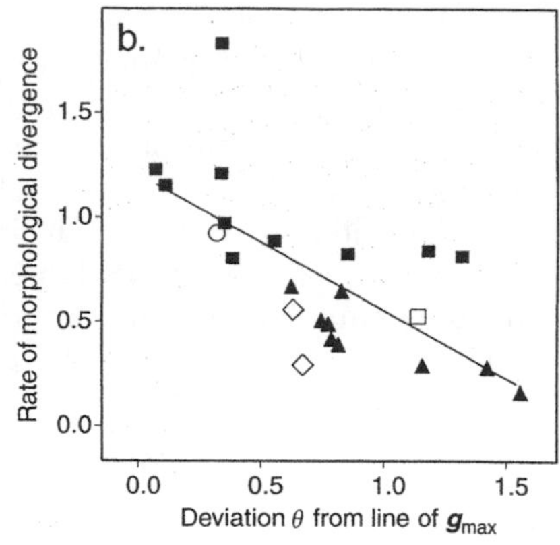


Figure 5.3 Evidence for evolution along the lines of least genetic resistance. (a) The angle ($\theta$) between direction of divergence and $\mathbf{g}_{max}$ as a function of Nei's genetic distance. The dotted line represents random expectation. (b) Rate of morphological divergence as a function of deviation from $\mathbf{g}_{max}$ among 23 pairs of sister taxa. Adapted from Schluter (2000) with permission from Oxford University Press.

1979). However, we suggest that additional tests, such as comparison of neutral marker (as measured in $F_{ST}$) and quantitative genetic (as measured in $Q_{ST}$) differentiation between species pairs, could be used to test for plausibility of drift as cause of the observed differentiation (review in Merilä and Crnokrak 2001). If drift as cause of population differentiation can be rejected, then Schluter's (1996) approach may provide a valid test for the role of integration constraining evolution by natural selection.

Returning to the issue of possible collinearity between $\mathbf{g}_{max}$ and $\omega_{max}$, there are currently too few studies evaluating how frequently this might be the case. However, at least a number of retrospective selection analyses—subject to numerous assumptions (see below)—indicate antagonistic selection (i.e., noncollinear selection in respect to genetic correlations) to be of frequent occurrence (e.g., Price et al. 1984; Price and Grant 1985; Lofsvold 1988; Merilä et al. 1994; Mitchell-Olds 1996). Etterson and Shaw (2001) found both antagonistic and reinforcing selection on genetic correlations on plant morphology, antagonistic selection being slightly less common (2 cases) than reinforcing selection (6 cases) when the plants were reared in their native environments, whereas when reared in nonnative environments, antagonistic (10 cases) and reinforcing selection (8 cases) were equally common. Likewise, current antagonistic selection has been implicated to constrain, for instance, the evolution of sexually dimorphic traits in two species of birds (Fig. 5.4). Here, positive genetic correlations between beak coloration among male and female zebra finches (*Taenopygia guttata;* Price 1996) and body size of male and female collared flycatchers (*Ficedula albicollis*) are opposed by selection on different directions in the two sexes. In both studies, it was demonstrated that the predicted selection responses over one generation would have been very different if the genetic correlation between the sexes had

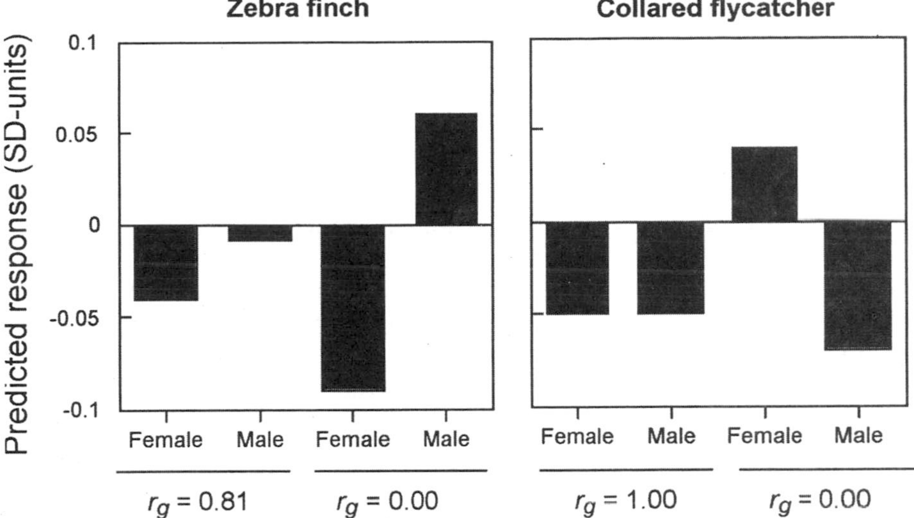


Figure 5.4 Effects of genetic correlations between the sexes in zebra finches (*Taenopygia guttata*) and collared flycatchers (*Ficedula albicollis*) on expected selection responses to empirically estimated selection. Bars represent expected selection responses over one generation using empirically derived genetic correlation estimates ($r_g$ = 0.81 and 1.0, respectively), and selection responses expected in the absence of genetic correlation ($r_g$ = 0) between the sexes. In both cases, the genetic correlation between the sexes (which differ in their optimal trait size as revealed by selection analyses; Price 1996; Merilä et al. 1998) will displace both sexes from their optimal response. Adapted from Merilä and Sheldon (2001) with permission from Plenum Press.

been absent, suggesting that genetic correlations between the sexes can retard evolution of sexual dimorphism and create a selection load that reduces fitness (Lande 1980; Price 1996; Merilä et al. 1998; Fig. 5.4). However, although these studies indicate that $\mathbf{g}_{max}$ and $\omega_{max}$ are not always aligned, there is currently neither any theory to predict, nor a review of empirical data to evaluate, how frequent antagonistic selection in the wild is. Hence, what is needed here are comparative studies of where patterns of population differentiation can be compared and contrasted with directions predicted by local selective regimes and **G**.

To our knowledge, there is only one study that has used actual measurements of selection in multiple populations to test whether the pattern of population differentiation matches better with trajectory predicted by local selection rather than integration as predicted by **G**. In a study of recently diverged populations of house finches (*Carpodacus mexicanus*), Badyaev and his coworkers (Badyaev and Hill 2000; Badyaev et al. 2000) demonstrated that the relatively recent population differentiation in the degree of sexual size dimorphism in this species was predictable from the local differences in selection pressures, rather than from the structure of **G**. In fact, the genetic correlation between the sexes was found to be close to 1 (Badyaev and Hill 2000), which should have posed considerable constraints for the observed divergence. It is not entirely clear how this differentiation has been possible in the face of the high genetic correlation (but see Badyaev and Martin 2000; Badyaev et al. 2002 for possible explanations), but the results suggest that in at least in this case, the constraints posed by integration have been overruled by local selection pressures.

Before moving on, it be may worth noting that the problem of differentiating between integration (cf. $\mathbf{g}_{max}$) versus adaptive landscape (cf. $\boldsymbol{\omega}_{max}$) as explanations for conservative patterns of morphological differentiation has been discussed extensively in a paleontological and macroevolutionary context as a possible explanation for prolonged morphological stasis (e.g., Eldredge 1995, 1999). Instead of viewing each species as subjects of selection *sensu* Lewontin (1983) and Sober (1984), it is also possible to view them as objects of selection (Lewontin 1983; Sober 1984). In other words, cross-species homogeneity in patterns of integration (e.g., Björklund 1991; Björklund and Merilä 1993), as well as the long-term stasis observed in paleontological records (e.g., Gould 2002), could be explained in terms of species choosing habitats into which they fit, rather than habitats selecting for changes in species only. Hence, in the face of selection pressures evoked by environmental changes (e.g., temperature changes), species may respond either evolutionarily or by tracking habitats to which they are already adapted (Eldredge 1995, 1999). Although habitat selection and range shifts/tracking occur, we are not aware of any empirical evidence for them to be evolutionarily more significant determinants (or reinforcers) of integration than selection arising from local environments.

## Additional Tests and Issues

Selection experiments provide a straightforward way of testing for the strength of genetic constraints for directions of evolution. For instance, Scheiner and Istock (1991) examined responses of pitcher-plant mosquitoes (*Wyeomyia smithii*) to correlational selection performed on all four combinations (= directions) of developmental time and propensity to diapause that were positively genetically correlated. They found that the response was rapid and consistent when performed in the direction of the correlation, whereas the response to selection performed orthogonal to correlation was slow and frequently in directions opposite to correlation (Scheiner and Istock 1991). Likewise, Dorn and Mitchell-Olds (1991) studied *Brassica campestris* in which flowering time and size exhibit a strong positive genetic correlation, and selected for different extreme combinations of these traits in a short-term experiment. As predicted by theory, responses were in the direction of selection only when the direction of selection was parallel to the direction of genetic correlation, but not when selection was in other directions. Unfortunately, this type of study is seldom possible except with some model organisms, and even then, typically only over relatively short periods of time (for which a sole knowledge of **G** and selection intensities may provide easier and equally good predictions; e.g., Grant and Grant 1995). Nevertheless, as illustrated by the work of Phillips et al. (2001), carefully planned and conducted experiments of this kind could provide important insights into how populations and integration evolve. For instance, although selection experiments have been used to evaluate the change in **G** under selection in *Drosophila* (Wilkinson et al. 1990), we are not aware of long-term experiments where the effects of differences in initial **G** on adaptation to some novel selection pressure would have been evaluated.

Genetic models of phenotypic evolution have generated testable predictions on how intra- and interspecific allometries should evolve under different evolutionary scenarios (e.g., Lande 1979; Zeng 1988; Riska 1989). For populations differentiating due to natural selection, there is no necessary relationship between multivariate patterns of covariation within and among populations (Lande 1979), except when the component of multivariate stabilizing (i.e., correlative) selection is less variable among populations than the directional component (Zeng 1988). In the latter case, theory predicts partial correspondence between patterns of covariation within and among populations, so that traits that are highly integrated within populations will also be so across populations (i.e., intra- and interspecific allometries similar), whereas this is not necessarily the case for less integrated traits (Zeng 1988). For populations differentiating under genetic drift and correlated responses models of evolution, multivariate patterns of covariation within and among populations are expected to be similar (Lande 1979; Zeng 1988). Consequently, a combined prediction that can be derived from these models is that if multivariate patterns of trait covariation within and among populations do not align, natural selection is a likely cause of differentiation, barring a few assumptions (see below). However, as should be obvious from above, the reverse is not true: similar patterns of integration can be produced by both selection and drift. In the present context, the utility of these predictions is that they can be informative in elucidating the role of genetic constrains in multivariate evolution. Namely, if the pattern of the population divergence—as summarized in the variance-covariance matrix (**L**) constructed from population means—does not follow the directions predicted by **G** (i.e., **G** and **L** are not proportional), this indicates that constraints dictated by **G** have been overruled by the action of natural selection (Merilä and Björklund 1999).

Surprisingly few studies have compared the orientation of within- and among-population/species patterns of integration. Most of them (Sokal and Riska 1981; Gibson et al. 1984; Lofsvold 1988; Voss et al. 1990; Armbruster 1991; Björklund and Merilä 1993; Mitchell-Olds 1996) have found that within- and among-population/species patterns of integration are similar, whereas only a few have concluded the opposite (Lofsvold 1988; Merilä and Björklund 1999; Badyaev and Hill 2000). In the two latter studies of greenfinches (*Carduelis chloris*) and house finches (*Carpodacus mexicanus*), respectively, it is worth noting that both of them considered relatively recently derived populations, and that additional evidence was provided to support the inference that patterns of divergence were caused by natural selection. In the case of the house finch, it was demonstrated that the pattern of population divergence was not predictable from **G** but was predictable from the variation in empirical measurements of local selection on the traits studied (Badyaev and Hill 2000; Badyaev et al. 2000), whereas in the greenfinch the divergence exceeded that to be expected under drift only (Merilä 1997; Merilä and Björklund 1999), and the largest dimension of nonallometric change was confined to beak traits likely to be under strong selection (Merilä 1997; Merilä and Björklund 1999).

Almost all studies comparing within- and among-population allometries up to date (but see Lofsvold 1988; Mitchell-Olds 1996; Badyaev and Hill 2000) have been based on purely phenotypic data and, hence, subject to the assumption that within-population genetic (**G**) and phenotypic (**P**) variance-covariance matrices can be equated, which is not always the case (Steppan et al. 2002; but see Roff 1995; Badyaev and Hill 2000). Likewise, as the mean trait values used to construct the among-population/species variance-covariance matrix (**L**) have not usually been estimated in a common garden situation (but see Lofsvold 1988; Mitchell-Olds 1996), nearly all studies assume implicitly that environmental and/or maternal effects have not been an important source of variation in population means. This is a strong assumption given the abundant evidence for the opposite (e.g., James 1983; Conover and Schultz 1995; Rossiter 1996). The study by (Badyaev and Hill 2000) is a case in point: while both matrix comparisons and empirical estimates of selection (Badyaev et al. 2000) strongly suggest that the divergence between the populations is the result of natural selection, the possibility of environmental induction cannot be ruled out as **L** was estimated from purely phenotypic data. In fact, a more recent study revealed strong nongenetic effects on studied morphological traits within each of the populations (Badyaev et al. 2002).

An additional assumption in studies aiming to predict long-term patterns of evolutionary dynamics of quantitative traits is that **G** has remained relatively stable over the time period considered (e.g., Lande 1979; Turelli 1988). A number of studies have performed pairwise comparisons of **G**-matrices to test the hypothesis about the constancy of **G** (reviews in Roff 1997, 2000; Arnold and Phillips 1999; Roff and Mousseau 1999; Steppan et al. 2002), the conclusion being that this cannot be assumed, but can be tested, albeit the statistical machinery for this also needs development (Steppan et al. 2002; Roff 2002). Here, it is worth pointing out the hypothesis about the constancy of **G**, which, if rejected, would itself provide evidence for natural selection overriding constraints. However, there are at least three distinct reasons why such a comparison may have limited utility in the current context. First, comparison of two **G**-matrices instead of **G** and **L** is likely to be a less informative test of the role of natural selection in population divergence, for the reason illustrated in Fig. 5.5. Second, the large sampling variance in matrix structure caused by drift can apparently produce all kinds of changes in **G** (Phillips et al. 2001; see also Armbruster and Schwaegerle 1996). However, as pointed out by Phillips et al. (2001), pooling over large number of populations will reduce the sampling variance, and provide an estimate of **G** that adequately represents that in the original base population. Third, mutations can also cause changes in the structure of **G** and, hence, their cumulative effects might be difficult to distinguish from effects of selection (Camara et al. 2000; Steppan et al. 2002).

Finally, apart from effects of selection (e.g., Wilkinson et al. 1990) and mutation (e.g., Camara et al. 2000), it is worth pointing out that the covariance elements of **G** might be subject to environmental modification. Although it is well known that patterns of expression of genetic variation can be highly

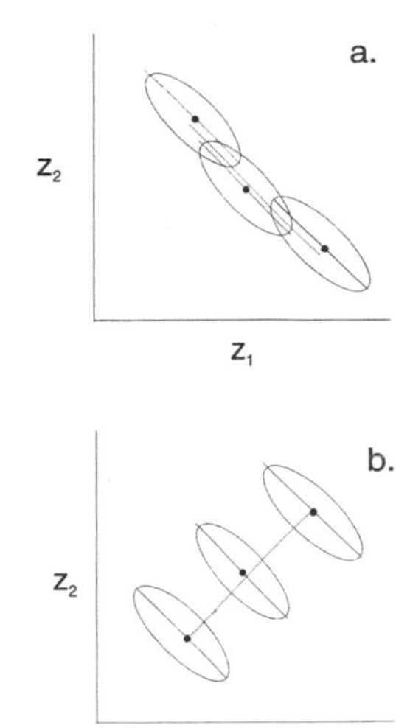


Figure 5.5 An illustration of distinction between patterns within and among population integration. (a) A case where comparison intra- and interpopulation allometries are equal and (b) a case where they differ. Each ellipse presents a bivariate ($z_1$, $z_2$) distribution of breeding values within populations, and the dots mark population mean breeding values. Note that in both cases, pairwise comparisons of intrapopulation allometries are equal. For this reason, pairwise comparisons of intrapopulation allometries are not as informative of significant evolutionary events as comparisons of comparison intra- and interpopulational allometries.

environment-dependent (Hoffmann and Merilä 1999), little attention in this context has been paid to covariance elements of **G** (but see Riska 1989). Service and Rose (1985) showed that the negative genetic correlation between fecundity and starvation time in *Drosophila melanogaster* subjected to a novel environment was reduced from −0.91 to −0.45. Similar change (from 0.85 to −0.97) in the sign of genetic correlation between beak and tarsus length in barnacle geese (*Branta leucopsis*) reared under good and poor environmental conditions was demonstrated by Larsson (1993). Further examples can be found in Giesel et al. (1982), Larsson (1993), Newman (1988), and Yampolsky and Ebert (1995). Consequently, the extent to which **G** acts to constraint the rate and path toward selective optima might be at least sometimes highly environment-dependent. To this end, comparative studies focusing on effects of environmental conditions on the structure of **G** are an obvious area where more research is needed.

Taken together, comparisons of **G** and **L** among relatively closely related taxa that are unlikely to have diverged considerably in **G**, can provide a valuable tool in evaluating how fast and frequently natural selection can overcome constraints posed by **G**. However, there are still far too few studies based on rigorous experimental and analytical methodology for us to leap to any broad conclusions in this respect.

## Integration as an Adaptive Feature

Until now, we have mostly emphasized the role of phenotypic integration as a potential constraint, with little reference to its potential constructive role in evolution. In the following sections we shall focus on the positive connotation of integration in evolution. In this context, it is important to distinguish between integration as an adaptation and integration as facilitating future adaptation.

## Integration as Adaptation

Phenotypic integration—as reflected in **G**—is itself thought to be a result of the action of natural selection which has favoured certain patterns of trait covariation over others (e.g. Lande 1980; Cheverud 1984). Evidence for a current adaptive value of integration comes from studies that demonstrate how certain patterns of trait covariation within populations are favored by natural selection over others (e.g., Björklund 1992; Brodie 1992; Björklund and Senar 2001), or that different patterns of integration match differences in selective regimes among populations (e.g., Schluter 1993; Relyea 2002). However, studies documenting correlative selection are still rare (Arnold et al. 2001; Kingsolver et al. 2001), let alone studies that have explicitly tested for within- and across-population concurrence between pattern of selection and integration as reflected in **G** (but see Jernigan et al. 1994). Hence, although it is fairly easy to come up with verbal and theoretical arguments to support the view that a given pattern of integration is an adaptation, there is currently very little hard evidence to support such a conjecture. Induced defences in amphibian larvae provide an example: larvae of many species develop a relatively smaller body and deeper tail as an induced response to predators (e.g., McCollum and Van Buskirk 1996; Lardner 2000; Laurila et al. 2002; Relyea 2002), and these responses are adaptive since the induced tadpoles survive better under predation risk, possibly as a consequence of their better swimming ability (Van Buskirk and McCollum 2000a, 2000b). Unfortunately, the genetics of these clearly adaptive responses have not been yet investigated.

Apart from the difficulty of obtaining relevant data (i.e., selection and **G**-matrix) to test the adaptive nature of integration, an additional potential shortcoming should be kept in mind: in order to call a certain pattern of integration an adaptation, one has to assume that the pattern of integration in question was established by selection for the current function. Of course, this need not be the case (cf. Gould and Lewontin 1979; Gould and Vrba 1982) as also indicated by the apparent—but still poorly documented and understood—time lags in how **G**-matrices evolve under changing environmental conditions (Steppan et al. 2002).

## Integration as Facilitating Adaptation

Once evolved, a certain pattern of integration can actually facilitate further evolution, and it has even been argued that developmental constraints are necessary to facilitate the evolution of functionally integrated traits (Bürger 1986; Wagner 1988a, 1988b). This is nicely illustrated by theoretical work by Bürger (1986) and Wagner (1988a), who modeled the rate of evolution as a function of phenotypic variance in various fitness landscapes. The essential result from their work was that if the variation in different traits is unconstrained (i.e., all traits have the same or at least constant ratios of variance), and the adaptive landscape has the shape of a ridge with increasing fitness in one trait in one direction and decrease in fitness in all other directions, then mean fitness will decrease despite positive selection. In general, mean fitness is negatively related to the variance of the other traits such that an upper limit to the rate of evolution is quickly reached as the number of traits increases (Bürger 1986; Wagner 1988a). The biological reason for this is that when variance is unconstrained in the functionally dependent traits, there is "too much" variation, that is, very few individuals have the optimal combination of trait values (but all sorts of other combinations), which results in very inefficient selection. This problem increases with an increasing number of dimensions so that if the number of traits is large, selection may have no variation on which to act (Wagner 1988a, 1988b). Hence, at least in certain types of realistic adaptive landscapes, phenotypic integration is expected to facilitate evolutionary transformations.

A similar argument was made by Björklund (1996a) using the **G**-matrix. The magnitude of response to selection is determined by the magnitude of genetic variation for the trait being selected. Consider two populations, one with strong positive correlations among a set of traits (strongly integrated), and one without any correlations at all (no integration). If we compare the **G**-matrices, we find that the length of the largest eigenvector ($\mathbf{g}_{max}$) in the integrated population is considerably larger than the length of the corresponding vector in the population with no integration. The response to any selection will be in terms of $\mathbf{g}_{max}$ in the first case, while in the other population the response will match selection in terms of direction. In the first case $\mathbf{g}_{max}$ will be interpreted as a general size vector, while in the other case no interpretation can be made. If selection acts on overall size (or a set of functionally dependent traits), or in other words, in a direction collinear with $\mathbf{g}_{max}$, then the response will be far greater in the case of a strongly integrated population than in the population with no integration. Thus, the correlations act to promote the response to selection.

Although it should be clear that phenotypic integration can have a facilitating role in evolution, it is not easy to find examples (but see Dorn and Mitchell-Olds 1991; Scheiner and Istock 1991) where integration has been demonstrated to be a facilitating factor in promoting speed or direction of evolution in any particular direction. Again, perhaps the best evidence comes from Schluter's (1996) study: rate of morphological evolution was fastest for the species pairs that had diverged mainly along $\mathbf{g}_{max}$, and slowest for those that had diverged in directions further away from $\mathbf{g}_{max}$ (Fig. 5.3). Similar studies at population level, and studies of integration in phylogenetic context (e.g., Goodin and Johnson 1992; Steppan 1997; Armbruster 2002), are obvious ways to gain more insight into this issue. Likewise, experimental studies of evolution with organisms (preferably populations) differing in their patterns of integration (**G**) subject to controlled selective regimes are an obvious way to test the facilitative role of integration for rates and directions of evolution.

Finally, moving from microevolutionary issues to the macroevolutionary level, it has been suggested that selection processes of entities higher than individuals may play a role in macroevolution (Vrba and Eldredge 1984; Jablonski 1987; Vrba 1989; Lloyd and Gould 1993). In particular, population-level traits such as variability seem likely candidates for higher-level selection processes because they interact with population fitness (survival; Lloyd and Gould 1993). Björklund (1994b) extended this theme, and argued that phenotypic integration—as reflected in **G** —can be subject to interpopulation selection. How this might work is best illustrated by a hypothetical example (Björklund 1994b). Consider the classical case of character displacement (Grant 1975): two allopatric species with the same niche become sympatric with the result that interspecific competition will favor differentiation in size. The species able to respond more rapidly to a given amount of selection for size change will be the more integrated one (more variation along $\mathbf{g}_{max}$). The less integrated species can achieve the same amount of change only through larger amounts of selective deaths. Hence, the highly integrated species should be less prone to suffer from interspecific competition and reduced population size, and thereby also have a lowered risk of extinction (Björklund 1994b). Again, although the idea of integration as a target of species-level selection is a logical possibility and is consistent with the macroevolutionary patterns observed in birds (Björklund 1994b), attempts at explicit tests are still lacking.

## Group Size Factors and Evolution of Modularity

Before closing the discussion about the adaptive nature of phenotypic evolution, one additional issue is worth considering, namely that relating to evolution modularity. Based on the results of his modeling work, Wagner (1988b) has suggested that there may exist an association between a constrained body plan and the evolution of complex functional adaptations. Progressive evolution may be preceded by or associated with adequate allocation of variance: for some traits to evolve, others must become constrained. This argument is closely related to a concept of "quasi-independence" (Lewontin 1978), according to which it must be possible to change and select for a given trait without at the same time changing large numbers of other traits. It has been suggested that this quasi-independence is achieved through modularization, where different functionally or developmentally related traits become associated to modular sets, each with a certain degree of evolutionary freedom (e.g., Cowley and Atchley 1990). This type of modularity in the phenotypic integration is illustrated in an analysis of morphometric variation in European greenfinches (Fig. 5.6a). Here, the traits were divided into four different a priori defined functional groups (viz., wing, leg, beak, and body traits) on the basis of their presumed functional and developmental proximity (Merilä and Björklund 1999). An analysis of residual correlations revealed a high degree of integration within these functional/developmental groups or complexes, but less so between them (Fig. 5.6). Reanalysis of the data with the method described by Magwene (2001) supports this conclusion: the greenfinch skeleton is modularized and consists of complexes of highly integrated traits, which in part are less loosely integrated with traits in other complexes.

This type of modularization ("group size factors" *sensu* Wright 1932; see also Riska 1985) is the basis for the theory of mosaic evolution (Simpson 1953; Wagner 1996). According to this theory, there are relatively independent functional complexes within an organismal structure that are subject to long-term stabilizing selection acting to preserve basic organismal structure. Hence, directional selection can favor rearrangements of these complexes without large cascading effects on the basic organismal structure. Consequently, long-term stabilizing selection on the basic structure in combination with short-term directional selection of functional complexes should lead to high integration of the functional complexes, which in turn are less integrated with each other (Simpson 1953; Berg 1960; Cheverud 1984; Zeng 1988; Wagner 1996; Cheverud et al. 1997). One pre-

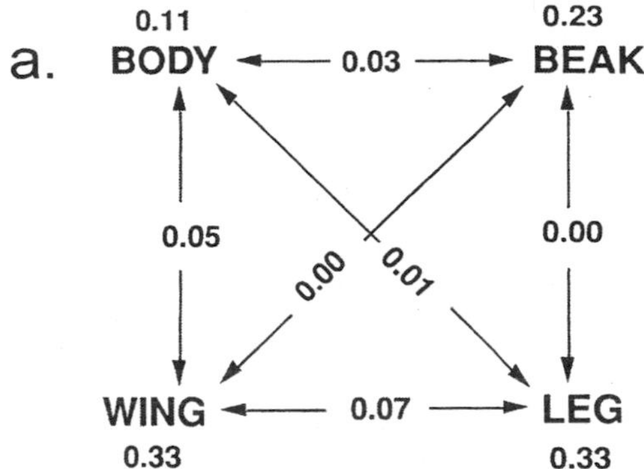


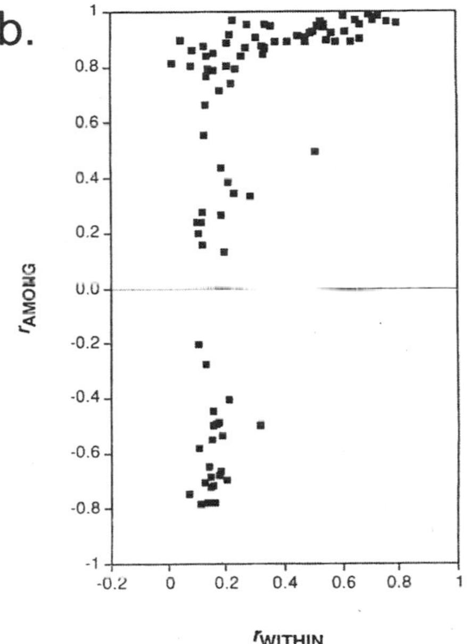


Figure 5.6 Morphometric integration in the greenfinch (*Carduelis chloris*). (a) Residual correlations between different a priori defined character complexes (group size factors). Residual correlation is defined as the correlation that exist between a pair of characters with the effects of other characters partialed out. (b) Among-population correlations between pairs of traits as a function of the same correlations within populations. Note that high within-population trait correlations are always associated with high among-population correlations, whereas low within-population correlations can be associated to between-population correlation of any magnitude. Adapted from Merilä and Björklund (1999) with permission from Blackwell Science.

diction from the mosaic theory of evolution is that those traits that are highly integrated within populations will be so also across populations, whereas traits that are weakly integrated within populations can be either weakly or highly integrated among populations (Zeng 1988). This is indeed a pattern seen in many studies (see above), and is illustrated in Fig. 5.6b with data from the greenfinch (Merilä and Björklund 1999).

Another prediction made by the mosaic theory of evolution is that functionally highly integrated traits should be less affected by environmental and genetic perturbations than functionally less integrated traits (Badyaev and Foresman 2000). Subjecting three species of shrews to environmental stress, and measuring the effects of stress on degree of integration in the shrew mandible, Badyaev and Foresman (2000) were able to verify this prediction: the stress-induced changes

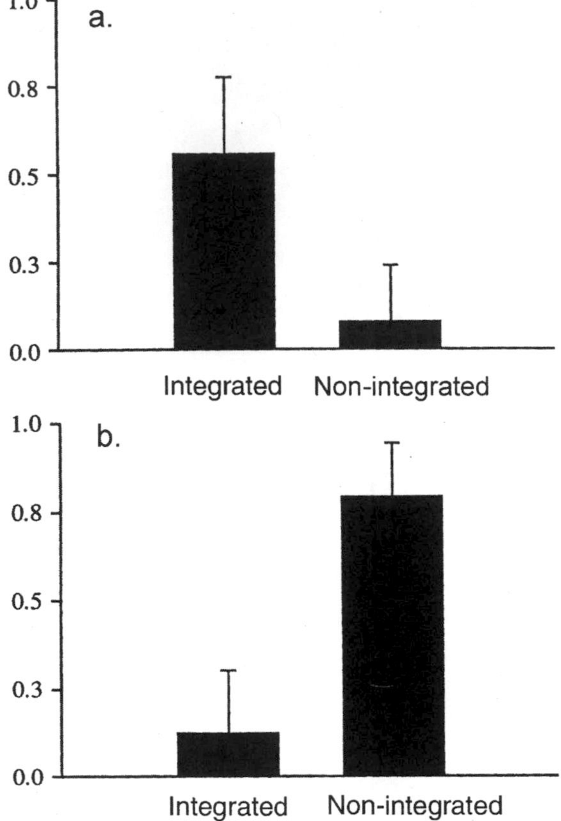


Figure 5.7 Vector correlations measuring the degree of concurrence in (a) between-species versus within-species and (b) between-species versus between-treatment (stressful versus nonstressful) patterns of variation for integrated and nonintegrated landmark displacements in shrews. Redrawn from Badyaev and Foresman (2000) with permission from the Royal Society of London.

were confined mostly to nonintegrated traits (Fig. 5.7). Perhaps more interestingly, when comparing the vector correlations of integrated and nonintegrated landmark displacements, they found that the species had diverged most in the same set of nonintegrated traits that were most displaced by stress within each species (Fig. 5.7). That the patterns of expression of environmental stress in the shrew mandible are concordant with patterns of evolutionary change suggests that low environmental canalization is associated with low genetic canalization.

## Optimal Integration?

Given the dual effect of integration for evolution, still another potentially useful perspective to the evolution of integration can be achieved by asking the obvious question: is there an optimal level of integration such that the constraining effect is minimized, and the adaptive feature is maximized? Is this achieved when all traits are correlated, or are fewer sets of correlations more optimal? This points to two different but interrelated questions: the number of integrated units (modules), and the optimal size of each module. This question has a close parallel to a question in theoretical physics, namely the stability-flexibility dilemma. In short, this can be described by noting the fact that large systems are flexible (in the sense of being able to perform many different tasks) but slow, while small systems are not flexible but fast. One attempt to look into this problem was made by Repsilber et al. (2003) by means of letting neural networks of different sizes (3–10 nodes) compete with each other using a genetic algorithm. The networks were exposed to situations with a number (3 or 8) of different environments that were encountered randomly and with equal frequency. The results were clear: there was a positive correlation between the number of different environments encountered and the size of the winning network. However, within each set of environments, smaller networks outcompeted larger ones in the short run, but as time passed, the advantage of larger networks increased and they eventually outcompeted smaller ones. However, network size did not increase linearly with time but seemed to reach an optimal size of around five nodes.This points to the following broad conclusions with regard to the patterns of integration. If environmental variation is small and varying over time, then we can expect a low level of integration, while if environmental variation is large and predictable over time, a higher level of integration can be expected (Repsilber et al. 2003). This can be translated in the context of constraint and adaptation such that when there are considerable changes in the environment over short time periods, integration acts more as a constraint and is consequently selected against. One the other hand, in predictable stable environmental situations, integration serves as an adaptation and there is consequently selection acting to maintain or strengthen the correlations, albeit within certain size limits. How general these conclusions are remains to be seen, but they provide at least a framework for the development of more specific biological models. Note also that the framework presented here allows for the analysis of the evolution of the number of modules as well as the number of parts within each module (Hartwell et al. 1999). Thus, the use of neural networks is a very flexible technique that may advance this field beyond the current level by, for example, allowing for a more detailed analysis of the genotype-phenotype mapping (Wagner and Altenberg 1996; Repsilber et al. 2001), which is an integral part of the understanding of integration.

## Conclusions

To sum up, while it is clear on theoretical grounds that integration—as summarized in **G**—can constrain rates and directions of evolutionary transformations at least in the short term, there is still a paucity of tests and evidence for the degree to which patterns of population and species differentiation reflect constraints posed by integration, and the degree of multivariate stabilizing (i.e., correlative) selection. The dual nature of integration in evolution becomes apparent if we consider the fact that the pattern of integration of a given taxon can act as a constraint and an adaptive feature at the same time. It is a constraint when the pattern of multivariate selection is in directions other than the directions of the correlations, but adaptive if selection and integration vectors are collinear. Furthermore, in a given population, the nature of integration may change from an adaptive to a constraining feature depending on the prevailing environmental conditions (i.e., selection), and on the possible direct effects of environment on genetic variances and covariances. This points to the important conclusion, namely, that any discussion about integration must by necessity be context-dependent—dependent on the pattern of selection, time scale, and environment.

This dual and context-dependent nature of the integration is nicely illustrated by two examples that also highlight the increasing concern about an organism's ability to adapt to environmental changes brought about by predicted climate changes. In a selection experiment conducted with blackcaps (*Sylvia atricapilla*), Pulido et al. (1996) demonstrated that incidence of migration (migratory/non-migratory) and migratory activity (strongly correlated with migratory distance) are genetically strongly positively correlated, such that selection on one trait will lead to strong correlated response in the other. A potentially important consequence of this is that selection favoring birds overwintering in areas close to breeding sites (i.e., short migration distance), and selection favoring nonmigratory individuals—both of which are expected under global warming (Berthold 1991)—will jointly increase the number of residents (Pulido et al. 1996). However, if the selection for residency causes a correlated response in migration distance to coincidence with migration to areas not suitable for overwintering, then the genetic correlation may oppose evolution of residency. Hence, the facilitative versus constraining role of the integration here depends on the ecological context. An example highlighting the time-scale-dependent nature of integration comes from the study by Etterson and Shaw (2001). In reciprocal transplant experiments of locally adapted annual legumes (*Chamaecrista fasciculata*), they demonstrated that patterns of integration that were concurrent with patterns of selection when the plants were grown in their native sites, sometimes became disconcordant when the plants were grown under conditions predicted by climate change scenarios (Etterson and Shaw 2001). Consequently, although patterns of integration might be adaptive today, they might not be so in the near future. In fact, Etterson and Shaw (2001) showed that the predicted rates of evolutionary response are much slower than the predicted rate of climate change. This underlines also the conjecture that although constraints posed by the genetic architecture might largely be transitory, they might have significant and enduring consequences even on shorter time scales.

Studies such as those presented by Schluter (1996), using multiple relatively recently diverged populations for which constancy of **G** is more likely to hold (e.g., Brodie 1993; Arnold and Phillips 1999), combined with empirically estimated selective regimes, provide attractive means to evaluate the relative roles of integration versus selection in explaining patterns of population differentiation. Likewise, comparisons of **G** and **L** are still amazingly rare, although this approach provides a theoretically sound framework to test for the importance of selection as an explanation of patterns of population differentiation. Although still far from optimal (Steppan et al. 2002), the recent advances in matrix comparison methods (e.g., Arnold and Phillips 1999; Roff 2002) have paved the way for such studies. However, given that the data collected from the wild species and populations are by definition of a correlative nature, it is always difficult—if not impossible—to differentiate between correlation and causation, however sophisticated analytical tools might be available. Therefore, long-term laboratory and field experiments where the effects of integration in evolutionary dynamics under different selective regimes can be assessed, could lead to a better understanding of the relative roles of selection versus constraints as determinants of the directions and pace of evolution. At the same time, they might give us important insights into how integration and modularity evolve.

*Acknowledgments* We thank Katherine Preston and Massimo Pigliucci for helpful comments on earlier version of this chapter. We were supported by the Academy of Finland (J.M.), the Nordic Institute for Advancement of Higher Education (NorFA; J.M.), and the Swedish Science Council (M.B.).

## *Literature Cited*

Armbruster, W. S. 1991. Multilevel analysis of morphometric data from natural plant populations: insights into ontogenetic, genetic, and selective correlations in *Dalechampia scandens*. Evolution 45:1229–1244.

Armbruster, W. S. 2002. Can indirect selection and genetic context contribute to trait diversification? A transition-probability study of blossom-colour evolution in two genera. J. Evol. Biol. 15:468–486.

Armbruster, W. S., and K. E. Schwaegerle 1996. Causes of covariation of phenotypic traits among populations. J. Evol. Biol. 9:261–276.

Arnold, S. J. 1992. Constraints on phenotypic evolution. Am. Nat. 140:S85-S107.

Arnold, S. J., and P. C. Phillips. 1999. Hierarchical comparison of genetic variance-covariance matrices. II. Coastal-inland divergence in the garter snake, *Thamnophis elegans*. Evolution 53:1516–1527.

Arnold, S. J., M. E. Pferender, and A. G. Jones. 2001. The adaptive landscape as a conceptual bridge between micro- and macroevolution. Genetica 112–113:9–32.

Badyaev, A. V., and K. R. Foresman. 2000. Extreme environmental change and evolution: stress-induced morphological variation is strongly concordant with patterns of evolutionary divergence in shrew mandibles. Proc. R. Soc. Lond. B 267:371–377.

Badyaev, A. V., and G. E. Hill. 2000. The evolution of sexual dimorphism in the house finch. I. Population divergence in the morphological covariance structure. Evolution 54:1784–1794.

Badyaev, A. V., and T. E. Martin. 2000. Individual variation in growth trajectories: phenotypic and genetic correlations in ontogeny of the house finch (*Carpodacus mexicanus*). J. Evol. Biol. 13:290–301.

Badyaev, A. V., G. E. Hill, A. M. Stoehr, P. M. Nolan, and K. J. McGraw. 2000. The evolution of sexual dimorphism in the house finch. II. Population divergence in relation to local selection. Evolution 54:2134–2144.

Badyaev, A. V., G. E. Hill, M. L. Beck, A. A. Dervan, R. A. Duckworth, K. J. McGraw, P. M. Nolan, and L. A. Whittingham. 2002. Sex-biased hatching order and adaptive population divergence in a passerine bird. Science 295:316–318.

Berg, R. L. 1960. The ecological significance of correlational pleiades. Evolution 14:171–180.

Berthold, P. 1991. Patterns of avian migration in light of current global "greenhouse" effects: a Central European perspective. Pp. 780–786 in B. D. Bell, R. O. Cossee, J. E. C. Flux et al. (eds.), Acta XX Congressus Internationalis Ornithologici. New Zealand Ornithological Trust, Wellington.

Björklund, M. 1991. Patterns of morphological variation among cardueline finches (Fringillidae: Carduelinae). Biol. J. Linn. Soc. 43:239–248.

Björklund, M. 1992. Selection of bill size proportions in the Common Rosefinch (*Carpodacus erythrinus*). Auk 109:637–642.

Björklund, M. 1993. Phenotypic variation of growth trajectories in finches. Evolution 47:1506–1514.

Björklund, M. 1994a. Allometric relationships in three species of finches (Aves, Fringillidae). J. Zool. 233:657–668

Björklund, M. 1994b. Species selection on organismal integration. J. Theor. Biol. 171:427–430.

Björklund, M. 1996a. The importance of evolutionary constraints in ecological time scales. Evol. Ecol. 10:423–431.

Björklund, M. 1996b. Similarity of growth among Great tits (*Parus major*) and Blue tits (*P. caeruleus*). Biol. J. Linn. Soc. 58:343–355.

Björklund, M. 1997. Variation in growth in the blue tit (*Parus caeruleus*). J. Evol. Biol 10:139–155.

Björklund, M., and J. Merilä. 1993. Morphological differentiation in *Carduelis* finches: adaptive vs constraint models. J. Evol. Biol. 6:359–373.

Björklund, M. and J. C. Senar. 2001. Sex differences in survival selection in the serin, *Serinus serinus*. J. Evol. Biol. 14:841–849.

Brodie, E. D. III. 1992. Correlational selection for color pattern and antipredator behavior in the garter snake *Thamnophis ordinoides*. Evolution 46:1284–1298.

Brodie, E. D. III. 1993. Homogeneity of the genetic variance-covariance matrix for anti-predator traits in two natural populations of the garter snake *Thamnophis ordinoides*. Evolution 47:844–854.

Bürger, R. 1986. Constraints for the evolution of functionally coupled characters: a non-linear analysis of a phenotypic model. Evolution 40:182–193.

Camara, M. D., C. A. Ancell, and M. Pigliucci. 2000. Induced mutations: a novel tool to study phenotypic integration and evolutionary constraints in *Arabidopsis thaliana*. Evol. Ecol. Res. 2:1009–1029.

Cheverud, J. M. 1984. Quantitative genetics and developmental constraints on evolution by selection. J. Theor. Biol. 110:155–171.

Cheverud, J. M. 1989. A comparative analysis of morphological variation patterns in the Papionins. Evolution 43:1737–1747.

Cheverud, J. M. 1996. Developmental integration and the evolution of pleiotropy. Amer. Zool. 36:44–50.

Cheverud, J. M., E. J . Routman, and D. J. Irschick. 1997. Pleitropic effects of individual gene loci on mandibular morphology. Evolution 51:2006–2016.

Conover, D. O., and E. D. Schultz. 1995. Phenotypic similarity and the evolutionary significance of countergradient variation. Trends Ecol. Evol. 10:248–252.

Cowley, D. E., and W. R. Atchley. 1990. Development and quantitative genetics of correlation structure among body parts of *Drosophila melanogaster*. Am. Nat. 135:242–268.

Dorn, L. A., and T. Mitchell-Olds. 1991. Genetics of *Brassica campestris*. 1. Genetic constraints on evolution of life-history characters. Evolution 45:371–379.

Eldredge, N. 1995. Reinventing Darwin: The Great Debate at the High Table of Evolution. John Wiley, New York.

Eldredge, N. 1999. The Pattern of Evolution. W. H. Freeman, New York.

Etterson, J. R., and R. G. Shaw. 2001. Constraint to adaptive evolution in response to global warming. Science 294:151–154.

Gibson, A. R., A. J. Baker, and A. Moeed. 1984. Morphometric variation in introduced populations of the common mynah (*Acridotheres tristis*). Syst. Zool. 33:217–237.

Giesel, J. T., P. A. Murphy, and M. N. Manlove. 1982. The influence of temperature on genetic interrelationships of life history traits in a population of *Drosophila melanogaster*: what tangled data sets we weave. Am. Nat. 119:464–479.

Gomulkiewicz, R., and M. Kirkpatrick. 1992. Quantitative genetics and the evolution of reaction norms. Evolution 46:390–411.

Goodin, J. T. and, M. S. Johnson 1992. Patterns of morphological covariation in *Parula*. Syst. Biol. 41:292–304.

Gould, S. J. 2002. The Structure of Evolutionary Theory. Belknap Press, Cambridge, MA.

Gould, S. J., and R. C. Lewontin. 1979. The spandrels of San Marco and the Panglossian paradigm: a critique of the adaptationist program. Proc. R. Soc. Lond. B 205: 581–598.

Gould, S. J., and E. S. Vrba 1982. Exaptation—a missing term in the science of form. Paleobiology. 8:4–15.

Grant, P. R. 1975. The classical case of character displacement. Evol. Biol. 8:237–337.

Grant, P. R., and B. R. Grant. 1995. Predicting microevolutionary responses to directional selection on heritable variation. Evolution 49:241–251.

Haldane, J. B. S. 1927. Possible Worlds and Other Essays. Chatto & Windus, London.

Hartwell, L. E., J. J. Hopfield, S. Leibler and A. W. Murray. 1999. From molecular to modular cell biology. Nature 402:C47–C52.

Hoffmann, A. A., and J. Merilä. 1999. Heritable variation and evolution under favourable and unfavourable conditions. Trends Ecol. Evol. 14:96–101.

Houle, D. 1992. Comparing evolvability and variability of quantitative traits. Genetics 130:195–204.

Jablonski, D. 1987. Heritability at the species level: analysis of geographic ranges of Cretaceous mollusks. Science 238:360–363.

James, F. C. 1983. Environmental component of morphological differentiation in birds. Science 221:184–186.

Jernigan, R. W., D. C. Culver, and D. W. Fong. 1994. The dual role of selection and evolutionary history as reflected in genetic correlations. Evolution 48:587–596.

Kingsolver, J. G., H. E. Hoekstra, J. M. Hoekstra, D. Berrigan, S. N. Vignieri, C. E. Hill, A. Hoang, P. Gilbert, and P. Beerlio. 2001. The strength of phenotypic selection in natural populations. Am. Nat. 157:245–261.

Kirkpatrick, M., and D. Lofsvold. 1992. Measuring selection and constraint in the evolution of growth. Evolution 46:954–971.

Lande, R. 1976. Natural selection and random genetic drift in phenotypic evolution. Evolution 30:314–334.

Lande, R. 1979. Quantitative genetic analysis of multivariate evolution, applied to brain body size allometry. Evolution 33:402–416.

Lande, R. 1980. The genetic covariance between characters maintained by pleiotropic mutations. Genetics 94:203–215.

Lardner, B. 2000. Morphological and life history responses to predators in larvae of seven anurans. Oikos 88:169–180.

Larsson, K. 1993. Inheritance of body size in the Barnacle Goose under different environmental conditions. J. Evol. Biol. 6:195–208.

Laurila, A., M. Järvi-Laturi, S. Pakkasmaa, P. A. Crochét, and J. Merilä. 2002. Predator-induced plasticity in early life history and morphology in two anuran amphibians. Oecologia 132:524–530.

Lewontin, R. C. 1978. Adaptation. Sci. Am. 239:212–230.

Lewontin, R. C. 1983. The organism as the subject and object of evolution. Scientia 118: 65–82.

Lloyd, E. A., and S. J. Gould. 1993. Species selection on variability. Proc. Natl. Acad. Sci. U.S.A. 90:595–599.

Lofsvold, D. 1988. Quantitative genetics of morphological differentiation in *Peromyscus*. II. Analysis of selection and drift. Evolution 42:54–67.

Lynch, M. and B. Walsh 1998. Genetics and Analysis of Quantitative Traits. Sinauer, Sunderland, MA.

Magwene, P. M. 2001. New tools for studying integration and modularity. Evolution 55:1734–1745.

Maynard Smith, J., R. Burian, S. Kauffman, P. Alberch, J. Campbell, B. Goodwin, R. Lande, D. Raup and L. Wolpert. 1985. Developmental constraints and evolution. Quart. Rev. Biol. 60:265–287.

McCollum, S. A., and J. Van Buskirk. 1996. Costs and benefits of a predator-induced polyphenism in the gray treefrog *Hyla chrysoscelis*. Evolution 50:583–593.

Merilä, J. 1997. Quantitative trait and allozyme divergence in the greenfinch (*Carduelis chloris*, Aves: Fringillidae). Biol. J. Linn. Soc. 61:243–266.

Merilä, J., and M. Björklund. 1999. Population divergence and morphometric integration in the greenfinch (*Carduelis chloris*)—evolution against the trajectory of least resistance? J. Evol. Biol. 12:103–112.

Merilä, J., and P. Crnokrak. 2001. Comparison of genetic differentiation at marker loci and quantitative genetic traits. J. Evol. Biol. 14:892–903.

Merilä, J., and B.C. Sheldon. 2001. Avian quantitative genetics. Curr. Ornithol. 16:179–255.

Merilä, J., M. Björklund, and L. Gustafsson. 1994. Evolution of morphometric differences under moderate genetic correlations as exemplified by flycatchers (*Ficedula*; Muscicapidae). Biol. J. Linn. Soc. 52:19–30.

Merilä, J., B. C. Sheldon, and H. Ellegren. 1998. Quantitative genetics of sexual size dimorphism in the collared flycatcher, *Ficedula albicollis*. Evolution 52:870–876.

Mitchell-Olds, T. 1996. Pleiotropy causes long term genetic constraints on life-history evolution in *Brassica rapa*. Evolution 50:1849–1858.

Mousseau, T. A., and D. A. Roff. 1987. Natural selection and the heritability of fitness components. Heredity 59:181–197.

Newman, R. A. 1988. Adaptive plasticity in development of *Scaphiopus couchii* tadpoles in desert ponds. Evolution 42:774–783.

Olson, E. C., and R. L. Miller 1958. Morphological Integration. University of Chicago Press, Chicago.

Phillips P. C., M. C. Whitlock, and K. Fowler. 2001. Inbreeding changes the shape of the genetic covariance matrix in *Drosophila melanogaster*. Genetics 158:1137–1145.

Price, D. K. 1996. Sexual selection, selection load and quantitative genetics of zebra finch bill colour. Proc. R. Soc. Lond. B 263:217–221.

Price, T. D., and P. R. Grant. 1985. The evolution of ontogeny in Darwin's finches: a quantitative genetic approach. Am. Nat. 125:169–188.

Price, T. D., P. R. Grant, and P. T. Boag. 1984. Genetic changes in the morphological differentiation of Darwin's ground finches. Pp. 49–66 in K. Wöhrmann and V. Loeschke (eds.), Population Biology and Evolution. Springer Verlag, Berlin.

Price, T. D., M. Turelli, and M. Slatkin. 1993. Peak shifts produced by correlated response to selection. Evolution 47:280–290.

Pulido, F., P. Berthold, and A. J. van Noordwijk. 1996. Frequency of migrants and migratory activity are genetically correlated in a bird population: evolutionary implications. Proc. Natl. Acad. Sci. U.S.A. 93:14642–14647.

Relyea, R. A. 2002. Local population differences in phenotypic plasticity: predator-induced changes in wood frog tadpoles. Ecol. Monogr. 72:77–93.

Repsilber, D., H. Liljenström, and S. Andersson. 2001. Reverse engineering of regulatory networks: simulation studies on a genetic algorithm approach for ranking hypotheses. Biosystems 66:31–41.

Repsilber, D., M. Björklund, and T. Martinez. 2003. Evolutionary dynamics of regulatory networks—optimal size for plasticity and adaptation as dependent on environmental heterogeneity and time scale of adaptation. In preparation.

Riska, B. 1985. Group size factors and geographic variation of morphometric correlation. Evolution 39:792–803.

Riska, B. 1989. Composite traits, selection response, and evolution. Evolution 43:1172–1191.

Roff, D. A. 1995. The estimation of genetic correlations from phenotypic correlations: a test of Cheverud's conjecture. Heredity 74:481–490.

Roff, D. A. 1997. Evolutionary Quantitative Genetics. Chapman & Hall, New York.

Roff, D. A. 2000. Evolution of the **G** matrix: Selection or drift? Heredity 84:135–142.

Roff, D. A. 2002. Comparing **G** matrices: a manova approach. Evolution 56:1286–1291.

Roff, D. A., and T. A. Mousseau. 1999. Does natural selection alter genetic architechture? An evaluation of quantitative genetic variation among natural populations of *Allonemobius socius* and *A. fasciatus*. J. Evol. Biol. 12:361–369.

Rossiter, M. 1996. Incidence and consequences of inherited environmental effects. Annu. Rev. Ecol. Syst. 27:451–476.

Scheiner, S. M., and C. A. Istock. 1991. Correlational selection on life-history traits in the pitcher-plant mosquito. Genetica 84:123–128.

Schlichting, C. D., and M. Pigliucci. 1998. Phenotypic Evolution: A Reaction Norm Perspective. Sinauer, Sunderland, MA.

Schluter, D. 1993. Adaptive radiation in sticklebacks: size, shape, and habitat use efficiency. Ecology 74:699–709.

Schluter, D. 1994. Adaptive radiation along lines of least resistance? J. Ornithol. 135:357.
Schluter, D. 1996. Adaptive radiation along genetic lines of least resistance. Evolution 50:1766–1774.
Schluter, D. 2000. The Ecology of Adaptive Radiation. Oxford University Press, Oxford.
Service, P. M., and M. R. Rose 1985. Genetic covariation among life-history components: the effect of novel environments. Evolution 39:943–945.
Simpson, G. G. 1953. The Major Features of Evolution. Simon & Schuster, New York.
Sober, E. 1984. The Nature of Selection. MIT Press, Cambridge, MA.
Sokal, R. R., and B. Riska. 1981. Geographic variation in *Pemphigus populitransversus* (Insects: Aphididae). Biol. J. Linn. Soc. 15:201–223.
Steppan, S. J. 1997. Phylogenetic analysis of phenotypic covariance structure. I. Contrasting results from matrix correlation and common principal component analysis. Evolution 51:571–586.
Steppan, S. J., P. C. Phillips, and D. Houle. 2002. Comparative quantitative genetics: evolution of the G matrix. Trends Ecol. Evol. 17:320–327.
Turelli, M. 1988. Phenotypic evolution, constant covariances, and the maintenance of additive genetic variance. Evolution 42:1342–1347.
Van Buskirk, J., and S. A. McCollum. 2000a. Influence of tail shape on tadpole swimming performance. J. Exp. Biol. 203:2149–2158
Van Buskirk, J., and S. A. McCollum. 2000b. Functional mechanisms of an inducible defence in tadpoles: morphology and behaviour influence mortality risk from predation. J. Evol. Biol. 13:336–347.
Voss, R., L. F. Marcus, and P. Escalante. 1990. Morphological evolution in muroid rodents. I. Conservative patterns of craniometric covariance and their ontogenetic basis in the neo-tropical genus *Zygodontomys*. Evolution 44:1568–1587.
Vrba, E. S. 1989. Levels of selection and sorting with special reference to the species level. Pp. 111–168 in P. H. Harvey and L. Partridge (eds.), Oxford Surveys in Evolutionary Biology. Oxford University Press, Oxford.
Vrba, E. S., and N. Eldredge. 1984. Individuals, hierarchies and processes: toward a more complete evolutionary theory. Paleobiology 10:146–171.
Wagner, G. P. 1984. On the eigenvalue distribution of genetic and phenotypic dispersion matrices: evidence for a nonrandom organization of quantitative character variation. J. Math. Biol. 21:77–95.
Wagner, G. P. 1988a. The influence of variation and of developmental constraints on the rate of multivariate phenotypic evolution. J. Evol. Biol. 1:45–66.
Wagner, G. P. 1988b. The significance of developmental constraints for phenotypic evolution by natural selection. Pp. 222–229, in G. de Jong (ed.), Population Genetics and Evolution. Springer-Verlag, Berlin.
Wagner, G. P. 1996. Homology, natural kinds, and the evolution of modularity. Am. Zool. 36:36–43.
Wagner, G. P., and L. Altenberg. 1996. Complex adaptations and the evolution of evolvability. Evolution 50:967–976.
Wagner, G. P., G. Booth, and H. Bacheri-Chaichian. 1997. A population genetic theory of canalization. Evolution 51:329–347.
Wilkinson, G. S., K. Fowler, and L. Partridge. 1990. Resistance of genetic correlation structure to directional selection in *Drosophila melanogaster*. Evolution 44:1990–2003.
Wright, S. 1932. General, group and special size factors. Genetics 17:603–619.
Yampolsky, L. Y., and D. Ebert. 1995. Variation and plasticity of biomass allocation in *Daphnia*. Funct. Ecol. 9:725–733.
Zeng, Z.-B. 1988. Long-term correlated response, interpopulation covariation, and interspecific allometry. Evolution 42:363–374.

# 6

# Evolvability, Stabilizing Selection, and the Problem of Stasis

THOMAS F. HANSEN
DAVID HOULE

> Organisms ... have not done nearly as much evolving as we should reasonably expect.
>
> *Williams (1992)*

> For a century we have been mesmerized by the successes of evolution. It is time now that we paid equal attention to its failures
>
> *Bradshaw (1991)*

Evolutionary stasis has a paradoxical position in current evolutionary thinking. Williams (1992) argued convincingly that stasis is one of the most important unsolved problems of evolutionary biology (see also Bradshaw 1991). Stasis is arguably the predominant mode of evolution (Gould and Eldredge 1993); the cry "stasis is data" (Gould and Eldredge 1977) echoes faintly in most neontologists' ears. Despite this, stasis is one of the most neglected theoretical problems in evolutionary biology. When population geneticists think of stasis at all, they usually regard it as an almost trivial consequence of stabilizing selection (e.g., Charlesworth et al. 1982; Maynard Smith 1983; Lande 1985, 1986). The fundamental problem of what may cause persistent stabilizing selection in changing environments is rarely addressed at length. It has been largely left to macroevolutionists to speculate on the microevolutionary underpinnings of stasis (e.g., Van Valen 1982; Wake et al. 1983; Williamson 1987; Hoffman 1989; Lieberman and Dudgeon 1996; Sheldon 1996; Eldredge 1999; Gould 2002).

The paradox of stasis has its roots in quantitative genetics. It is the abundant variation in quantitative traits that makes stasis so difficult to explain on current thinking. Quantitative genetic experiments and simple population genetical theory seem to indicate that most characters should be very evolvable (i.e., have a high capacity to evolve), and there are many examples of rapid microevolutionary change that seem to confirm this ability (Hendry and Kinnison 1999; Kinnison and Hendry 2001). For example, the neutral theory for the evolution of quantitative genetic characters (Lynch and Hill 1986; Lynch 1993) shows that very rapid changes are expected on a geological time scale even in the absence of any selection, simply due to drift and random fixation of new mutations. In fact, it is rare to find characters that evolve faster than the neutral expectation on macroevolutionary time scales, and very easy to find characters that are much too conservative for it (Lynch 1990). The need for phylogenetic comparative methods is another simple illustration of how conservative evolution can be.

This conflict between stasis and abundant genetic variation has an all-too-obvious solution in stabilizing selection. Alternatively, the abundance of genetic variation may be illusory, and some form of variational constraint limits the evolvability of characters, precluding them from tracking environmental changes. In this chapter we discuss the logic of these two hypotheses. We argue that stabilizing selection is too readily accepted, while constraints are too readily dismissed. In particular, we suggest that conceiving of an organism as an integrated phenotypic and/or genotypic entity suggests two classes of variational explanation for stasis. Although there is often abundant genetic variation in quantitative characters, it is the quality, and not the quantity, of variation that is important for evolvability. The quality of variation may be reduced by either integration of characters (pleiotropic constraints) or by specific types of integration among genes (epistatic constraints). We suggest several ways to operationalize these notions.

## Stasis in Insect Wings

Our thinking on this problem is best introduced with an example. Insect wings are conservative characters. Wing characters are usually reliable taxonomic indicators on the level of families, and qualitative differences are rare within genera, although coloration is sometimes a striking exception. Quantitative differences in shape occur, but are limited in extent. A good example is provided by variation in the genus *Drosophila*. Figure 6.1 summarizes the size-adjusted positions of vein intersections in 19 species of *Drosophila*, plus four representatives of closely related genera in the family Drosophilidae (Galpern 2000). While the wings differ in length by an order of magnitude, the size-adjusted positions of the intersections are remarkably conservative. This conservatism is also readily detected in the behavior of *Drosophila* in flight. The slow, hovering "cargo helicopter" flight mode is instantly recognizable to the experienced eye. Despite this conservatism, discriminant function analysis of wing shape shows that nearly all of the specimens in Fig. 6.1 can be correctly identified to species, as shown in Table 6.1. This seeming paradox is resolved when we note that it is the result of low variation within species and not high variation among species.

The relative conservatism of wing shape in *Drosophila* is particularly remarkable because the genus is thought to be at least 50 million years old (Powell 1997). As expected in such an old group, there is substantial ecological diversity in the genus, with flies inhabiting both temperate and tropical habitats from rain forest to deserts. Their larvae feed on the microflora in a wide range of substrates, including decaying fruits, flowers, wood, leaves, fungi, or carrion. A few exceptional species mine leaves, feed on pollen, or prey on other insects (Powell 1997; Kambysellis and Craddock 1997). As wings are often used in elaborate courtship

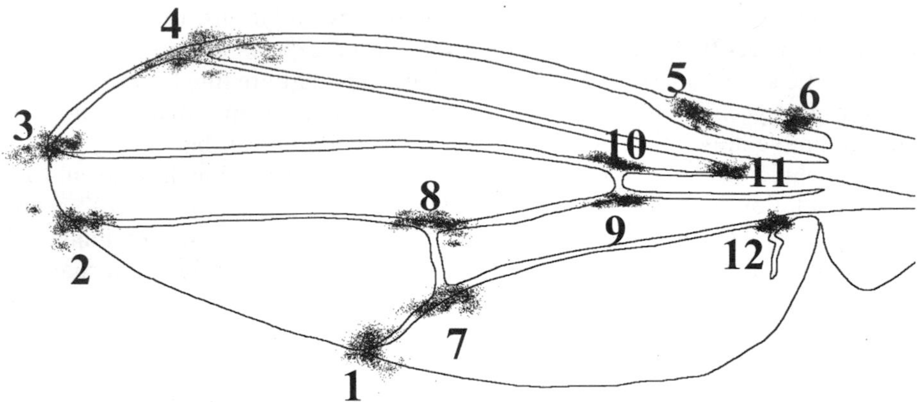


Figure 6.1 The wing vein landmark coordinates of Drosophilid flies when centroid size is removed and the data are aligned to a consensus configuration (Galpern 2000). The data consist of 2774 individuals of 23 species: 18 *Drosophila* species (*busckii, nebulosa, willistoni, saltans, sturtevanti, melanogaster, simulans, algonquin, athabasca, immigrans, sulfurigaster, falleni, guttifera, micromelanica, robusta, americana, virilis, hydeii,* and *repleta*), *Scaptodrosophila lebanonensis* and *S. stonei*, *Hirtodrosophila pictiventris*, and *Chymomyza procnemis*.

Table 6.1 Discriminant function analysis of Drosophilid wing shape data. The discriminant function was estimated based on 2/3 of the data, and tested on the remaining 1/3 using S-Plus routine lda (Venables and Ripley 1994, pp. 315–316).

| | Number of specimens | | | |
|---|---|---|---|---|
| Species | Trained | Tested | Classification errors | Error rate |
| *algonquin* | 43 | 21 | 1 | 0.05 |
| *athabasca* | 52 | 26 | 5 | 0.19 |
| *hydeii* | 119 | 61 | 13 | 0.21 |
| *melanogaster* | 128 | 65 | 3 | 0.05 |
| *simulans* | 78 | 40 | 9 | 0.22 |
| 18 other species | 1177 | 592 | 0 | 0.00 |
| Total | 1597 | 805 | 31 | 0.036 |

displays, the diversity in mating systems, including lek breeding, solitary territoriality, assault mating, and scrambles, is particularly noteworthy.

Even more remarkable is that wings throughout the Acalyptratae, the huge subordinal clade that includes *Drosophila*, are quite similar to those of the family Drosophilidae. The acalyptrates comprise over 22,000 species and 65 of the 113 families recognized in the Diptera (McAlpine 1989). There are a few key differences: the two costal breaks used as landmarks in Figure 6.1 are missing in most other acalyptrate families, and *Drosophila* lack a basal crossvein that most other acalyptrates have. Otherwise, the overall shape and placement of the common veins is usually conserved, despite the even greater range of ecological variation in the wider group.

Wing size and shape are highly heritable in *Drosophila* populations (e.g., Robertson and Reeve 1952; Robertson 1959; Cavicchi et al. 1981), making this a typical example of the paradox of stasis. Mutation produces novel phenotypic variation in wings at typical rates (Santiago et al. 1992; Houle, unpublished). Variation among populations has been demonstrated in both natural (e.g., Coyne and Beecham 1987; Gilchrist et al. 2000) and laboratory populations (Cavicchi et al. 1985), as has rapid divergence under natural selection (Huey et al. 2000; Gilchrist et al. 2001 for field studies and Cavicchi et al. 1985 for laboratory populations). Particularly compelling evidence for heritable variation in shape comes from the artificial selection experiments of Weber (1990, 1992; Weber et al. 1999), who obtained responses of up to 20 standard deviations in seven different arbitrary measures of wing shape in a relatively small number of generations.

## Stabilizing Selection as an Explanation of Stasis

Stabilizing selection has a somewhat odd position in evolutionary biology. On the one hand, there are few detailed empirical studies, and the direct evidence for stabilizing selection in the wild is far from overwhelming (Endler 1986; Travis 1989; Kingsolver et al. 2001). On the other hand, some of the most successful research programs in evolutionary biology are based on the assumption that traits are maintained at local optima by stabilizing selection (Mitchell and Valone 1990; Parker and Maynard Smith 1990). This is a fundamental assumption in behavioral ecology, life-history theory, and functional biology. These fields could not be as successful as they are if stabilizing selection is not the norm.

There appears to be fairly general acceptance of the idea that stabilizing selection is the cause of stasis. Even Gould, who may have been the most prominent champion of constraint hypotheses, recently acknowledged stabilizing selection based on niche tracking as a likely explanation of stasis (Gould 2002, pp. 880–885). From the theoretical point of view, stabilizing selection is a potent conservative force, but its relation to stasis is not simple. Explaining stasis with stabilizing selection requires not just that stabilizing selection is common, but that the selective optimum varies only within a narrow range. A stable optimum may seem plausible for a trait affected by a single selective factor, but the fitness functions of most quantitative traits are likely the result of a compromise among a large number of selective factors, any of which may be affected by changes in the environment. Dipteran wing shapes are certainly under stabilizing selection, but are the optimal shapes likely to be nearly the same in thousands of species of widely different size, living under widely different conditions with respect to temperature, humidity, and wind conditions? Why does not allocation to wing mass and muscle depend on the relative importance of flight to energetic constraints? Should not shape depend on this allocation? Should not the importance of wings for mate choice have substantial effects on their optimal shape? Should not males and females with differently shaped bodies have wings more different in shape? And if there really is one global optimum that fits all these conditions, why then would thousands of similarly sized hymenopterans have such different wings?

Mammalian body temperatures (BTs) provide another good illustration of the problems involved in assuming a constant optimum (Williams 1992). Almost all placental mammals keep their operating BT between 37 and 38°C, and, equally puzzling, they keep their testicular temperatures 1°C below that. If this is to be explained in terms of direct selection on BT, we need to show that an arctic lemming and an African elephant have similar ecologically determined temperature optima. This seems next to impossible in view of the huge differences in ambient temperature, heat exchange, metabolic needs, and energetic constraints. An explanation based on internal selection where BT is "burdened" (Riedl 1978) by interactions with other traits is perhaps more plausible (see fuller discussion below), but we agree with Williams (1992) that mammalian BT seems inexplicably conservative under a stabilizing-selection hypothesis.

Thus, just evoking stabilizing selection to explain stasis is insufficient; we need to explain why selective optima themselves should be conservative. Surprisingly little work has been devoted to this problem. Only a few sketches of candidate hypotheses have been proposed, such as niche tracking, population averaging, and ecological equilibration.

Niche tracking is perhaps the best-known mechanism proposed for maintaining stable selective environments (e.g., Eldredge 1999). To varying extents, all organisms are able to seek out favorable living conditions by behavioral means. In doing so, they will also stabilize many selective factors. More generally, Wake et al. (1983) suggested that any sort of plasticity (behavioral, physiological, or developmental) in one set of traits will tend to allow stasis in the remaining set of characters. There are, however, problems with these propositions. Note that this explanation just shifts the problem from one set of traits to another, for example from morphology to behavior. Why should a given behavioral "habitat" preference remain optimal in a changing environment? Indeed, the idea of the "Baldwin effect" is that plasticity, including behavioral plasticity, facilitates adaptive shifts (Baldwin 1896; Robinson and Dukas 1999), precisely the opposite of niche tracking. It is thus unclear if adaptive plastic responses will generally stabilize or disrupt selection pressures. Explicit models may help to clarify when plasticity helps or hinders further evolutionary change (e.g., Ancel 2000), but more work is needed in this area.

Lieberman and Dudgeon (1996) suggest that stasis is a result of averaging over many semi-independent populations that separately track fluctuating optima.

This simply lifts the problem to a higher hierarchical level. Why should the separately fluctuating optima exactly cancel, and why would the many environmental conditions that undoubtedly affect the entire metapopulation remain constant?

Williams (1992) suggests that stasis could be explained by the existence of hyperstable niches. These are core sets of environmental conditions that are always present somewhere. Unstable "niches" come and go as the environment changes, and species that adapt to them will tend to go extinct with these shifts. What we observe in the fossil record are thus the forms that reside in these hyperstable niches, which also must be the common niches for the forms that reside in them to dominate the fossil record.

Sheldon (1996) suggests a similar lineage-selection mechanism. The species that survive environmental fluctuations are those that are least affected by them, typically generalists. But Sheldon's hypothesis does not explain why some species should be immune to the fluctuations. Similarly, Williams's hypothesis does not explain why we should expect niches to be hyperstable. How can we test the hypotheses that Dipteran wings and mammalian BT are conserved by stabilizing selection in a hyperstable niche? The ecological niche concept itself may not be sufficiently operational to allow a direct test, as this would require very precise descriptions of niches. At least these hypotheses predict that we should find numerous short-lived taxa that do deviate from the norm. We believe this can be rejected in the case of both Dipteran wings and mammalian BT, but Williams suggests that repeated fresh-water radiations of sticklebacks from a stable marine form (Bell 1989) may provide an example.

A theoretical mechanism that may favor niche conservation has been proposed by Holt and Gaines (1992; Holt 1996) and Kawecki (1995). The idea is that selection for adaptation to a core niche is stronger than selection for adaptation to any marginal niche or habitat, since individuals living under conditions to which they are well adapted have higher reproductive output than individuals living under marginal (sink) conditions. In other words, more individuals are affected by selection in the core habitat and selection in this habitat is therefore more important. This helps maintain adaptation to the core habitat and makes adaptation to alternative habitats more difficult. Although a significant theoretical observation, this does not solve the stasis problem. First, the argument still depends on the core niche conditions remaining constant, and second, the argument is only valid if the core habitat is more abundant than any marginal habitat the species may encounter. In fact, these models may as well predict that shifting habitat abundances should be a powerful driver of adaptive shifts.

Ecology is no doubt essential in understanding stasis. Stenseth and Maynard Smith (1984) developed a model of community dynamics that was able to predict stasis or gradual (Red Queen) evolution depending on the strength of ecological interactions. As with Sheldon's hypothesis, this is still not a complete explanation of stasis within lineages, as it simply assumes that evolution will come to a halt in a stable environment.

A different type of explanation for stasis focuses on the complexity of forces that affect optima. Adaptation in one focal trait to one "primary" selective factor may be hindered by the need to stay adapted to a myriad of "secondary" factors, including other traits that have been tuned to the previous state of the focal trait (Simpson 1944; Hansen 1997). Change is not impossible in this scenario, but may be slow, as any new adaptation needs to be coordinated with a host of secondary adaptations. Consistent with this scenario, related species often deviate from an adaptive prediction in the same direction, leading to phylogenetic correlations. Although not stasis per se, this is direct evidence of evolutionary inertia. Hansen (1997) suggested that evolutionary inertia may be governed by the dynamics of the secondary factors, and developed a phylogenetic comparative method around this assumption. A skeptic may, however, argue that it is just as plausible that the secondary factors provide more opportunity for environmental changes to nudge the optimal state around. Theoretical work is necessary to determine whether, or under what conditions, a complex system of interrelated traits and selection forces will resist change.

In short, all proposed mechanisms for preserving optima are ultimately based on shifting the problem elsewhere, be it to other traits or to other levels of ecological organization. Thus, we join Arnold et al. (2001) in suggesting that work is urgently needed on the estimation and dynamics of adaptive landscapes. We simply do not have enough empirical evidence to conclude that landscapes are stable, nor do we have any compelling theoretical justification for assuming such stability. Given this situation, the alternative notion that stasis is due to constraints should be entertained as a valid possibility.

## Constraints and Evolvability

A constraint is any mechanism that may limit or bias the response to selection (see Wagner 1986; Arnold 1992; Houle 2001 for review). We make a distinction between variational and selective constraints. Variational constraints are due to limitation and biases in the variability of characters. Developmental constraints are sometimes also defined in this way (e.g., Maynard Smith et al. 1985), but we prefer the more general term, as character variability need not be a consequence of development, as in the case of many cellular or biochemical traits. Selective "constraints" derive from conflicting selection pressures, and are constraints only from the perspective of achieving specific adaptations, and not from the perspective of optimizing the fitness of the organism as a whole. The shape of the fitness landscape itself as a selective constraint has also been widely discussed (e.g., Kauffman 1993; Fear and Price 1998; Arnold et al. 2001), but we shall not consider this further here. As we shall see, the distinction between selective and variational constraints is sometimes a matter of perspective.

The most common objection to constraint as an explanation of stasis is the notion that quantitative characters are very evolvable because they show ample amounts of standing additive genetic variance and new mutational variation. In the following we shall show that this variation is not sufficient to ensure evolvability. Levels of additive genetic and mutational variation have traditionally not been measured in a way that is operationally linked to evolvability. Furthermore, it is the quality and not the quantity of variation that is important. Finally, the evolution of variability itself must also be taken into account.

## Genetic Constraints and Short-Term Evolvability

In an important contribution, Bradshaw (1991) reviewed cases where a failure to adapt in the face of unambiguous evidence for selection is plausibly due to a lack of appropriate genetic variation. For example, although the evolution of heavy-metal tolerance in plants is a celebrated textbook example of rapid evolution, Bradshaw pointed out that there are many plant populations that have failed to adapt in this way, and he showed that this is linked to an absence of variation for metal tolerance in the candidate populations.

Still, most traits exhibiting stasis appear to be genetically variable. This leads us to a short-term version of the paradox of stasis. There are now several well-documented examples of traits under directional selection in the field, such as clutch size in birds, that show no evolutionary response despite demonstrable heritability (Price and Liou 1989; Cooke et al. 1990; Frank and Slatkin 1992; Merilä et al. 2001). A number of plausible explanations can be evoked, including soft selection, confounding effects of condition, poor estimates of selection or genetic variance, G $\times$ E interactions, environmental deterioration, and a failure to account for selection throughout the life cycle. Despite this we believe that it is worthwhile to take a closer look at the evolvability of quantitative traits, and ask whether such cases of evolutionary failure may also be caused by a lack of useful genetic variation.

We have argued (Houle 1992; Hansen et al. 2003a) that the use of heritability, $h^2$, as a measure of evolvability is misleading. One reason is that there is a strong correlation between additive genetic and phenotypic variation, which means that the heritability is poorly correlated with additive genetic variation (see Fig. 6.2a for an example). A second reason is that heritability is not independent of its corresponding measure of selection strength, the selection differential. Under directional selection, the selection differential, $S$, is proportional to the phenotypic variance, which also enters in the denominator of the heritability. Thus, if $h^2$ is high, we may well expect a proportionally smaller $S$ for the same fitness function. This makes heritability highly suspect as an a priori predictor of evolvability.

Thus, $h^2$ and $S$, the components of the breeder's equation ($R = h^2 S$), do not constitute proper measures of evolvability and selection, as is often implicitly assumed. This suggests that the theoretically preferred separation should be based on the selection gradient, $\beta$, and the additive genetic variance, $G$, as in the Lande (1976, 1979; Lande and Arnold 1983) equation ($R = G\beta$). The selection gradient is a descriptor of the adaptive landscape and thus of the causal basis of selection. In our view, evolvability should be seen as the ability to respond to an externally imposed selection regime as represented by a fitness function.

Following Houle (1992), Hansen et al. (2003a) showed that $I_A$, the additive genetic variance scaled with the square of the trait mean, is an operational measure of evolvability, as it can be interpreted as expected percent change per generation per unit strength of directional selection. This holds for traits on a ratio scale, and requires the use of mean-scaled selection gradients, or fitness elasticities, as measures of selection strength (see van Tienderen 2000). Mean-scaled selection gradients have a natural unit as they measure the strength of selection on fitness itself as 1. Thus, $I_A$ is interpretable as the expected proportional

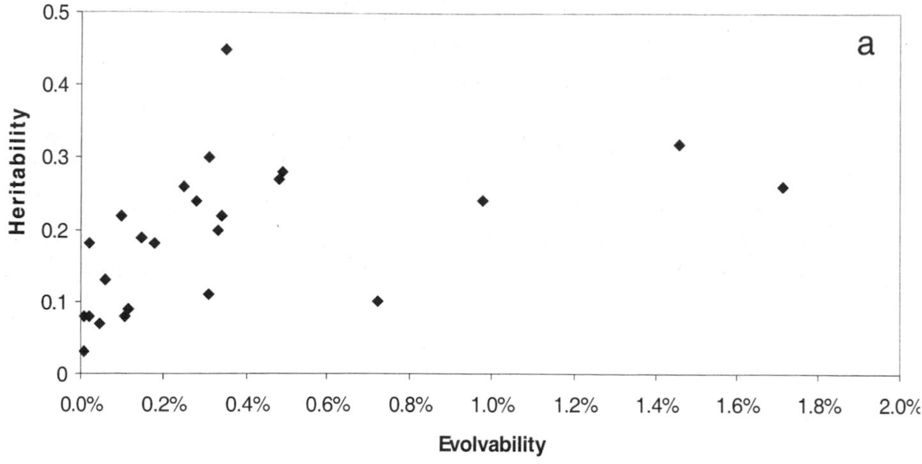


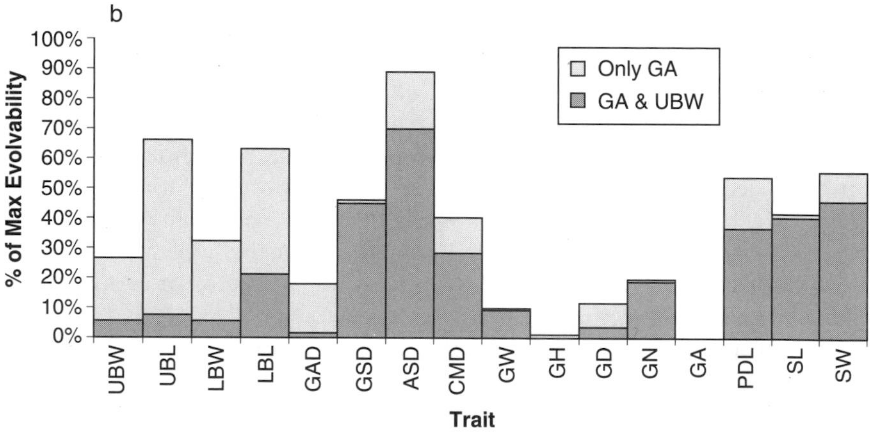


Figure 6.2 (a) Relationship between heritabilities and "evolvabilities," measured as $I_A$, in a set of floral traits from a population of *Dalechampia scandens*. An "evolvability" of 1% means that the response to directional selection of the same strength as on fitness itself would be 1% per generation. Generally the heritabilities and evolvabilities are unrelated, with the exception of a few shape characters in the lower left part of the plot that were practically devoid of genetic variation. (b) Effects of conditioning evolvabilities on two functionally important traits (GA is size of a gland that produce resins as a pollinator reward, and UBW is the size of an involucral bract with an important protective function; see also Armbruster et al., Chapter 2, this volume for a description of traits). The graph shows reduction in evolvability due to conditioning on either GA alone, or on both traits together. From Hansen et al. (2003a, 2003b).

response if selection was to be causally as strong as on fitness itself. The poor correlation between $I_A$ and $h^2$ (e.g., Houle 1992) means that judgments of evolvability based on the latter are irrelevant. Houle (1992) found that the conclusion that life-history traits were less evolvable than morphological traits (e.g., Gustafsson 1986; Mousseau and Roff 1987) was turned on its head if evolvability was measured by coefficients of additive genetic variation ($CV_A = \sqrt{I_A}$) instead of heritability. Hansen et al. (2003a) found that the evolvabilities ($I_A$) of floral traits in a *Dalechampia scandens* population were typically less than half a percent (i.e., response to selection as strong as on fitness itself would be less than half a percent per generation), a significant constraint on adaptation to potentially rapidly changing pollination regimes. This limitation is not apparent from $h^2$ alone.

## Genetic Constraints and Long-Term Evolvability

Regardless of how variation is measured, it might appear that the constraint hypothesis is falsified for traits such as Dipteran wing shape or mammalian body temperature, which are conserved on higher taxonomic levels, as there arguably are detectable levels of additive genetic variation in each case (see Lynch 1994 for BT and references above for wings). At sufficiently long time scales, any level of replenishable variation could be turned into monumental changes. On longer time scales it is variability, the capacity of traits to vary (Wagner and Altenberg 1996), rather than the standing level of variation that dictates evolvability. The requirement that new genetic variation is produced by mutation is fulfilled for at least some aspects of wing variation in *Drosophila* (Santiago et al. 1992; Houle, unpublished). More generally, studies of a variety of traits in many organisms indicate that a fair amount of mutational variability is generated each generation (Lynch 1988; Houle et al. 1996; Lynch et al. 1999).

As with additive genetic variation, mutational variation can be measured on a variance-standardized scale (i.e., mutational heritability) or on a mean-standardized scale (Houle et al. 1996; Houle 1998). It is instructive to consider the level of evolvability in wing morphology that can be maintained through mutation. A mutation-accumulation study by Santiago et al. (1992) showed that the mutational heritability of wing traits was a typical 0.001; however, estimates of mean-standardized mutational variance for wing morphology was $\sim 2 \times 10^{-6}$ (0.0002%), an unusually low value (cf. Houle et al. 1996). This means that long-term directional selection of similar strength as selection on fitness would only be able to change the traits by 0.0002% per generation once standing variation from the base population is exhausted. This suggests fairly strong genetic constraints on wing shape. Once again, mutational heritabilities give a misleading picture of potential evolvability.

It is thus debatable to what time scales genetic constraints may extend. It seems likely that a lack of mutational and standing genetic variation may sometimes limit evolution on ecological time scales, but it is also reasonably clear that stasis on a true geological time scale, comprising millions of generations, cannot be explained in this way. Doubling of wing trait value at the low rate of evolution calculated above would take only 150,000 generations, or perhaps 15,000 years, for a *Drosophila melanogaster* population. If stasis in macroevolution is caused by constraints, these have to render even the low level of evolvability revealed by $I_A$ evolutionarily irrelevant. The crucial question to which we now turn is whether genetic variation can be translated into adaptive changes without compromising other aspects of organismal function.

## Variational Constraints Due to Functional Architecture

The variational properties of a character are determined by its genotype-phenotype map—its underlying functional architecture. The functional architecture is the collection of pathways that lead from the genes to the character (Houle 1991, 2001). Thus, a deeper understanding of constraint must come from an understanding of the functional architecture. This includes the relationships among different characters, and the way in which genes are combined in the mapping from genotype to phenotype (termed functional epistasis by Hansen and Wagner 2001). Below, we discuss in more detail two forms of constraints due to functional architecture. Pleiotropic constraints stem from interactions among characters; epistatic constraints stem from interactions among genes. In each case we evaluate their potential as explanations of stasis, and suggest ways to operationalize the concepts to make them empirically accessible.

### Pleiotropic Constraints and Conditional Evolvability

All too many assessments of character evolvability are done in isolation. In such situations it is customary to point out that the character in question is, or is likely to be, genetically variable. One of the assumptions of this line of argument is that the character has a property that Lewontin (1978) called *quasi-independence,* meaning that it is possible to change the character without unduly disturbing other aspects of the organism.

Such independence may be compromised by pleiotropy, where the genes that create variation in a focal character also create variation in other characters. If the normal state of affairs is that most characters are under stabilizing selection most of the time, the evolvability of the focal character will often be compromised by pleiotropy. This possibility may also be viewed as a potential cause of apparent stabilizing selection, as pointed out above. There is an urgent need to assess how severe such pleiotropic constraints may be, both for standing and mutational variation. To do this we need to devise measures of evolvability that can be used to assess how much of the variation is effectively available to produce particular adaptations.

The concept of conditional evolvability is such an attempt (Hansen et al. 2003b; Hansen 2003). Conditional evolvability is defined as the evolvability of a character $y$ in the event that a set of constraining characters $x$ is not allowed to change. Under the usual assumptions of the Lande equation, conditional evolvability is determined by the conditional genetic variance (i.e., the residual variance when the genetic value of $y$ is regressed on the genetic value of $x$). It can be further shown that the conditional variance is approximately valid as a predictor of evolvability regardless of the strength of stabilizing selection on $x$ (Hansen

2003). When directional selection is placed on the focal character, the constraining characters are displaced from their optima by a small amount, and after a few generations the strength of stabilizing selection only affects the size of this displacement and not the degree of constraint on the focal character.

The conditional evolvability is always relative to the particular set of constraining characters that are included in $x$. It may for example be used to assess the constraining effect of particular characters that are believed to be under stabilizing selection. In Fig. 6.2b we show how the evolvabilities of the floral traits in the *Dalechampia scandens* population mentioned above could be further reduced by conditioning on two functionally important floral characters that may be under stabilizing selection (Hansen et al. 2003b; see also Armbruster et al., Chapter 2, this volume).

Conditional evolvability may also be assessed on some measure of background fitness. This will efficiently capture the organism-level pleiotropic constraints generated by a large number of characters that are not directly studied. Weber (1996) provided an interesting example of this sort of reasoning when he assessed the fitness consequences of his artificial-selection experiment on wind-tunnel flight performance in *Drosophila*. Selection over 100 generations resulted in a response of over 30 standard deviations in performance (see Weber's Fig. 2), but competitive fitness dropped by only 6%, although the confidence limits on this estimate are wide. Weber concludes that performance in *Drosophila* is relatively unconstrained by pleiotropy. It certainly seems unlikely that such a moderate fitness reduction provides an insurmountable constraint. As for body temperature, it is unfortunate that no one has yet taken up Williams's (1992) excellent suggestion of selecting on body temperature in mice.

Conditioning on background fitness may be particularly useful in assessing the adaptive potential of mutational variation. Mutational variation is likely to hold a much higher fraction of useless variation than standing genetic variation, as it is not yet filtered by selection (Houle et al. 1996; Houle 1998). It is possible that a large majority of mutational variation is due to mutations in housekeeping and general regulatory genes that are expressed in a multitude of tissues and circumstances. This may appear as variability in individual characters, but is hardly good material for building adaptations. Galis (1999) suggests an intriguing example, where apparent genetic variability in the number of mammalian neck vertebrae is rendered useless by pleiotropic effects that greatly elevate cancer risk (see also Galis and Metz 2001; Galis et al. 2001 for further examples). Stern (2000) made the interesting suggestion that adaptively useful mutations are largely limited to regulatory elements that have temporally and spatially restricted effects, which may be less pleiotropically constrained than other mutations. At present the quality of mutational variation is basically unknown. Assessing the amount of conditional character evolvability provided by mutation is essential to evaluate the pleiotropic-constraint hypothesis.

The idea of conditional evolvability is related to a variety of proposals concerning "internal" selection pressures (see Wagner and Schwenck 2000 for review). The logic of these hypotheses is that the complexity of development links the variational properties of characters to each other. Some key characters may then become so entrenched in development that they completely lose their variational freedom. Riedl (1977, 1978) proposed entrenchment, or "developmental burden," as an explanation for deep invariant homologies such as the vertebra in vertebrates.

Mammalian body temperature (BT) is a plausible example of this sort of entrenchment. Since the function of every protein is likely to be affected by a change in BT, evolution of a different BT might require adaptation throughout the genome. This may equally well be viewed as a selective constraint, as there is no conceptual difference between adaptation to the internal and the external environment. It is, however, not clear whether this mechanism can generate an absolute constraint on the position of the optimum. Mammalian tissues are tolerant to minor variations in temperature, and are able to continue operation within a few degrees of the normal BT. In other words, the physiological or ecological constraints are not absolute. This means that the optimum should be at least somewhat negotiable. A large change in, say, external temperature should produce at least a small shift in the BT optimum. After this, the physiology should be able to adapt to the new optimum, making possible another shift in the direction of ecological adaptation, and so on. In general, continuous variation combined with plasticity may make adaptive optima inherently responsive to external change. Again, formal models need to be developed to assess the potential for entrenchment of quantitative characters due to interdependence with numerous secondary factors that may be either ecological (external) or physiological (internal).

The idea of entrenchment is based on the assumption that complex interdependence among traits is a constraint. This assumption has its theoretical basis in Fisher's (1930, pp. 38–41) geometrical model of adaptation. Fisher's model was an attempt to formalize the notion that a random change in a complex apparatus is less likely to improve function than a random change in a relatively simple one. Fisher showed that the probability of a mutation being advantageous was a decreasing function of the dimensionality of the trait. This result, however, needs qualification. If we take the perspective of the evolvability of an individual trait, pleiotropic links to other characters may act as a constraint, but also serve to increase the mutational target size and therefore the evolvability of the trait. Hansen (2003) investigated this tradeoff between selective constraints and evolvability in formal models and showed that an intermediate level of pleiotropy was likely to maximize the independent evolvabilities of characters.

A neglected approach to the detection of pleiotropic constraints is the estimation of the dimensionality of phenotypic variation (reviewed in Houle 2001; Steppan et al. 2002). Multivariate variation may exist for only a few kinds of change in form, and be absent in all others. In such a case, evolution could occur only along these few variable directions. Quantitatively it may be far more likely for evolution to proceed along relatively variable aspects of the phenotype, the "genetic lines of least resistance," than less variable ones (Schluter 1996). Indeed, several studies have found that a large amount of multivariate genetic variation within and among species is concentrated in only a few aspects of form (Kirkpatrick and Lofsvold 1992; Björklund 1996; Schluter 2000, chap. 9). For wing shape in *Drosophila melanogaster*, however, genetic variance in wing shape is distributed across all measured aspects of form (Mezey and Houle, unpublished). This perhaps renders pleiotropic constraint less plausible as an explanation of stasis in wing shape.

## Epistatic Constraints and the Evolution of Evolvability

In an influential nineteenth-century critique of Darwin, Fleeming Jenkin argued that although selection may bring a species out toward the limit of its natural variation, it would get stuck, as there would be no more useful variation in the appropriate direction (see Gould 2002, p.142). Jenkin thus imagined a species as a fixed sphere of variation around some immutable essence. While Jenkin's reasoning is a reflection of a flawed typological species concept, the hypothesis that the range of possible variation is limited is still plausible. Indeed such limits are inherent in any mutation model based on a finite number of alleles with fixed effects (e.g., Zeng et al. 1989). Empirically, Mackay et al. (1995) found that the divergence of *Drosophila* in bristle number between lines accumulating spontaneous mutation reached a plateau after less than two hundred generations. This notion of limits can be contrasted with its opposite extreme, the additive model of quantitative genetics (e.g., Lande 1976). In the additive model, the distribution of the effects of new mutations on the phenotype is independent of the genetic background. Thus, when selection moves a character toward more extreme values, the variational properties of the genome stay the same. This implies that new variation can thus be produced ad infinitum, and the sphere of variation moves along with the population mean.

In between these extremes of fixed and unlimited ranges of potential variation lies the probably more realistic idea that the range of variation is itself evolvable. The genotype-phenotype map is a very complex function, where the correct functioning of large numbers of genes is necessary to achieve a particular phenotype. This implies that functional epistatic interactions among genes must exist. By definition, epistasis means that the effects of alleles at a focal locus will change with changes in the rest of the genotype, the genetic background. Any sort of evolutionary change in the genetic background may affect the variation expressed by variants at a particular locus (Hansen and Wagner 2001). Epistasis thus makes evolvability evolvable. Note, however, that this evolution could either enhance or diminish the potential for adaptation. Thus another potential explanation for stasis is that epistatic interactions tend to restrict variation under selection.

If genes interact in such a way as to diminish each other's effects as the population is selected in a particular direction, then evolvability will be diminished when natural selection pushes the population in that direction. We call this negative epistasis. In contrast, if genes interact positively by mutually reinforcing each other, evolvability will be enhanced. If negative epistasis dominates in all directions in phenotype space, then a form of Jenkin's sphere may be generated, as the variability will be diminished when the population moves to more extreme character values. We call this an epistatic constraint. Note that a population subject to an epistatic constraint might possess ample genetic variance and mutational variability in each trait, but the additive and mutational variance will be scaled down or become biased as selection changes the character mean.

Formal equations describing these effects in the case of multilinear epistasis are given in Hansen and Wagner (2001). Epistasis may generally be described by a set of epistasis factors, $f$, that describes how a change in the genetic background changes the effect of any particular gene substitution. If $\delta$ is the effect of the gene substitution in some reference genotype, then $f\delta$ is the effect in an alternative genetic background. The epistasis factor is thus a function of the genetic background. If $f < 1$ with respect to a particular perturbation, then epistasis is negative, while $f > 1$ corresponds to positive epistasis. The average values of the epistasis factors with respect to particular genetic pertubances are therefore the parameters we need to predict the evolution of evolvability.

The hypothesis of epistatic constraints is thus operational, and can be tested by estimation of epistasis factors in real genetic architectures. We are currently preparing to do this for fly wing characters. Estimation of epistasis factors requires the creation of different genetic backgrounds followed by a comparison of the effects of known genetic elements in the different backgrounds. To test the epistatic-constraint hypothesis, backgrounds with extreme character values need be compared to backgrounds with less extreme values. These backgrounds may be created through artificial selection, for example.

The question that emerges from the epistatic-constraint hypothesis is whether we should expect genetic architectures to exhibit negative epistasis. We may distinguish two subquestions. First, are there system-theoretic reasons why functional architectures should produce particular patterns of epistasis? Second, does evolution lead to functional architectures with particular patterns of epistasis? Despite the novelty of these questions, there are some relevant observations.

With regard to the first question, there are certainly aspects of functional architectures that lead to particular patterns of interaction among components (e.g., Kauffman 1993). For example, if there is widespread redundancy in the genotype-phenotype map, we may expect negative epistasis, as there are many subsystems that can produce adaptively relevant variation but when change is achieved through one of these, the variability of the others becomes adaptively irrelevant. The theoretical study of epistasis in model architectures, as exemplified by Szathmáry's (1993) study of epistasis in metabolic pathways or Gibson's (1996) study of epistasis in gene regulation, may prove illuminating here.

The evolution of functional architectures has usually been addressed in terms of the evolution of canalization or robustness. The phenomenon of genetic assimilation, first demonstrated in fly wing morphology (Waddington 1953), shows that increased evolvability can evolve when directional selection is applied in a novel environment. Indeed, when we conclude that the wild type is genetically canalized, we are saying that the epistasis factors are positive with respect to perturbations of the wild-type genotype. The epistatic-constraint hypothesis thus runs into the difficulty of positing a (relatively) decanalized wild type, so that there is room to diminish the variability when selection perturbs the population. Evidence for canalization of the wild type has come from a series of empirical studies demonstrating that severe mutational or environmental disturbances can release "hidden" genetic variation (reviewed in Scharloo 1991; Moreno 1994; Gibson and Wagner 2000; Rutherford 2000). The changes in genetic background that produce such decanalization are usually extreme, however, and likely to be associated with severe deleterious side effects, making them less relevant to microevolutionary changes. Similarly, the effects detected in such studies are themselves very large, leaving it unclear whether alleles of small effect are also unmasked by decanalizing perturbations.

Since Waddington it has been thought that stabilizing selection may lead to the evolution of canalization. This prediction is supported by some models (Wagner et al. 1997; Rice 1998), but the effects are weak. Hermisson et al. (2003) have further shown that these results may not hold with more realistic representations of multi-locus dynamics. Although there is selection for canalization at individual loci, the interactions among loci means that many loci will still become decanalized. Loci with the highest mutation rates are usually canalized, but the net phenotypic effect depends on how genes interact and stabilizing selection may well produce a somewhat decanalized phenotype with high mutational variability.

The epistatic-constraint hypothesis is thus at least tenable, but needs to be further evaluated with modeling and empirical studies of functional epistasis within the range of "normal" variation. It is particularly important to consider the effects of functional epistatic architecture on the evolution of pleiotropic patterns, that is, on the evolution of pleiotropic constraints. It has been proposed that the functional architecture may evolve to resemble the adaptive landscape, either through correlated stabilizing selection (Olson and Miller 1958; Cheverud 1984, 2001), or through correlated shifting directional selection (Wagner 1996). The mechanisms behind these hypotheses need more study and this will necessarily involve the study of patterns of epistasis.

## Conclusions

Character conservation remains an important challenge for evolutionary theorists. None of the many hypotheses we have considered provides a completely satisfactory explanation for stasis in our two examples: fly wing shape and mammalian body temperature. Stabilizing selection undoubtedly contributes to stasis, but alone is an insufficient explanation for it. Selective explanations of stasis will need to focus on the dynamics of adaptive landscapes (Simpson 1944; Arnold et al. 2001), and it will be necessary to further develop and test ideas for how adaptive optima themselves can be stabilized. Currently, variational constraints can neither be generally rejected nor generally accepted as an explanation of stasis. The constraint hypothesis has suffered from a certain vagueness, and perhaps from its association with the untenable notion that evolution only happens in association with speciation. The recent interest in evolvability (e.g., Raff 1996; Wagner and Altenberg 1996; Gerhart and Kirschner 1997) provides a new perspective on constraints that may prove helpful. As outlined above, the theory of evolvability is in the process of being operationalized on several levels, and may eventually provide the tools for assessing the prevalence and power of constraint in both micro- and macroevolution.

*Acknowledgments* We thank Ashley Carter, Massimo Pigliucci, and Katherine Preston for helpful comments on this chapter.

*Literature Cited*

Ancel, L. W. 2000. Undermining the Baldwin expediting effect: does phenotypic plasticity accelerate evolution? Theor. Pop. Biol. 58:307–319.

Arnold, S. J. 1992. Constraints on phenotypic evolution. Am. Nat. 140:S85–S107.

Arnold, S. J., M. E. Pfrender, and A. G. Jones. 2001. The adaptive landscape as a conceptual bridge between micro- and macroevolution. Genetica 112/113: 9–32.

Baldwin, J. M. 1896. A new factor in evolution. Am. Nat. 30:354–451.

Bell, M. A. 1989. Stickleback fishes: bridging the gap between population biology and paleobiology. Trends Ecol. Evol. 3:320–325.

Björklund, M. 1996. The importance of evolutionary constraints in ecological time scales. Evol. Ecol. 10:423–431.

Bradshaw, A. D. 1991. Genostasis and the limits to evolution. Phil. Trans. R. Soc. Lond. B 333:289–305.

Cavicchi, S., G. Giorgi, and C. Pezzoli. 1981. Effects of environmental temperature on developmental patterns in *Drosophila melanogaster* (Meig). Boll. Zool. 48:227–231.

Cavicchi, S., D. Guerra, G. Giorgi, and C. Pezzoli. 1985. Temperature-related divergence in experimental populations of *Drosophila melangaster*. I. Genetic and developmental basis of wing size and shape variation. Genetics 109:665–689.

Charlesworth, B., R. Lande, and M. Slatkin. 1982. A neodarwinian commentary on macroevolution. Evolution 36:474–498.

Cheverud, J. M. 1984. Quantitative genetics and developmental constraints on evolution by selection. J. Theor. Biol. 110:155–171.

Cheverud, J. M. 2001. The genetic architecture of pleiotropic relations and differential epistasis. Pp. 411–433 in G. P. Wagner (ed.), The Character Concept in Evolutionary Biology. Academic Press, San Diego.

Cooke, F., P. D. Taylor, C. M. Francis, and R. F. Rockwell. 1990. Directional selection and clutch size in birds. Am. Nat. 136:261–267.

Coyne, J. A., and E. Beecham. 1987. Heritability of two morphological characters within and among natural populations of *Drosophila melanogaster*. Genetics 117: 727–737.

Eldredge, N. 1999. The Pattern of Evolution. Freeman, New York.

Eldredge, N., and S. J. Gould. 1972. Punctuated equilibria: an alternative to phyletic gradualism. Pp. 82–115 in T. J. M. Schopf (ed.), Models in Paleobiology. Freeman, San Francisco.

Endler, J. A. 1986. Natural Selection in the Wild. Monographs in Population Biology 21. Princeton University Press, Princeton, NJ.

Fear, K. K., and T. Price. 1998. The adaptive surface in ecology. Oikos 82:440–448.

Fisher, R. A. 1930. The Genetical Theory of Natural Selection. Oxford University Press, London. A complete variorum edition, 1999.

Frank, S. A., and M. Slatkin. 1992. Fisher's fundamental theorem of natural selection. Trends Ecol. Evol. 7:92–95.

Galis, F. 1999. Why do almost all mammals have seven cervical vertebrae? Developmental constraints, Hox genes, and cancer. J. Exper. Zool. (MDE) 285:19–26.

Galis, F., and J. A. J. Metz. 2001. Testing the vulnerability of the phylotypic stage: on modularity and evolutionary conservation. J. Exper. Zool. (MDE) 291:195–204.

Galis, F., J. M. van Alphen, and J. A. J. Metz. 2001. Why five fingers? Evolutionary constraint on digit numbers. Trends Ecol. Evol. 16:637–646.

Galpern, P. 2000. The use of common principal component analysis in studies of phenotypic evolution: an example from the Drosophilidae. M.S. thesis, Department of Zoology, University of Toronto, Toronto, Canada.

Gerhart, J., and M. Kirschner. 1997. Cells, Embryos and Evolution: Toward a Cellular and Developmental Understanding of Phenotypic Variation and Evolutionary Adaptability. Blackwell, Oxford.

Gibson, G. 1996. Epistasis and pleiotropy as natural properties of transcriptional regulation. Theor. Pop. Biol. 49:58–89.

Gibson, G., and G. P. Wagner. 2000. Canalization in evolutionary genetics: a stabilizing theory? BioEssays 22:372–380.

Gilchrist, G. W., R. B. Huey, and L. Serra. 2001. Rapid evolution of wing size clines in *Drosophila subobscura*. Genetica 112/113:273–286.

Gilchrist, A. S., R. B. R. Azevedo, L. Partridge, and P. O'Higgins. 2000. Adaptation and constraint in the evolution of *Drosophila melanogaster*. Evol. and Dev. 2:114–124.

Gould, S. J. 2002. The Structure of Evolutionary Theory. Belknap Press, Cambridge, MA.

Gould, S. J., and N. Eldredge. 1977. Punctuated equilibria: the tempo and mode of evolution reconsidered. Paleobiology 3:115–151.

Gould, S. J., and N. Eldredge. 1993. Punctuated equilibrium comes of age. Nature 366:223–227.

Gustafsson, L. 1986. Lifetime reproductive success and heritability: empirical support for Fisher's fundamental theorem. Am. Nat. 158:761–764.

Hansen, T. F. 1997. Stabilizing selection and the comparative analysis of adaptation. Evolution 51:1341–1351.

Hansen, T. F. 2003. Is modularity necessary for evolvability? Remarks on the relationship between pleiotropy and evolvability. Biosystems 69:83–94.

Hansen, T. F., and G. P. Wagner. 2001. Modeling genetic architecture: a multilinear model of gene interaction. Theor. Pop. Biol. 59:61–86.

Hansen, T. F., C. Pelabon, W. S. Armbruster, and M. L. Carlson. 2003a. Evolvability and constraint in *Dalechampia* blossoms: components of variance and measures of evolvability. J. Evol. Biol. 16:754–765.

Hansen, T. F., W. S. Armbruster, M. L. Carlson, and C. Pelabon. 2003b. Evolvability and genetic constraint in *Dalechampia* blossoms: genetic correlations and conditional evolvability. J. Exper. Zool. (MDE) 296B:23–39.

Hendry, A. P., and M. T. Kinnison. 1999. The pace of modern life: measuring rates of contemporary microevolution. Evolution 53:1637–1653.

Hermisson, J., T. F. Hansen, and G. P. Wagner. 2003. Epistasis in polygenic traits and the evolution of genetic architecture. Am. Nat. 161:708–734.

Hoffman, A. 1989. Arguments on Evolution: A Paleontologist's Perspective. Oxford University Press, New York.

Holt, R. D. 1996. Demographic constraints in evolution: toward unifying the evolutionary theories of senesence and niche conservativism. Evol. Ecol. 10:1–11.

Holt, R. D. and M. S. Gaines. 1992. Analysis of adaptation in heterogeneous landscapes: implications for the evolution of fundamental niches. Evol. Ecol. 6:433–447.

Houle, D. 1991. Genetic covariance of fitness correlates: what genetic correlations are made of and why it matters. Evolution 45:630–648.

Houle, D. 1992. Comparing evolvability and variability of quantitative traits. Genetics 130:195–204.

Houle, D. 1998. How should we explain variation in the genetic variance of traits? Genetica 102/103:241–253.

Houle, D. 2001. Characters as the units of evolutionary change. Pp. 109–140 in G. P. Wagner (ed.), The Character Concept in Evolutionary Biology. Academic Press, San Diego.

Houle, D., B. Morikawa, and M. Lynch. 1996. Comparing mutational variabilities. Genetics 143:1467–1483.

Huey, R. B., G. W. Gilchrist, M. L. Carlson, D. Berrigan, and L. Serra. 2000. Rapid evolution of a geographic cline in size in an introduced fly. Science 287:308–309.

Kambysellis, M. P., and E. M. Craddock. 1997. Ecological and reproductive shifts in the diversification of the endemic Hawaiian *Drosophila*. Pp. 475–509 in T. J. Givnish and K. J. Sytsma (eds.), Molecular Evolution and Adaptive Radiation. Cambridge University Press, Cambridge.

Kauffman, S. A. 1993. The Origins of Order: Self-Organization and Selection in Evolution. Oxford University Press, New York.

Kawecki, T. J. 1995. Demography of source-sink populations and the evolution of ecological niches. Evol. Ecol. 9:38–44.

Kingsolver, J. G., H. E. Hoekstra, J. M. Hoekstra, D. Berrigan, S. N. Vignnieri, C. E. Hill, A. Hoang, P. Gibert, and P. Beerli, P. 2001. The strength of phenotypic selection in natural populations. Am. Nat. 157:245–261

Kinnison, M. T., and A. P. Hendry. 2001. The pace of modern life. II. From rates of contemporary microevolution to pattern and process. Genetica 112/113:145–164.

Kirkpatrick, M., and D. Lofsvold. 1992. Measuring selection and constraint in the evolution of growth. Evolution 46:954–971.

Lande, R. 1976. The maintenance of genetic variability by mutation in a polygenic character with linked loci. Genet. Res. 26:221–235.

Lande, R. 1979. Quantitative genetic analysis of multivariate evolution, applied to brain:-body size allometry. Evolution 33:402–416.

Lande, R. 1985. Expected time for random genetic drift of a population between stable phenotypic states. Proc. Natl. Acad. Sci. U.S.A. 82:7641–7645.

Lande, R. 1986. The dynamics of peak shifts and the pattern of morphological evolution. Paleobiology 12:343–354.

Lande, R., and S. J. Arnold. 1983. The measurement of selection on correlated characters. Evolution 37:1210–1226.

Lewontin, R. C. 1978. Adaptation. Sci. Am. 239:212–231.

Lieberman, B. S., and S. Dudgeon. 1996. An evaluation of stabilizing selection as a mechanism for stasis. Paleaeogeogr., Paleaeoclimatol., Palaeoecol. 127:229–238.

Lynch, C. B. 1994. Evolutionary inferences from genetic analyses of cold adaptation in laboratory and wild populations of the house mouse. Pp. 278–304 in C. R. B. Boake (ed.), Quantitative Genetic Studies of Behavioral Evolution. University of Chicago Press, Chicago.

Lynch, M. 1988. The rate of polygenic mutation. Genet. Res. Camb. 51:137–148.

Lynch, M. 1990. The rate of morphological evolution in mammals from the standpoint of the neutral expectation. Am. Nat. 136:727–741.

Lynch, M. 1993. Neutral models of phenotypic evolution. Pp. 86–108 in L. Real (ed.), Ecological Genetics. Princeton University Press, Princeton, NJ.

Lynch, M., and W. G. Hill. 1986. Phenotypic evolution by neutral mutation. Evolution 40:915–935.

Lynch, M., J. Blanchard, D. Houle, T. Kibota, S. Schultz, L.Vassilieva, and J. Willis. 1999. Perspective: Spontaneous deleterious mutation. Evolution 53:645–663

Mackay, T. F. C., R. F. Lyman, and W. G. Hill. 1995. Polygenic mutation in *Drosophila melanogaster*: non-linear divergence among unselected strains. Genetics 139:849–859.

Maynard Smith, J. 1983. The genetics of stasis and punctuation. Annu. Rev. Genet. 17:11–25.

Maynard Smith, J., R. Burian, S. Kauffman, P. Alberch, J. Campbell, B. Goodwin, R. Lande, D. Raup, and L. Wolpert. 1985. Developmental constraint and evolution. Quart. Rev. Biol. 60:265–287.

McAlpine, J. F. 1989. Phylogeny and classification of the Muscomorpha. Pp. 1397–1518 in J. F. McAlpine and D. M. Wood (eds.), Handbook of Nearctic Diptera, Vol. 3. Research Branch of Agriculture Canada, Monograph No. 32, Ottawa.

Merilä, J., B. C. Sheldon, and L. E. B. Kruuk. 2001. Explaining stasis: microevolutionary studies in natural populations. Genetica 112/113:199–222.

Mitchell, W. A., and T. J. Valone. 1990. The optimization research program: studying adaptations by their function. Quart. Rev. Biol. 65:43–52.

Moreno, G. 1994. Genetic architecture, genetic behavior, and character evolution. Annu. Rev. Ecol. Syst. 25:31–45.

Mousseau, T. A., and D. A. Roff. 1987. Natural selection and the heritability of fitness components. Heredity 59:181–197.

Olson, E. C. and Miller, R. L. 1958. Morphological Integration. University of Chicago Press, Chicago.

Parker, G. A., and J. Maynard Smith. 1990. Optimality theory in evolutionary biology. Nature 348:27–33.

Powell, J. R. 1997. Progress and Prospects in Evolutionary Biology: The *Drosophila* Model. Oxford University Press, New York.

Price T., and L. Liou. 1989. Selection on clutch size in birds. Am. Nat. 134:950–959.

Raff, R. A. 1996. The Shape of Life: Genes, Development, and the Evolution of Animal Form. University of Chicago Press, Chicago.

Rice S. H. 1998. The evolution of canalization and the breaking of von Baer's laws: modeling the evolution of development with epistasis. Evolution 52:647–656.

Riedl, R. J. 1977. A systems-analytical approach to macro-evolutionary phenomena. Quart. Rev. Biol. 52:351–370.

Riedl, R. J. 1978. Order in Living Organisms: A Systems Analysis of Evolution. Wiley, New York.

Robertson, F. W. 1959. Studies in quantitative inheritance. XII. Cell size and number in relation to genetic and environmental variation of body size in *Drosophila*. Genetics 44:869–896.

Robertson, F. W., and E. C. R. Reeve. 1952. Studies in quantitative inheritance. I. The effects of selection for wing and thorax length in *Drosophila melanogaster*. J. Genet. 50:416–448.

Robinson, B. W., and R. Dukas. 1999. the influence of phenotypic modifications on evolution: the Baldwin effect and modern perspectives, Oikos 85:582–89.

Rutherford, S. L. 2000. From genotype to phenotype: buffering mechanisms and the storage of genetic information. BioEssays 22:1095–1105.

Santiago, E., J. Albornoz, A. Dominguez, M. A. Toro, and C. López-Fanjul. 1992. The distribution of spontaneous mutations on quantitative traits and fitness in *Drosophila melanogaster*. Genetics 132:771–781.

Scharloo, W. 1991. Canalization: genetic and developmental aspects. Annu. Rev. Ecol. Syst. 22:65–93.

Schluter, D. 1996. Adaptive radiation along genetic lines of least resistance. Evolution 50:1766–1774.

Schluter, D. 2000. The Ecology of Adaptive Radiation. Oxford University Press, Oxford.

Sheldon, P. R. 1996. Plus ça change—a model for stasis and evolution in different environments. Paleaeogeog., Paleaeoclimatol., Palaeoecol. 127:209–227.

Simpson, G. G. 1944. Tempo and Mode in Evolution. Columbia University Press, New York.

Stenseth, N. C., and J. Maynard Smith. 1984. Coevolution in ecosystems: red queen evolution or stasis? Evolution 38:870–880.

Steppan, S. J., P. C. Phillips, and D. Houle. 2002. Comparative quantitative genetics: evolution of the G matrix. Trends Ecol. Evol. 17:320–327.

Stern, D. 2000. Perspective: Evolutionary developmental biology and the problem of variation. Evolution 54:1079–1091.

Szathmáry, E. 1993. Do deleterious mutations act synergistically? Metabolic control theory provides a partial answer. Genetics 133:127–132

Travis, J. 1989. The role of optimizing selection in natural populations. Annu. Rev. Ecol Syst. 20:279–296.

van Tienderen, P. M. 2000. Elasticities and the link between demographic and evolutionary dynamics. Ecology 81:666–679.

Van Valen, L. M. 1982. Integration of species, stasis and biogeography. Evol. Theory 6:99–112.

Venables, W. N., and B. D. Ripley. 1994. Modern Applied Statistics with S-Plus. Springer-Verlag, New York.

Waddington, C. H. 1953. Genetic assimilation of an acquired character. Evolution 7:118–126

Wagner, G. P. 1986. The systems approach: an interface between development and population genetic aspects of evolution. Pp. 149–165 in D. M. Raup and D. Jablonski (eds.), Patterns and Processes in the History of Life. Springer-Verlag, Berlin.

Wagner, G. P. 1996. Homologues, natural kinds and the evolution of modularity. Am. Zool. 36:36–43.

Wagner, G. P., and L. Altenberg. 1996. Complex adaptations and evolution of evolvability. Evolution 50:967–976.

Wagner, G. P., and K. Schwenk. 2000. Evolutionarily stable configurations: functional integration and the evolution of phenotypic stability. Evol. Biol. 31:155–217.

Wagner, G. P., G. Booth, and H. Bagheri-Chaichian. 1997. A population genetic theory of canalization. Evolution 51:329–347.

Wake, D. B., G. Roth, and M. Wake. 1983. On the problem of stasis in organismal evolution. J. Theor. Biol. 101:211–224.

Weber, K. E. 1990. Selection on wing allometry in *Drosophila melanogaster*. Genetics 126:975–989.

Weber, K. E. 1992. How small are the smallest selectable domains of form? Genetics 130:345–353.

Weber, K. E. 1996. Large genetic changes at small fitness cost in large populations of *Drosophila melanogaster* selected for wind tunnel flight: rethinking fitness surfaces. Genetics 144:205–213.

Weber, K., R. Eisman, L. Morey, A. Patty, J. Sparks, M. Tausek, and Z.-B. Zeng. 1999. An analysis of polygenes affecting wing shape on chromosome 3 in *Drosophila melanogaster*. Genetics 153:773–786.

Williams, G. C. 1992. Natural Selection: Domains, Levels, and Challenges. Oxford University Press, Oxford.

Williamson P. G. 1987. Selection or constraint? A proposal on the mechanism for stasis. Pp. 129–142 in M. F. Campbell (ed.), Rates of Evolution. Allen & Unwin, London.

Zeng, Z.-B., H. Tachida, and C. C. Cockerham. 1989. Effects of mutation on selection limits in finite populations with multiple alleles. Genetics 122:977–984.

## PART II

# PHENOTYPIC PLASTICITY AND INTEGRATION

Phenotypic plasticity is conceptually distinct from phenotypic integration, yet research on these two aspects of organismal complexity has in many ways evolved along parallel trajectories. While both ideas are very old (the concept of plasticity goes back to the beginning of the twentieth century: Woltereck 1909; that of integration to the middle of the same century: Clausen et al. 1940; Olson and Miller 1958), they have both largely been ignored or have played a background role in evolutionary theory and practice until the last few decades (Schlichting and Pigliucci 1998, chaps. 3 and 7).

In some sense, this long gestation was to be expected: both plasticity and integration concern complex aspects of the phenotypes of organisms, and even the first steps toward addressing them require a sophisticated set of conceptual, analytical, and investigative tools. Unless earlier evolutionary ecologists and geneticists had sharpened their thinking on simpler problems, such as the genetic basis and evolution of relatively straightforward Mendelian or simple quantitative traits, we would surely not have the vocabulary to include either plasticity or integration into the modern evolutionary synthesis.

In both plasticity and integration studies, recent progress has been made possible by the introduction of new conceptual frameworks as well as innovative statistical techniques (both topics that, as far as integration is concerned, will be discussed in Part V of this book). For example, we had to wait until the development of the so-called new morphometrics (Rohlf and Marcus 1993; Marcus et al. 1996) to be able to apply quantitative analyses to the beautiful idea of "grids of morphological change" introduced by D'Arcy Thompson at the beginning of the twentieth century (Thompson 1917). Similarly, it was not until Wagner's introduction of concepts such as parcellation or the distinction between variation and variability (Wagner 1995; Wagner and

Altenberg 1996) that it has been possible to place studies of integration firmly in the logical space of efforts to understand both micro- and macroevolution.

This section of the book explores the relationship between plasticity and integration, for example through the treatment of the plasticity of integration (Schlichting 1989), that is, the phenomenon that whatever measure of integration we devise (e.g., the pattern of genetic covariances among a given set of traits), it can be affected by the specific environment in which we measure it, that is, it is plastic. Phenotypic plasticity becomes important when one focuses on the concerted response of a suite of traits to ecologically relevant environmental conditions, as in the case of shade avoidance in plants (e.g., Schmitt et al. 1999; Weinig 2000; Donohue et al. 2001), or predator-induced morphological shifts in animals (e.g., Lively 1986; Reznick et al. 2001; Van Buskirk 2002), some instances of which are discussed below in Chapter 8 by Relyea. In other words, what is particularly fruitful to study is the plasticity of functionally integrated characters. Examples include diet-induced alterations in character correlations in animals (e.g., Greene 1989; Emlen 1997; Thompson 1999), as well as the plasticity of components of growth habit in plants (e.g., van Tienderen and Hinsberg 1996).

Part II comprises two chapters: the first one (Chapter 7) by one of the editors, Massimo Pigliucci, the other (Chapter 8) by Rick Relyea. Pigliucci's goal is to discuss the use of model systems, such as the plant *Arabidopsis thaliana* in the study of integration. He does so through the presentation of two specific case studies that illustrate the advantages and limits of model systems in research on the evolution and ecology of complex phenotypes: the instance of "touch response" to mechanical stimulation, and the case of response to multiple simultaneous stresses such as flood and low light levels. The same chapter also incorporates a more broadly ranging discussion of the very concept of hypothesis testing, so crucial not just to research on integration, but more in general to all quantitative biology. The idea is to stimulate thinking and discussion on the very foundations of the tools we use for data analysis, the limitations of which become increasingly clear precisely when their application is stretched by problems that are more complex than usual, such as phenotypic integration data sets.

Relyea's chapter uses predator-induced plasticity in the patterns of phenotypic integration as a model system. He focuses on the sort of insights one can gather by concentrating on a particular aspect of integration regarding which we have a wealth of functional ecological and developmental information. Relyea advocates moving away from attention to only one or a few characters at a time, so convenient for both logistical and conceptual reasons. One of the welcome consequences of this is the ability to solve puzzles that would otherwise have been difficult to put together simply because we were missing important corollary information on underlying tradeoffs between traits, as well as on the costs of exhibiting plasticity under certain environmental circumstances. Interestingly, Relyea makes the point that it actually becomes *easier* to appreciate the adaptive significance of certain traits and of their plasticity when these are put into the more complex context of the several other characteristics that play into antipredator defense or avoidance.

Clearly, both chapters are also about adaptation and constraints, and in that sense they could have been inserted in Part I as well. However, given the special emphasis on plasticity and the historical parallels between the two fields of inquiry, we felt that a separate treatment was warranted. This does not mean, of course, that many of the issues raised earlier about the relationship between selection and genetic/developmental constraints do not apply to studies of the plasticity of integration patterns. On the contrary, they do so all the more because additional parameters are thrown into the system by taking into consideration the environmentally induced flexibility of correlations within functionally adaptive character complexes.

### *Literature Cited*


Clausen, J., D. Keck, and W. M. Hiesey. 1940. Experimental studies on the nature of plant species. I. Effect of varied environment on western north American plants. Carnegie Institution, Washington, DC.

Donohue, K., D. Messiqua, E. H. Pyle, M. S. Heschel, and J. Schmitt. 2001. Evidence of adaptive divergence in plasticity: density- and site-dependent selection on shade-avoidance responses in *Impatiens capensis*. Evolution 55:1956–1968.

Emlen, D. J. 1997. Diet alters male horn allometry in the beetle *Onthophagus acuminatus* (Coleoptera: Scarabaeidae). Proceedings of the Royal Society of London 264:567–574.

Greene, E. 1989. A diet-induced developmental polymorphism in a caterpillar. Science 243:643–646.

Lively, C. M. 1986. Predator-induced shell dimorphism in the acorn barnacle *Chthamalus anisopoma*. Evolution 40:232–242.

Marcus, L. F., M. Corti, A. Loy, G. J. P. Naylor, and D. E. Slice. 1996. Advances in Morphometrics. Plenum Press, New York.

Olson, E. C., and R. L. Miller. 1958. Morphological Integration. University of Chicago Press, Chicago.

Reznick, D., M. J. Butler IV, and H. Rodd. 2001. Life-history evolution in guppies. VII. The comparative ecology of high- and low-predation environments. American Naturalist 157:126–140.

Rohlf, F. J., and L. F. Marcus. 1993. A revolution in morphometrics. Trends in Ecology and Evolution 8:129–132.

Schlichting, C. D. 1989. Phenotypic plasticity in *Phlox*. II. Plasticity of character correlations. Oecologia 78:496–501.

Schlichting, C. D., and M. Pigliucci. 1998. Phenotypic Evolution: A Reaction Norm Perspective. Sinauer, Sunderland, MA.

Schmitt, J., S. Dudley, and M. Pigliucci. 1999. Manipulative approaches to testing adaptive plasticity: phytochrome-mediated shade avoidance responses in plants. American Naturalist 154:S43–S54.

Thompson, D. 1917. On Growth and Form. Cambridge University Press, Cambridge.

Thompson, D. B. 1999. Genotype-environment interaction and the ontogeny of diet-induced phenotypic plasticity in size and shape of *Melanoplus femurrubrum* (Orthoptera: Acrididae). Journal of Evolutionary Biology 12:38–48.

Van Buskirk, J. 2002. Phenotypic lability and the evolution of predator-induced plasticity in tadpoles. Evolution 56:361–370.

van Tienderen, P. H., and A. V. Hinsberg. 1996. Phenotypic plasticity in growth habit in *Plantago lanceolata*: how tight is a suite of correlated characters? Plant Species Biology 1:87–96.

Wagner, G. P. 1995. Adaptation and the modular design of organisms. Pp. 317–328 in F. Moran, A. Moreno, J. J. Merelo and P. Chacon, eds. Advances in Artificial Life. Springer, Berlin.

Wagner, G. P., and L. Altenberg. 1996. Complex adaptations and the evolution of evolvability. Evolution 50:967–976.

Weinig, C. 2000. Differing selection in alternative competitive environments: shade-avoidance responses and germination timing. Evolution 54:124–136.

Woltereck, R. 1909. Weiterer experimentelle Untersuchungen über Artveränderung, speziell über das Wesen quantitativer Artunterschiede bei Daphniden. Verhandlungen der Deutschen Zoologischen. Gesselleschaft 19:110–172.

# 7

# Studying the Plasticity of Phenotypic Integration in a Model Organism

MASSIMO PIGLIUCCI

## The Problem: Ecology of Phenotypic Integration

It should be clear from a survey of the heterogeneous literature on phenotypic integration that biologists still have to agree on exactly what integration is to begin with. Indeed, one of the aims of this book is to clarify the discussion on this essential aspect of the problem. Since we do not have a clear definition of integration, it may seem premature to ask other questions about it, such as how it evolves, what are its molecular underpinnings, and in what ecological context it needs to be framed. Yet, as the philosopher Ludwig Wittgenstein (1953/1973) has pointed out, most complex concepts are defined not by some essential property, but by a network of properties that intersect in different ways, depending on the viewpoint from which a given question is asked. Concepts whose exact nature eludes easy definition, often referred to as "family resemblance" concepts, may include phenotypic integration, and that may be why it is difficult to put a finger on what exactly the latter is. Moreover, scientists—by definition of what they do—cannot afford to be shy about investigating things that they cannot define in an essentialist fashion (just think of the amount of work done on speciation, despite the lack of a generally agreed-upon definition of species: Barraclough and Nee 2001; Schluter 2001). It therefore makes perfect sense to ask how to go about investigating the ecology of phenotypic integration even though we may disagree on the very definition of the concept.

This said, I still need to provide at least a general sense in which I am using both the terms "phenotypic integration" and "ecology" within the scope of this chapter. By *integration* I simply mean whatever set of evolutionary and developmental

an observable network of multivariate relationships among the
ıat define the morphology and life history of a living organism.
ıis book deal with the problem of identifying the best ways to
ıships, but here I shall use a variety of statistical approaches
oyed in the past to summarize the characteristics of multi-
traits. In my experience, the exact choice of a metric is far less
rtant then the choice of traits and the sample size used to study them. By *ecology* I mean the relationship between the networks of traits in question and the external environment—biotic or abiotic—in which the organisms that we study live. In other words, I am interested in what one can think of as the environmental contingency, and degree of phenotypic plasticity, of integration.

Phenotypic plasticity is a term normally used in reference to individual traits to summarize (by means of so-called reaction norm diagrams) how a given genotype (or population, or species) responds to a series of different environmental conditions by producing a more or less varied array of phenotypes (Fig. 7.1, left). It is therefore a logical extension of the concept to trace the same genotype-environment interaction in the case of complex measures of multivariate phenotypes, including whatever quantification of the amount and pattern of integration one wishes to use (Fig. 7.1, right). Consequently, most of the concepts, questions, and methodological approaches usually associated with the study of phenotypic plasticity (Pigliucci 2001) translate naturally to research on phenotypic integration (Schlichting 1989a, 1989b). This is important, because it provides us with a ready and tested toolbox from which we can choose to help build our understanding of phenotypic integration.

Now that we have set things up so that we are thinking of the ecology of phenotypic integration in a sense similar to the usual approach to genotype-environment interactions, we need to ask ourselves what it is that we want to know about the plasticity of integration. As in the case of phenotypic plasticity of individual traits, there is a risk of adopting simplistic null hypotheses (more on this below) and asking essentially trivial or irrelevant questions. For example, we

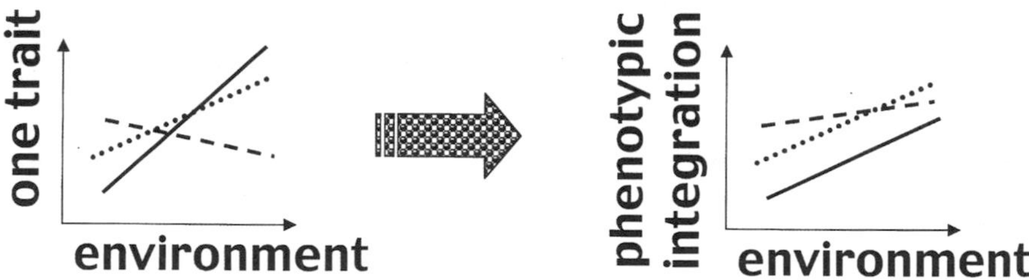


Figure 7.1 The conceptual shift from studying phenotypic plasticity of individual traits (left) to that of measures of associations among multiple characters (right) leaves unaltered the basic concepts, questions, and methodologies used in the study of phenotypic plasticity in general. The individual lines within each graph represent genotypes, populations, or species in whose plasticity one is interested. On the right, the vertical axis may represent a measure of the quantity or pattern of phenotypic integration in a group of organisms (population, species, etc.).

could wonder if there is genetic differentiation for the plasticity of integration among different populations and species. The answer ought to be yes, otherwise there would have to be a striking pattern of parallel covariances among traits in different species and environments that would be hard to miss even after a superficial glance at the natural world. Trying to be a bit more sophisticated, we could ask the same question for specific environmental factors and particular species, such as "Does the multivariate network of traits in population X of *Drosophila melanogaster* change in response to temperature?" But this would be only slightly better than the first case. The answer would probably be yes again (there are very few characters in very few species that do not show a degree of plasticity to at least some environmental factor) and it would therefore be of very little value for our understanding of the phenomenon of the modulation of integrated phenotypes in response to environmental changes. In all these cases, the null hypotheses (e.g., there is no plasticity of integration of traits X, Y, Z in response to environmental factors A, B, C) is just too simple for its (likely) rejection to be informative.

What I would rather suggest is that we need to start formulating more specific hypotheses that lead to more probing questions than we have largely been able to do so far. As any practicing organismal biologist will quickly recognize, however, this is much easier said than done. What sort of guidelines can one adopt to pursue this more informative search? How do we generate more interesting hypotheses that can make us feel that thousands of hours spent in the greenhouse are really worth the effort? I suggest that the answer lies along two distinct lines: on the one hand, we need to alter radically our ideas on hypothesis testing in general; on the other hand, we need to make intelligent use of what we already know about the ecology and genetics of certain species, especially of so-called model systems such as *Drosophila* or *Arabidopsis*. In the rest of this chapter I shall tackle first the philosophical and methodological issues related to hypothesis testing (which have a much broader application than just the study of phenotypic integration), then consider the use of a particular model system, its advantages and inevitable shortcomings, and conclude by examining in more detail two case studies in progress to illustrate some of the key concepts discussed in the chapter.

## A Brief Detour on the Concept of Testing Hypotheses

So, what is wrong with standard hypothesis testing? A great deal, and evolutionary biologists (not just those interested in phenotypic plasticity or integration) have been unaware of a long discussion that has recently started to bear fruit, yielding significant changes in the *modus operandi* of several researchers in other fields (especially the social sciences). In what follows, I shall provide only a brief overview of the problem and of some of the proposed alternatives, since the topic of this book is phenotypic integration, not statistics or philosophy of science. Nevertheless, the references cited here should be very helpful to students of complex phenotypes whenever they find themselves at the point of formulating hypotheses, evaluating experimental designs, and conducting statistical analyses.

As unbelievable as it may sound, the standard statistical approach so common in the quantitative sciences is highly problematic and does not provide a very useful tool for scientific (as opposed to statistical) inference. The main problem with the classical approach is that it involves testing the probability of the observed data given a rather arbitrarily constructed "null" hypothesis. That is, we typically set up statistical tests to tell us if, for example, the correlation coefficient between variables $X$ and $Y$ is zero (null hypothesis) or different from zero (any other value). We then collect the data and estimate the probability of getting those data if the null hypothesis were correct. If such probability is low, we happily "reject" the null hypothesis and proclaim the acceptance of the alternative hypothesis. Otherwise, we say that there is no evidence to reject the null, and sometimes even use statements that imply that the null has been confirmed (i.e., there is no effect of $X$ on $Y$).

There are many problems with this (Gregson 1997; Dixon and O'Reilly 1999). For one thing, we are actually interested in determining the probability of a series of alternative hypotheses given the data, not vice versa. In other words, scientific inference is about comparing hypotheses on the basis of data, not the other way around (which is what we do when we use the standard statistical approach). Also, the null hypotheses are often uninteresting (why did you bother starting the experiment if you really thought that there was no correlation between $X$ and $Y$?), and almost certain to be rejected given a large enough sample size. The latter is because the null hypothesis is stated very precisely (e.g., $r_{xy} = 0$, the correlation between $X$ and $Y$ is zero). In reality, what we mean to test is whether the correlation, measure of phenotypic integration, or whatever other value we are interested in, is close enough to zero to be scientifically uninteresting. But how close is close enough? Furthermore, the standard approach does not take into account any prior knowledge one may have about the system (like the expectation that not only will there be a correlation between $X$ and $Y$, but that it will be positive, for example); but scientists do not start each project with a blank slate. It should also be obvious that the so-called alternative hypothesis is in reality an infinite family of hypotheses, which means that concluding that the result does not accord with the null is not very informative. Indeed, good science proceeds by considering several alternative hypotheses (Chamberlain 1897), not just two: the more alternatives, and the better defined, the more we can design informative experiments and make progress (Platt 1964).

There are serious problems also with the companion concept of "$p$-values," since in the standard statistical approach they are often misinterpreted in a variety of ways (Loftus 1993; Gregson 1997). The most important thing to understand is that the $p$-value is neither a measure of $P(H|D)$ (the probability of a hypothesis given the data, which is what is of interest to the scientist) nor (exactly) of $P(D|H_0)$ (the probability of the data given the null hypothesis, what we allegedly test with the standard approach). In fact, it is the cumulative probability of the point-null hypothesis and of all values more extreme than the one observed. This is related to $P(D|H_0)$, but clearly is not the same. Furthermore, it is absolutely incorrect to interpret a $p$-value as an estimate of the strength of the effect under investigation (as in "the smaller the $p$-value the more 'significant' the results are": significant in what sense, statistically or scientifically?). This is because for a fixed effect size (large or small), the $p$-value is a function of the sample size. One can repeat the experiment with a much larger sample size, obtain a much smaller *p*-value associated with the null hypothesis, and yet the effect size (e.g., the strength of a correlation coefficient) can be unchanged (if it were properly estimated before).

The reader will notice throughout this discussion that many of the objections and alternatives to the standard approach to hypothesis testing are compatible with a Bayesian framework for scientific and statistical inference (Howson and Urbach 1991; Jefferys and Berger 1992). However, I shall not take that route here (even though I do consider it valid) for two reasons: first, there are several interesting philosophical problems raised by the adoption of a Bayesian framework (such as the estimation of prior probabilities) that I do not have space to discuss in this chapter. Second, it is currently very difficult to utilize the Bayesian insight in practice because of the lack of general-purpose software. One has instead to write custom programs tailored to each specific problem, something that people have started doing in a variety of fields within organismal biology, from systematics (Huelsenbeck et al. 2000), to genetics (Shoemaker et al. 1999), to evolutionary biology (Rudge 1998), to conservation biology (Wade 2000). Here I shall instead briefly examine some alternatives to the standard approach that are of immediate application for any researcher who is familiar with widely available statistical packages.

One powerful alternative practice to the standard approach is simply to think in terms of multiple hypotheses, none of which gets to play default. Interesting and informative scientific papers are those that consider two (or more) plausible hypotheses and can actually conclude for an increase or decrease in likelihood of one or more of them (Chamberlain 1897). Most scientists of course do realize the distinction between statistical and scientific inference but, because of the constraints (imposed by editors and reviewers, as well as by available textbooks and software) favoring standard statistical analyses, they end up writing rather schizophrenic papers in which the Results section is presented in terms of uninformative *p*-values while the Introduction and Discussion sections are where the real theoretical action lies. There are also better ways of reporting the relationship between the actual results one obtains and the predictions of the alternative hypotheses. A simple but often-neglected approach is to plot the data and their confidence intervals. As Loftus (1993) puts it, "a picture is worth a thousand *p*-values." Loftus provides examples from the primary literature from which it is clear that a graph of means and standard errors gives us all the information that a *p*-value provides, and then some. Moreover, the really important information can be gathered only from the graph, while the rest is not that interesting scientifically. In particular, graphs of means and their associated measures of dispersion provide us with a visual estimate of effect sizes, of the power of the analysis (which is inversely proportional to the standard errors), and with an immediate understanding of the patterns identified by the data. Loftus points out that these quantities are completely invisible in a table of *p*-values, where all we are told is that certain, largely irrelevant, null hypotheses are to be rejected. Furthermore, even in those cases in which the null is not rejected, Loftus reminds us that we can immediately discern from a plot if this is because there is little power in the analysis (large standard errors when compared to the differences between means), or because the population means are really well estimated, and indeed close to each other. This distinction is important, because in the first case we deduce that we need more data, in the latter that we really have no reason to think there is a scientifically relevant difference among means.

The next alternative to consider combines the insights of the Bayesian approach (*sans* the difficulties related to the estimation of priors) and the simplicity of already available statistical approaches (and therefore of computer packages): maximum likelihood ratios (Dixon and O'Reilly 1999). A maximum likelihood ratio is a quantity that compares the likelihood of the data based on one model with the likelihood of the data based on a second, competing, model (comparisons can also be made among multiple models, pairwise or cumulatively). The ratio therefore provides a measure of the relative match of models and data. Most sources suggest that a ratio of 10:1 is equivalent to a *p*-value of 0.05, that is, it should be the minimum ratio at which one should consider one model better than another. However, remember the fact that any such threshold (including the magical *p*-value of 0.05) is arbitrary and that there is no shortcut to using one's own judgment. The maximum likelihood ratio is obviously conceptually (and mathematically; see Dixon and O'Reilly 1999) related to the Bayesian approach, but it is also conveniently close to the standard statistics we are all familiar with, which makes it a powerful practical alternative. There are several ways to calculate likelihood ratios from standard statistics, which are explained in detail by Dixon and O'Reilly: all one needs are standard estimates of variances obtainable from ANOVA or regression output tables.

Yet another approach to the solution of the same problems concerning model selection is provided by the use of the Akaike Information Criterion (AIC). This is a measure based on information theory and derived from the concept of entropy in physics (Anderson et al. 2000). Briefly, AIC measures the fit of the data to a given model, when the model is penalized in proportion to the number of parameters it employs. This is necessary because, other things being equal, a model with more parameters will always fit the data better than one with fewer parameters, but the improved fit will not always be scientifically relevant. The Akaike Information Criterion and its derivates (which, like likelihood ratios, can be calculated from standard statistical quantities) lend themselves very easily to the comparison of several alternative models, while not being at all "tests" in the sense of requiring the calculation of associated *p*-values. The AIC also benefits from solid theoretical and philosophical foundations given its relationship to information theory.

At the end of this brief tour one obvious question comes to mind: should we throw away everything that has been published using the standard statistical methods? Fortunately, no: it turns out that in many cases inferences based on the classical *p*-value-based approach are similar to those arrived at using the alternative methods described here. However, there are some important differences that are especially useful to quantitative biologists, particularly in fields such as phenotypic integration where hypotheses are complex and sample sizes may vary dramatically from study to study. First, there are cases in which the two approaches will yield very different results. In particular, as we have seen, *p*-values are problematic when sample sizes are either too small or very large. In the first case, a failure to reject the null hypothesis may actually be premature, since the problem may be the low power inherent in small sample sizes. The *p*-values do not carry any information about this problem, and one needs to plot the data and their confidence intervals to get an idea, as mentioned above (just reporting the sample sizes will not do, because that information does not provide clues to the relevant measures of dispersion of the data). In the second case, *p*-values may lead us to reject the null when the difference between effects is scientifically irrelevant, though technically nonzero. As we have seen, this is because p-values refer to a point-hypothesis, which is never of interest to scientists; any small, but systematic, unaccounted source of variation will produce deviations from the point-null which will be detectable for large enough sample sizes. In this case again, plotting the data will reveal more readily both the statistical and the scientific significance of the results.

The second reason to opt for alternatives to standard hypothesis testing is a matter of training ourselves to think along realistic, and inevitably complex, lines: typically, we are simply not interested in the hypotheses tested with the standard methods. Again, good science actually proceeds by comparing the relative fitness (using the data) of a series of reasonable alternative models (hypotheses), something that standard hypothesis testing simply cannot do.

It is now time to turn to the second source of help that students of integration can use to make some progress in this murky field: model systems.

## The Use of Model Systems in Evolutionary Ecology

The virtues and benefits of the use of model systems in biological research have been discussed (Kellogg and Shaffer 1993), and a reasonable position to take is that while they provide certain advantages that are impossible to gather from the majority of potential experimental organisms, there are some serious limitations to the generalizations one can make from research conducted on *Drosophila*, *Caenorhabditis*, *Arabidopsis* and the few other species of choice that have emerged over the last few decades.

On the one hand, it is undeniable that remarkable progress has been made, especially in molecular, but also in organismal biology, with the use of model systems. It is simply not possible to imagine modern genetics without *Drosophila*, and we owe much of our understanding of flower development to the famous ABC model developed in *Arabidopsis* (Pidkowich et al. 1999). The very idea of a "model system" is central to that part of biology most closely resembling research in physics and chemistry: molecular biology is the search for universals analogous to the structure of the atom and the principles of thermodynamics. In this framework, it is logical to focus the attention on a few, easily manipulated systems, and attempt to generalize to the rest of the living world.

On the other hand, organismal biology (ecology and evolutionary biology) is a much messier business, dominated by the importance of historical events, where organisms differ from each other in important respects, and where the focus is on understanding the differences, not the similarities. In this context, concentrating our attention on a few (often unrepresentative) species can be seen as a fatal flaw that will lead us into the false security of very narrow understanding.

While this distinction between ahistorical and historical science is indeed of fundamental interest, and does have important consequences for our understanding of the very nature of science (Dupré 1993), I think that a useful middle ground can be found. I shall illustrate the advantages of such an intermediate position for the study of phenotypic integration in the remainder of this chapter using research on *Arabidopsis thaliana*, from which discoveries can easily be extended to its close phylogenetic relatives.

*Arabidopsis* has been a model system in molecular, developmental, and cell biology for quite some time now (Griffing and Scholl 1991; Pyke 1994; Anderson and Roberts 1998), but several researchers have recently also been proposing it as an interesting tool for quantitative genetics, ecology, and evolutionary biology (Pigliucci 1998; Alonso-Blanco and Koornneef 2000; Mitchell-Olds 2001). While *A. thaliana* is a highly selfing plant, several of its relatives are not, which allows us to study the evolution of mating systems and of their relationship to the ecology of a group of taxa. Furthermore, from an ecological perspective *A. thaliana* can be described as an opportunistic weed, a plant of ruderal habitats (Ratcliffe 1965; Napp-Zinn 1985). It grows along a huge latitudinal span (from northern Africa to the polar circle), and it is found in a surprising number of soil types in varying association with other vegetation. The ecology of the other species of this group is much less understood, but the clade includes alpine species, annuals, biennials, and perennials, as well as species with a different base chromosome number and even hybrids and polyploids (Mummenhoff and Hurka 1995; Lee and Chen 2001). For a long time the systematics of this group was confusing (Price et al. 1994; O'Kane and Al-Shehbaz 1997), but recent morphological and especially molecular research has significantly cleared up the relationships among many of the more or less close relatives of *A. thaliana* (Al-Shehbaz et al. 1999; Koch et al. 1999, 2000, 2001). Currently a robust phylogenetic hypothesis is available (Fig. 7.2) to provide an invaluable comparison base for research on the evolution of phenotypes in this clade using standard phylogenetic comparative methods (Harvey and Pagel 1991; Huelsenbeck et al. 2000; Martins 2000). Examples of this in *Arabidopsis* have already appeared in the literature (Pigliucci et al. 1999, 2003).

Having made my case for a judicious use of *Arabidopsis* as a model system in organismal biology, it is now time to turn to a few examples of ongoing research on phenotypic integration that illustrate several of the concepts I have been discussing so far.

## Case Study 1: Bushy Phenotypes in Response to Touch

An interesting and understudied type of phenotypic plasticity affecting several aspects of the phenotype is the so-called thigmomorphogenic response, that is, the reaction that many plants display to mechanical stimulation (Biddington 1986; Jaffe and Forbes 1993; Wisniewski 1996). The ecological context of thigmomorphogenesis can be varied and complex in itself, as this plasticity can be a reaction to abiotic mechanical stimulation as imposed by wind, rain, or snow (Braam and Davis 1990; Telewski and Pruyn 1998; Cordero 1999), to contact

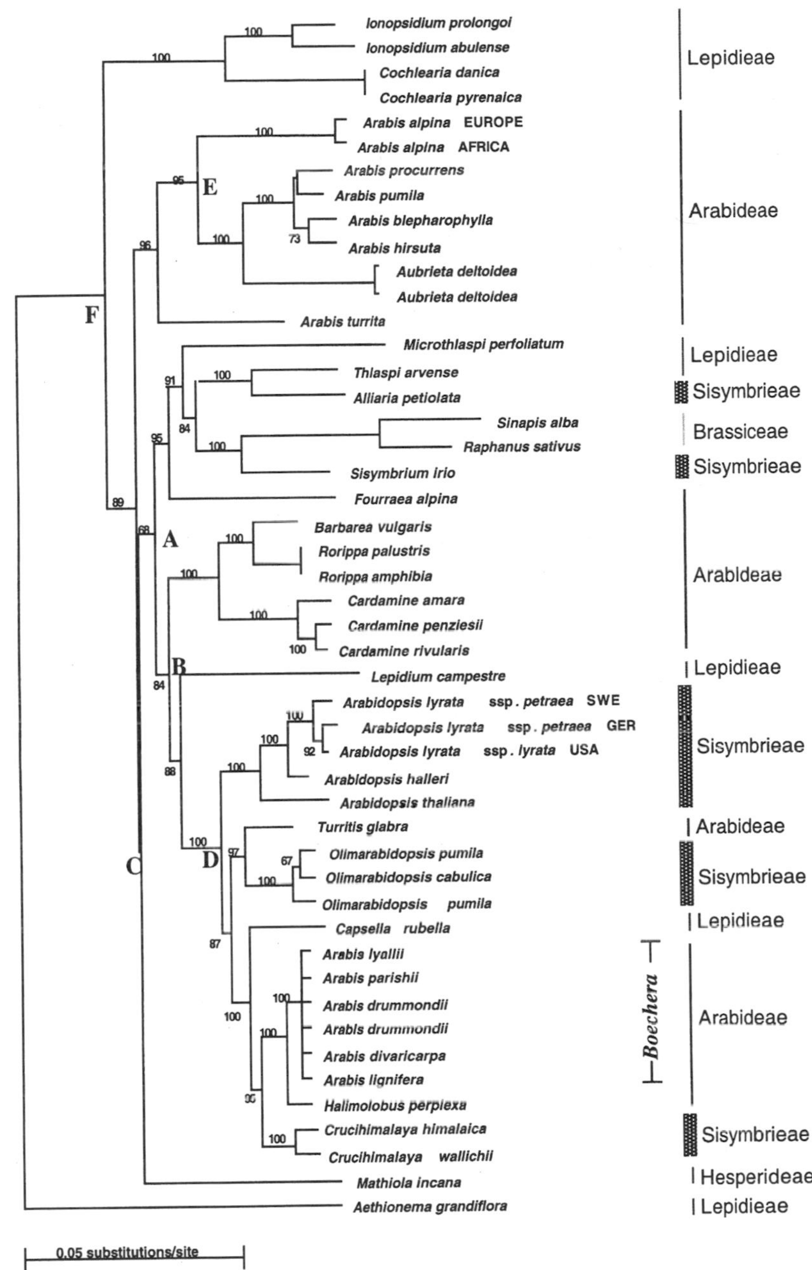


Figure 7.2 Neighbor-joining distance tree of *Arabidopsis* and some of its close relatives based on *matK* and *Chs* sequences. Bootstrap values in percent from 1000 replicates are shown above branches. Tribal assignments are given in the right margin. Approximate divergence dates (million years before present) for nodes A–F are: A, 16–21; B, 13–19; C, 19–25; D, 10–14; E, 15–17; F, 26–32. Several important taxa are not shown in this figure: *Arabidopsis arenosa* and *A. suecica* are closely related to *A. lyrata*. *Brassica* is near *Raphanus* and *Sinapis*. Finally, *Leavenworthia* is related to *Barbarea*. Reproduced with permission from Koch et al. (2001).

by insects (van Emden et al. 1990; Cipollini 1997), or to contact by other plants in crowded stands (Emery et al. 1994). It has even been suggested that touch-response plasticity may have practical applications, since in some plants it increases resistance to pests (Cipollini 1997).

The molecular basis of touch-response plasticity has been investigated, especially in *Arabidopsis*, and a series of relevant genes has been identified and is currently under study (Braam and Davis 1990; Sistrunk et al. 1994; Xu et al. 1995, 1996; Mizoguchi et al. 1996; Antosiewicz et al. 1997; Braam et al. 1997). Braam and collaborators (1997) found that the *TCH* (touch) gene products in *Arabidopsis* include a xyloglucan endotransglycosylase and several calmodulin and calmodulin-related proteins. They found that the functions of these proteins comprise cell and tissue expansion under mechanical strain, and hypothesize that a subset of *TCH* genes may be involved in cell wall biogenesis, all of which makes sense in light of their relationship to phenotypic plasticity triggered by mechanical stimulation.

I studied thigmomorphogenesis in response to winds sustained for different periods of time in 11 populations of *A. thaliana* to determine the degree of genetic differentiation among populations in plasticity to mechanical stimulation, as well as to test the hypothesis that recognizable "wind-specialized" populations exist and are characterized by an identifiable pattern of phenotypic integration (Pigliucci 2002b). More specifically, I hypothesized that when plants are exposed to sustained winds, populations can be grouped by their multivariate response into "bushy" and "nonbushy." If true, the clustering of populations could then be used to predict which populations are actually found in high-wind environments in the field (a piece of information that is still currently missing from the puzzle).

I did find that 8 of the 11 populations responded to sustained winds, especially by increasing their branching intensity, which made them look more "bushy." The remaining 3 did not respond at all to the environmental stimulus (Fig. 7.3). More interestingly, the multivariate pattern of phenotypic integration—measured by the correlation matrices among all traits—clearly separated bushy from nonbushy phenotypes (Fig. 7.4). This result confirmed the existence of a recognizable phenotypic syndrome related to mechanical stimulation under sustained winds, but not under wind-free conditions. Since I had previous experience with *A. thaliana*, I had also set up a series of three alternative models predicting the patterns of phenotypic integration based on my knowledge of the biology of this species and on basic considerations of plant physiology. The three hypotheses were set up as matrices of character correlations which where then compared with the observed matrices using standard matrix correlation tests (Cheverud et al. 1989). The hypothetical matrices assumed one of three scenarios: (1) integration within trait classes, where characters within vegetative, architectural, and reproductive classes are hypothesized to be highly positively correlated, but independent of characters in the other classes; (2) developmental integration with no tradeoffs, in which traits expressed contiguously (in time) during the life cycle are assumed to be positively correlated with each other, but the correlation decays with time (i.e., between traits expressed very early and very late); and (3) developmental integration with tradeoffs, similar to scenario (2) but where tradeoffs (manifested as negative correlations) are predicted between vegetative and reproductive traits, as well as between different reproductive traits (such as main versus

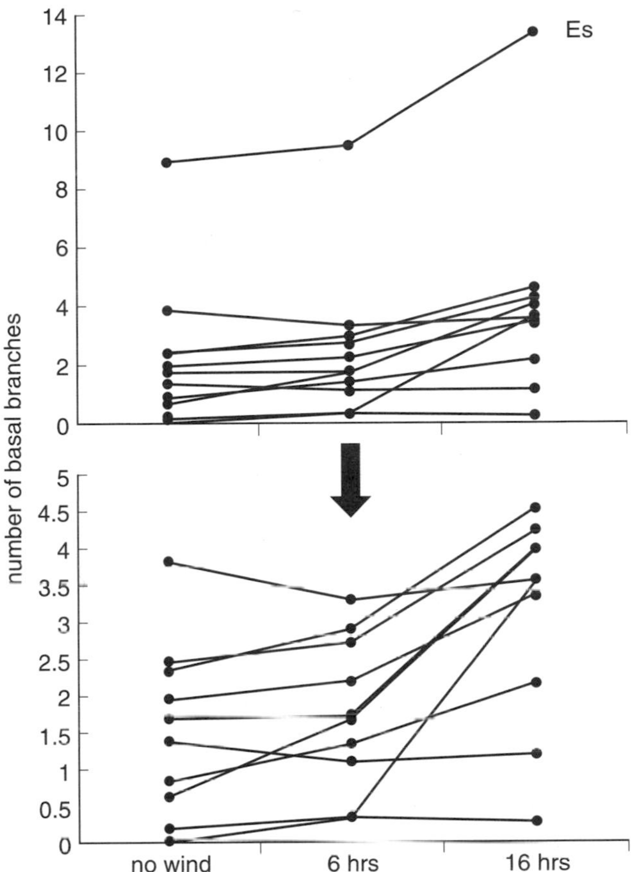


Figure 7.3 Reaction norms of individual populations of *Arabidopsis thaliana* graphing the response of branching to exposure to wind for increasing amounts of time. Each line represents a different population. The lower graph is a zoom into the lower level of the upper one with the exclusion of the late-flowering population Es (Espoo, from Finland), which produced many more branches than the others regardless of environmental conditions. (From Pigliucci, 2002b.)

basal branching). Models were compared by means of likelihood ratios, one of the alternative statistical approaches mentioned above. I found that under all three environmental conditions the third model was the best predictor of the observed patterns of integration, with amounts of explained variance ranging from 63% under no wind to 74% for 6-hour wind and 52% for 16-hour wind.

This study pointed out a series of interesting things about the plasticity of phenotypic integration when the response of *Arabidopsis* to mechanical stimulation is concerned. First, it is possible to predict the general pattern of phenotypic integration as reflected in character correlations, if one has a minimal knowledge of the life cycle and physiological characteristics of the organism in question. (This has been done in other systems for which prior knowledge of the developmental trajectory was available: Cheverud 1996; Mezey et al. 2000). Second,

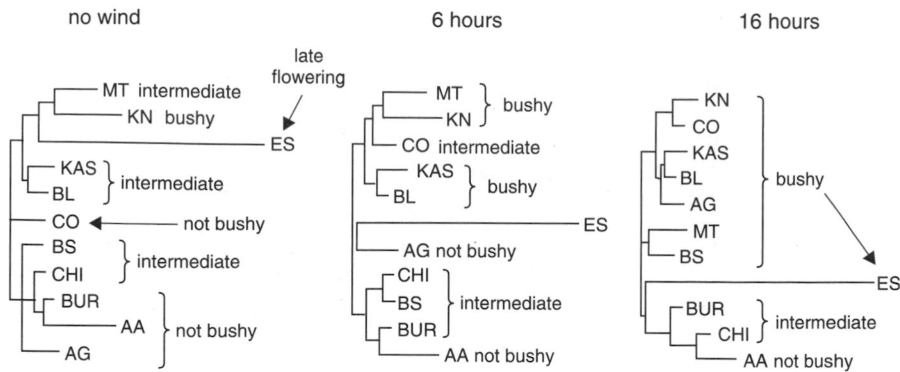


Figure 7.4 Multivariate clustering of 11 populations of *Arabidopsis thaliana* according to their patterns of phenotypic integration (measured as matrices of character correlations) within each of three environmental treatments. Notice how the populations characterized by the plastic response that alters the degree of their branching are scattered throughout the dendrogram on the left (no wind) but become increasingly separated as a group with more and more sustained winds (especially under 16 hours of exposure, right). The clustering was obtained by computing additive trees based on Euclidean distances calculated on phenotypic data. (From Pigliucci, 2002b.)

phenotypic integration can nevertheless be substantially altered by environmental changes. Third, if the environmental stimuli can be connected even roughly with the field ecology of the species under investigation, one can in fact predict the existence of specific patterns of integration, predictions that can later be used to test for the existence of a link between phenotypic syndromes expressed under controlled conditions and prevalent environmental regimes in the field. To show that making sense of phenotypic integration in this way can be applied to other circumstances, I shall now turn to our second case study, in which the same species (but different populations) is examined in response to two more environmental factors relevant to its ecology.

## Case Study 2: No Stress, One Stress, Two Stresses

Two of the most common sources of environmental variation in most plants are heterogeneities in the levels of water and light. *Arabidopsis thaliana* occasionally grows under a relatively dense canopy of short grass, and more often has to deal with competition for light from shorter neighbors of the same or others species (such as *Trifolium*, pers. obs.). As far as water is concerned, while certainly not likely to experience prolonged conditions under flood, *A. thaliana* nevertheless lives in habitats such as ditches along roads where water can accumulate to the point of saturating the soil and causing anoxia in the roots and possibly the leaves.

As in the case of mechanical stimulation, there is a large literature on the molecular basis of responses to light intensity (Yanovsky et al. 1997; Whitelam and Devlin 1998; Lasceve et al. 1999) and water (Wang et al. 1995; Ishitani et al.

1997; Zhang et al. 2000) in *Arabidopsis*. As far as responses to light are concerned, *Arabidopsis* is equipped with a battery of photoreceptors, some sensitive to red and far-red light (phytochromes), some to the blue and UV components of the spectrum (so-called cryptochromes and UV-receptors). Phytochromes, especially phytochrome A and B (there are five in *Arabidopsis*), are the entry point for a series of distinctive responses known as the very-low-fluence, the low-fluence, and the high-irradiance responses (Yanovsky et al. 1997). The blue-light-activated signal transduction pathways themselves are of at least four different kinds, each regulated by a distinct photoreceptor as the environmental signal is received by the plant (Lasceve et al. 1999). Furthermore, we are beginning to unravel the complexity of the transduction pathways connected to these two large families of photoreceptors, and the emerging picture is one that includes a battery of repressors (the so-called *COP*/*DET*/*FUS* genes) acting downstream of multiple photoreceptors, which in turn regulate the cellular levels of hormones such as cytokinins and brassinosteroids (Whitelam and Devlin 1998). Concerning the molecular basis of responses to levels of water, Zhang et al. (2000) have demonstrated with the use of transgenic *Arabidopsis* that endogenously produced cytokinin increases flood tolerance in this species, probably in part by extending the length of the part of the life cycle between flowering and senescence, thereby allowing plants to mature more seeds under stressful conditions. Ishitani and coworkers (1997) have identified a number of mutants that affect osmosis, and therefore the reaction to water stress (drought, in this case) in *Arabidopsis*, and have argued for the existence of extensive cross-talk between pathways that are dependent of abscisic acid and those that are independent of this hormone, contrary to what was previously thought.

In the context of phenotypic integration, Ania Kolodynska and I set out to test several hypotheses about how patterns of phenotypic integration in *A. thaliana* could be altered by simultaneously varying light and water levels. In particular, we were interested in the multivariate phenotype expressed under normal levels of both environmental factors, as opposed to situations in which one or the other stress (low light or high water) was experienced, and then to a scenario in which both stresses were applied simultaneously (Pigliucci and Kolodynska, in preparation). We predicted, in agreement with other suggestions published in the integration literature (Schlichting 1986, 1989a; Chapin 1991) that the highest degrees of stress would result in a more "tight" phenotype, that is, in the expression of a higher number of correlations of large magnitude between different traits and reproductive fitness. We also made a number of specific predictions about the behavior of individual traits in response to the four imposed combinations of light and water levels. While the latter were largely confirmed, I shall focus here only on the results pertinent to phenotypic integration and not on those concerning individual traits. We obtained very clear support for our hypothesis of a positive relationship between degree of stress (as perceived by the plant and manifested in decreased fruit production) and phenotypic integration as gauged by the number of interactions between phenotypic traits and reproductive fitness, itself tested by fitting different models to the data and choosing based on the corresponding likelihood ratios (Fig. 7.5). While under nonstressful conditions the only variable that seemed to affect fitness was shoot biomass, under moderate stress (either low light

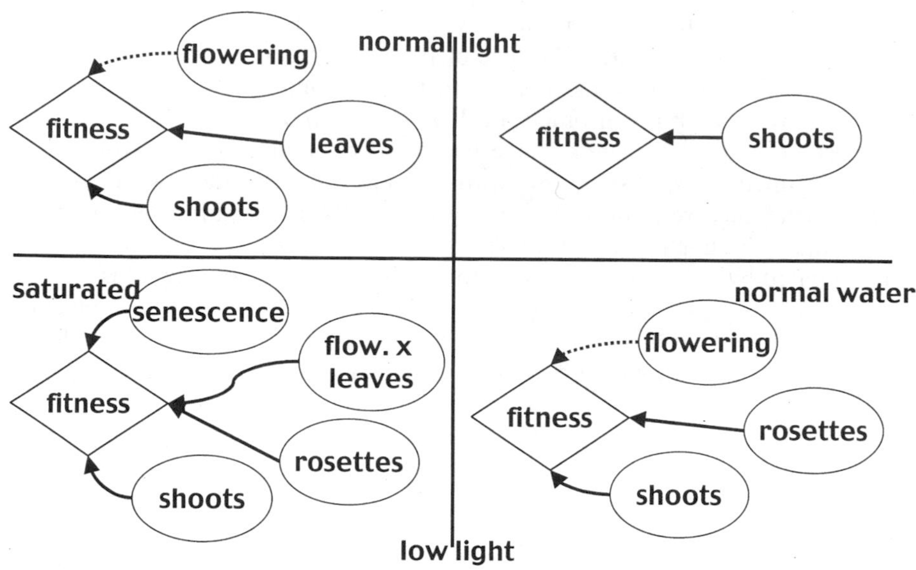


Figure 7.5 Patterns of phenotypic integration in 22 populations of *Arabidopsis thaliana* in response to four combinations of water (horizontal axis) and light (vertical axis) levels. Each quadrant shows the best model that fits the data according to a maximum likelihood ratio analysis of several alternative models. Notice how the number of traits affecting reproductive fitness (number of fruits produced) increases from no stress (upper right) to moderate stress (upper left and lower right) to high stress (lower left). Broken lines indicate a negative relationship between a given trait and fitness. "Flow. × leaves" indicates an interaction between leaf production and flowering time.

or water-saturated soil) flowering time, and either leaf number or size, were important. When the stress was highest, new traits came into play (such as the time to senescence), as well as higher-order interactions among individual characters (e.g., the particular combination of flowering time and leaf number became important).

While we do not have a good enough understanding of the ecophysiology of *A. thaliana* to allow us to predict in advance which combinations of characters will be more or less relevant under certain conditions, some of the changes in the patterns of integration are readily explainable. More importantly, such explanations can be submitted to further experimental tests. For example, if the stress was induced by water, with light at normal levels, leaf number was more important than leaf size; however, under low light it was the size, not the number, of leaves that was important. This is understandable if one thinks that under low light an increase in photosynthetic area is a more decisive parameter, because it compensates directly for reduced photosynthetic activity without costing the commitment of additional meristems to leaf production. It is the combination of such first principles of physiology and knowledge of the ecology of the particular species that it is more likely to generate detailed and testable hypotheses concerning phenotypic integration. The road ahead of us, however, is still fraught with many difficulties, as I shall discuss in the concluding section of this chapter.

## What We Need to Do

For a long time molecular and organismal biology have proceeded largely independently of each other, and have managed to make significant progress (Pigliucci in press). In recent years, however, a significant amount of interdisciplinary research has been carried out, largely centering on model systems such as *Drosophila*, *Caenorhabditis*, and *Arabidopsis* (Schlichting and Pigliucci 1998, chap. 8). This multi-pronged approach is likely to be useful also within the field of phenotypic integration, providing us with valuable insights into the molecular basis of complex phenotypes. Examples of these approaches include QTL mapping and the study of the pleiotropic effects of candidate genes for important regulatory functions (Juenger et al. 2000; Mezey et al. 2000), and robust phylogenetic hypotheses to be used in comparative studies of the evolution of phenotypic integration (Losos and Miles 1994; Ackerly and Donoghue 1998). For example, the phylogenetic hypothesis of the *Arabidopsis* and closely related clades resulting from extensive work by Mitchell-Olds's laboratory (Koch et al. 1999, 2000, 2001; see Fig. 7.2) will allow countless detailed comparisons of evolutionary patterns in this group to be carried out in a fashion that is informed by the likely evolutionary history of these species.

This optimism notwithstanding, there are some cautionary considerations to be made. First, it should be obvious to every evolutionary biologist that our discipline in general (not just the study of phenotypic integration) is a strange mixture of historical and experimental science (Cleland 2001). We can carry out experiments on phenotypic integration under controlled conditions (in the laboratory, greenhouse, or field), but the results will be highly dependent on which particular species, populations, and genotypes we have chosen. The quest to make evolutionary biology into a "hard" science on the model of physics or chemistry (or even, to some extent, molecular biology) is simply a misguided one (Pigliucci 2002a).This means that not only the study of model systems, but the study of *any* particular group of organisms, will not likely yield results that can be generalized beyond fairly narrow taxonomic or life-history categories. As evolutionary biologists we should embrace this as a natural outcome of the diversity of life, not as a limitation on our ability to conduct science.

Similarly, perhaps talk of "phenotypic integration" is itself misleading, as is talk about "phenotypic plasticity," without reference to specific combinations of traits and environments. Bradshaw (1965) pointed out decades ago that plasticity is not a meaningful property of a genotype, but it can be different for different traits, and even for the same trait when expressed under different environmental circumstances. Similarly, phenotypic integration may not intelligibly be attributed to the totality of an organism's traits, but make sense only for certain groups of characters (modules) that have been subjected to joint selection, or that have recently ceased to be coupled functionally and are now evolving independently. There is no reason to think that natural selection works on the full phenotypic variance-covariance matrix of an organism any more than we are well grounded in thinking that it operates simultaneously to coordinate the plasticities of all traits (and of course, there is the additional problem of what exactly a trait is, a discussion that recently has deserved a whole book in itself: Wagner 2001). Rather, we should focus our attention on groups of characters about which we have reasonable grounds to think that they are functionally integrated (Klingenberg and Zaklan 2000; Mezey et al. 2000; Wagner and Schwenk 2000; Klingenberg et al. 2001; Magwene 2001), or that may have evolved in concert as a result of the particular genetic architecture of the organism under study. Generalized measures of integration that are calculated across the whole organism are likely to be uninformative because they result from averaging the effects of distinct processes. It would be as if we calculated an "average plasticity" for a given genotype.

This problem has already emerged clearly within that part of phenotypic integration research that deals with the study of genetic variance-covariance matrices (Jernigan et al. 1994; Roff 1996; Lynch 1999). Authors write about the evolution of "**G**-matrices" as if these were a homogeneous biological class (what philosophers of science would call a natural category). But they are clearly not, because **G**-matrices can be measured in phylogenetically highly divergent taxa, and different authors concentrate on distinct subsets of traits, which can in no biologically meaningful sense be considered "samples" of the whole **G**-matrix. The result is that comparative studies of **G**-matrices among disparate biological entities tend to be highly uninformative (Roff 1996; Roff and Mousseau 1999). Rather, one should concentrate on closely related taxa and on groups of characters that are chosen because of the specific biology of those taxa (e.g., Kassen and Bell 2000; Begin and Roff 2001; Ferguson and Fairbairn 2001). Once again, misguided quests for ahistorical generalizations are a waste of time at best in evolutionary biology research.

This leads us to a reconsideration of the phylogenetic comparative method already discussed. Unfortunately, things are not easy for students of phenotypic integration even *if* a robust phylogenetic hypothesis is available for their group of interest. This is because of a simple reason pointed out by Westoby et al. (1995) and discussed in detail by Ackerly (2000). Most phylogenetic comparative analyses include some sort of "correction" (usually in the form of independent contrasts: Felsenstein 1985; Abouheif 1999) intended to "control" for historical events (which causes statistical nonindependence of the data points represented by the terminal taxa in a phylogeny), and uncover instead ecological patterns. The problem can readily be understood by glancing at Fig. 7.6: it is possible (indeed, likely) that phylogenetic niche conservatism (i.e., the tendency of daughter taxa to retain the ecology typical of their parent species) is present in the group being studied. This causes an overlap between the components of variance explainable by phylogeny and by ecology. Thinking about it in terms of standard analysis of variance, this would be a "phylogeny-by-ecology" interaction, analogous to the genotype-by-environment interaction in plasticity studies. If we factor out any variance related to the phylogenetic effect, this may throw away valuable information on the ecology and, in extreme cases, leave us with apparently no ecological signal when in fact there is one.

There is only one solution to this conundrum, and that is to gather as detailed an ecological database as one has on the phylogenetic relatedness

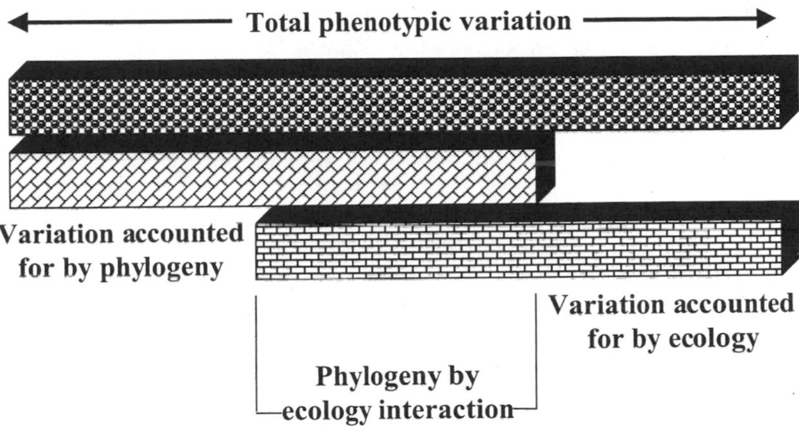


Figure 7.6 The total phenotypic variance observed in a group of phylogenetically related taxa (top) is usually thought of as the result of two components, one accounted for by historical commonality (middle), the second by the specifics of the ecology of each taxon (bottom). Standard "phylogenetic corrections" tend to eliminate the historical effect with the aim of uncovering the ecological specificities. However, if there is phylogenetic niche conservatism (i.e., daughter taxa maintain the ecological characteristics of their parent species), there may be a more or less large amount of phylogeny-by-ecology interaction that risks being thrown away by phylogenetic corrections. The only way around this problem is the difficult and tedious gathering of actual ecological information to complement the now relatively easily obtained phylogenetic one. (Based on a concept by Westoby et al. 1995).

and on the phenotypic variation. The problem identified by Westoby and colleagues stems from the fact that we normally use only the latter two types of data and infer ecological effects by difference. This clearly will not do because of the possibility of tantalizing interactions between causal agents such as selection and genetic constraints on one hand and drift and migration on the other. Unfortunately, it has been historically rather easy to gather copious (albeit not always informative) phenotypic data (Gould 1966), and recently it has been increasingly cheaper to obtain molecular data allowing us to reconstruct phylogenetic relationships. Conversely, to gather satisfactory ecological data is tedious and extremely time-consuming, and it is the sort of basic research that most funding agencies do not find sexy enough to support adequately. As a result, we may find that one of the major stumbling blocks to our understanding of phenotypic integration in the near future is the simple lack of solid field ecology for the taxa being studied.

*Acknowledgments* I wish to thank Ania Kolodynska for collaborating in some of the actual research summarized here and to the National Science Foundation (grant DEB-0089493) for partial funding. I should also like to thank Carl Schlichting and Gunter Wagner for many discussions on the concept of phenotypic integration and Katherine Preston for her insightful comments on this chapter.

*Literature Cited*

Abouheif, E. 1999. A method for testing the assumption of phylogenetic independence in comparative data. Evolutionary Ecology Research 1:895–909.

Ackerly, D. D. 2000. Taxon sampling, correlated evolution, and independent contrasts. Evolution 54:1480–1492.

Ackerly, D. D., and M. J. Donoghue. 1998. Leaf size, sapling allometry, and Corner's rules: phylogeny and correlated evolution in maples (*Acer*). American Naturalist 152:767–791.

Alonso-Blanco, C., and M. Koornneef. 2000. Naturally occurring variation in *Arabidopsis*: an underexploited resource for plant genetics. Trends in Plant Science 5:22–29.

Al-Shehbaz, I. A., S. L. O'Kane, Jr., and R. A. Price. 1999. Generic placement of species excluded from *Arabidopsis* (Brassicaceae). Novon 9:296–307.

Anderson, D. R., K. P. Burnham, and W. L. Thompson. 2000. Null hypothesis testing: problems, prevalence, and an alternative. Journal of Wildlife Management 64:912–923.

Anderson, M., and J. A. Roberts. 1998. *Arabidopsis*. Sheffield Academic Press, Sheffield, UK.

Antosiewicz, D. M., M. M. Purugganan, D. H. Polisenksy, and J. Braam. 1997. Cellular localization of *Arabidopsis* xyloglucan endotransglycosylase-related proteins during development and after wind stimulation. Plant Physiology 115:1319–1328.

Barraclough, T. G., and S. Nee. 2001. Phylogenetics and speciation. Trends in Ecology and Evolution 16:391–399.

Begin, M., and D. A. Roff. 2001. An analysis of G matrix variation in two closely related cricket species, *Gryllus firmus* and *G. pennsylvanicus*. Journal of Evolutionary Biology 14:1–13.

Biddington, N. L. 1986. The effects of mechanically-induced stress in plants—a review. Plant Growth Regulation 4:103–123.

Braam, J., and R. W. Davis. 1990. Rain-, wind-, and touch-induced expression of calmodulin and calmodulin-related genes in *Arabidopsis*. Cell 60:357–364.

Braam, J., M. L. Sistrunk, D. H. Polisensky, W. Xu, M. M. Purugganan, D. M. Antosiewicz, P. Campbell, and K. A. Johnson. 1997. Plant responses to environmental stress: regulation and functions of the *Arabidopsis TCH* genes. Planta 203:S35-S41.

Bradshaw, A. D. 1965. Evolutionary significance of phenotypic plasticity in plants. Advances in Genetics 13:115–155.

Chamberlain, T. C. 1897. The method of multiple working hypotheses. Science 15:92–96.

Chapin, F. S., III. 1991. Integrated responses of plants to stress. BioScience 41:29–36.

Cheverud, J. M. 1996. Quantitative genetic analysis of cranial morphology in the cotton-top (*Sanguinus oedipus*) and saddle-back (*S. fuscicolis*) tamarins. Journal of Evolutionary Biology 9:5–42.

Cheverud, J. M., G. P. Wagner, and M. M. Dow. 1989. Methods for the comparative analysis of variation patterns. Systematic Zoology 38:201–213.

Cipollini, D. F., Jr. 1997. Wind-induced mechanical stimulation increases pest resistance in common bean. Oecologia 111:84–90.

Cleland, C. E. 2001. Historical science, experimental science, and the scientific method. Geology 29:987–990.

Cordero, R. A. 1999. Ecophysiology of *Cecropia schreberiana* saplings in two wind regimes in an elfin cloud forest: growth, gas exchange, architecture and stem biomechanics. Tree Physiology 19:153–163.

Dixon, P., and T. O'Reilly. 1999. Scientific versus statistical inference. Canadian Journal of Experimental Psychology 53:133–149.

Dupré, J. 1993. The Disorder of Things: Metaphysical Foundations of the Disunity of Science. Harvard University Press, Cambridge, MA.

Emery, R. J. N., D. M. Reid, and C. C. Chinnappa. 1994. Phenotypic plasticity of stem elongation in two ecotypes of *Stellaria longipes*: the role of ethylene and response to wind. Plant, Cell and Environment 17:691–700.

Felsenstein, J. 1985. Phylogenies and the comparative method. American Naturalist 125:1–15.

Ferguson, I. M., and D. J. Fairbairn. 2001. Estimating genetic correlations from measurements of field-caught waterstriders. Evolution 55:2126–2130.

Gould, S. J. 1966. Allometry and size in ontogeny and phylogeny. Biological Review 41:587–640.

Gregson, R. A. M. 1997. Signs of obsolescence in psychological statistics: significance versus contemporary theory. Australian Journal of Psychology 49:59–63.

Griffing, B., and R. L. Scholl. 1991. Qualitative and quantitative genetic studies of *Arabidopsis thaliana*. Genetics 129:605–609.

Harvey, P. H., and M. D. Pagel. 1991. The Comparative Method in Evolutionary Biology. Oxford University Press, Oxford.

Howson, C., and P. Urbach. 1991. Scientific Reasoning: the Bayesian Approach. Open Court, La Salle, IL.

Huelsenbeck, J. P., B. Rannala, and J. P. Masly. 2000. Accomodating phylogenetic uncertainty in evolutionary studies. Science 288:2349–2350.

Ishitani, M., L. Xiong, B. Stevenson, and J.-K. Zhu. 1997. Genetic analysis of osmotic and cold stress signal transduction in *Arabidopsis*: interactions and convergence of abscisic acid-dependent and abscisic acid-independent pathways. Plant Cell 9:1935–1949.

Jaffe, M. J., and S. Forbes. 1993. Thigmomorphogenesis: the effect of mechanical perturbation on plants. Plant Growth Regulation 12:313–324.

Jefferys, W. H., and J. O. Berger. 1992. Sharpening Ockam's razor on a Bayesian strop. American Scientist 80:64–72.

Jernigan, R. W., D. C. Culver, and D. W. Fong. 1994. The dual role of selection and evolutionary history as reflected in genetic correlations. Evolution 48:587–596.

Juenger, T., M. Purugganan, and T. F. C. Mackay. 2000. Quantitative trait loci for floral morphology in *Arabidopsis thaliana*. Genetics 156:1379–1392.

Kassen, R., and G. Bell. 2000. The ecology and genetics of fitness in *Chlamydomonas*. X. The relationship between genetic correlation and genetic distance. Evolution 54:425–432.

Kellogg, E. A., and H. B. Shaffer. 1993. Model organisms in evolutionary studies. Systematic Biology 42:409–414.

Klingenberg, C. P., and S. D. Zaklan. 2000. Morphological integration between developmental compartments in the *Drosophila* wing. Evolution 54:1273–1285.

Klingenberg, C. P., A. V. Badyaev, S. M. Sowry, and N. J. Beckwith. 2001. Inferring developmental modularity from morphological integration: analysis of individual variation and asymmetry in bumblebee wings. American Naturalist 157:11–23.

Koch, M., J. Bishop, and T. Mitchell-Olds. 1999. Molecular systematics and evolution of *Arabidopsis* and *Arabis*. Plant Biology 1:529–537.

Koch, M., B. Haubold, and T. Mitchell-Olds. 2000. Comparative evolutionary analysis of chalcone synthase and alcohol dehydrogenase loci in *Arabidopsis*, *Arabis*, and related genera (Brassicaceae). Molecular Biology and Evolution 17:1483–1498.

Koch, M., B. Haubold, and T. Mitchell-Olds. 2001. Molecular systematics of the Brassicaceae: evidence from coding plastidic *matK* and nuclear *Chs* sequences. American Journal of Botany 88:534–544.

Lasceve, G., J. Leymarie, M. A. Olney, E. Liscum, J. M. Christie, A. Vavasseur, and W. R. Briggs. 1999. *Arabidopsis* contains at least four independent blue-light-activated signal transduction pathways. Plant Physiology 120:605–614.

Lee, H.-S., and Z. J. Chen. 2001. Protein-coding genes are epigenetically regulated in *Arabidopsis* polyploids. Proceedings of the National Academy of Sciences USA 98:6753–6758.

Loftus, G. R. 1993. A picture is worth a thousand *p* values: on the irrelevance of hypothesis testing in the microcomputer age. Behavior Research Methods, Instruments and Computers 25:250–256.

Losos, J. B., and D. B. Miles. 1994. Adaptation, constraint, and the comparative method: phylogenetic issues and methods. Pp. 60–98 in P. C. Wainwright and S. M. Reilly (eds.), Ecological Morphology: Integrative Organismal Biology. University of Chicago Press, Chicago.

Lynch, M. 1999. Estimating genetic correlations in natural populations. Genetical Research 74:255–264.

Magwene, P. M. 2001. New tools for studying integration and modularity. Evolution 55:1734–1745.

Martins, E. P. 2000. Adaptation and the comparative method. Trends in Ecology and Evolution 15:296–299.

Mezey, J. G., J. M. Cheverud, and G. P. Wagner. 2000. Is the genotype-phenotype map modular? A statistical approach using mouse Quantitative Trait Loci data. Genetics 156:305–311.

Mitchell-Olds, T. 2001. *Arabidopsis thaliana* and its wild relatives: a model system for ecology and evolution. Trends in Ecology and Evolution 16:693–699.

Mizoguchi, T., K. Irie, T. Hirayama, N. Hayashida, K. Yamaguchi-Shinozaki, K. Matsumoto, and K. Shinozaki. 1996. A gene encoding a mitogen-activated protein kinase kinase kinase is induced simultaneously with genes for a mitogen-activated protein kinase and an S6 ribosomal protein kinase by touch, cold, and water stress in *Arabidopsis thaliana*. Proceedings of the National Academy of Sciences USA 93:765–769.

Mummenhoff, K., and H. Hurka. 1995. Allopolyploid origin of *Arabidopsis suecica* (Fries) Norrlin: evidence from chloroplast and nuclear genome markers. Botanica Acta 108:449–456.

Napp-Zinn, K. 1985. *Arabidopsis thaliana*. Pp. 492–503 in A. H. Halevy (ed.), CRC Handbook of Flowering. CRC, Boca Raton, FL.

O'Kane, S. L., Jr., and I. A. Al-Shehbaz. 1997. A synopsis of *Arabidopsis* (Brassicaceae). Novon 7:323–327.

Pidkowich, M. S., J. E. Klenz, and G. W. Haughn. 1999. The making of a flower: control of floral meristem identity in *Arabidopsis*. Trends in Plant Science 4:64–70.

Pigliucci, M. 1998. Ecological and evolutionary genetics of *Arabidopsis*. Trends in Plant Science 3:485–489.

Pigliucci, M. 2001. Phenotypic Plasticity: Beyond Nature and Nurture. Johns Hopkins University Press, Baltimore, MD.

Pigliucci, M. 2002a. Are ecology and evolutionary biology "soft" sciences? Annales Zoologici Finnici 39:87–98.

Pigliucci, M. 2002b. Touchy and bushy: phenotypic plasticity and integration in response to wind stimulation in *Arabidopsis thaliana*. International Journal of Plant Science 163:399–408.

Pigliucci, M. 2003. From molecules to phenotypes? The promise and limits of integrative biology. Basic and Applied Ecology 4:297–306.

Pigliucci, M., K. Cammell, and J. Schmitt. 1999. Evolution of phenotypic plasticity: a comparative approach in the phylogenetic neighborhood of *Arabidopsis thaliana*. Journal of Evolutionary Biology 12:779–791.

Pigliucci, M., H. Pollard, and M. Cruzan. 2003. Comparative studies of evolutionary responses to light environments in *Arabidopsis*. American Naturalist 161:68–82.

Platt, J. R. 1964. Strong inference. Science 146:347–353.

Price, R. A., J. D. Palmer, and I. A. Al-Shehbaz. 1994. Systematic relationships of *Arabidopsis*: a molecular and morphological perspective. Pp. 7–20 in E. M. Meyerowitz and C. R. Somerville (eds.), *Arabidopsis*. Cold Spring Harbor Laboratory Press, Cold Spring Harbor, NY.

Pyke, K. 1994. *Arabidopsis*—its use in the genetic and molecular analysis of plant morphogenesis. New Phytologist 128:19–37.

Ratcliffe, D. 1965. The geographical and ecological distribution of *Arabidopsis* and comments on physiological variation. Web address: www.arabidopsis.org.

Roff, D. A. 1996. The evolution of genetic correlations: an analysis of patterns. Evolution 50:1392–1403.

Roff, D. A., and T. A. Mousseau. 1999. Does natural selection alter genetic architecture? An evaluation of quantitative genetic variation among populations of *Allenomobius socius* and *A. fasciatus*. Journal of Evolutionary Biology 12:361–369.

Rudge, D. W. 1998. A Bayesian analysis of strategies in evolutionary biology. Perspectives on Science 6:341–360.

Schlichting, C. D. 1986. The evolution of phenotypic plasticity in plants. Annual Review of Ecology and Systematics 17:667–693.

Schlichting, C. D. 1989a. Phenotypic integration and environmental change. BioScience 39:460–464.

Schlichting, C. D. 1989b. Phenotypic plasticity in *Phlox*. II. Plasticity of character correlations. Oecologia 78:496–501.

Schlichting, C. D., and M. Pigliucci. 1998. Phenotypic Evolution: A Reaction Norm Perspective. Sinauer, Sunderland, MA.

Schluter, D. 2001. Ecology and the origin of species. Trends in Ecology and Evolution 16:372–380.

Shoemaker, J. S., I. S. Painter, and B. S. Weir. 1999. Bayesian statistics in genetics: a guide for the uninitiated. Trends in Genetics 15:354–358.

Sistrunk, M. L., D. M. Antosiewicz, M. M. Purugganan, and J. Braam. 1994. *Arabidopsis TCH3* encodes a novel $Ca^{2+}$ binding protein and shows environmentally induced and tissue-specific regulation. Plant Cell 6:1553–1565.

Telewski, F. W., and M. L. Pruyn. 1998. Thigmomorphogenesis: a dose response to flexing in *Ulmus americana* seedlings. Tree Physiology 18:65–68.

van Emden, H. F., R. J. Macklin, and S. Staunton-Lambert. 1990. Stroking plants to reduce aphid populations. Entomologist 109:184–188.

Wade, P. R. 2000. Bayesian methods in conservation biology. Conservation Biology 14:1308–1316.

Wagner, G. 2001. The Character Concept in Evolutionary Biology. Academic Press, San Diego.

Wagner, G. P., and K. Schwenk. 2000. Evolutionarily stable configurations: functional integration and the evolution of phenotypic stability. Evolutionary Biology 31:155–217.

Wang, H., R. Datla, F. Georges, M. Loewen, and A. J. Cutler. 1995. Promoters from *kin1* and *cor6.6*, two homologous *Arabidopsis thaliana* genes: transcriptional regulation and gene expression induced by low temperature, ABA, osmoticum and dehydration. Plant Molecular Biology 28:605–617.

Westoby, M., M. R. Leishman, and J. M. Lord. 1995. On misinterpreting the "phylogenetic correction." Journal of Ecology 83:531–534.

Whitelam, G. C., and P. F. Devlin. 1998. Light signalling in *Arabidopsis*. Plant Physiology and Biochemistry 36:125–133.

Wisniewski, M. 1996. Recent advances in plant responses to stress. HortScience 31:30–57.

Wittgenstein, L. 1953/1973. Philosophical Investigations. Macmillan, New York.

Xu, W., M. M. Purugganan, D. H. Polisensky, D. M. Antosiewicz, S. C. Fry, and J. Braam. 1995. *Arabidopsis TCHA*, regulated by hormones and the environment, encodes a xyloglucan endotransglycosylase. Plant Cell 7:1555–1567.

Xu, W., P. Campbell, A. K. Vargheese, and J. Braam. 1996. The *Arabidopsis XET*-related gene family: environmental and hormonal regulation of expression. Plant Journal 9:879–889.

Yanovsky, M. J., J. J. Casal, and J. P. Luppi. 1997. The *VLF* loci, polymorphic between ecotypes Landsberg *erecta* and Columbia, dissect two branches of phytochrome A signal transduction that correspond to very-low-fluence and high-irradiance responses. Plant Journal 12:659–667.

Zhang, J., T. Van-Toai, L. Huynh, and J. Preiszner. 2000. Development of flooding-tolerant *Arabidopsis thaliana* by autoregulated cytokinin production. Molecular Breeding 6:135–144.

# 8

# Integrating Phenotypic Plasticity When Death Is on the Line

## *Insights from Predator–Prey Systems*

RICK A. RELYEA

Environmental heterogeneity is the norm of nature. As a result, many organisms have evolved the ability to alter their phenotypes to track changing environments and improve their fitness. During the past century, investigators have studied phenotypic plasticity with a primary focus on single traits (see discussion in Schlichting and Pigliucci 1998). This has provided us with an excellent collection of case studies that have documented the species that possess plastic traits, the types of traits that are plastic, and the types of environments that induce the trait changes. This focus on single traits has extended to models of plasticity evolution, primarily because single-trait models are considerably easier to handle mathematically than are multi-trait models (Levins 1963; Via and Lande 1985, 1987; Lively 1986; van Tienderen 1991; Gomulkiewicz and Kirkpatrick 1992; Moran 1992; Padilla and Adolph 1996). However, while the focus of theoretical and empirical work has been largely on single traits, it is becoming increasingly clear that we need to examine multiple traits and determine how they work together in producing an integrated organism. In this chapter, I argue that while it requires more time, money, and energy to examine multiple traits, it is imperative that we do so if we wish to understand the evolution and ecology of phenotypic plasticity.

In animal systems, one of the most extensively studied types of reactions to the environment is predator-induced plasticity (Dill 1987; Sih 1987; Lima and Dill 1990; Kats and Dill 1998; Tollrian and Harvell 1999). Thus, it can serve as a good system from which to draw examples of the importance of integrating multiple plastic traits. Over the past decade, there has been a growing interest in how predators induce changes in prey phenotypes. A simply query of *Biological*

*Abstracts* during the past decade (keywords = predator and (plasticity or behavior or morphology or physiology or life history)) indicates that the number of studies addressing the effects of predators on prey has increased from approximately 200 per year in 1992 to nearly 400 per year in 2001 (Fig. 8.1). During that same time, the number of studies that have specifically identified their work with the larger framework of phenotypic plasticity has increased more than fourfold (Fig. 8.1), but the percentage of these studies remains relatively low (4–9% from 1992 to 2001, respectively). This low percentage may be the result of many researchers believing that certain inducible traits do not qualify as phenotypic plasticity (e.g., behavioral or physiological traits). A more inclusive view of plasticity that combines predator-induced changes in behavior, physiology, morphology, and life history into a single conceptual framework is undoubtedly a more profitable way to understand how organisms integrate traits.

A cursory glance at the literature indicates that prey plasticity studies vary widely in the degree to which they address trait integration. Using the subset of studies produced by the query of "predator and plasticity" (189 studies from 1992 to 2001 in *Biological Abstracts*), I quantified how frequently these studies focused on one or more prey traits and how often they examined more than one type of trait (categorized as either behavior, morphology, life history, or physiology). In 125 of the studies, I was able to identify the number and types of traits from the abstracts. More than one-third of the studies examined only one trait, and 73% of the studies examined three or fewer traits (Fig. 8.2a). Through the decade, there

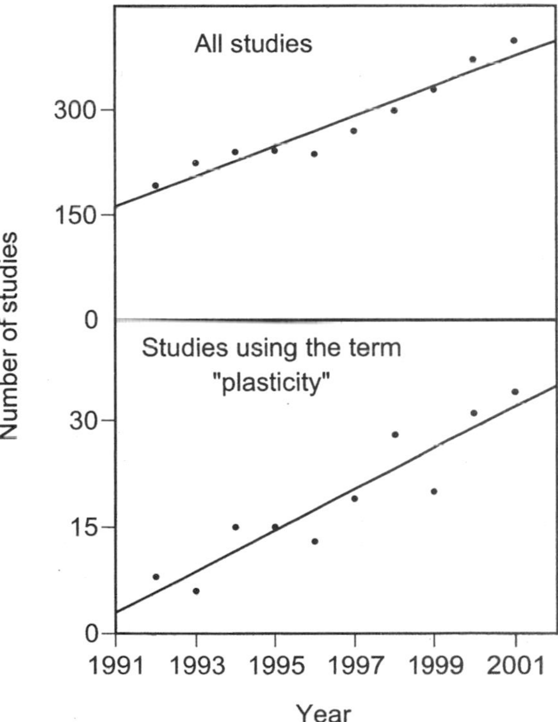


Figure 8.1 The results of a ten-year literature review (1992–2001) to determine the total number of studies that have examined the effect of predators on prey. The upper panel contains all studies that have examined predator-induced behavior, morphology, physiology, or life history. The lower panel contains the subset of those studies that have generalized their work into a framework of plasticity.

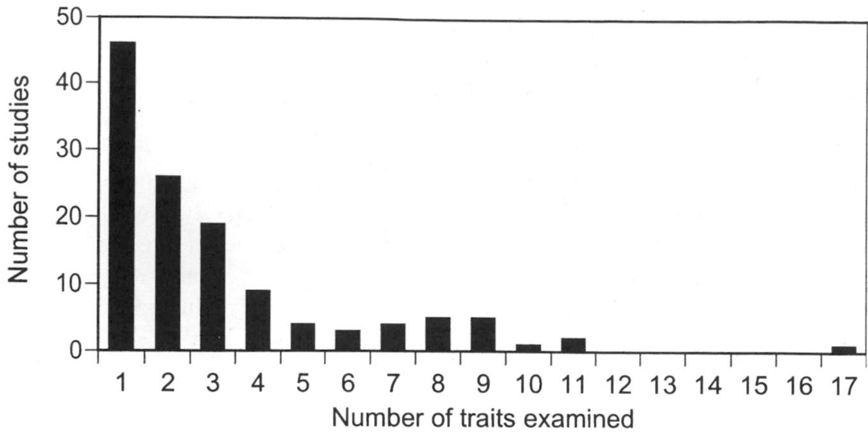


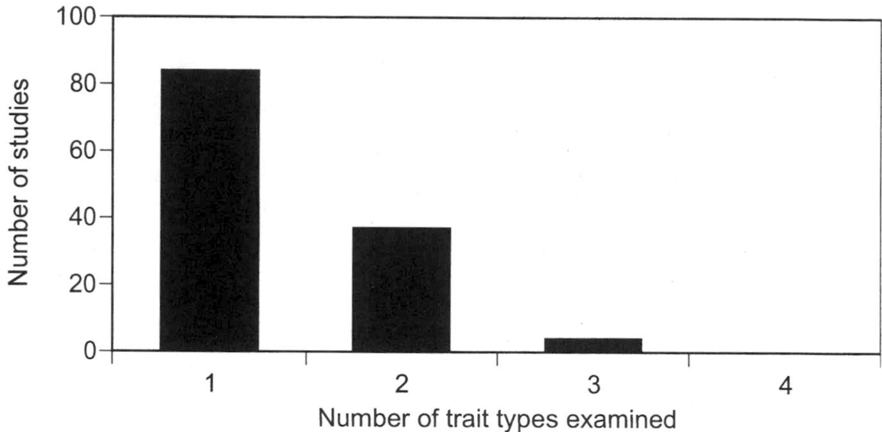


Figure 8.2 Predator-induced plasticity studies from 1992 to 2001. The upper panel represents the number of traits examined by the 125 studies. The lower panel represents the number of trait types that were examined (either behavior, morphology, life history, or physiology).

has been little change in the mean ($\pm 1$ SE) number of traits examined ($3.0 \pm 0.4$ for 1992–1996 versus $3.2 \pm 0.3$ for 1997–2001). There were similar emphases on behavioral traits (26%), morphological traits (41%), and life-historical traits (32%), but little attention to physiological traits (1%). Most studies examined only one type of trait (67%); nearly all studies examined two or fewer trait types (97%; Fig. 8.2b). In short, it appears that investigators of predator-induced plasticity have focused primarily on only a few traits, and typically we have focused on either behavior, morphology, or life history, but rarely have we integrated across different types of traits. The central question is whether we can actually learn more about predator-induced plasticity (and phenotypic plasticity in general) if we were to take a more integrated approach that did incorporate more traits and more types of traits.

## Prey Can Change a Lot of Different Traits

If we wish to understand predator–prey interactions, both from an ecological and an evolutionary perspective, it is imperative that we examine the full suite of traits that prey can change when predators are present and examine how these multiple defenses combine to produce a coordinated defense. Prey can employ numerous antipredator behaviors, including reduced foraging activity, increased spatial avoidance of areas containing predators, increased density to better detect the predator (e.g., schooling in fish, herding in mammals), pursuit-deterrents that signal prey health, and mobbing of predators (Dill 1987; Sih 1987; Lima and Dill 1990; Werner and Anholt 1993; Kats and Dill 1998). Prey also can alter numerous morphological traits in response to predators. Some morphological changes make prey more difficult to capture, including increased tail depth in larval anurans and salamanders (Smith and Van Buskirk 1995; McCollum and Van Buskirk 1996; Relyea 2001a; Van Buskirk and Schmidt 2000), and the production of "winged" morphs in protozoans (Kuhlmann and Heckmann 1985; Kusch 1995). Other morphological changes make prey more difficult to consume after being captured, including increased spine length in cladocerans (Krueger and Dodson 1981; Dodson and Havel 1988), increased body depth in fish (Brönmark and Miner 1992; Nilsson et al. 1995), and changes in shell shape and thickness in snails (DeWitt 1998; Trussell 2000a, 2000b). Predators can also induce prey to alter their life history, including changes in growth, development, size at metamorphosis, and size at sexual maturity (Skelly and Werner 1990; Riessen 1999; Barry 2000; Johnson 2001; Warkentin et al. 2001). Although studying the full suite of plastic prey traits is complex, it is this complexity that makes research on predator-induced plasticity simultaneously challenging and exciting. Below, I argue that empirically embracing this diversity of responses can have a profound effect on how we interpret our experiments and how empirical results provide feedback to our theoretical models.

## Determining the Antipredator Strategies of Prey

It is logical to hypothesize that prey responses to predators are adaptive, in that they make prey more resistant to predation. Although not all plasticity is adaptive, the vast majority of observed prey responses do appear to be in adaptive directions (Lima and Dill 1990). Extending this logic, most researchers have further hypothesized that prey should exhibit more extreme phenotypic responses in the presence of more dangerous predators. A lot of evidence supports this hypothesis. Most of the evidence is based upon studies that compare a prey's response to two predators (one that poses a low risk of prey death and another that poses a high risk) and studies that examine a single trait (usually a behavioral trait; Sih 1987). However, as discussed above, prey typically alter several traits. To rigorously test the hypothesis that prey should exhibit more extreme phenotypic responses in the presence of more dangerous predators, we should examine many traits and many categories of traits. For example, our studies of nine behavioral and morphological responses of six tadpole species to five different predators reject the hypothesis that more dangerous predators generally induce more extreme phenotypic responses (Relyea 2001a, 2001b).

So why is it that we can arrive at such different conclusions when we examine two predators and single traits than when we examine multiple predators and multiple traits? The answer lies in two important points. First, using only high- and low-risk predators is similar to using high-risk predators and no predators (given that few prey respond to low-risk predators). In reality, most prey face predation risks from a diverse assemblage of predators that have different hunting strategies. Second, most prey can alter multiple traits (Tollrian and Harvell 1999), and different trait changes are effective against different predator strategies (Relyea 2001b). By focusing on only one high-risk predator and only one trait, we may miss some of the complexity of such responses. For example, hiding may be an effective strategy against a hunting predator; rapid locomotion and unpalatability may be an effective strategy against an ambush predator; and rapid growth into a size refuge may be an effective strategy against a gape-limited predator. In other words, different predators have evolved numerous ways to kill prey and the latter can potentially evolve a variety of options to counter each predator's strategy. As a result, the best prey strategy may not simply be a scaled response of multiple defenses against different predators (i.e., correlated trait changes).

Because so few traits are typically observed in prey plasticity studies, we know little about the extent to which prey employ correlated trait changes against different predators. When trait changes are highly correlated, different predators induce a suite of traits to varying magnitudes. Under what conditions would we expect highly correlated defenses? If there are no genetic or developmental constraints that prevent traits from being correlated, prey should use correlated defenses when each defensive trait helps to improve survival against any predator (e.g., Scrimgeour and Culp 1994; Eklöv 2000). For example, if decreased activity, increased locomotory ability, and longer spines all help to reduce predation risk against all predators, then the prey could simply scale the magnitudes of the responses to the mortality risk posed by each predator. Extreme expression of defensive traits would not occur with low-risk predators because most antipredator phenotypes are costly (Werner et al. 1983; Woodward 1983; Leibold 1990; Neill 1990; Bollins and Frost 1991; Ringelberg 1991; Short and Holomuzki 1992; Semlitsch 1993; Van Buskirk and Relyea 1998; Relyea and Werner 1999). In contrast, if one trait is disproportionately effective against a given predator, then prey should only use the more effective trait. In this case, prey would use unique combinations of traits against each predator, resulting in uncorrelated prey responses across different predator species. Uncorrelated responses to different predators appear to be common in the limited set of studies that are available (Havel 1985; Peckarsky and McIntosh 1998; Turner et al. 2000; Relyea 2001a; Fig. 8.3).

While quantifying multiple traits is important for understanding how prey respond to individual predators, it is particularly important for understanding how prey respond to combinations of predators (the much more natural scenario for prey: Sih et al. 1998). Nearly two dozen studies of combined predator effects have been conducted during the past two decades, and these studies have demonstrated that a focus on a single trait is frequently insufficient to understand how prey respond to combined predators (reviewed in Relyea 2003a). For example,

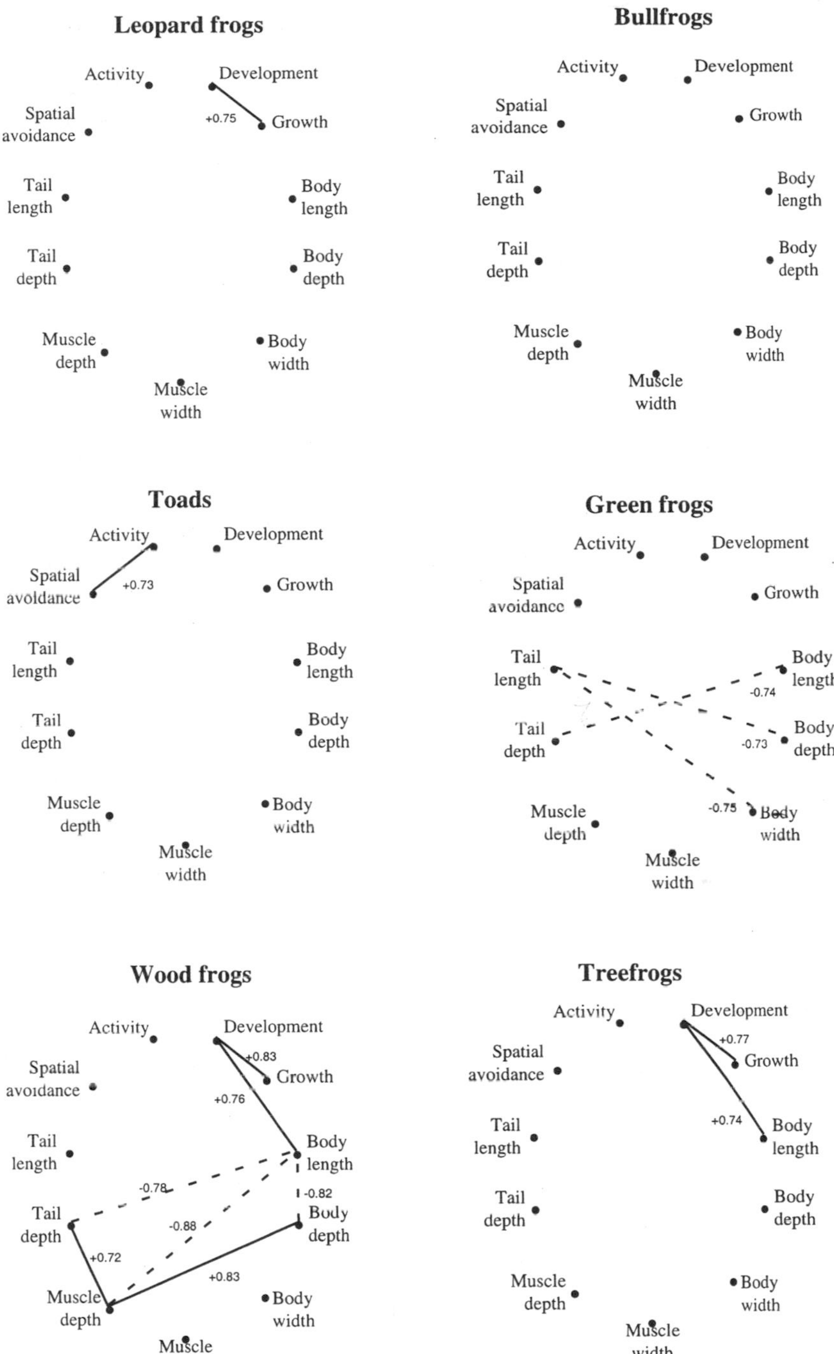


Figure 8.3 The integration of behavioral, morphological, and life-historical traits in six species of larval anurans when reared in five different predator environments. Solid lines indicate positive trait correlations, whereas dashed lines indicate negative trait correlations. Data are from Relyea (2001a).

predatory birds induce prey fish (*Leiostomus xanthurus*) to move into deeper water, while predatory flounder (*Paralichthys lethostigma*) induce prey fish to move into shallow water. When both predators are present, the prey fish do not exhibit the expected intermediate habitat use because the prey begin employing a second plastic response of increased schooling (Crowder et al. 1997). Similarly, fishing spiders (*Dolomedes vittatus*) scare water striders into open water, while sunfish scare water striders to the pool edge. When the two predators are combined, the predicted intermediate habitat use is not observed, because the prey add a second defense—activity reduction (Krupa and Sih 1998). Had these researchers not taken a more integrated trait approach to understand prey responses to combined predators, they would have been left with confusing conclusions rather than insightful interpretations.

## Searching for the Costs and Benefits of Alternative Phenotypes

Theoretical models of plasticity evolution are built on the premise that no single phenotype is superior in all environments. Empirically, this premise has been translated into the "adaptive plasticity hypothesis" (Dudley and Schmitt 1996), which states that an induced phenotype should experience higher fitness than alternative phenotypes when placed into the environment that caused the induction. In contrast, the same phenotype should experience lower fitness than alternative phenotypes when placed into alternative environments. While this operational definition is by no means the only way to achieve adaptive plasticity (e.g., a less plastic genotype could experience higher fitness across all environments), it has inspired an explosion of studies in search of these expected costs and benefits in predator-induced traits in animals and herbivore-induced traits in plants. This search for costs and benefits has met with rather mixed results (Karban and Baldwin 1997; Tollrian and Harvell 1999).

Given our recurrent bias toward examining only single (or few) traits, we may be inadvertently hindering our ability to identify the costs and benefits of alternative phenotypes. In some instances, focusing on a single trait may be sufficient. For example, most animals reduce their time spent foraging in the presence of predators. Reduced foraging reduces the probability of being detected by a predator, but less food is consumed (Gerritsen and Strickler 1977; Werner 1991; Juliano and Reminger 1992; Grill and Juliano 1996; Relyea and Werner 1999). Thus, the same trait directly determines both the organism's fitness benefit and fitness cost. Identifying costs and benefits can become more problematic when the predators alter multiple prey traits and we focus on only a single prey trait (Spitze 1992; Spitze and Sadler 1996). For example, aquatic predators simultaneously induce relatively large tails and small bodies in tadpoles. A focus on only the tail readily identified the benefit of the predator-induced phenotype (an increased ability to escape predator strikes: Van Buskirk et al. 1997; Van Buskirk and Relyea 1998). However, the cost of being predator-induced in a predator-free environment was not immediately evident. Early work found slower growth of predator-induced tadpoles in caged-predator environments (McCollum and Van

Buskirk 1996; Van Buskirk 2000), but caged-predator environments are not predator-free environments (they simply lack *lethal* predators). Consequently, these early results did not answer the theoretical question: why not exhibit the predator-induced phenotype in a predator-free environment? Looking across the suite of morphological traits induced by predators led us to hypothesize that the smaller tadpole body might translate into reduced growth ability in a predator-free environment (because a tadpole's body is dominated by intestines). This hypothesis was recently confirmed (Relyea 2002a): the larger tail directly produces the fitness benefit, but it is correlated with a smaller body, which produces the fitness cost. This sort of work demonstrates that when we examine a large number of traits, we are more likely to correctly identify the costs and benefits of alternative phenotypes.

## Searching for the Costs of Carrying Plasticity

While there has been a large effort toward identifying the costs and benefits of alternative phenotypes, there has been considerably less effort toward detecting the costs of simply carrying the ability to be plastic. These costs are incurred over and above the fitness effects of producing a particular phenotype (Newman 1992; DeWitt et al. 1998). If carrying plasticity were costly, then a plastic genotype exhibiting the same phenotype as a nonplastic genotype would experience lower fitness. The existence of plasticity costs is important to our knowledge of how plasticity evolves. If carrying greater plasticity reduces fitness, then the evolutionary outcome becomes a compromise between the benefits of exhibiting the optimal phenotype and the costs of carrying the plasticity. Possible costs of carrying plasticity include: (1) maintenance costs of sustaining the sensory and response pathways responsible for plastic responses (not required for nonplastic genotypes); (2) production costs above those necessary for a nonplastic individual to build the same phenotype; (3) genetic costs, in which plasticity genes interact with other genes via linkage, pleiotropy, or epistasis and reduce fitness; (4) developmental instability costs; (5) information acquisition costs that are required to detect changes in the environment (e.g., prey putting themselves at risk to detect a predator); and (6) limits on the organism's ability to assess its environment (Charnov and Bull 1977; Reeve 1989; Getty 1996; DeWitt et al. 1998).

Early models of plasticity evolution included costs of carrying plasticity, but there has been little empirical work on the question (Via and Lande 1985, 1987; van Tienderen 1991; Moran 1992). The first investigations frequently used predator-induced traits as model systems and were conducted using one (Nguyen et al. 1989; Krebs and Feder 1997), two (DeWitt 1998), or three traits (Scheiner and Berrigan 1998). In all cases, the evidence for plasticity costs was either weak or nonexistent. In an investigation using wood frog tadpoles, we examined eight behavioral and morphological traits and found that detecting the costs of carrying plasticity was highly dependent on which traits we examined (Relyea 2002b). Possessing greater plastic ability could cause decreased fitness (body length), increased fitness (tail depth), or no effect on fitness (activity). Surprisingly, theoretical models have never considered the possibility that simply carrying plasticity (whether expressed or not) could increase fitness. By casting a wide experimental net across a larger number of traits, we can better appreciate the complex relationships between predator-induced plasticity and fitness to provide critical feedback to theoretical models of plasticity evolution.

## Observing Antipredator Strategies over Ontogeny

As prey grow, their vulnerability to a given predator can either increase or decrease, depending on the size selection preferences of the predator (Brooks and Dodson 1965; Werner 1988; Persson and Eklöv 1995). Therefore, a prey's antipredator strategy should change over ontogeny. For reasons that are not clear, the ontogeny of predator-induced plasticity remains poorly studied, yet the few studies available suggest that it is indeed an intriguing phenomenon; phenotypic responses to predators can change throughout a prey's lifetime (Krueger and Dodson 1981; Harvell 1991; Pettersson et al. 2000). In two studies that have taken an integrated approach that combined both behavioral and morphological responses to predators, researchers have found that only behavioral defenses were used against predators early in ontogeny. Later in ontogeny, prey began to develop morphological defenses and gradually terminated their use of behavioral defenses (Pettersson et al. 2000; Relyea 2003b). Had the researchers only examined the behavioral defenses, they would have concluded that prey do not exhibit predator-induced defenses later in ontogeny and, therefore, must by that time be invulnerable to predators. Such incorrect conclusions were avoided by integrating across multiple prey traits.

## Identifying the Mechanistic Causes of Trait-Mediated Effects in a Community

While hundreds of studies have documented that predators induce prey to alter their phenotypes (e.g., Lima and Dill 1990; Travis 1994; Kats and Dill 1998), the impact of environmentally induced phenotypes goes well beyond the level of the individual. Not long ago, ecologists argued that, because an individual's traits determine an individual's performance (e.g., competitive ability or predator resistance), predator-induced changes in prey traits should alter prey performance and, as a result, alter the per-capita interaction rates between prey and other species in the community (Abrams 1983; Werner 1992). In short, predator-induced changes in prey traits should cause "trait-mediated indirect effects" (Abrams 1995) that alter how species interact with each other and ultimately change the structure of ecological communities. These effects are distinct from the numerically mediated indirect effects (changes in the number of individuals via predation or competition) that have been the traditional focus of ecological studies (e.g., Paine 1966; Lubchenco 1978; Menge 1995). In the past decade, numerous investigators have confirmed that trait-mediated effects can alter the interactions between predators and prey, between prey and their resources, and between two species of competing prey (Power et al. 1985; Mittelbach and Chesson 1987; Turner and Mittelbach 1990; Wootton 1992; Wissinger and McGrady 1993; Diehl and Eklöv 1995; Persson and Eklöv 1995; Werner and Anholt 1996; Peacor and Werner 1997; Schmitz et al. 1997; Relyea and Yurewicz 2002).

Similar to the history of examining predator-induced traits in general, studies of trait-mediated indirect effects have rarely taken an integrated approach. They have focused almost exclusively on behavioral traits and typically on a single behavioral trait. While this bias may have occurred because behavioral studies were the most obvious places to begin searching for trait-mediated effects, there is no inherent reason to believe that other types of plastic traits could not also cause trait-mediated indirect effects. For example, predator-induced morphological changes also can affect a prey's performance by increasing predator resistance and decreasing competitive ability. As a result, predator-induced changes in morphology should be able to cause trait-mediated effects. For example, when larval wood frogs and leopard frogs are raised together in a pond, wood frogs are the superior competitors. However, if we add caged fish or caged dragonfly predators, wood frogs produce a more extreme defensive morphology than leopard frogs at the cost of becoming the inferior competitors. In short, the caged predators (which cannot alter the numbers of either tadpole species) induce differential trait changes that can reverse the outcome of competition (Relyea 2000). Trait-mediated effects also can occur with physiological traits. For example, Martinsen et al. (1998) found that when beaver (*Castor canadensis*) consume aspen trees, the aspens produce sprouts with elevated antimammal chemicals that are sequestered by herbivorous beetles; these sequestered chemicals subsequently reduce the per-capita predation rate between beetles and their mammalian predators. Thus, the antiherbivore physiological response of aspen caused a trait-mediated effect between beetles and beetle predators. These examples make it clear that although the focus has remained largely on predator-induced behaviors, trait-mediated effects can be generated by many different types of predator-induced responses.

Because predators can induce numerous traits and types of traits, we face the challenge of determining which trait changes are responsible for the trait-mediated indirect effects that we observe. This is a relatively trivial question if all of the traits change in directions that affect performance in the same way (e.g., changes in behavior, physiology, and morphology that all make the prey more resistant to predation), although we may wish to know the relative contribution of each trait. The situation is much more challenging if a predator alters a suite of traits and each trait affects performance in different ways (e.g., behavioral changes that decrease competitive ability, physiological changes that have no effect on competitive ability, and morphological changes that increase competitive ability). Because there have been so few studies incorporating multiple traits in trait-mediated effects, it remains to be seen if the latter scenario ever occurs.

## The Challenges That Lie Before Us

The suite of possible plastic traits that prey can possess is indeed impressive and underscores the need for investigators to consider a greater diversity of traits and to determine how prey integrate multiple traits to form a cohesive defense against the assemblage of predators that they face. It is not enough to simply catalog how numerous traits change in the presence of predators. The challenge is to understand how each trait functions and how traits interact with each other to interpret how prey make their antipredator decisions. As future studies will undoubtedly focus on more and more traits, we must not lose sight of the need to put such "dissected" prey back together into a cohesive whole, for it is the whole organism that is the ecological and evolutionary unit of interest. At the same time, because of practical constraints, researchers cannot realistically measure every conceivably plastic trait in preys. Rather, we must rely on our knowledge of natural history and our empirical experience in deciding which traits to examine, while continually considering integrating additional traits into our research.

As we continue to take a more integrated approach, several specific research avenues need to be pursued. First, we need more studies that examine diverse prey responses across multiple predators to determine whether prey commonly employ highly correlated suites of traits against predators or noncorrelated, predator-specific responses. Such studies could investigate both responses to separate predators as well as several combinations of predators. Second, while the search for benefits of predator-induced phenotypes has been fruitful, in many systems the hunt for the costs that must be associated to these phenotypes has been less productive. We need a more extensive approach that incorporates more traits to provide empirical clues as to the traits responsible for costs and the conditions under which these costs might be expressed. Third, the fitness effects of simply carrying the ability to be plastic is unknown in most systems, preventing any general sense of how often it is costly or beneficial to carry the plasticity option. Fourth, although we have hundreds of studies that examine predator-induced trait changes at a given developmental stage, an integrated approach needs to be taken over prey ontogeny, a point that has been made repeatedly in the past (e.g., Schlichting and Pigliucci 1998). Finally, investigators need to explore the wide variety of traits that can cause trait-mediated effects. When investigators use their knowledge of natural history combined with a holistic phenotypic perspective to select the traits to be examined, we are more likely to make progress in understanding the ecology and evolution of predator-induced plasticity.

*Acknowledgments* I should like to thank M. Pigliucci and K. Preston for their reviews of this chapter. I thank the National Science Foundation for their continued support of my work.

## *Literature cited*

Abrams, P. A. 1983. Arguments in favor of higher-order interactions. American Naturalist 121:887–891.

Abrams, P. A. 1995. Implications of dynamically variable traits for identifying, classifying, and measuring direct and indirect effects in ecological communities. American Naturalist 146:112–134.

Barry, M. J. 2000. Inducible defences in *Daphnia*: responses to two closely related predator species. Oecologia 124:396–401.

Bollins, S. M., and B. W. Frost. 1991. Diel vertical migration in zooplankton: rapid individual response to predators. Journal of Plankton Research 13:1359–1365.

Brönmark, C., and J. G. Miner. 1992. Predator-induced phenotypical change in body morphology in Crucian carp. Science 258:1348–1350.

Brooks, J. L., and S. I. Dodson. 1965. Predation, body size, and composition of plankton. Science 150:28–35.

Charnov, E. L., and J. Bull. 1977. When is sex environmentally determined? Nature 266:828–830.

Crowder, L. B., D. D. Squires, and J. A. Rice. 1997. Nonadditive effects of terrestrial and aquatic predators on juvenile estuarine fish. Ecology 78:1796–1804.

DeWitt, T. J. 1998. Costs and limits of phenotypic plasticity: tests with predator-induced morphology and life history in a freshwater snail. Journal of Evolutionary Biology 11:465–480.

DeWitt, T. J., A. Sih, and D. S. Wilson. 1998. Costs and limits of phenotypic plasticity. Trends in Ecology and Evolution 13:77–81.

Diehl, S., and P. Eklöv. 1995. Effects of piscivore-mediated habitat use on resources, diet, and growth of perch. Ecology 76:1712–1726.

Dill, L. M. 1987. Animal decision making and its ecological consequences: the future of aquatic ecology and behavior. Canadian Journal of Zoology 65:803–811.

Dodson, S. I. and J. E. Havel. 1988. Indirect prey effects: some morphological and life history responses of *Daphnia pulex* exposed to *Notonecta undulata*. Limnology and Oceanography 33:1274–1285.

Dudley, S. A., and J. Schmitt. 1996. Testing the adaptive plasticity hypothesis: density-dependent selection on manipulated stem length in *Impatiens capensis*. American Naturalist 147:445–465.

Eklöv, P. 2000. Chemical cues from multiple predator-prey interactions induce changes in behavior and growth of anuran larvae. Oecologia 123:192–199.

Gerritsen, J., and J. R. Strickler. 1977. Encounter probabilities, and the community structure in zooplankton: a mathematical model. Journal of the Fisheries Board of Canada 34:73–82.

Getty, T. 1996. The maintenance of phenotypic plasticity as a signal detection problem. American Naturalist 148:378–385.

Gomulkiewicz, R., and M. Kirkpatrick. 1992. Quantitative genetics and the evolution of reaction norms. Evolution 46:390–411.

Grill, C. P., and S. A. Juliano. 1996. Predicting species interactions based on behaviour: predation and competition in container-dwelling mosquitoes. Journal of Animal Ecology 65:63–76.

Harvell, C. D. 1991. Coloniality and inducible polymorphism. American Naturalist 138:1–14.

Havel, J. E. 1985. Predation of common invertebrate predators on long- and short-featured *Daphnia retrocurva*. Hydrobiologia 124:141–149.

Johnson, J. B. 2001. Adaptive life-history evolution in the livebearing fish *Brachyrhaphis rhabdophora*: genetic basis for parallel divergence in age and size at maturity and a test of predator-induced plasticity. Evolution 55:1486–1491.

Juliano, S. A., and L. Reminger. 1992. The relationship between vulnerability to predation and behavior of larval treehole mosquitoes: geographic and ontogenetic differences. Oikos 63:465–476.

Karban, R, and I. T. Baldwin. 1997. Induced Responses to Herbivory. University of Chicago Press, Chicago.

Kats, L. B., and L. M. Dill. 1998. The scent of death: chemosensory assessment of predation risk by prey animals. Ecoscience 5:361–394.

Krebs, R. A., and M. E. Feder. 1997. Natural variation in the expression of the heat-shock protein HSP70 in a population of *Drosophila melanogaster* and its correlation with tolerance of ecologically relevant thermal stress. Evolution 51:173–179.

Krueger, D. A., and S. I. Dodson. 1981. Embryological induction and predation ecology in *Daphnia pulex*. Limnology and Oceanography 26:219–223.

Krupa, J .J., and A. Sih. 1998. Fishing spiders, green sunfish, and a stream-dwelling water strider: male-female conflict and prey responses to single versus multiple predator environments. Oecologia 117:258–265.

Kuhlmann, H., and K. Heckmann. 1985. Interspecific morphogens regulating prey-predator relationships in protozoa. Science 227:1347–1349.

Kusch, J. 1995. Adaptation of inducible defenses in *Euplotes diadaleos* (Ciliophora) to predation risks by various predators. Microbial Ecology 30:79–88.

Leibold, M. 1990. Resources and predators can affect the vertical distributions of zooplankton. Limnology and Oceanography 35:938–944.

Levins, R. 1963. Theory of fitness in a heterogeneous environment. II. Developmental flexibility and niche selection. American Naturalist 47:75–90.

Lima, S. L., and L. M. Dill. 1990. Behavioral decisions made under the risk of predation: a review and prospectus. Canadian Journal of Zoology 68:619–640.

Lively, C. M. 1986. Canalization versus developmental conversion in a spatially-variable environment. American Naturalist 128:561–572.

Lubchenco, J. 1978. Plant species diversity in a marine intertidal community: importance of herbivore food preference and algal competitive abilities. American Naturalist 112:23–39.

Martinsen, G. D., E. M. Driebe, and T. G. Whitham. 1998. Indirect interactions mediated by changing plant chemistry: beaver browsing benefits beetles. Ecology 79:192–200.

McCollum, S. A., and J. Van Buskirk. 1996. Costs and benefits of a predator-induced polyphenism in the gray tree frog *Hyla chrysocelis*. Evolution 50:583–593.

Menge, B. A. 1995. Indirect effects in marine rocky intertidal interaction webs: patterns and importance. Ecological Monographs 65:21–74.

Mittelbach, G. G., and P. L. Chesson. 1987. Predation risk: indirect effects on fish populations. Pp. 315–332 in W. C. Kerfoot and A. Sih, eds. Predation: Direct and Indirect Impacts on Aquatic Communities. University Press of New England, Hanover, NH.

Moran, N. A. 1992. The evolutionary maintenance of alternative phenotypes. American Naturalist 139:971–989.

Neill, W. E. 1990. Induced vertical migration in copepods as a defence against invertebrate predation. Nature 345:524–526.

Newman, R. A. 1992. Adaptive plasticity in amphibian metamorphosis. BioScience 42:671–678.

Nguyen, T. N. M., Q. G. Phan, L. P. Duong, K. P. Bertrand, and R. E. Lenski. 1989. Effects of carriage and expression of the Tn10 tetracyline-resistance operon on the fitness of *Escherichia coli* K12. Molecular Biology and Evolution 6:213–225.

Nilsson, P. A., C. Brönmark, and L. B. Pettersson. 1995. Benefits of a predator-induced morphology in crucian carp. Oecologia 104:291–296.

Padilla D. K., and S. C. Adolph 1996. Plastic inducible morphologies are not always adaptive: the importance of time delays in a stochastic environment. Evolutionary Ecology 10:105–117.

Paine, R. T. 1966. Food web complexity and species diversity. American Naturalist 100: 65–75.

Peacor, S. D., and E. E. Werner. 1997. Trait-mediated indirect interactions in a simple aquatic food web. Ecology 78:1146–1156.

Peckarsky, B. L., and McIntosh, A. R. 1998. Fitness and community consequences of avoiding multiple predators. Oecologia 113:565–576.

Persson, L. and P. Eklöv. 1995. Prey refuges affecting interactions between piscivorous perch and juvenile perch and roach. Ecology 76:70–81.

Pettersson, L. B., P. A. Nilsson, and C. Brönmark. 2000. Predator recognition and defence strategies in crucian carp, *Carassius carassius*. Oikos 88:200–212.

Power, M. E., W. J. Matthews, and A. J. Stewart. 1985. Grazing minnows, piscivorous bass, and stream algae: dynamics of a strong interaction. Ecology 66:1448–1456.

Reeve, H. K. 1989. The evolution of conspecific acceptance thresholds. American Naturalist 133:407–435.

Relyea, R. A. 2000. Trait-mediated effects in larval anurans: reversing competitive outcomes with the threat of predation. Ecology 81:2278–2289.

Relyea, R. A. 2001a. Morphological and behavioral plasticity of larval anurans in response to different predators. Ecology 82:523–540.

Relyea, R. A. 2001b. The relationship between predation risk of different predators and the predator-induced, phenotypic plasticity of their prey. Ecology 82:541–554.

Relyea, R. A. 2002a. Competitor-induced plasticity in tadpoles: consequences, cues, and connections to predator-induced plasticity. Ecological Monographs 72:523–540.

Relyea, R. A. 2002b. Costs of phenotypic plasticity. American Naturalist 159:272–282.

Relyea, R. A. 2003a. How prey respond to combined predators: a review and an empirical test. Ecology 84:1827–1839.

Relyea, R. A. 2003b. Predators come and predators go: the reversibility of predator-induced traits. Ecology 84:1840–1848.

Relyea, R. A., and E. E. Werner. 1999. Quantifying the relation between predator-induced behavior and growth performance in larval anurans. Ecology 80:2117–2124.

Relyea, R. A., and K. L. Yurewicz. 2002. Predicting community outcomes from pairwise interactions: integrating density- and trait-mediated effects. Oecologia 131:569–579.

Riessen, H. P. 1999. Predator-induced life history shifts in *Daphnia*: a synthesis of studies using meta-analysis. Canadian Journal of Fisheries and Aquatic Sciences 56:2487–2494.

Ringelberg, J. 1991. Enhancement of the phototactic reaction in *Daphnia hyalina* by a chemical mediated by juvenile perch (*Perca fluviatilis*). Journal of Plankton Research 13:17–25.

Scheiner, S. M., and Berrigan, D. 1998. The genetics of phenotypic plasticity. VIII. The cost of plasticity in *Daphnia pulex*. Evolution 52:368–378.

Schlichting, C. D., and M. Pigliucci. 1998. Phenotypic Evolution: A Reaction Norm Perspective. Sinauer, Sunderland, MA.

Schmitz, O. J., A. P. Beckerman, and K. M. O'Brien. 1997. Behaviorally mediated trophic cascades: effects of predation risk on food web interactions. Ecology 78:1388–1399.

Scrimgeour, G. J., and J. M. Culp. 1994. Foraging and evading predators: the effect of predator species on a behavioural trade-off by a lotic mayfly. Oikos 69:71–79.

Semlitsch, R. D. 1993. Effects of different predators on the survival and development of tadpoles from the hybridogenetic *Rana esculenta* complex. Oikos 647:40–46.

Short, T. M., and J. R. Holomuzki. 1992. Indirect effects of fish on foraging behaviour and leaf processing by the isopod *Lirceus fontinalis*. Freshwater Biology 27:91–97.

Sih, A. 1987. Predators and prey lifestyles: an evolutionary and ecological overview. Pp. 203–224 in W. C. Kerfoot and A. Sih, eds. Predation: Direct and Indirect Impacts on Aquatic Communities. University Press of New England, Hanover, NH.

Sih, A., G. Englund, and D. Wooster. 1998. Emergent impacts of multiple predators on prey. Trends in Ecology and Evolution 13:350–355.

Skelly, D. K., and E. E. Werner. 1990. Behavioral and life-historical responses of larval American toads to an odonate predator. Ecology 71:2313–2322.

Smith, D. C., and J. Van Buskirk. 1995. Phenotypic design, plasticity, and ecological performance in two tadpole species. American Naturalist 145:211–233.

Spitze, K. 1992. Predator-mediated plasticity of prey life-history and morphology: *Chaoborus americanus* predation on *Daphnia pulex*. American Naturalist 139:229–247.

Spitze, K. and T. D. Sadler. 1996. Evolution of a generalist genotype: multivariate analysis of the adaptiveness of phenotypic plasticity. American Naturalist 148:S108–S123.

Tollrian, R., and D. Harvell. 1999. The Ecology and Evolution of Inducible Defenses. Princeton University Press, Princeton, NJ.

Travis, J. 1994. Evaluating the adaptive role of morphological plasticity. Pp. 99–122 in P. C. Wainwright and S. M. Reilly, eds. Ecological Morphology. University of Chicago Press, Chicago.

Trussell, G. C. 2000a. Phenotypic clines, plasticity, and morphological trade-offs in an intertidal snail. Evolution 54:151–166.

Trussell, G. C. 2000b. Predator-induced plasticity and morphological trade-offs in latitudinally separated populations of *Littorina obtusata*. Evolutionary Ecology Research 2:803–822.

Turner, A. M., and G. G. Mittelbach. 1990. Predator avoidance and community structure: interactions among piscivores, planktivores, and plankton. Ecology 71:2241–2254.

Turner, A. M., R. J. Bernot, and C. M. Boes. 2000. Chemical cues modify species interactions: the ecological consequences of predator avoidance by freshwater snails. Oikos 88:148–158.

Van Buskirk, J. 2000. The costs of an inducible defense in anuran larvae. Evolution 81:2813–2821.

Van Buskirk, J., and R. A. Relyea. 1998. Natural selection for phenotypic plasticity: predator-induced morphological responses in tadpoles. Biological Journal of the Linnean Society 65:301–328.

Van Buskirk, J., and B. R. Schmidt. 2000. Predator-induced phenotypic plasticity in larval newts: trade-offs, selection, and variation in nature. Ecology 81:3009–3028.

Van Buskirk, J., S. A. McCollum, and E. E. Werner. 1997. Natural selection for environmentally-induced phenotypes in tadpoles. Evolution 52:1983–1992.

van Tienderen, P. H. 1991. Evolution of generalists and specialists in spatially heterogeneous environments. Evolution 45:1317–1331.

Via, S., and R. Lande. 1985. Genotype-environment interaction and the evolution of phenotypic plasticity. Evolution 39:502–522.

Via, S., and R. Lande. 1987. Evolution of genetic variability in a spatially heterogeneous environment: effects of genotype-environment interaction. Genetic Research 49:147–156.

Warkentin, K. M., C. R. Currie, and S. A. Rehner. 2001. Egg-killing fungus induces early hatching of red-eyed treefrog eggs. Ecology 82:2860–2869.

Werner, E. E. 1988. Size, scaling, and the evolution of complex life cycles. Pp. 60–81 in B. Ebenman and L. Persson, eds. Size-Structured Populations. Springer-Verlag, Berlin.

Werner, E. E. 1991. Nonlethal effects of a predator on competitive interactions between two anuran larvae. Ecology 72:1709–1720.

Werner, E. E. 1992. Individual behavior and higher-order species interactions. American Naturalist 140:S5–S32.

Werner, E. E., and B. R. Anholt. 1993. Ecological consequences of the trade-off between growth and mortality rates mediated by foraging activity. American Naturalist 142:242–272.

Werner, E. E., and B. R. Anholt. 1996. Predator-induced behavioral indirect effects: consequences to competitive interactions in anuran larvae. Ecology 77:157–169.

Werner, E. E., J. F. Gilliam, D. J. Hall, and G. G. Mittelbach. 1983. An experimental test of the effects of predation risk on habitat use in fish. Ecology 64:1540–1548.

Wissinger, S., and J. McGrady. 1993. Intraguild predation and competition between larval dragonflies: direct and indirect effects on shared prey. Ecology 74:207–218.

Woodward, B. D. 1983. Predator-prey interactions and breeding-pond use of temporary-pond species in a desert anuran community. Ecology 64:1549–1555.

Wootton, J. T. 1992. Indirect effects, prey susceptibility, and habitat selection: impacts of birds on limpets and algae. Ecology 73:981–991.

PART III

# GENETICS AND MOLECULAR BIOLOGY OF PHENOTYPIC INTEGRATION

The question of the genetics and molecular basis of biologically relevant processes is one that often plays a central role in organismal biology, especially given the panoply of techniques that have recently become available to molecular and developmental geneticists. However, one needs to be clear on what exactly "the question" is in cases such as these.

There are two general meanings that we can associate with the quest for the genetic bases of organismal characteristics, such as phenotypic integration or plasticity. On the one hand, one can ask if there is a particular sort of genetic mechanism underlying a particular sort of biological process. For example, we can pursue the genetic bases of phenotypic plasticity (Scheiner 1993; Schlichting and Pigliucci 1993; Via 1993). On the other hand, one can ask a much more specific sort of question, along the lines of: what are the genetic bases of *this particular instantiation* of phenomenon X (where X can be plasticity, integration, allometry, or whatever else one might be interested in)?

We think that asking the question in the first sense is essentially meaningless in evolutionary biology, regardless of the fact that it is done as a matter of normal course. It is becoming clear from the mounting evidence coming from molecular genetics itself that—broadly speaking—the genetic bases of *any* high-level phenotype are essentially the same: we will always find some regulatory switch that channels development down one particular pathway (sometime the choice is between two or more, and sometimes the choice is determined by a receptor of environmental signals); there will always be a series of "transduction elements," often arranged in complex networks rather than in a nice linear sequence (Wilkins 2002, chap. 4, esp. fig. 4.5); finally, there will be a large number of "structural" genes whose products will actually do all the work that the high-level regulatory elements will instruct them to do and that results in the particular phenotype object of our attentions. Above the basic molecular level, all of this can be recast as intracellular interactions and intercellular morphogenic gradients, and at a still higher level of analysis this will result in the observation of certain genetic variance-covariance structures among the traits involved. There will always be, for any complex phenotype, genes with "large effects" and a variable number of "modifiers," which will be characterized by more or less complex pleiotropic and epistatic effects. This general story will hold no matter what the phenotype or the organism. So, if one asks, "What are the genetic bases of phenotypic integration?" in this generic sense, the answer is going to be: pretty much the same as those of any other complex aspect of the phenotype.

In the second sense, on the other hand, the question does have sensible and interesting answers, but they may not fit the aspirations at reaching general conclusions held by some evolutionary biologists. If what we mean when we inquire about the genetic bases of trait X is which particular genetic elements, biochemical pathways, and developmental processes underlie the production of the *specific* phenotype X, then the answer is going to be intricate and fascinating. Indeed, the answer will often require a lot of painstaking work, and will yield increasing amounts of detail, eventually resulting in the completion of a satisfyingly engaging puzzle. However, that is exactly how it will look to anybody else who was not interested in that particular phenotype: we have solved another puzzle, but we have gained little in terms of general results that can be applied across the board to the problems of our science.

The philosopher of science Thomas Kuhn (1970) pointed out that most of science is indeed a puzzle-solving activity, and that only rarely do we experience one of those great moments of turmoil that reshape our thinking about nature (what Kuhn called a "scientific revolution"). Biology has already had several of these earthquakes, starting with Darwin's discovery of natural selection, and continuing with the elucidation of the structure of DNA. No doubt, there will be many more in future centuries. But most of what we do as practicing scientists falls into the rather more modest, but no less engaging, activity of puzzle solving.

One other way to consider the difference between the first and second sense of the "What are the genetic bases of X?" question is that evolutionary biology is a rather strange beast as far as sciences go. By its nature, it is a mix of historical contingency, accessible only through indirect observation, and general patterns (such as Mendel's laws, or the structure of DNA), on which one can conduct repeatable experimental research. Results of general interest tend to come from the second aspect of this mix, though occasionally they do come from historical studies. The latter, though, provide that magnificent repertoire of case studies that usually attracts organismal biologists to their discipline to begin with. As evolutionary biologists, we are destined to balance our work between the two extremes of arriving at general, but rare, conclusions on the one hand, and a cornucopia of specific cases, which, however, require mostly local explanations.

The three chapters in this section reflect the above considerations to some extent. We start with Chapter 9 by Courtney Murren and Paula Kover, who address the use of quantitative trait loci techniques in studying the genetics of phenotypic integration. After a brief summary of how QTL research is conducted in general, they discuss the specific statistical and methodological problems posed by the functional integration of several traits. They then consider several examples from the literature, focusing on situations in which the environment plays a determinant role (plasticity of integration), those in which we wish to follow the changing genetic bases of a given trait throughout its development, and finally the study of quantitative trait loci underlying major evolutionary shifts such as the evolution of maize from teosinte.

In Chapter 10, Christian Klingenberg gives us an overview of how in fact it is possible to study phenotypic integration at multiple levels, from the molecular to the morphological. He starts out by considering the function of intercellular signals and the conceptual role played by the always relevant idea of "morphogen." The latter leads him into a discussion of morphogenetic fields from the modern standpoint of molecular and developmental biology, eventually arriving at the idea that integration results from developmental interactions that manifest themselves in the observable covariation among the parts of the finished structure. This covariation can then be tackled with modern versions of the morphometric approach, to which Klingenberg has provided many valuable contributions and examples himself.

Chapter 11, the last in this section, is by one of the editors, Massimo Pigliucci, and it deals with the organismal end of the question of the genetics of phenotypic integration. In particular, Pigliucci addresses the experimental investigation of Wagner's (1995) idea that both current and short-term future genetic variation (what he refers to respectively as variation and variability) need to be estimated in order to be able to predict the likely response of a population to natural selection. After examining specific implementations of this idea, the chapter ends with a broader philosophical discussion of the degree of usefulness of adopting a metaphor of "evolutionary forces" in guiding theoretical and empirical research on the relationship between selection and "constraints."

## *Literature Cited*

Kuhn, T. 1970. The Structure of Scientific Revolutions. University of Chicago Press, Chicago.

Scheiner, S. M. 1993. Plasticity as a selectable trait: reply to Via. American Naturalist 142:371–373.

Schlichting, C. D., and M. Pigliucci. 1993. Evolution of phenotypic plasticity via regulatory genes. American Naturalist 142:366–370.

Via, S. 1993. Adaptive phenotypic plasticity: target or by-product of selection in a variable environment? American Naturalist 142:352–365.

Wagner, G. P. 1995. Adaptation and the modular design of organisms. Pp. 317–328 in F. Moran, A. Moreno, J. J. Merelo and P. Chacon, eds. Advances in Artificial Life. Springer, Berlin.

Wilkins, A. S. 2002. The Evolution of Developmental Pathways. Sinauer, Sunderland, MA.

# 9

# QTL Mapping

## *A First Step Toward an Understanding of Molecular Genetic Mechanisms Behind Phenotypic Complexity/ Integration*

COURTNEY J. MURREN
PAULA X. KOVER

The challenge of the study of phenotypic integration is to comprehend both how a single trait evolves, and how combinations of traits evolve together to produce and maintain a functional phenotype. Olson and Miller (1958), in their now classic work, define morphological integration as the interdependence of morphological traits that produces an organized functional organism. Empirical studies demonstrate that selection on multiple traits throughout ontogeny and across environments can result in character correlations (Schlichting and Pigliucci 1998). Ecological genetic approaches suggest that gene regulation, pleiotropy, and linkage may provide the genetic mechanisms of phenotypic correlations (Cheverud 1988; Schlichting 1989; Schlichting and Pigliucci 1998). However, further progress in unraveling the genetic basis of phenotypic integration will require moving a step closer to the genes that are involved in determining the traits and their interactions.

Investigations into the genetic basis of phenotypic integration and the plasticity of integration have historically used one of two approaches. The first method is based on the use of isolated mutants that affect a well-established phenotypic correlation between traits across environments. The second method involves examination of gene or protein expression across environments, developmental stages, and so on (Pigliucci 1996). Although very useful, these methods are restricted to traits for which candidate genes are already well documented and for which single genes of major effect contribute to the observed phenotype. Since the genes underlying the traits of interest are usually unknown, an alternative is to use quantitative trait loci (QTL) analysis. The QTL approach is an effective method for mapping portions of the genome associated with a trait or multiple traits. Additionally, it allows for further statistical inquiry to examine the relative importance of pleiotropy, modularity, development, and the environment in the genetics of integration among traits of interest.

Here, we shall discuss how QTL analysis may assist in the identification of genes involved in phenotypic integration and in distinguishing between different genetic mechanisms proposed to be involved in phenotypic integration. We shall begin by briefly describing the methods behind QTL analyses. We then review several investigations that use QTL to explicitly address questions about the genetic basis of phenotypic integration. Next, we shall discuss methods that promise to advance our understanding of how the genetic architecture behind trait correlations may change across environments and through development. We shall also explore the importance of genetic architecture in influencing major shifts in evolution. Our final section will address some of the limitations and highlight the potential future uses of QTL studies to answer questions of phenotypic integration and plasticity of integration.

## QTL Mapping

The objective of QTL mapping is to identify the number, position, and effects of genetic factors that can contribute to phenotypic variation observed for a trait that varies quantitatively. This is accomplished by testing for statistical associations between the genotype of molecular markers distributed across the genome and the phenotypic value of the trait of interest. QTL mapping can also be used to analyze interactions among QTL, and between QTL and the environment. The concept of marker-based mapping is an old one (e.g., Mather 1938; Breese and Mather 1957), but the approach has become easier to implement due to the advancement of PCR-based molecular markers (Kearsey and Farquhar 1998), new statistical procedures (Lander and Botstein 1989; Zeng 1993), and the availability of software programs that can perform the required analysis in reasonable time (Basten et al. 1994; Manly and Olson 1999; Sen and Churchill 2001).

The simplest experimental design uses an $F_2$ cross, where two inbred lines that differ for the trait(s) of interest are used as parents (see Fig. 9.1). In the $F_2$ design, the $F_1$ between the different inbred lines are selfed or crossed through brother-sister mating to produce an $F_2$ population where the trait(s) of interest as well as genetic markers that differ between the two parental lines segregate. In this $F_2$ population there will still be significant linkage disequilibrium, so that QTL correlated to the phenotypic distribution can be identified through their statistical association with scorable genetic markers (Lander and Botstein 1989). Only genetic markers for which the two parental lines differ can be used for the analysis. The number of QTL that can be identified is constrained by the number of markers available, the size of the $F_2$ population, and the number of QTL that differ between the parental lines. Once each of the $F_2$ individuals has been scored for its phenotype and genotype at each polymorphic marker, associations between marker genotype and phenotype can be identified. The most straightforward approach is to use a single regression analysis at each marker location (Soller et al. 1976). However, when markers are more than a few centimorgans apart,

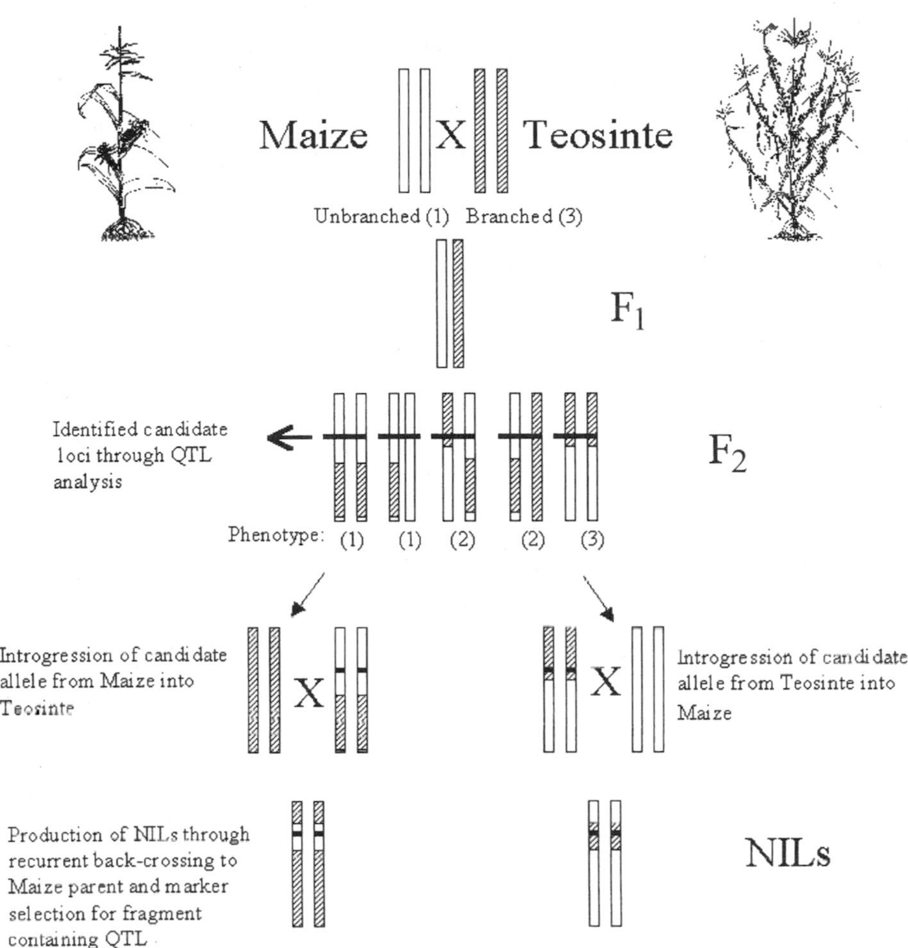


Figure 9.1 Schematic representation of the process through which QTL analysis on an $F_2$ progeny is performed. Maize and teosinte phenotypic differences in terms of numbers of branches are used as an example. This scheme also shows how NILs (Near Isogenic Lines) can be obtained for further dissection of regions containing identified QTL (details in the text). Line drawing of maize and teosinte courtesy of J. Doebley.

interval mapping is preferable, as it takes into consideration the rate of recombination between marker and QTL (Lander and Botstein 1989). Furthermore, other statistical variations have been developed in which the effect of other segregating markers is taken into account. For example, composite interval mapping analysis (Jansen 1993; Zeng 1993) or multiple interval mapping analysis (Zeng 1994) were developed to better identify the position of a QTL by removing the effect of other markers as covariates.

Many other crossing designs have been developed that are more appropriate for different species and specific questions (see, e.g., Lynch and Walsh 1998). Advances in statistical analysis have made it possible to take into consideration multiple traits, clonal designs, and other covariates. For a more detailed review of these methods see Broman (2001), Haley and Knott (1992), and Lander and Botstein (1989). Follow-up experiments using Near Isogenic Lines (NILs, discussed below and illustrated in Fig. 9.1) can offer further insight into the positioning of QTL of interest. These molecular genetic and statistical approaches allow many questions to be addressed regarding the number and influence of particular QTL on a trait or an array of traits, and these methods can be valuable in gaining an understanding of the genetic basis of phenotypic integration.

## The Genetic Basis of Phenotypic Integration

Central to the concept of phenotypic integration is the idea that functionally related traits have higher genetic correlations within a functional unit than across units. Early studies of phenotypic integration addressing this issue demonstrate that phenotypic correlations are indeed stronger and more numerous among functionally than nonfunctionally related traits (e.g., Berg 1960; Clausen and Heisey 1960; see also Chapter 2 by Armbruster et al. in this volume, and Murren, 2002). However, the actual details of the molecular genetic mechanisms behind such observations remain mostly unknown. It has been hypothesized that the observed trait correlations can be explained in part by pleiotropy, although in many cases it cannot be distinguished from close linkage (e.g., Berg 1960; Cheverud 1982; Zelditch 1988; Schlichting 1989). Conceptual models developed to explore the genetic basis of integration hypothesize that a functionally integrated phenotypic unit (a phenotypic module) should also have genetic elements associated into genetic modules (Wagner 1996; Wagner and Altenberg 1996; Winther 2001). A genetic module is defined by two conditions: interaction among genes acting on traits within a phenotypic functional unit (modularity), and dissociation of the genes underlying a specific phenotypic module from genes associated with traits of phenotypic modules with other functions (parcelation: Wagner 1996; Mezey et al. 2000). Pleiotropy and physical linkage act as genetic mechanisms of modularity. A reduction in the number of pleiotropic effects among genes influencing different phenotypic modules is a measure of parcelation. A QTL approach allows for the investigation of these theoretical considerations, and additionally offers an excellent starting point to explore the molecular genetic details of pleiotropy.

The mouse mandible is an ideal system in which the genetic basis of phenotypic integration can be explored, as there is a wealth of data from morphological, developmental, and functional studies on the mouse skeletal system (e.g., Cheverud 1982; Atchley et al. 1985; Bailey 1985, 1986; Cheverud et al. 1991). The mouse mandible is composed of two major functional sections that need to be integrated to insure proper mastication: the alveolar region and the ascending ramus. The alveolar region primarily supports the tooth roots, whereas the ascending ramus has several structures to which the muscles of mastication insert (Atchley and Hall 1991; see Fig. 9.2). Cheverud et al. (1997) pursued a QTL analysis as a first step toward uncovering the importance of pleiotropy as a component of the genetic basis of integration. A series of measurements of two-

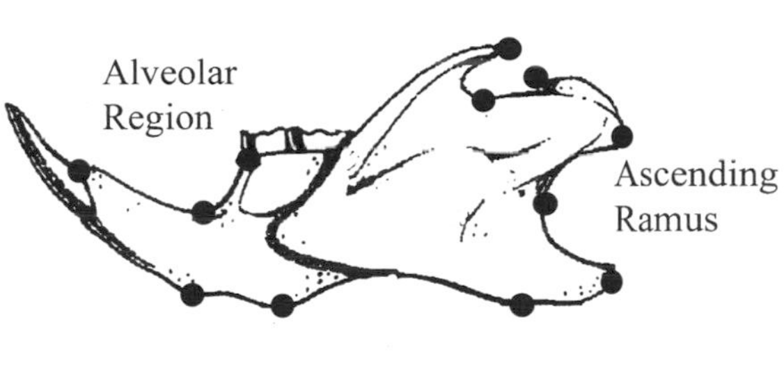


Figure 9.2 Line drawing of the mouse mandible (adapted from Cheverud et al. 1996) showing the two functional regions studied through QTL analysis as described in the text. The alveloar region (left) and the ascending ramus (right) are involved in mastication. The alveolar region primarily supports the tooth roots, while the ascending ramus has several structures to which the muscles of mastication insert. Circles on the drawing represent landmarks used in the QTL and morphological studies by Cheverud and colleagues.

dimensional coordinates composed of 15 landmarks (positions on the mandible—such as bone processes—that can be compared among specimens) and 21 linear distances among the coordinates of the mandible were made on a large set of $F_2$ progeny created from a cross between large- and small bodied inbred mice. The landmarks and distances were classified into the two major sections (Fig. 9.2) and subclassified into localized mandibular regions. Taken together, the measurements were used to describe the overall size and shape of the mandible, and analyzed for their association with a large set of molecular markers to locate QTL corresponding to these mandibular characteristics. Cheverud and colleagues found that the two major functional units of the mouse mandible mostly appear to have unique genetic control, but a few QTL are associated with more than one localized mandibular region.

As a first pass at identifying possible pleiotropy or linkage, intervals that influenced more than two inter-landmark distances (traits) were identified. Of those, 70% affected more than two traits, and no single QTL explained more than 20% of the phenotypic variance of a trait. Half of the QTL that affected more than two traits were associated only with the ascending ramus, whereas 27% were associated with the alveolus and a smaller proportion (23%) of QTL affected the entire mandible. The results showed that five QTL were significantly associated with the ascending ramus and two with the alveolar region. However, 60% of the QTL were significantly associated with more localized mandibular regions. From this evidence, it appears that pleiotropy among traits within a phenotypic module plays an important role in morphological integration.

Similar data suggesting the importance of pleiotropy or linkage for functionally or developmentally related traits have been reported in a number of studies that did not explicitly set out to examine phenotypic integration. For example, in a study of the plant *Senecio vulgaris* evidence was presented suggesting either pleiotropy or linkage among several traits: time to first flower bud, height, and leaf number (Moritz and Kadereit 2001). In *Arabidopsis thaliana*, data suggestive of pleiotropy have been shown for traits including developmental timing, rosette morphology, inflorescence architecture, and reproductive fitness (e.g., Ungerer et al. 2002). A growing number of QTL studies have supported the idea that pleiotropy is the genetic mechanism behind trait correlations, and these data offer excellent starting points for future fine-scale mapping to further identify the genes and details of the genetic architecture of phenotypically correlated traits.

To explore the relative amount of association among genetic elements of phenotypic traits, Mezey et al. (2000) took advantage of the same mouse mandible data set described above (see also Housworth et al. 2001). First, these authors defined an integration statistic ($M_I$):

$$M_I = \frac{\sum_{j=1}^{m} (\text{number of traits affected by } T_1 \text{ affected by QTL } j) - m}{m(k-1)}$$

where $m$ is the number of QTL that affect at minimum one trait in a set of traits (a phenotypic module) ($T_1$), $k$ is the number of phenotypic traits, and $j$ is a specific QTL. The authors describe $M_I$ as the difference in the total number of traits in the first phenotypic module affected by a particular QTL and the total number of QTL affecting traits in that phenotypic module. This difference was then divided by the maximum number of traits these QTL could affect. On the other hand, $M_P$ (parcelation) was defined by a Pearson's chi square:

$$M_P = \sum_{j=1}^{n} \sum_{i=1}^{2} \frac{(O_{ji} - E_{ji})^2}{E_{ji}}$$

where $n$ is the total number of QTL that have effects on at least two traits, $i$ refers to trait set $T_1$ or $T_2$ (the phenotypic modules), $O$ is then the observed number of traits affected in either of the two trait sets by each QTL $j$, and $E$ is the expected number of traits affected (for more details on these two statistics see Mezey et al. 2000). This parcelation statistic compares the observed number of QTL affecting the two modules to the expected number of QTL. The expected number of QTL was defined as the marginal distribution of effects among all QTL and all traits. If the null hypothesis for the chi square was rejected, then QTL were considered to be clustered within a hypothesized phenotypic module. Both $M_I$ and $M_P$ are sensitive to how the phenotypic modules ($T_1$ and $T_2$) are defined. The data showed support for both modularity and parcelation of the genetics of the alveolar region of the mandible. For the ascending ramus, however, there was only evidence for parcelation. In other words, the effects of the QTL generally were restricted to one of the modules, yet there was significant dissociation between the QTL affecting the two modules. Nevertheless, differentiation of the QTL associated with characters of the mouse mandible lent evidence for the existence of distinct genetic modules behind these two functional units.

A conceptually different approach to understanding the genetic basis of phenotypic integration is to emphasize the differences in size and shape from the perspective of a phenotypic module. Klingenberg et al. (2001) applied a Procrustes method to determine whether groups of QTL map to similar size or shape features of the mouse mandible, which would suggest a complex of genes that together regulate shared developmental processes resulting in an integrated mandible. The Procrustes method is similar to the analyses of other skeletal morphology studies in that it uses distances between landmarks. However, Procrustean superimposition differs in that it examines both the distances between landmarks and the deviance from the centroid (the mean of all $x$ and $y$ coordinates of all landmarks) (see other examples of new morphometrics: Zelditch et al. 1992; Rohlf and Marcus 1993). The Procrustean superimposition method first scales two given mandibles to the same units size, then overlaps the centroids, which is followed by a rotation to optimize the alignment of the centroids. The resulting measures of shape change can be used as traits in a QTL analysis (Klingenberg et al. 2001). In the mouse mandible, 25 separate QTL for shape were located on 16 out of 19 chromosomes. More QTL were detected for shape than for size. In particular, the relative positions of two specific landmarks overwhelmingly influenced the shape of the mouse mandible.

Workman et al. (2002), using these same statistical methods and experimental mouse population, examined the QTL of another set of characters, molar shape and size. They found a large number of QTL for shape and argue that this result is not surprising given that molar development is the result of a complex interaction of cells from multiple distinct tissue types. In contrast, size is largely regulated by the endocrine system. Both the mouse mandible and molar studies found that there were continuous distributions of QTL effects for shape variation and no evidence for distinct modules; rather, the mandible and molars actually shared a number of QTL. Overall, the results from the analyses of shape may reflect the importance of shared gene regulators among traits of a phenotypic module that orchestrate the development of the overall complex phenotype.

Analyses of pleiotropy (Cheverud et al. 1997), modularity (Mezey et al. 2000; Housworth et al. 2001), and shape (Klingenberg et al. 2001) generally detected QTL located within the same marker intervals. The shape-based analyses are set apart from the other studies as they suggest that perhaps the QTL are not restricted to either the ascending ramus or alveolar regions of the mandible, and instead their evidence suggests that QTL likely affected traits across multiple mandible regions. Overall, the studies of the mouse skull have opened up our understanding of the genetics of phenotypic integration, highlighting the importance of pleiotropy and linkage, which prior to these analyses had been theorized. Although the mouse mandible studies demonstrated the difficulty in defining the genetic architecture underlying complex phenotypes, they have implemented a set of statistical tools that can be used to further investigate the genetics of modularity in other taxa. We argue, though, that future investigations should follow these preliminary studies with fine-scale mapping to further test the relative contributions of linkage and pleiotropy to phenotypic integration.

### Across Environments

Functionally related traits have to be integrated in a particular way to produce a functional organism; thus, changes in trait correlations in response to environmental conditions can be critical for survival and reproduction. For example, the relationships among thallus traits of the giant kelp changed according to water flow conditions (Johnson and Koehl 1994). Phenotypic plasticity across environments has been carefully characterized for many morphological, life-history, anatomical, and architectural traits across a wide group of taxa (Schlichting and Pigliucci 1998; Pigliucci 2001). However, the genetic mechanisms by which the phenotype is integrated and how this integration may change or be maintained over a range of environmental conditions is a budding area of research. At this stage, we are beginning to shed light on the possible molecular genetic mechanisms of phenotypic integration across environments by the investigation of the genetics of plasticity for single traits.

Two main hypotheses have been put forward as to the genetic basis of plasticity. The "allelic sensitivity/pleiotropy" hypothesis suggests that differential phenotypic expression of the same trait in different environments results from allelic sensitivity of the structural genes responsible for the trait expression (Smith-Gill 1983; Via and Lande 1985; Schlichting and Pigliucci 1998). In contrast, the "environmentally dependent gene regulation" hypothesis suggests that environmentally sensitive regulatory genes affect the phenotypic expression of a suite of traits through epistatic interactions (Bradshaw 1965; Schlichting and Pigliucci 1993, 1995). In terms of the genetic mechanisms behind plasticity of trait correlations the discussion is less advanced. One hypothesis is that regulatory genes with environmentally dependent expression might explain correlation of plasticities among traits (Schlichting and Pigliucci 1998). Although we are not aware of QTL studies that have specifically investigated the genetic mechanisms of phenotypic integration across environments, there are a growing number of studies on the environmental dependence of QTL that have developed new statistical methodologies that may lead to novel insights about integration. We discuss these in more detail below.

Testing the two possible alternatives outlined above for the genetic basis of plasticity would be facilitated by knowledge of the identity of all genes in all environments that affect a trait. Molecular genetics approaches have allowed the identification of many genes that affect qualitative traits with Mendelian segregation. However, the genes underlying quantitative traits, which most often show environment-specific expression, remain generally unknown (but for a recent example see Braam and Davis 1990 and Pigliucci 2002). Thus, QTL analysis can be helpful in allowing the identification of candidate genes that affect plastic traits. However, actual identification of a gene is difficult since the confidence intervals for QTL positioning are typically around 10 cM (which can include large numbers of genes). More fruitful methodologies directly test for the environmental and epistatic effects of QTL. These methods take advantage of Recombinant Inbred Lines (RILs), which allow the same genotype to be grown in replicates over many environmental conditions. RILs are commonly obtained by performing the typical $F_2$ cross for QTL analysis (Fig. 9.1), followed by many generations of selfing for each of the $F_2$ progeny, such that the resulting population is a set of fully homozygous individuals that segregate for alleles and traits. Using replicates of the RILs in different environments, it is possible to test directly for QTL by environment (QTL $\times$ E) interactions. For example, Gurganus et al. (1998) identified QTL that are differentially expressed across environments in

*Drosophila* RILs. They identified 19 markers that showed a main effect on the number of sternopleural bristles present, that is, they affected the trait independent of the environment. They also identified five QTL with a significant QTL × E interaction, which indicated that the QTL effect on bristle number depended on the environment in which the flies were being raised. Furthermore, four of the five markers that showed QTL × E had no significant main effect, suggesting that these might be regulatory genes that are detected only through their epistatic interactions.

Comparing the effect of QTL within and between environments can more specifically test the role of regulatory versus structural genes in plasticity. Structural genes are expected to have an effect on the plastic trait independent of the environmental conditions. Gene expression, however, can be environmentally sensitive, which will influence the across-environment covariance of its effect. On the other hand, a gene that explains changes in expression, but does not have a main effect on the trait in either environment, indicates a regulatory role of the gene. Wu (1998) tested for the role of regulatory genes in plastic responses of poplar trees using three replicates of each of 90 $F_2$ individuals growing in two different locations. He tested separately for QTL effects on growth traits within each of the environments, and then tested for QTL that show significant QTL by environment interactions. He also calculated the genetic variation explained by each QTL in each environment and the across-environment genetic correlation ($rg_{Qxy}$, where rg stands for the genetic correlation and $_{Qxy}$ indicates that it is for a QTL across environments $x$ and $y$). Since a perfect correlation of QTL effects across environment indicates that they are responding equally to both environments, QTL for which $rg_{Qxy}$ was not significantly different from 1 were not considered to affect trait plasticity. Thus, environmentally sensitive genes were the ones for which $rg_{Qxy}$ was significantly different from 1. Among those, regulatory genes were identified indirectly by comparing the amount of genetic variation explained in each environment by a QTL. When QTL for which $rg_{Qxy}$ were significantly different from 1, and explained significantly different amounts of genetic variance in the two environments, this indicated environmental-dependent expression mediated by regulatory loci. A second method used to identify the role of regulatory genes determines which QTL affects the environmental response (i.e., has a significant QTL × E effect) without having any direct effect on the trait when mapped within an environment. Using a combination of these methods, Wu (1998) concluded that gene regulation plays a prevailing role in determining the plastic response of stem growth in poplar trees.

The methods described above are limited to identifying QTL that affect phenotypic responses across two discrete environments, which inherently assume a linear function for the norm of reaction. In contrast, under natural conditions organisms experience a range of environments, which often vary continuously, and not necessarily linearly. To address this problem, Stratton (1998) proposed a method to search for QTL that affect the shape of the norm of reaction across an environmental gradient. Of the seven QTL found to have an effect on the average flowering time of *Arabidopsis thaliana* calculated over a light gradient, none was sensitive to light changes. In contrast, two of the eight QTL found to affect the number of leaves present at flowering significantly affected the linear and the quadratic effect describing the shape of the norm of reaction. These two QTL explained a smaller proportion of the total genotype by light interaction than the QTL on the main effect (i.e., QTL that affect flowering independent of light conditions) explained of the total genetic variation for number of leaves at flowering. However, the two QTL identified by their effect on the shape of the reaction norm had little influence on the mean leaf number, suggesting that they might be regulatory genes. This approach also offers promise as a direct test for the genetic basis of phenotypic integration across environments. Specifically, it may be used to test the hypothesis that regulatory genes can explain the correlation between norms of reactions of a pair of traits. It should be possible to compare QTL that affect norms of reactions of functionally related traits and whether these QTL affect trait correlations.

## Throughout Development

Different stages of development can be seen as different internal environmental conditions (Schlichting and Pigliucci, 1998). Given that trait correlations can depend on the environment, trait correlations can also change throughout stages of development. There are still very few studies that specifically map traits associated with a plant's or animal's development. To date, QTL studies have been generally restricted to single traits (but see the approaches of Mezey et al. 2000). However, we foresee these approaches expanding to multiple correlated traits in the near future.

Three different approaches have been taken to understand development of a single trait using QTL. First, the most common strategy is to map QTL related to the timing of the onset of a trait of interest. For example, QTL have been mapped for the timing of transition from vegetative to reproductive stage in *Arabidopsis thaliana* (e.g., Alonso-Blanco et al. 1998; Juenger et al. 2000; Ungerer et al. 2002), the time of bud development in *Brassica oleracea* (Lan and Paterson 2000), and the time to hatch in salmonids (Robison et al. 2001). Leamy et al. (1999) showed that QTL for the cranial vault and the face group of the mouse skull were generally distinct, which corresponds to the difference in developmental timing of these two sets of characters. These studies have begun to highlight the important details of genetic architecture behind the functioning of developmental switches.

A second approach has been to map QTL of growth or growth trajectories directly (e.g., Wu et al. 2002). The third approach, rarely implemented, examines the same trait at two developmental stages. This is similar to the analysis of phenotypes in multiple discrete environments (discussed above), with the added complexity of dependence between the two times measured. For example, in winter wheat, at least four QTL were found to be important in the genetic control of resistance to mildew in two different adult populations. Three QTL for resistance were found in the seedling stage. A single QTL was common to both life stages, but it described a large proportion of the genetic variation for resistance to mildew. By identifying QTL important at different times during development that result in functionally integrated phenotypes, these studies begin to expand our understanding of the genotype to phenotype mapping process (*sensu* Wagner and Altenberg 1996), in which development plays a central role.

Although development serves a key function in the evolution of phenotypic integration (Cheverud 1982; Zelditch 1988; Klingenberg et al. 2001), we are unaware of any studies that use QTL analysis specifically to address questions of phenotypic integration at different points during development. In part, the limitation on the number of studies of development or developmentally related characters is due to the logistics of large sample sizes. These are needed both to adequately map QTL by environment interactions and also to take into account development (Schlichting and Pigliucci 1998). However, the time and cost of measuring developmental characters in a large population can be tremendous. Despite these difficulties, a QTL approach has great potential for documenting the combined importance of the genetic and developmental basis of phenotypic integration.

We envision two kinds of studies. The first would measure a number of phenological, morphological, and architectural traits for a set of phenotypic modules (e.g., rosette versus reproductive traits of *Arabidopsis thaliana*), to test whether QTL associated with timing traits had pleiotropic effects across both modules or were restricted to a single module (as opposed to the morphological measurements most commonly taken). A single QTL mapping to the same marker interval for a number of phenological traits may suggest a regulatory element with important pleiotropic effects through developmental time and across phenotypic modules. Once identified, this QTL could be examined more explicitly via other molecular genetic approaches for its regulatory role.

A second type of study would focus on understanding the pattern of modularity of QTL action through both time and space. For example, we know that time to flower varies widely among *Arabidopsis* accessions and that it can be affected by, among other factors, day length, cold treatment of the seed, and rosette size (Karlsson et al. 1993; Nordborg and Bergelson 1999). Jermstad et al. (2001) found in Douglas fir that QTL determining time of bud flush through development were the same. However, geographic location strongly affected the timing of bud emergence, and there were some year by site interactions. To explore these issues we suggest a study that uses the Mezey et al. (2000) approach to test for genetic modularity or parcelation of QTL and expands their methods to an examination of these parameters at several points throughout ontogeny. If this analysis could be completed both across developmental time and across environments, it could contribute significantly to our understanding of the changes in importance of QTL through time and space and the multifaceted nature of modules (as described by Raff 1996). As in the case of the external environment, development could be considered as a continuous gradient. Implementation of Stratton's (1998) analyses, modified for the dependence of trait measurements at different developmental times, would increase our understanding of the changing influence of QTL through time. We envision that these types of studies would demonstrate the dynamic nature of the genetic components that influence quantitative traits through the lifetime of organisms living in heterogeneous environments.

### Through Evolutionary Time

Phenotypic integration also plays an important role in the evolution of multiple traits that are functionally related. Systems for which we already know a great deal about the genetic basis of multitrait evolution are domesticated plants. When compared with their wild ancestors, they usually have undergone dramatic shifts in many aspects of the phenotype for which no variation is observed within the wild or domesticated varieties. Examples of such shifts during domestication are the plant and flower architecture of maize (Doebley et al. 1990) and of pearl millet (Poncet et al. 2000). Since, on an evolutionary scale, cereal domestication is a recent event (initiated 5000 to 10 000 years ago: Poncet et al. 2000), it is remarkable that so many traits have changed so dramatically in numerous domesticated species.

One of the best-studied cases of rapid evolutionary shifts in multiple traits is the domestication of maize from its ancestor teosinte. Maize has strong apical dominance, leading to short lateral branches where female flowers (the corn ears) are produced. In contrast, teosinte usually has many long lateral branches with male inflorescences at their apex. The fruit in teosinte has a single row of kernels that disarticulate upon maturity, while maize has many rows of kernels that do not disarticulate (Doebley et al. 1990). Using QTL analysis, Doebley and colleagues have dissected the genetic basis of maize domestication (see below). Their results so far indicate that genes of large effect were key to this dramatic architectural shift. Furthermore, they obtained remarkable data that suggested that the way in which few genes can accomplish such major phenotypic changes is by the alteration of regulator genes that act epistatically across many traits.

Originally, two hypotheses had been proposed to explain the changes observed between maize and teosinte. The first contemplated that independent genes of large effect accomplished each of the major phenotypic changes (although modifiers of smaller effects may also exist: Beadle 1939). According to this hypothesis, five major genes were invoked to explain the changes in ear morphology. The second hypothesis proposed that all changes could be explained by the feminization of the lateral meristems, and that this syndrome would be under the control of many genes of small effect (Iltis 1983).

To distinguish between these two hypotheses, Doebley et al. (1990) performed a QTL study using an $F_2$ progeny array from an $F_1$ cross between maize and teosinte. For each of nine traits analyzed, they found between four and eight significant QTL, indicating that these traits support Iltis's (1983) polygenic hypothesis. However, many of the QTL explained very little of the variation and others were found in close proximity to each other (indicating that it may in fact be the same gene underlying these QTL). Taking into consideration only the QTL of larger effect, they identified five discrete regions of the genome that explained most of the variation for all the nine traits. This finding supported Beadle's hypothesis, assuming the genes of smaller effect acted as modifiers. More importantly, some of these regions (like the long arm of chromosome 1) affected five of the nine traits studied. This result suggests that these major QTL act pleiotropically, and that they might be regulatory genes responsible for whole organism integration; although it is also possible that, instead of one single gene of large effect, these regions have many genes in tight linkage within one small region, each of them independently affecting a different trait.

To dissect these chromosomal regions further, Doebley and colleagues used fine-scale QTL mapping (Dorweiler et al. 1993; Doebley et al. 1995). Using backcrosses and marker selection, Doebley et al. (1995) built Near Isogenic Lines (NILs) where the significant segment found in the long arm of chromosome 1 (segment 1L) from maize was transferred into a teosinte line and the same segment from teosinte was introduced into a maize line. They then crossed the introgressed line with the recurrent parent (the parent that is providing the background), so that when they performed a QTL analysis on the selfed $F_2$ progeny, only the segment of interest was segregating in a homozygous background. For example, in the QTL analysis of the $F_2$ progeny from a cross between teosinte and the constructed teosinte line with segment 1L from maize, they found a single QTL that affected eight of the nine traits measured. This QTL had a very large effect, changing some traits' phenotypic mean by as much as 50%. Using mapping and complementation tests, they established that this QTL corresponded to the gene *tb1*, previously identified as a recessive mutant in maize, which affect many aspects of corn morphology. These results favored the hypothesis of pleiotropy of a single regulator within the reproductive module instead of tight linkage as a mechanism behind the phenotypic correlation of traits.

Given the information obtained from the QTL analysis, Doebley and colleagues hypothesized that the change from a branched architecture into the maize architecture with strong apical dominance occurred through the wiring of *tb1*. They believed that *tb1* was once under environmental control, allowing plants to developed a branched architecture when the environment was not competitive, or increasing apical dominance in crowded situations. If this model is correct, they expected the expression of the *tb1* allele from teosinte to be plastic and to respond to plant density, but not the *tb1* allele from maize. To test this alternative model they examined whether the teosinte allele of *tb1* provided greater phenotypic plasticity when NILs with the maize background and either the teosinte or the maize allele of *tb1* were grown in low or high density (Lukens and Doebley 1999). As predicted, they found that the lines containing the *tb1* allele from teosinte expressed plasticity for more traits than pure corn lines. To develop this model further they are pursuing gene expression experiments to determine possible mechanisms through which *tb1* affect branching, internode length, and sex determination in flowers and acts as a global regulator across many modules.

Further evidence for the role of regulatory genes in rapid evolutionary shifts corresponding to domestication, which may alter patterns of phenotypic integration, is found on the transition from the small fruit of *Lycopersicon pennellii,* the ancestral species, to the large fruit size of the domesticated tomato, *L. esculentum* (Grandillo and Tanksley 1996; Frary et al. 2000). These studies also support the idea that regulatory genes are important not only for the integration of adult morphological traits, but also through the evolutionary changes of complex traits that require integration of many parts.

## Caveats and Future Endeavors

It is likely that in the next few years, QTL analyses will continue to be the best first step in the dissection of the molecular genetic basis of complex traits and of phenotypic integration. However, as more QTL studies are being published, we are becoming more aware of the limitations of this method (e.g., Beavis 1994; Doerge et al. 1997) and of the importance of further molecular genetic experiments to uncover details of the observed patterns of genetic architecture.

One problem is the large size of the confidence intervals associated with detected QTL. For example, Ungerer et al. (2002) point out that an average interval between markers on the *Arabidopsis* QTL mapping population includes between 375 and 783 genes when the interval was compared to the fully sequenced genome. The best way to address this problem is to increase the number of recombinations either by increasing the number of $F_2$ progeny scored or using an Advance Intercross Design where the $F_2$ progeny are randomly crossed for one or more generations before they are used for QTL mapping.

Another limitation of QTL experiments is that they only suggest the occurrence of pleiotropy. However, fine-scale mapping can improve our ability to distinguish between pleiotropy and linkage, or confirm the homology of gene effects underlying QTL across environments. Once individual genes are identified and cloned, their pleiotropic effects or homology can be more directly assessed (but see Callahan et al. 1999). Although these methods can be an expensive and long-term effort, it has been accomplished in tomato and maize (Doebley et al. 1997; Frary et al. 2000), and we expect that it will get easier as the protocols become better established.

Other developments, which are likely to improve the ability of QTL analyses to reveal the genetic basis of phenotypic integration, are new statistical models. For example, testing hypotheses of genetic modularity using QTL and associated molecular genetic approaches are in their statistical infancy. A limitation of the model described by Mezey et al. (2000) requires that QTL be attributed to either one of two phenotypic modules. A test that does not constrain the traits to belong to either one of the two modules would be a desirable extension to this analysis (e.g., see the approach by Magwene 2001, although the extensions to QTL analyses have not yet been attempted). The combination of fine-scale mapping with newly developed statistical procedures will further add to the power of these studies. For example, the ability of QTL studies to detect epistasis is significantly improved by the use of NILs or previously identified candidate loci (Cheverud and Routman 1995; Lukens and Doebley 1999).

We predict that our understanding of the molecular genetic basis of phenotypic integration will accelerate when QTL studies are combined with high-resolution mapping and microarray experiments. The results of QTL analyses to date offer an excellent foundation for further work to identify the genetic mechanisms of phenotypic integration. When studies of the molecular genetic basis of integration are combined with whole organism studies, we shall gain a greater understanding of how to define a complex phenotypic module through shared functional developmental, and genetic mechanisms (see also Chapter 10 by Klingenberg in this volume).

*Acknowledgments* We thank Massimo Pigliucci and Katherine Preston for the invitation to write this chapter and editorial comments on an earlier version. We are grateful for the comments of Michele Dudash, Matt Rutter, and Jason Wolf, which greatly improved the chapter. Also, Courtney Murren wishes to thank Carl Schlichting for inspiration on the topic of integration.

## *Literature Cited*

Alonso-Blanco, C., S. E.-D. El-Assal, G. Coupland, and M. Koornneef. 1998. Analysis of natural allelic variation at flowering time loci in the Landsberg erecta and Cape Verde Islands ecotypes of *Arabidopsis thaliana*. Genetics 149:749–764.

Atchley, W. R., and B. K. Hall. 1991. A model of development and evolution of complex morphological structures. Biological Review 66:101–157.

Atchley, W. R., A. A. Plummer, and B. Riska. 1985. Genetics of mandible form in the mouse. Genetics 111:555–577.

Bailey, D. 1985. Genes that affect the shape of the murine mandible: congenic strain analysis. Journal of Heredity 76:107–114.

Bailey, D. 1986. Genes that affect morphogenesis of the murine mandible: recombinant-inbred strain analysis. Journal of Heredity 77:17–25.

Basten, C., B. S. Weir, and Z. B. Zeng. 1994. Zmap—a QTL cartographer. Pp. 65–66 in C. Smith et al., eds. Proceedings of the 5th World Congress on Genetics Applied to Livestock Production: Computing Strategies and Software, Guelph, Ontario, Canada.

Beadle, G. W. 1939. Teosinte and the origin of maize. Journal of Heredity 30:245–247.

Beavis, W. 1994. The power and deceit of QTL experiments: lessons from comparative QTL studies. Pp. 250–266 in Proceedings of the 49th Annual Corn and Sorghum Research Conference, Chicago.

Berg, R. L. 1960. The ecological significance of correlation pleiades. Evolution 14:171–180.

Braam, J., and R. W. Davis. 1990. Rain-, wind-, and touch-induced expression of calmodulin and calmodulin-related genes in *Arabidopsis*. Cell 60:357–364.

Bradshaw, A. D. 1965. Evolutionary significance of phenotypic plasticity in plants. Advances in Genetics 13:115–155.

Breese, E. L., and K. Mather. 1957. The organization of polygenic variation within a chromosome in *Drosophila*. I. Hair characters. Heredity 11:373–395.

Broman, K. W. 2001. Review of statistical methods for QTL mapping in experimental crosses. Lab Animal 30:44–52.

Callahan, H. S., C. L. Wells, and M. Pigliucci. 1999. Light-sensitive plasticity genes in *Arabidopsis thaliana*: mutant analysis and ecological genetics. Evolutionary Ecology Research 1:731–751.

Cheverud, J., and E. Routman. 1995. Epistasis and its contribution to genetic variance. Genetics 139:1455–1461.

Cheverud, J. M. 1982. Phenotypic, genetic and environmental morphological integration in the cranium. Evolution 36:499–516.

Cheverud, J. M. 1988. The evolution of genetic correlation and developmental constraints. Pp. 94–101 in G. de Jong, ed. Population Genetics and Evolution. Springer-Verlag, Berlin.

Cheverud, J. M., S. E. Hartman, J. T. Richtsmeier, and W. R. Atchley. 1991. A quantitative genetic analysis of localized morphology of inbred mice using finite element scaling. Journal of Craniofacial Genetics Development and Biology 11:122–137.

Cheverud, J. M., E. J. Routman, and D. J. Irschick. 1997. Pleiotropic effects of individual gene loci on mandibular morphology. Evolution 51:2006–2016.

Clausen, J., and W. M. Heisey. 1960. The balance between coherence and variation in evolution. Proceedings of the National Academy of Sciences, USA 46:494–506.

Doebley, J., A. Stec, J. Wendel, and M. Edwards. 1990. Genetic and morphological analysis of a maize-teosinte F2 population: implication for the origin of maize. Proceedings of the National Academy of Sciences USA 87:9888–9892.

Doebley, J., A. Stec, and G. Gustus. 1995. *teosinte branched1* and the origin of maize: evidence for epistasis and the evolution of dominance. Genetics 141:333–346.

Doebley, J., A. Stec, and L. Hubbard. 1997. The evolution of apical dominance in maize. Nature 386:485–488.

Doerge, R. W., Z.-B. Zeng, and B. S. Weir. 1997. Statistical issues in the search for genes affecting quantitative traits in experimental populations. Statistical Science 12:195–219.

Dorweiler, J., A. Stec, K. Kermicle, and J. Doebley. 1993. Teosinte glume architecture. *1*: A genetic locus controlling a key step in maize evolution. Science 262:233–235.

Frary, A., T. Nesbitt, S. Grandillo, E. van der Knaap, B. Cong, J. Liu, J. Meller, R. Elber, K. Alpert, and S. Tanksley. 2000. *fw2.2*: a quantitative trait locus key to the evolution of tomato fruit size. Science 289:85–88.

Grandillo, S., and S. D. Tanksley. 1996. QTL analysis of horticultural traits differentiating the cultivated tomato from the closely related species *Lycopersicon pimpinellifolium*. Theoretical and Applied Genetics 92:935–951.

Gurganus, M. C., J. D. Fry, S. V. Nuzhdin, E. G. Pasyukova, R. F. Lyman, and T. F. C. Mackay. 1998. Genotype-environment interaction at quantitative trait loci affecting sensory bristle number in *Drosophila melanogaster*. Genetics 149:1883–1898.

Haley, C., and S. Knott. 1992. A simple regression method for mapping quantitative trait loci in line crosses using flanking markers. Heredity 69:315–324.

Housworth, E., J. G. Mezey, J. M. Cheverud, and G. P. Wagner. 2001. The test distribution of modularitiy statistics: a correction and a clarification. Genetics 158:1381.

Iltis, H. H. 1983. From teosinte to maize: the catastrophic sexual transmutation. Science 222:886–894.

Jansen, R. C. 1993. Interval mapping of multiple quantitative trait loci. Genetics 135:205–211.

Jermstad, K. D., D. L. Bassoni, K. S. Jech, N. C. Wheeler, and D. B. Neale 2001. Mapping of quantitative trait loci controlling adaptive traits in coastal Douglas-fir. I. Timing of vegetative bud flush. Theoretical and Applied Genetics 102:1142–1151.

Johnson, A. S., and M. Koehl. 1994. Maintenance of dynamic strain similarity and environmental stress factor in different flow habitats: thallus allometry and material properties of a giant kelp. Journal of Experimental Biology 195:381–410.

Juenger, T., M. Purugganan, and T. F. C. Mackay. 2000. Quantitative trait loci for floral morphology in *Arabidopsis thaliana*. Genetics 156:1379–1392.

Karlsson, B. H., G. R. Sills, and J. Nienhuis. 1993. Effects of photoperiod and vernalization on the number of leaves at flowering in 32 *Arabidopsis thaliana* (Brassicaceae) ecotypes. American Journal of Botany 80:646–648.

Kearsey, M. J., and A. G. L. Farquhar. 1998. QTL analysis in plants: where are we now? Heredity 80:137–142.

Klingenberg, C. P., L. J. Leamy, E. J. Routman, and J. M. Cheverud. 2001. Genetic architecture of mandible shape in mice: effects of quantitative trait loci analyzed by geometric morphometrics. Genetics 157:785–802.

Lan, T.-H., and A. H. Paterson. 2000. Comparative mapping of quantitative trait loci sculpting the curd of *Brassica oleracea*. Genetics 155:1927–1954.

Lander, E. S., and D. Botstein. 1989. Mapping mendelian factors underlying quantitative traits using RFLP linkage maps. Genetics 121:185–199.

Leamy, L. J., E. J. Routman, and J. M. Cheverud. 1999. Quantitative trait loci for early and late developing skull characters in mice: a test of the genetic independence model of morphological integration. American Naturalist 153:201–214.

Lukens, L. N., and J. Doebley. 1999. Epistatic and environmental interactions for quantitative trait loci involved in maize evolution. Genetical Research 74:291–302.

Lynch, M., and B. Walsh. 1998. Genetics and Analysis of Quantitative Traits. Sinauer Associates, Sunderland, MA.

Magwene, P. M. 2001. New tools for studying integration and modularity. Evolution 55:1734–1745.

Manly, K. F., and J. M. Olson. 1999. Overview of QTL mapping software and introduction to Map Manager QT. Mammalian Genome 10:327–334.

Mather, K. 1938. The Measurement of Linkage in Heredity. Methuen, London.

Mezey, J. G., J. M. Cheverud, and G. P. Wagner. 2000. Is the genotype to phenotype map modular? A statistical approach using mouse quantitative trait loci data. Genetics 156:305–311.

Moritz, D. M. L., and J. W. Kadereit. 2001. The genetics of evolutionary change in *Senecio vulgaris* L.: a QTL mapping approach. Plant Biology 3:544–552.

Murren, C. J. 2002. Phenotypic integration in plants. Plant Species Biology 17:89–99.

Nordborg, M., and J. Bergelson. 1999. The effect of seed and rosette cold treatment on germination and flowering time in some *Arabidopsis thaliana* (Brassicaceae) ecotypes. American Journal of Botany 86:470–475.

Olson, E. C., and R. L. Miller. 1958. Morphological Integration. University of Chicago Press, Chicago.

Pigliucci, M. 1996. How organisms respond to environmental changes: from phenotypes to molecules (and vice versa). Trends in Ecology and Evolution 11:168–173.

Pigliucci, M. 2001. Phenotypic Plasticity: Beyond Nature and Nurture. Johns Hopkins University Press, Baltimore, MD.

Pigliucci, M. 2002. Touchy and bushy: phenotypic plasticity and integration in response to wind stimulation in *Arabidopsis thaliana*. International Journal of Plant Science 163:399–408.

Poncet, V., F. Lamy, K. M. Devos, M. D. Gale, A. Sarr, and T. Robert. 2000. Genetic control of domestication traits in pearl millet (*Pennisetum glaucum* L., Poaceae). Theoretical and Applied Genetics 100:147–159.

Raff, R. A. 1996. The shape of Life: Genes, Development and the Evolution of Animal Form. University of Chicago Press, Chicago.

Robison, B. D., P. A. Wheeler, K. Sundin, P. Sikka, and G. H. Thorgaard. 2001. Composite interval mapping reveals a major locus influencing embryonic development rate in rainbow trout (*Oncorhynchus mykiss*). Journal of Heredity 92:16–22.

Rohlf, F. J. and L. F. Marcus. 1993. A revolution in morphometrics. Trends in Ecology and Evolution. 8:129–132.

Schlichting, C. D. 1989. Phenotypic integration and environmental change. BioScience 39:460–464.

Schlichting, C. D., and M. Pigliucci. 1993. Control of phenotypic plasticity via regulatory genes. American Naturalist 142:366–370.

Schlichting, C. D., and M. Pigliucci. 1995. Lost in phenotypic space: environmental dependent morphology in *Phlox drummondii* (Polemoniaceae). International Journal of Plant Science 156:542–546.

Schlichting, C. D., and M. Pigliucci. 1998. Phenotypic evolution: a reaction norm perspective. Sinauer Associates, Sunderland, MA.

Sen, S., and G. Churchill. 2001. A statistical framework for quantitative trait mapping. Genetics 159:371–387.

Smith-Gill, S. J. 1983. Developmental plasticity: developmental conversion versus phenotypic modulation. American Zoologist 23:47–55.

Soller, M., et al. 1976. On the power of experimental designs for the detection of linkage between marker loci and quantitative loci in crosses between inbred lines. Theoretical and Applied Genetics 47:35–39.

Stratton, D. A. 1998. Reaction norm functions and QTL-environment interactions for flowering time in *Arabidopsis thaliana*. Heredity 81:144–155.

Ungerer, M. C., S. S. Halldorsdottir, J. L. Modliszewski, T. F. C. Mackay, and M. Purugganan. 2002. Quantitative trait loci for inflorescence development in *Arabidopsis thaliana*. Genetics 160:1133–1151.

Via, S., and R. Lande. 1985. Genotype-environment interaction and the evolution of phenotypic plasticity. Evolution 39:505–522.

Wagner, G. P. 1996. Homologues, natural kinds, and the evolution of modularity. American Zoologist 36:36–43.

Wagner, G. P., and L. Altenberg. 1996. Perspective: Complex adaptations and the evolution of evolvability. Evolution 50:967–976.

Winther, R. G. 2001. Varieties of modules: kinds, levels, origins, and behaviors. Journal of Experimental Zoology (Molecular Development and Evolution) 291:116–129.

Workman, M. S., L. J. Leamy, E. J. Routman, and J. M. Cheverud. 2002. Analysis of quantitative trait locus effects on the size and shape of mandibular molars in mice. Genetics 160:1573–1586.

Wu, R. 1998. The detection of plasticity genes in heterogeneous environments. Evolution 52:967–977.

Wu, R. L., C. X. Ma, J. Zhu, and G. Casella. 2002. Mapping epigenetic quantitative trait loci (QTL) altering a developmental trajectory. Genome 45:28–33.

Zelditch, M. L. 1988. Ontogenetic variation in patterns of phenotypic integration in the laboratory rat. Evolution 42:28–41.

Zelditch, M. L., F. L. Bookstein, and B. L. Lundrigan. 1992. Ontogeny of integrated skull growth in the cotton rat *Sigmodon fulviventer*. Evolution 46:1164–1180.

Zeng, Z. B. 1993. Theoretical basis for separation on multiple linked gene effects in mapping quantitative trait loci. Proceedings of the National Academy of Sciences USA 90:10972–10976.

Zeng, Z. B. 1994. Precision mapping of quantitative trait loci. Genetics 136:1457–1468.

# 10

# Integration, Modules, and Development

## *Molecules to Morphology to Evolution*

CHRISTIAN PETER KLINGENBERG

Living organisms are intricately organized developmental systems, which at the same time are very flexible but also highly robust. They are flexible to respond to environmental conditions by changing developmental processes and the resulting phenotype accordingly, but they are also robust in that all these developmental changes in different parts are coordinated and the end result is an integrated, functional organism. Similarly, there is considerable flexibility for evolutionary change of specific parts, while robustness of the overall body plan ensures continued integration of multiple organismal functions.

Research in recent years has identified the modular architecture of organisms as one of the major principles that underlies this simultaneous flexibility and robustness (e.g., Raff 1996; Kirschner and Gerhart 1998; von Dassow and Munro 1999; Bolker 2000; Winther 2001). Modules are units that are made internally coherent by manifold interactions of their parts, but are relatively autonomous from other such units with which they are connected by fewer or weaker interactions (Fig. 10.1). Modules are therefore "individualized" to some extent and can be delimited from their surroundings. They are units that can function in different contexts and can undergo developmental and evolutionary change separately. Modular organization has been found at many levels of organization, from the molecular structure of individual genes to the body plans of whole organisms. At the molecular level, *cis*-regulatory sequences of single genes are subdivided into distinct modules that control expression of the gene in different locations or at different times in development (Yuh et al. 1998; Davidson 2001). Modularity is also found in gene regulatory networks, where the interactions among genes tend to be concentrated in particular "clusters"

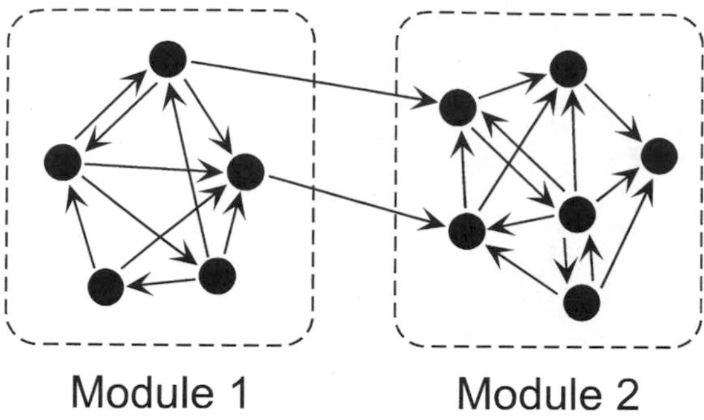


Figure 10.1 Definition of modules by developmental interactions. Component parts within modules are interconnected by many interactions, whereas there are fewer interactions between modules.

that are stable in themselves, and where such modules can be flexibly deployed in different developmental contexts (von Dassow et al. 2000; Wilkins 2002, pp. 348–350). The most apparent manifestation of modularity, of course, is in the structural parts that make up the bodies of organisms, where modules can originate as developmentally distinct parts or perform different functions (Cheverud et al. 1997; Klingenberg et al. 2001). These examples show how different the domains are to which the principle of modularity can be applied. Modules can be tangible material units, as in the examples of morphological parts or of *cis*-regulatory modules that are specific stretches of DNA sequence, or they can be abstract, as in the example of gene regulatory networks, where modularity resides in the regulatory relationships among genes. In modular systems at all levels, however, the primary criterion for identifying modules is the strong internal coherence and connectivity of modules coupled with their relative independence from other parts of the system (for additional discussion, see Bolker 2000).

Phenotypic studies address modularity primarily at the morphological level. The body parts of organisms behave as modules because they are internally coherent and show some degree of mutual autonomy corresponding to their developmental origins and functions (Cheverud 1996; Wagner 1996). Developmental biologists have long considered modules of this kind under the concepts of morphogenetic fields or embryonic fields (e.g., Davidson 1993; Gilbert et al. 1996; Wilkins 2002, pp. 255–258). These modules are internally coherent due to signaling interactions that are part of the patterning processes that generated the structure. In the resulting body parts this coherence is manifest as morphological integration (e.g., Olson and Miller 1958; Cheverud 1996). Therefore, developmental integration and modularity are amenable to quantitative study with morphometric methods (Pimentel 1979; Bookstein 1991; Dryden and Mardia 1998).

In this chapter, I review the developmental origins of integration and modularity from molecular mechanisms to their morphological manifestation. Consideration of these issues reveals that morphological variation originating from different sources intrinsic or extrinsic to the organism can be analyzed to infer the developmental origins of integration and to delimit the spatial extent of morphological modules. I describe this morphometric approach for identifying developmental modules and briefly review the few available case studies. I also examine the implications of this developmental perspective on morphological integration for evolutionary quantitative genetics, where it can shed new light on the evolution of pleiotropy and genetic covariances.

## Modularity and Integration by Intercellular Signals

Modules are units within a developmental system that are defined by their internal coherence and relative independence from other parts of the system. They are made internally coherent by manifold interactions among their component parts, and the nature of those interactions is therefore a defining property of the modules themselves. Different kinds of modules are based on different kinds of interactions, but they are recognizable because there are numerous and strong interactions within modules and fewer or weaker interactions between a module and the rest of the system.

In a morphological context, developmental modules are spatially delimited domains of developing organisms within which signaling interactions take place that organize patterning and morphogenesis of the resulting adult structures. The interactions that give coherence to these modules must therefore act over the spatial scale of the module. Probably the most widespread mechanism for such interactions over a distance is signaling via morphogens (Neumann and Cohen 1997; Kerszberg 1999; Podos and Ferguson 1999; Gurdon and Bourillot 2001). The signaling molecules can be proteins, for instance of the FGF, Hedgehog, Wnt, or TGF-$\beta$ families, or other molecules such as retinoic acid (e.g., Begemann and Meyer 2001). Spatial patterning by morphogen gradients is a process consisting of two main steps: the establishment of the gradient and its interpretation by cells (Kerszberg 1999; Gurdon and Bourillot 2001).

Morphogen molecules are secreted by some cells and diffuse or are transported to others that may be several cell diameters away. The distances depend on the specific signal molecule, as there are short-range as well as long-range morphogens. Transport can occur via "bucket brigades" of membrane-bound receptor molecules that can carry signaling molecules along the cell surface and from one cell to another, but relay mechanisms involving sequential uptake and re-release by cells have also been shown (Kerszberg 1999; Gurdon and Bourillot 2001). Morphogen transport, and therefore the shape of the gradient, can be influenced by the binding to receptors and interactions with antagonistic proteins, as well as degradation of the morphogen. As a result, many factors can at least potentially influence gradient shape (e.g., Entchev et al. 2000; Teleman and Cohen 2000), and there may also be ample opportunity for evolutionary changes.

Interpretation of morphogen concentration by cells occurs through cellular signal transduction pathways that are activated when morphogen molecules bind to receptors on the cell surface. Because the response to signaling is usually a change of the cell's transcriptional activity, the signal is transmitted from the activated receptors at the cell surface to the nucleus by signal transduction molecules. At least in the particularly well-studied case of activin signaling in *Xenopus* blastula cells, it has been shown that the absolute number of occupied receptors, and not the ratio of occupied to unoccupied receptors on the cell surface, determines the response of the cell (Dyson and Gurdon 1998), and that the transduction system operates linearly, that is, a threefold difference in the number of occupied receptors translates into a threefold difference in activated cellular transduction molecules (Shimizu and Gurdon 1999). The transduction proteins can interact with *cis*-regulatory elements of downstream genes and activate or repress their transcription. To the extent that cells are homogeneous in their interpretation of morphogen concentrations, morphogens will have a coordinating effect and integrate variation across the domain of signaling, thereby promoting the coherence of the module.

A highly simplified model of genetic control for a morphogen gradient and threshold response, when the phenotypic outcome was analyzed with the methods of quantitative genetics, produced complex outcomes including additive genetic variation, dominance, and epistasis among the components of the model (Nijhout and Paulsen 1997; Klingenberg and Nijhout 1999; Gilchrist and Nijhout 2001). A more realistic model including details of transcriptional control of a target gene by the concentration of a transcription factor yielded similarly complex results (Gibson 1996). Given that many gene products are involved in setting up and interpreting morphogen gradients, it is clear that these systems offer a substantial potential for evolutionary change in signaling. Such evolutionary flexibility of signaling processes also provides the potential for changes in the spatial extent, patterning, and integration within developmental modules.

## Morphogenetic Fields

Developmental modules that are spatially defined units giving rise to specific body parts have been discussed in developmental biology in relation to the classical concept of morphogenetic fields (e.g., Gilbert et al. 1996). Morphogenetic fields (also termed secondary embryonic fields) are embryonic regions that are precursors of specific parts of the developing organism, which, once they have been established, have considerable autonomy from the development of other parts of the embryo (e.g., Wilkins 2002, pp. 255–258). This concept has recently been refined in the light of new information on the molecular mechanisms that establish and delimit the fields (Davidson 1993, 2001, chap. 4; Gilbert et al. 1996; Carroll et al. 2001; Wilkins 2002, pp. 302–305). A critical factor for the initial establishment of fields is intercellular signaling, in which cells that receive the initiating signal are set apart from neighboring cells to organize the prospective module. The distinctness of the field is usually assured by the expression of one or more transcription factors that act as field-specific selector genes (Carroll et al. 2001, pp. 26–28) and commit the cells to fates specific to the prospective body part. Once a field is specified, further signaling steps are activated, which mediate the patterning processes leading to further subdivision and specification within the field.

The cells within a morphogenetic field are not necessarily homogeneous, but there may be internal boundaries delimiting cell populations with different properties. For instance, the wing imaginal disc of *Drosophila* is divided into compartments, which are distinct cell lineages because cells normally do not cross the boundary to move from one compartment into another (Dahmann and Basler 1999; Irvine and Rauskolb 2001; Held 2002). Moreover, the compartments are also characterized by the expression of specific selector genes; for instance, the posterior compartments of *Drosophila* imaginal discs express *engrailed*. The compartment boundaries are not just inert division lines separating distinct populations of cells, but they are themselves active signaling centers. For instance, in the *Drosophila* wing, perpendicular morphogen gradients of the Decapentaplegic and Wingless emanate from the anterior–posterior and dorsal–ventral compartment boundaries and set up a coordinate system of positional values throughout the imaginal disc (Lawrence and Struhl 1996). These signals have a double function. On the one hand, through the different expression patterns of target genes that differ in the concentrations required for transcriptional activation, the morphogen gradients define the further subdivision of the field into domains corresponding to specific portions of the final body part (e.g., Lecuit et al. 1996; Nellen et al. 1996; Held 2002). On the other hand, because the signals are transported to both sides of the respective compartment boundary, they are contributing to integration across the compartments.

The partitioning of the field into subdomains creates new boundaries where populations of cells expressing different regulatory genes are juxtaposed to each other. These new boundaries can in turn be the origin of signaling through morphogens. Through sequential rounds of intercellular signaling and division of transcription domains, the initial pattern of the morphogenetic field can be elaborated (Davidson 1993, 2001, chap. 4). Because this process usually proceeds while the field itself is growing, signals that travel over a constant distance, as measured in cell diameters, will act at a successively smaller scale relative to the field as a whole. To make this stepwise elaboration of preexisting pattern elements more intuitive, Coen (1999, pp. 131–143) has used the metaphor of an artist painting on an imaginary canvas that is expanding while the strokes of the paintbrush always have the same width: at first, the coarse outlines of the overall composition are laid out, whereas the later brush strokes add successively finer details.

The iterative patterning through successive rounds of signaling and establishment of transcription domains specifies the overall topology and pattern elements of the body part that will arise from the morphogenetic field. This specification of the prospective structure is by a combinatorial code of selector genes, whose transcription domains will overlap to various degrees, depending on the sequence of subdivisions. The organization of patterning processes is therefore hierarchical, where overall integration is expected to result from the early signaling steps with morphogen gradients extending across the entire field, but where later patterning steps would generate only local integration within progressively finer subdomains.

The regional code of selector genes in the morphogenetic field influences the patterning of the prospective structure by locally variable rates of cell proliferation and directional alignment of new cells (e.g., Resino et al. 2002). This prepattern is translated into the geometry of the final structure by differentiation of tissues and by morphogenetic movements of parts, for instance, deformations such as stretching, folding, and distal outgrowth. These processes do not all need to reflect the hierarchical fashion in which the domains for pattern elements originally were laid down in the field, and may even obscure some of the original localized structure by overall deformations that force different parts of the field to fit together. On the whole, these late morphogenetic events are not nearly as well understood as the early patterning processes, but they clearly have the potential to influence patterns of integration of morphological structures decisively.

## Morphological Integration

Integration resulting from developmental interactions can be studied by analysis of covariation among the parts of the fully formed structure. However, developmental connection is not the only cause of covariation, because genetic and environmental factors also may contribute to simultaneous variation of multiple parts. It is therefore helpful to examine briefly how covariation between morphological traits can arise (see also Klingenberg 2002a). Covariation is the regular association of variation between different traits. Therefore, if one trait deviates in a particular way from its average value, there is an expectation that a different trait will also deviate from its average in a specific direction. What is required for covariation is a source of variation and a mechanism that generates a regular association between the traits. The source of variation may be linked to the mechanism that generates the association, but this is by no means necessary. Associations between traits are generated primarily in two different manners: by direct connections between the developmental pathways that produce the traits, or by parallel variation of separate pathways that respond to the same extrinsic factors (Fig. 10.2; for a detailed discussion of the concept of developmental pathways, see Wilkins 2002, chap. 4).

In the preceding sections, I have discussed developmental signaling as a source of covariation, where signals originating from a restricted area such as a compartment boundary are transmitted through a much larger expanse of a morphogenetic field. Variation arising at the origin of the signal is therefore transmitted over a distance and can affect large parts of the developing structure simultaneously, generating systematic covariation. The patterning processes that subdivide the field into subdomains, and therefore define the spatial organization of the prospective body part, also rely on these signals. Riska (1986) examined a series of models in which developmental precursors are partitioned into parts that give rise to different traits (Fig. 10.2A). Variation in growth before the fission will result in positive correlations between the resulting parts, whereas variation in the proportions allocated to the parts will generate negative associations. These elementary mechanisms are involved in complex developmental processes such as the growth, partitioning, and migration of cell populations, for example in the neural crest, where they are critical determinants of patterning (e.g., Köntges and Lumsden 1996; Hall 1999). Therefore, processes like these can mold the associations among the resulting adult traits.

Signals from one pathway to another, often localized in distinct portions of the developmental field, are another mechanism that can generate covariation between the resulting traits (Fig. 10.2B). Such signaling has also been referred

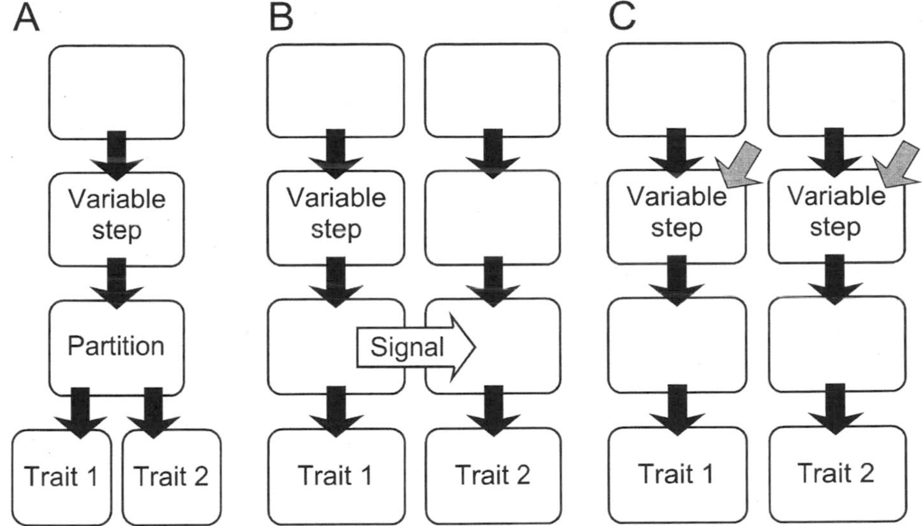


Figure 10.2 Origins of covariation between morphological traits (modified after Klingenberg 2002a). A. Direct connection between pathways of two traits due to partitioning of a common developmental precursor. The variation existing in the pathway before or at the partition is transmitted and can manifest itself as covariation between the traits. B. Direct connection by signaling between pathways. Variation is transmitted from the pathway containing the source of the signal, and therefore can jointly affect the traits that arise from both pathways. C. Parallel variation of two separate developmental pathways. Because there is no transmission of variation between pathways, covariation relies entirely on the simultaneous effects on both developmental pathways by an extrinsic source of variation (gray arrows).

to as epigenetic control (Atchley and Hall 1991; Cowley and Atchley 1992; Hall 1999, chap. 7). Signals may even originate from adjoining structures outside the field itself, such as signaling from the endoderm to the cephalic neural crest and later between elements derived from them (Hall 1999; Couly et al. 2002), but will still cause covariation when the signal from one source has effects over an extended domain where it is received. Although these signaling mechanisms are likely to be the predominant source of interactions within modules that give rise to integration, there are also other processes that can result in direct transmission of variation between pathways.

All these effects are associations due to the direct developmental interactions among the developmental pathways that give rise to the parts concerned. These interactions can transmit variation originating in a single developmental pathway to multiple others, that is, variation from one source is transmitted to multiple pathways via the interactions among them, and can manifest itself as covariation among all the resulting traits.

The origin of the variation does not matter in this context: whatever the source of the variation in a given pathway, the variation will be transmitted to the other pathways if the variable step in the pathway precedes (is "upstream of") the developmental connection between pathways (Fig. 10.2A,B). If the variation is of genetic origin, its transmission to multiple traits generates pleiotropy (relational pleiotropy of Hadorn 1945; Pyeritz 1989; Wilkins 2002, pp. 117–118). If the variation is environmentally induced, it will result in coordinated patterns of phenotypic plasticity. Even for random variation arising spontaneously within the developmental system itself (e.g., McAdams and Arkin 1999; Klingenberg 2002b), the connections of developmental pathways will result in patterned morphological variation.

There is another possible origin of covariation among traits, however, which is not based on direct connections between developmental pathways. This is the parallel variation of separate developmental pathways in response to extrinsic sources of variation that affect the pathways simultaneously (Fig. 10.2C). Joint variation of the morphological traits is produced by an outside factor that affects a step in each pathway and thereby elicits responses in all of them. The developmental effects of this extrinsic variation are transmitted in parallel along each of the developmental pathways, but not from one to another. Moreover, because no direct exchange between pathways occurs, the developmental precursors of the traits that covary in this way are not necessarily adjacent to each other—there is no need for any particular spatial relationship between them. Possible sources of variation include environmental factors such as temperature changes and nutrition. Allelic variation in genes that affect multiple developmental processes also can produce covariation in this manner, which is a form of pleiotropy, because the gene products are involved in multiple pathways that are otherwise independent (mosaic pleiotropy of Hadorn 1945; Wilkins 2002, pp. 117–118). An example of such a gene is *Distal-less* in butterflies, where it is involved in the development of the distal parts of limbs as well as in the later specification of the colored eyespots on the wings (Carroll et al. 1994; Panganiban et al. 1994).

An important consequence of variation in separate pathways is that perturbations arising within the developmental pathway of one trait cannot be transmitted to other pathways and traits in this manner. To produce covariation by parallel variation, the source of variation must be extrinsic to the pathways themselves, and will usually be outside the developing organism as well. This is particularly clear for environmental variation, which affects the developing organism from outside. Genetic variation, although perhaps in a less obvious way, is also extrinsic to the developing organism, because it consists of differences in the genotypes among individuals that are already established at the zygote stage, but can of course affect the later development.

It is important to distinguish the two components of mechanisms that produce integration among morphological traits: on the one hand the source of variation, on the other hand the processes by which the variation is channeled into patterns of association between traits, that is, the processes that manifest the variation multiple traits simultaneously. Both are necessary for covariation between morphological traits to arise, but they play different roles in the mechanisms that produce covariation. Direct connections between developmental pathways generate regular associations among morphological traits by acting as conduits for variation regardless of its origin. For parallel variation of separate developmental pathways, however, the regularity of the association arises from the source of variation itself, which generates covariation through its simultaneous effects on multiple pathways.

The theoretical framework of Cowley and Atchley (1992) distinguishes the effects of developmental interactions among traits as epistatic effects from the intrinsic pleiotropic effects that genes exert on separate traits simultaneously. Their concept of epistatic effects of a gene on multiple traits approximately corresponds to pleiotropy by direct connection between developmental pathways. Likewise, their notion of intrinsic pleiotropy is more or less equivalent to pleiotropy by parallel effects of genes on separate developmental pathways. Cowley and Atchley make this framework amenable for statistical analysis by assuming that the effects are additive, meaning that the system is linear. In general, however, developmental processes are nonlinear, and often extremely so, and it cannot be assumed that any rescaling of phenotypic values is able to linearize the effects of all processes simultaneously. In a developmental system of multiple nonlinear and interdependent processes that is not known completely, therefore, it is unlikely that epigenetic effects and intrinsic pleiotropy among traits as proposed by Cowley and Atchley (1992) can be separated by statistical means. This theoretically elegant approach will therefore not be practical for empirical studies of developmental integration.

## A Morphometric Approach to Delimit Developmental Modules

Developmental modules can be recognized as those spatial domains of organisms within which there is strong integration through direct developmental interactions, and which are relatively independent of other such domains. Therefore, to identify developmental modules from morphological data, covariation due to direct connection of developmental pathways is informative, but not covariation from parallel variation of separate pathways (see also Klingenberg 2002a). To isolate covariation due to direct connection of developmental pathways, it is desirable to control rigorously for environmental and genetic variation, because that would eliminate the variation leading to parallel variation of separate pathways.

A straightforward biometric protocol that contains an inherent control for genetic and environmental factors is to analyze patterns of covariation in fluctuating asymmetry. Fluctuating asymmetry refers to small random differences between corresponding parts on the left and right body sides of each individual (e.g., Palmer and Strobeck 1986; Palmer 1994; Møller and Swaddle 1997). The left and right body sides share the same genome and in most organisms also very nearly the same environment. Because they are "held constant" between the body sides of each individual, genotype and environment cannot produce left–right asymmetries, nor can genotype × environment interactions. This argument assumes that phenomena like somatic mutation and somatic recombination are rare, and it may not apply to sessile organisms located in an environmental gradient, but it should hold at least for most mobile animals (Klingenberg 2002b). Therefore, the structures arising on either body side are replicates of each other that develop separately under nearly identical conditions, and, in a completely deterministic system, would be identical mirror images of each other. Development is not strictly deterministic, however, and there are small random perturbations during development leading to differences between corresponding morphological structures on the left and right body sides. Random variation from many developmental processes can generate such fluctuating asymmetry, because the dynamics of most cellular processes are inherently stochastic (McAdams and Arkin 1999; Klingenberg 2002b), but it must originate within the developmental system itself.

Covariation in fluctuating asymmetry between traits can only arise through direct connections between their developmental pathways. Because the perturbations responsible for the asymmetry originate within the pathways themselves, they can generate covariation of asymmetry only if the perturbations themselves are transmitted between pathways through direct connection. Completely separate pathways also can show fluctuating asymmetry, but the asymmetries are uncorrelated because perturbation cannot be transmitted among pathways. Therefore, the analysis of covariation in fluctuating asymmetry is a way to isolate the contribution of direct connections between developmental pathways to the integration among traits. Comparing the patterns of covariance in asymmetry to the patterns of covariance among individuals, which also includes a contribution from parallel variation of separate pathways, will then make it possible to assess the importance of both ways of generating morphological integration (Klingenberg and Zaklan 2000; Klingenberg et al. 2001; Klingenberg 2002a).

These analyses focus on the covariation of fluctuating asymmetries, that is, the joint variation of asymmetry in multiple variables around the average asymmetry. Therefore, such analyses automatically correct also for directional asymmetry, the systematic difference between the averages of traits on the left and right sides, as it is commonly found in subtle form even in structures that superficially appear symmetric (Klingenberg et al. 1998; Klingenberg 2002b).

These analyses of covariation of fluctuating asymmetry for studying the developmental basis of morphological integration differ in important ways from other analyses of fluctuating asymmetry in multiple traits (e.g., Lens and van Dongen 1999; Leung et al. 2000). Those analyses examine whether individuals differ consistently in the amount of asymmetry in different traits, reflecting variation in the organism-wide capacity to buffer against developmental perturbation. Therefore, those analyses consider traits that are developmentally independent of one another, so that different traits can be used as independent sources of information. Those studies also use the absolute values of asymmetries (unsigned asymmetry), because it is the magnitude and not the direction of asymmetry that is of interest. In contrast, to identify developmental modules, it is essential that signed asymmetries are analyzed (e.g., raw right–left differences for each variable), because the directions of asymmetries are of critical importance for analyzing the covariation among traits (Klingenberg 2002a).

Covariation of signed asymmetries for linear distance measures has long been documented (e.g., Jolicoeur 1963; Leamy 1984, 1993; Hallgrímsson 1998), but these studies did not specifically examine the developmental relationships among traits (but for a partial attempt, see Sakai and Shimamoto 1965). In recent years, the methods of geometric morphometrics have been adapted to study left–right asymmetry (e.g., Klingenberg and McIntyre 1998; Auffray et al. 1999; Klingenberg et al. 2002). This approach offers a particular potential for delimiting developmental modules, because it explicitly takes into account the geometry of patterns of variation, and therefore facilitates their interpretation in the anatomical context of the structure under study. These geometric methods have been applied for the comparison of covariance patterns between fluctuating asymmetry and individual variation (Klingenberg and McIntyre 1998; Debat et al. 2000; Klingenberg et al. 2002) and specifically for delimiting developmental modules (Klingenberg and Zaklan 2000; Klingenberg et al. 2001).

Only a few studies have used this approach so far, but they have confirmed the feasibility of the method and have produced some first results (for a more detailed review, see Klingenberg 2002a). A morphometric study of *Drosophila* wings (Klingenberg and Zaklan 2000) examined the question whether the entire wing is a single module or whether the anterior and posterior compartments, which are separate cell lineages from the inception of the wing imaginal discs (Held 2002, pp. 87–91), are distinct modules. The study found that fluctuating asymmetry is almost completely integrated throughout the wing, because the component of variation shared between the two compartments accounted for nearly all the variation across the entire wing. These results indicated that the entire wing is a single coherent module, and that the anterior and posterior wing compartments are not separate modules (Klingenberg and Zaklan 2000). This agrees well with results from developmental biology indicating that the boundary between the anterior and posterior compartments is the source of signals that are critical for patterning in both compartments (Held 2002, chap. 6). The boundary is therefore not an inert delimiter between compartments, but is itself an active center of integration throughout the wing.

A further study showed that the fore- and hindwings of bees are each an integrated module and clearly separated from one another (Klingenberg et al. 2001). Accordingly, in flies and bees, each entire wing constitutes a module, which presumably relates to the fact that each wing is derived from a separate imaginal disc and that the signaling interactions or other processes taking place within each disc provide strong integration.

Moreover, the studies of fly and bee wings have also found good agreement between the covariance patterns for fluctuating asymmetry and variation among individuals (Klingenberg and McIntyre 1998; Klingenberg and Zaklan 2000; Klingenberg et al. 2001). This agreement suggests that the same processes may be responsible for covariation of asymmetry as well as of variation among individuals, and in particular, that direct connections among developmental pathways also may have an important or even dominant role in shaping genetic and environmental components of covariance in insect wings.

In contrast, a study of mouse skulls found considerable discrepancies between the covariance patterns for fluctuating asymmetry and individual variation, and suggested that different processes were responsible for each (Debat et al. 2000). A comparable result with no similarity between covariance patterns for fluctuating asymmetry and variation among individuals was also obtained in a small study of pharyngeal jaws in a species of cichlid fish (Klingenberg et al. 2002). In that case, the dominant pattern for interindividual variation may correspond to phenotypic plasticity associated with the trophic polymorphism of this fish, and thus represents an outside factor fundamentally different from the developmental processes controlling integration for fluctuating asymmetry. Clearly, the few studies that are available so far are not sufficient for a generalization of the results. Nevertheless, some interesting patterns have emerged, which indicate that further research, particularly on more complex structures such as whole skulls, will be worthwhile.

## Modules, Integration, and the Evolution of Pleiotropy

Integration and modularity have often been discussed in an evolutionary context, frequently with the connotation that they are both adaptive themselves. Integration evolves to ensure that different parts and organ systems are coordinated into a whole functioning organism. Modularity, however, allows for evolution in some body parts without effects on others, and thereby provides an escape from the universal tradeoffs between organismal functions as they would exist in a completely integrated organism. Therefore, an important question is how patterns of integration themselves evolve. Some authors have argued that genetic covariance matrices evolve to reflect the multivariate selection regime and the functional relationships of the morphological traits (Cheverud 1984, 1996; Wagner 1996; Wagner and Altenberg 1996). So far there are no empirical studies, however, that clearly document the adaptive evolution of patterns of variation and rule out nonadaptive alternatives. In these considerations, a crucial issue is the evolution of the patterns of pleiotropy for the genes involved.

This chapter offers a new perspective on this issue, which emphasizes development by distinguishing the different ways in which pleiotropy can originate (Hadorn 1945; Pyeritz 1989; Hodgkin 1998; Wilkins 2002, pp. 117–118). A gene can have simultaneous effects on multiple traits either by direct connection between developmental pathways or by their parallel effects on multiple separate pathways (Fig. 10.2). Clearly, both mechanisms can cause genes to have similar effects on the phenotype, since both can produce pleiotropy. A quite different question, however, is whether these two distinct developmental sources of pleiotropy also have the same potential for evolutionary change (see also Cowley and Atchley 1992).

With direct connection between developmental pathways, any gene that affects the pathway upstream of the connection ("variable step" in Figs. 10.2A,B) will have a pleiotropic effect on all the descendant traits because allelic effects are transmitted between pathways. Moreover, provided that allelic variation leads to differences in the activities of relevant gene products that can be transduced through the pathway, multiple "upstream" loci will have congruent patterns of pleiotropy due to the same connection of pathways. Evolution of the patterns of pleiotropy must therefore occur by changing the linkage among pathways itself, for example, by changes in signaling or the mechanism of partitioning a developmental field. These changes may have profound effects on the resulting morphological structure and its function. In other words, it is likely that these changes in signaling mechanisms will often be under stabilizing selection and that patterns of pleiotropy through direct connection of developmental pathways will be fairly conservative. Because direct connections between developmental pathways occur primarily within modules and only to a lesser degree between modules, the evolutionary conservatism of the resulting patterns of pleiotropy will contribute to the evolutionary inertia of the modular organization itself. If a change in the connection of developmental pathways is selectively advantageous, it can be a source of morphological innovation. Such an evolutionary transition to a novel interaction between pathways could then lead to a complete reorganization of the spatial pattern of the module, and therefore to a concerted change in the patterns of pleiotropy for all the genes upstream of the link between pathways.

In contrast, pleiotropy by parallel effects of a gene on multiple developmental pathways relies entirely on the activity of that gene alone. Because transcriptional control of genes is itself generally modular, the expression of the gene in each separate developmental context is normally controlled by one or a few separate enhancer elements (Davidson 2001). In order to exert a joint effect on two different pathways, allelic differences must lie either in a *cis*-regulatory element that is activated in both pathways, or they must affect the transcript itself (either by a difference in the protein-coding sequence or in untranslated regions affecting posttranscriptional processing and the control of translation). Therefore, pleiotropy by this mechanism requires a particular kind of allelic variation to exist in a population. There can be a great diversity of patterns of pleiotropy, however, because every allele of a gene can have a distinct combination of effects on different developmental pathways. These patterns of pleiotropy can be modified by mutations that affect the expression of the gene; for instance, any regulatory changes that lead to reduced expression of the gene in a subset of pathways can reduce the pleiotropic effects of allelic variation at that locus. Because the relevant changes in *cis*-regulatory regions can occur rapidly (Stone and Wray 2001), pleiotropy by parallel effects of a gene on multiple pathways is likely to evolve readily under natural selection.

## Evolution of Genetic Covariances

Just as the developmental origin of morphological covariation makes a difference for the evolution of the pleiotropic effects of single loci, it can also affect changes of genetic variances and covariances in natural populations, which are due to the aggregate effects of all segregating loci. The evolution of genetic variances and covariances among traits is an important issue in evolutionary quantitative genetics, because long-term predictions of response to selection or of random drift depend on the genetic covariance matrix (reviewed by Roff 1997, and in Chapter 16, this volume). Genetic covariances can be due to pleiotropic effects of individual loci, but they also can arise from genetic linkage among loci that affect different traits (e.g., Lynch and Walsh 1998). As outlined above, pleiotropic effects can originate through direct connection or parallel variation of developmental pathways. The origin of covariance by genetic linkage is a special case of parallel variation, in which different developmental pathways are affected by different loci whose effects are associated statistically by the genetic linkage.

If direct developmental linkages between developmental pathways contribute most of the covariation between traits, shifts in allele frequencies will have relatively small effects on the patterns of covariance. Because the connections of pathways act as a common conduit for the effects of multiple "upstream" genes, the patterns of genetic covariance will be similar regardless of the specific allelic differences and allele frequencies in a population. Direct links among developmental pathways will therefore contribute to the constancy of covariances among traits. Because of the strong direct interactions among the parts of a developmental module, this reasoning suggests that patterns of genetic covariances among traits within a module should be relatively stable, even over evolutionary time scales.

In contrast, if covariation among traits arises primarily by parallel variation of separate developmental pathways, the patterns of covariation will be more labile. Because every allele can have different combinations of pleiotropic effects, genetic covariances will depend strongly on allele frequencies in the population. Because genetic linkage is also subject to change in natural populations, the genetic covariances produced by it will also be evolutionarily fluid. Patterns of genetic covariances due to parallel variation of separate developmental pathways are therefore likely to undergo substantial evolutionary transformations by selection and drift.

## Conclusions

This chapter has reviewed the developmental origin of morphological integration and examined its implications for evolution. Morphological integration reflects the fact that organisms and their development are organized into modules. Most adult body parts arise from distinct morphogenetic fields within which spatial pattern is established by direct developmental interactions, which also integrate the components of the module into a coherent unit. The resulting morphological integration is manifest in genetic as well as nongenetic components of variation. The spatial extent of modules can be delimited by analyzing the patterns of covariation for fluctuating asymmetry, which indicate the domains within which there is integration by direct developmental interactions. Although it may seem paradoxical at first, it is possible to use this approach that is based on variation of nongenetic origin to study the developmental basis of pleiotropy and genetic integration.

Developmental integration by direct interactions within modules is one of the prime factors determining patterns of pleiotropy. It is likely that these patterns of pleiotropy are evolutionarily conservative, because to change them would require fundamental alterations of the developmental processes involved. In contrast, whole-organism integration across modules, by parallel variation of separate developmental pathways, relies on a different mechanism and requires an extrinsic source of variation. It is likely that pleiotropy due to this process can evolve easily by regulatory changes in the genes responsible. Similarly, the developmental origins of genetic covariances at the population level are important determinants for their evolution. Again, it is likely that patterns of genetic covariance that are due to direct developmental interactions within modules are more robust evolutionarily than covariances due to parallel variation of separate developmental pathways.

Clearly, the ideas and hypotheses presented here need to be developed further and tested empirically, but they have the potential to provide a new perspective on the role of development for genetic and phenotypic integration among traits. A developmental perspective offers a framework for obtaining a unified understanding of morphological variation, from molecular mechanisms to phenotypic manifestation. Inclusion of information on gene regulation, signaling, and the molecular basis of growth and differentiation has much to offer to evolutionary quantitative genetics.

## *Literature Cited*


Atchley, W. R., and B. K. Hall. 1991. A model for development and evolution of complex morphological structures. Biol. Rev. 66:101–157.

Auffray, J.-C., V. Debat, and P. Alibert. 1999. Shape asymmetry and developmental stability. Pp. 309–324 in M. A. J. Chaplain, G. D. Singh, and J. C. McLachlan, eds. On Growth and Form: Spatio-Temporal Pattern Formation in Biology. Wiley, Chichester, UK.

Begemann, G., and A. Meyer. 2001. Hindbrain patterning revisited: timing and effects of retinoic acid signalling. BioEssays 23:981–986.

Bolker, J. A. 2000. Modularity in development and why it matters to evo-devo. Am. Zool. 40:770–776.

Bookstein, F. L. 1991. Morphometric Tools for Landmark Data: Geometry and Biology. Cambridge University Press, Cambridge.

Carroll, S. B., J. Gates, D. N. Keys, S. W. Paddock, G. E. F. Panganiban, J. E. Selegue, and J. A. Williams. 1994. Pattern formation and eyespot determination in butterfly wings. Science 265:109–114.

Carroll, S. B., J. K. Grenier, and S. D. Weatherbee. 2001. From DNA to Diversity: Molecular Genetics and the Evolution of Animal Design. Blackwell Science, Malden, MA.

Cheverud, J. M. 1984. Quantitative genetics and developmental constraints on evolution by selection. J. Theor. Biol. 110:155–171.

Cheverud, J. M. 1996. Developmental integration and the evolution of pleiotropy. Am. Zool. 36:44–50.

Cheverud, J. M., E. J. Routman, and D. J. Irschick. 1997. Pleiotropic effects of individual gene loci on mandibular morphology. Evolution 51:2006–2016.

Coen, E. 1999. The Art of Genes: How Organisms Make Themselves. Oxford University Press, Oxford.

Couly, G., S. Creuzet, S. Bennaceur, C. Vincent, and N. M. Le Douarin. 2002. Interactions between Hox-negative cephalic neural crest cells and the foregut endoderm in patterning the facial skeleton in the vertebrate head. Development 129:1061–1073.

Cowley, D. E., and W. R. Atchley. 1992. Quantitative genetic models for development, epigenetic selection, and phenotypic evolution. Evolution 46:495–518.

Dahmann, C., and K. Basler. 1999. Compartment boundaries: at the edge of development. Trends Genet. 15:320–326.

Davidson, E. H. 1993. Later embryogenesis: regulatory circuitry in morphogenetic fields. Development 118:665–690.

Davidson, E. H. 2001. Genomic Regulatory Systems: Development and Evolution. Academic Press, San Diego.

Debat, V., P. Alibert, P. David, E. Paradis, and J.-C. Auffray. 2000. Independence between developmental stability and canalization in the skull of the house mouse. Proc. R. Soc. Lond. B Biol. Sci. 267:423–430.

Dryden, I. L., and K. V. Mardia. 1998. Statistical Analysis of Shape. Wiley, Chichester, UK.

Dyson, S., and J. B. Gurdon. 1998. The interpretation of position in a morphogen gradient as revealed by occupancy of activin receptors. Cell 93:557–568.

Entchev, E. V., A. Schwabedissen, and M. González-Gaitán. 2000. Gradient formation of the TGF-$\beta$ homolog Dpp. Cell 103:981–991.

Gibson, G. 1996. Epistasis and pleiotropy as natural properties of transcriptional regulation. Theor. Popul. Biol. 49:58–89.

Gilbert, S. F., J. M. Opitz, and R. A. Raff. 1996. Resynthesizing evolutionary and developmental biology. Dev. Biol. 173:357–372.

Gilchrist, M. A., and H. F. Nijhout. 2001. Nonlinear developmental processes as sources of dominance. Genetics 159:423–432.

Gurdon, J. B., and P.-Y. Bourillot. 2001. Morphogen gradient interpretation. Nature 413:797–803.

Hadorn, E. 1945. Zur Pleiotropie der Genwirkung. Arch. Klaus-Stift. Verebungsforsch. Supplement 20:82–95.

Hall, B. K. 1999. Evolutionary Developmental Biology. Kluwer, Dordrecht, The Netherlands.

Hallgrímsson, B. 1998. Fluctuating asymmetry in the mammalian skeleton: evolutionary and developmental implications. Evol. Biol. 30:187–251.

Held, L. I., Jr. 2002. Imaginal Discs: The Genetic and Cellular Logic of Patterns Formation. Cambridge University Press, Cambridge.

Hodgkin, J. 1998. Seven types of pleiotropy. Int. J. Dev. Biol. 42:501–505.

Irvine, K. D., and C. Rauskolb. 2001. Boundaries in development: formation and function. Annu. Rev. Cell Dev. Biol. 17:189–214.

Jolicoeur, P. 1963. Bilateral symmetry and asymmetry in limb bones of *Martes americana* and man. Rev. Can. Biol. 22:409–432.

Kerszberg, M. 1999. Morphogen propagation and action: toward molecular models. Semin. Cell Dev. Biol. 10:297–302.

Kirschner, M., and J. Gerhart. 1998. Evolvability. Proc. Natl. Acad. Sci. USA 95:8420–8427.

Klingenberg, C. P. 2002a. Developmental instability as a research tool: using patterns of fluctuating asymmetry to infer the developmental origins of morphological integration. Pp. 427–442 in M. Polak, ed. Developmental Instability: Causes and Consequences. Oxford University Press, New York.

Klingenberg, C. P. 2002b. A developmental perspective on developmental instability: theory, models and mechanisms. Pp. 14–34 in M. Polak, ed. Developmental Instability: Causes and Consequences. Oxford University Press, New York.

Klingenberg, C. P., and G. S. McIntyre. 1998. Geometric morphometrics of developmental instability: analyzing patterns of fluctuating asymmetry with Procrustes methods. Evolution 52:1363–1375.

Klingenberg, C. P., and H. F. Nijhout. 1999. Genetics of fluctuating asymmetry: a developmental model of developmental instability. Evolution 53:358–375.

Klingenberg, C. P., and S. D. Zaklan. 2000. Morphological integration between developmental compartments in the *Drosophila* wing. Evolution 54:1273–1285.

Klingenberg, C. P., G. S. McIntyre, and S. D. Zaklan. 1998. Left-right asymmetry of fly wings and the evolution of body axes. Proc. R. Soc. Lond. B Biol. Sci. 265:1255–1259.

Klingenberg, C. P., A. V. Badyaev, S. M. Sowry, and N. J. Beckwith. 2001. Inferring developmental modularity from morphological integration: analysis of individual variation and asymmetry in bumblebee wings. Am. Nat. 157:11–23.

Klingenberg, C. P., M. Barluenga, and A. Meyer. 2002. Shape analysis of symmetric structures: quantifying variation among individuals and asymmetry. Evolution 56:1909–1920.

Köntges, G., and A. Lumsden. 1996. Rhombencephalic neural crest segmentation is preserved throughout craniofacial ontogeny. Development 122:3229–3242.

Lawrence, P. A., and G. Struhl. 1996. Morphogens, compartments, and patterns: lessons from *Drosophila*? Cell 85:951–961.

Leamy, L. 1984. Morphometric studies in inbred and hybrid house mice. V. Directional and fluctuating asymmetry. Am. Nat. 123:579–593.

Leamy, L. 1993. Morphological integration of fluctuating asymmetry in the mouse mandible. Genetica 89:139–153.

Lecuit, T., W. J. Brook, M. Ng, M. Calleja, H. Sun, and S. M. Cohen. 1996. Two distinct mechanisms for long-range patterning by Decapentaplegic in the *Drosophila* wing. Nature 381:387–393.

Lens, L., and S. van Dongen. 1999. Evidence for organism-wide asymmetry in five bird species of a fragmented afrotropical forest. Proc. R. Soc. Lond. B Biol. Sci. 266:1055–1060.

Leung, B., M. R. Forbes, and D. Houle. 2000. Fluctuating asymmetry as a bioindicator of stress: comparing efficacy of analyses involving multiple traits. Am. Nat. 155:101–115.

Lynch, M., and B. Walsh. 1998. Genetics and Analysis of Quantitative Traits. Sinauer, Sunderland, MA.

McAdams, H. H., and A. Arkin. 1999. It's a noisy business! Genetic regulation at the nanomolecular scale. Trends Genet. 15:65–69.

Møller, A. P., and J. P. Swaddle. 1997. Asymmetry, Developmental Stability, and Evolution. Oxford University Press, Oxford.

Nellen, D., R. Burke, G. Struhl, and K. Basler. 1996. Direct and long-range action of a DPP morphogen gradient. Cell 85:357–368.

Neumann, C. J., and S. M. Cohen. 1997. Morphogens and pattern formation. BioEssays 19:721–729.

Nijhout, H. F., and S. M. Paulsen. 1997. Developmental models and polygenic characters. Am. Nat. 149:394–405.

Olson, E. C., and R. L. Miller. 1958. Morphological Integration. University of Chicago Press, Chicago.

Palmer, A. R. 1994. Fluctuating asymmetry analyses: a primer. Pp. 335–364 in T. A. Markow, ed. Developmental Instability: Its Origins and Implications. Kluwer, Dordrecht, The Netherlands.

Palmer, A. R., and C. Strobeck. 1986. Fluctuating asymmetry: measurement, analysis, patterns. Annu. Rev. Ecol. Syst. 17:391–421.

Panganiban, G., L. Nagy, and S. B. Carroll. 1994. The role of the *Distal-less* gene in the development and evolution of insect limbs. Curr. Biol. 4:671–675.

Pimentel, R. A. 1979. Morphometrics: The Multivariate Analysis of Biological Data. Kendall/Hunt, Dubuque, IA.

Podos, S. D., and E. L. Ferguson. 1999. Morphogen gradients: new insights from DPP. Trends Genet. 15:396–402.

Pyeritz, R. E. 1989. Pleiotropy revisited: molecular explanations of a classic concept. Am. J. Med. Genet. 34:124–134.

Raff, R. A. 1996. The Shape of Life: Genes, Development and the Evolution of Animal Form. University of Chicago Press, Chicago.

Resino, J., P. Salama-Cohen, and A. García-Bellido. 2002. Determining the role of patterned cell proliferation in the shape and size of the *Drosophila* wing. Proc. Natl. Acad. Sci. USA 99:7502–7507.

Riska, B. 1986. Some models for development, growth, and morphometric correlation. Evolution 40:1303–1311.

Roff, D. A. 1997. Evolutionary Quantitative Genetics. Chapman & Hall, New York.

Sakai, K.-I., and Y. Shimamoto. 1965. A developmental-genetic study on panicle charaters in rice, *Oryza sativa* L. Genet. Res. 6:93–103.

Shimizu, K., and J. B. Gurdon. 1999. A quantitative analysis of signal transduction from activin receptor to nucleus and its relevance to morphogen gradient interpretation. Proc. Natl. Acad. Sci. USA 96:6791–6796.

Stone, J. R., and G. A. Wray. 2001. Rapid evolution of *cis*-regulatory sequences via local point mutations. Mol. Biol. Evol. 18:1764–1770.

Teleman, A. A., and S. M. Cohen. 2000. Dpp gradient formation in the *Drosophila* wing imaginal disc. Cell 103:971–980.

von Dassow, G., and E. Munro. 1999. Modularity in animal development and evolution: elements of a conceptual framework for EvoDevo. J. Exp. Zool. (Mol. Dev. Evol.) 285:307–325.

von Dassow, G., E. Meir, E. M. Munro, and G. M. Odell. 2000. The segment polarity network is a robust developmental module. Nature 406:188–192.

Wagner, G. P. 1996. Homologues, natural kinds and the evolution of modularity. Am. Zool. 36:36–43.

Wagner, G. P., and L. Altenberg. 1996. Complex adaptations and the evolution of evolvability. Evolution 50:967–976.

Wilkins, A. S. 2002. The Evolution of Developmental Pathways. Sinauer, Sunderland, MA.

Winther, R. G. 2001. Varieties of modules: kinds, levels, origins, and behaviors. J. Exp. Zool. (Mol. Dev. Evol.) 291:116–129.

Yuh, C.-H., H. Bolouri, and E. H. Davidson. 1998. Genomic *cis*-regulatory logic: experimental and computational analysis of a sea urchin gene. Science 279:1896–1902.

# 11

# Studying Mutational Effects on G-Matrices

MASSIMO PIGLIUCCI

## Variation versus Variability, G versus M

Wagner and Altenberg (1996) introduced the concept of variability as distinct from variation in the study of the evolution of complex phenotypes, a concept that has parallels in complexity theory and that concerns the "genetic neighborhood" of a given genotype and its role in evolution on rugged adaptive landscapes (Kauffman and Levin 1987; Kauffman 1993). Their basic idea is that evolution depends not just on the extant genetic *variation* as studied by classical quantitative genetics, but also on the underlying genetic *variability* that quantifies the range of immediately possible genotypes that can occur in a given population in the not too distant future.

In evolutionary theory, the **G**-matrix, which summarizes the additive variances of and covariances among different traits of interest, quantifies the amount of extant genetic variation. **G** enters into the multivariate version of the "breeders' equation," which is taken to be a sufficient description of evolution by natural selection, given certain assumptions:

$$\Delta\overline{\mathbf{Z}} = \mathbf{GP}^{-1}\mathbf{S}$$

where $\Delta\overline{\mathbf{Z}}$ is the vector of change in trait means, **G** is the matrix of additive genetic variances and covariances, **P** is the matrix of phenotypic variances and covariances, and **S** is a vector of selection differentials. One problem with the use of the breeders' equation is that there are both theoretical (Turelli 1988) and empirical (Roff and Mousseau 1999) reasons to think that **G** does not stay constant or even proportional over evolutionary time (see also Hansen and Houle, Chapter 6, this volume, for a different angle), contrary to what is assumed by common evolutionary quantitative genetics theory. Hence, the breeders' equation can work only approximately, on the short run of a few generations (despite the common practice of projecting it over tens of thousands of generations: e.g., Via 1987).

Wagner and Altenberg's, as well as Kauffman's, idea is then that we need to conceive of an expansion of **G**, often referred to as **M** (for mutation, the ultimate source of new genetic variation), which estimates the *variability* potentially harbored by a population, and that can give us a better idea of the medium-capacity for evolutionary responses. I have conceptualized the difference in Fig. 11.1: suppose we measure selection in a natural population using the standard multiple regression approach (Lande and Arnold 1983; Rausher 1992) and find that there is a push to shift the position of the population in phenotypic space from where it currently is to the area identified by the star in the diagram. However, from a study of the quantitative genetics of the situation, we also uncover a fairly strong genetic correlation between these same two traits (a component of **G**, indicated by the narrow ellipse). Our conclusion, based on the breeders' equation, would be that the resultant of the two forces (natural selection and the existing genetic constraint) will actually deflect the evolutionary trajectory pretty far from the optimum favored by selection. This conclusion, of course, depends on usually unstated assumptions concerning **M**, which is normally not estimated empirically and cannot be derived from first principles. For example, it is possible that the genetic architecture of the traits in question is such that, if we allow for a sufficient amount of time to pass so that mutation and recombination will translate some of the underlying *variability* into actual *variation*, the resulting genetic correlation will be much less tight (larger ellipse in the figure). This, in turn, may allow the population to get significantly closer to the selected optimum than we might at first have surmised.

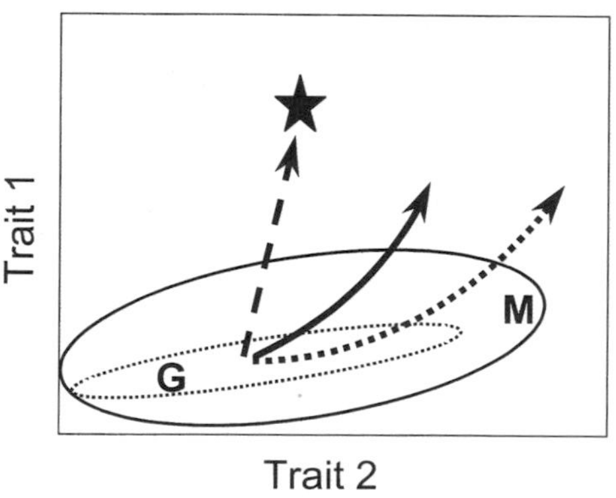


Figure 11.1 The difference between a **G**- and an **M**-matrix and the impact it can have on evolutionary change. The scenario considers the simplified situation of just two traits related by a genetic correlation, represented by the small ellipse. Assuming that natural selection pushes the population in the direction of phenotypic space identified by the star, the presence of a genetic correlation between the two traits instead generates a highly deflected evolutionary trajectory. However, if the **M**-matrix (represented by the larger ellipse) allows for more leeway in the relationship between the two characters, then the balance between selection and genetic constraints may move the population in an intermediate area of phenotypic space (solid arrow).

This view of the evolutionary process seems to open up several avenues of research. In the first place, it provides us with a new conceptual handle on the complex problem of the relationship between natural selection and genetic constraints, something that has been the focus of a sustained series of efforts (e.g., Antonovics 1976; Cheverud 1984; Maynard-Smith et al. 1985; Kirkpatrick and Lofsvold 1992; van Tienderen and Koelewijn 1994; Shaw et al. 1995; Schlichting and Pigliucci 1998; Barton and Partridge 2000). Second, considering **M** and not just **G** appears to offer a more realistic view of the evolutionary process itself, one that can be more comfortably extended to stretches of time beyond which the breeders' equation is of little use, and that are actually most interesting to evolutionary biologists. Third, the concept of variability readily allows us to see how to extend current techniques to empirically estimate components of **M**, just as we already do in the case of **G**. Furthermore, studies of the mutational properties of individual quantitative traits (e.g., Houle et al. 1994; Mackay et al. 1994; Fry et al. 1995, 1996; Fernandez and Lopez-Fanjul 1997; Keightley and Ohnishi 1998) can provide the starting point for analogous research into the effect of mutations on genetic covariances.

However, thinking in terms of **M** also immediately carries a slew of difficulties. First, despite Wagner's referring to the new matrix as **M**, it is obvious that in the short run the translation of variability into variation will more likely be the result of other genetic mechanisms, such as recombination (at least in outbreeding species), than of mutations. This brings into the equation complications related to the enormous natural variation in breeding systems, not to mention the necessity of estimating the frequencies of crossing-over events at the molecular level. Second, while **G** is clearly temporally defined (it can be observed *now*), it is not quite as clear over what time interval one needs to study **M**, since the more time passes, the more recombination and mutation will affect its structure. Third, at the moment, and despite Kauffman's valiant efforts in exploring the properties of *N*-*K* models (where *N* is the number of loci underlying a trait and *K* the number of interactions among them), we have very little theoretical grounding for studying the properties of the **M**-matrix or, more importantly, for making any predictions about its relationship to the more easily observable **G**. Finally, while it is cumbersome enough to estimate components of **G** via what often become logistically challenging experimental designs, it is even more difficult to obtain empirical estimates of components of **M**, since one has to take into account the effect of recombination and of new mutations on the currently existing **G**.

In the following pages I shall concentrate on the results of some attempts at studying components of **M** carried out in my laboratory during the last few years using the weedy herbaceous plant *Arabidopsis thaliana* as a model system (Pigliucci 1998; Alonso-Blanco and Koornneef 2000; Mitchell-Olds 2001; Simpson and Dean 2002). These examples are meant to provide an appreciation for both the insights yielded by this approach as well as the difficulties inherent in it. I then summarize the common threads emerging from our efforts in the hope of stimulating other students to engage in similar research in this and other systems, and conclude with a somewhat provocative discussion of the very conceptualization of selection and constraints as "forces" (as is done, for example, in Fig. 11.1).

## Evolution from a "Lab Rat"

Several of the *A. thaliana* genotypes used in genetic research have evolved under highly artificial conditions for a considerable amount of time, and have been more or less consciously selected for a very short life cycle (Redei 1992). Such selection has reshaped several aspects of the phenotype of this plant, not limited to its life history. For example, the Landsberg "lab rat" version of *A. thaliana* not only flowers and senesces much earlier than field-collected conspecifics, but produces many fewer leaves, tends to be smaller, is significantly less branched (if at all), and has a markedly reduced reproductive fitness (measured by fruit and seed production).

An interesting question, therefore, concerns the possibility and limits of reevolving a normal-looking *A. thaliana* from the starting point of the highly specialized Landsberg genotype, which is what I set out to do as a way to begin to study the characteristics of **M**-matrices in this species. It is important to note at the outset that Landsberg, properly speaking, does not *have* a **G**-matrix. This is because **G** is a population-level concept, and Landsberg is a single genotype, for which genetic correlations are simply undefined. This, of course, does not mean that Landsberg's characteristics are not genetically constrained, but simply that such constraints cannot be quantified as components of **G**. Nevertheless, the first question we asked was if a mutation-selection protocol could allow descendants of Landsberg to evolve a higher fecundity and, if so, by means of what changes in the phenotype of the starting genetic background (Pigliucci et al. 1998).

We used EMS- (ethyl methane sulfonate) mutagenized seeds of Landsberg, which we grew for two generations, the first of which experienced an episode of selection imposed on the now genetically variable population. Instead of selecting directly on fecundity, we favored an increase in leaf production because we wished to see if a commonly observed genetic correlation between leaf number and flowering time in *A. thaliana* (Mitchell-Olds 1996) would appear in the descendants of our isogenic line. Leaf production in our mutagenized populations turned out to be highly positively correlated with both fecundity and time to senescence. The results of the mutation-selection protocol were astounding, considering that we applied selection for only one generation (choosing plants with a mean leaf production more than three standard deviations from that of the baseline strain): the number of leaves produced by the mutants varied from around 6 (the average for Landsberg) to 39; correspondingly, the time to flower was extended from about 33 days to as much as 86. Interestingly, however, the mutants and Landsberg lined up pretty tightly to show a genetic (mutational) correlation of +0.86 between these two traits (Fig. 11.2a), essentially reproducing the known association between the same characters that has been described in many early-flowering populations of *A. thaliana*. Perhaps of equal interest was the fact that other traits were not genetically correlated in the descendants of Landsberg, as they usually are not in other populations of this species: for example, time to senescence and size of the main inflorescence (measured as plant height) were completely unrelated among the mutants, even though all plants senesced much later than the baseline strain (Fig 11.2b).

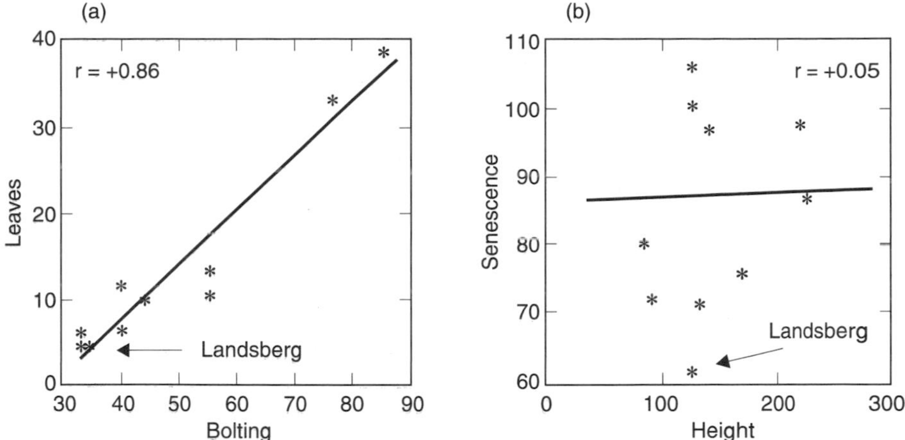


Figure 11.2 Experimental evolution of new genotypes of *A. thaliana* from a baseline inbred line. (a) High genetic correlation between leaf production and bolting (flowering) time in the baseline and mutants. (b) Low genetic correlation between height and time to senescence in the same population. Data from Pigliucci et al. (1998).

One can think of this experiment as a successful attempt at expanding the initial range of genetic variation of a base population (which in this case happened to be essentially zero, given that we started with an inbred line) and exploring the limitations intrinsic in the structure of the underlying **M**-matrix. It was possible to obtain the evolution of phenotypes far removed from the initial one, but for some combinations of traits this was clearly restricted to only certain preferential directions. It seems that some characters in *A. thaliana* can evolve only along "lines of least resistance" (Schluter 1996).

## The Importance of Genetic Background and Mutation Pressure

While the previous experiment provided a first glimpse into components of the **M**-matrix constraining multivariate trait evolution in *Arabidopsis thaliana*, one obvious follow-up question is how much the particular genetic background of the baseline population matters. This is a rather difficult question to answer because there are so many possible backgrounds, some of which are representative of populations with markedly distinct ecological niches (e.g., winter annual versus spring annual populations of *A. thaliana*: Donohue 2002). Furthermore, this is a question that opens up a logistical Pandora's box, since it is not clear what the consistency of mutagenic results is even for the same genetic background: it is possible that a given mutagenic correlation between two traits, like the one discussed above concerning flowering time and leaf production, will actually manifest itself only some of the times in which one attempts to bring out the variability of a population using the experimental approach advocated here.

In part to address these questions, Camara and I (Camara and Pigliucci 1999) examined the effect of EMS mutagenesis on three genetic backgrounds, one of which was the same as used in the first experiment: Landsberg, Dijon, and Wassilewskija. We found that the mutagenic treatment did not alter significantly any trait mean (which it was not expected to do, unlike the case of the previous experiment, which used a mutation-selection protocol). However, it did dramatically affect trait covariances, as was clear from a series of principal components analyses (Fig. 11.3). The detailed constitution of each eigenvector (i.e., the weights or loadings of each original variable on a given principal component) were highly heterogeneous among the three genetic backgrounds, indicating that the underlying **M**-matrix was different, or at least that the patterns of variability we were able to translate into actual variation in our mutant populations depended on the particular genetic background used. Notice in particular what happened to the genetic correlation between flowering time (vegetative period) and leaf production discussed above: in the Landsberg background it was still prominent, thereby confirming it as a major feature of the **M**-matrix of this line (consider the two tallest bars on the positive side of PC2 in Fig. 11.3 for the Landsberg genotype). However, the same relationship was much less marked in the Wassilewskija and Dijon backgrounds, where it showed up to a reduced extent only on the third principal component. Moreover, this relationship appeared in a different association with other traits; the negative relationship between the two focal traits and number of nonelongated inflorescences—a measure of potential additional

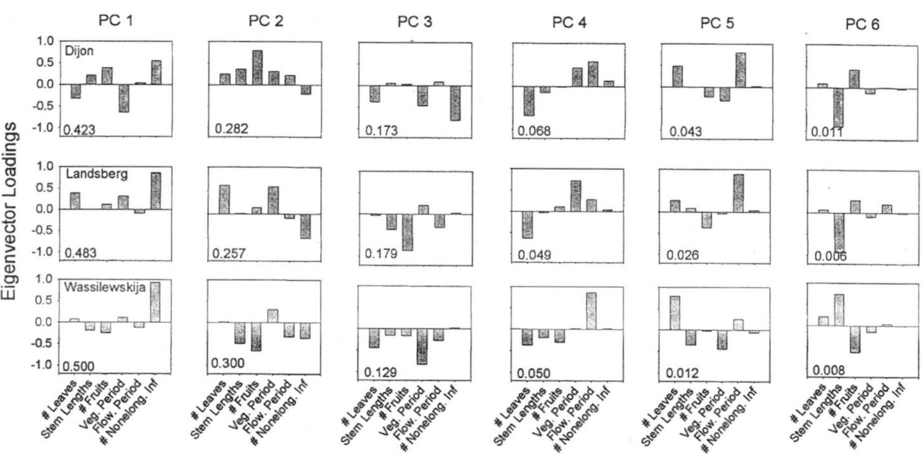


Figure 11.3 The six principal components summarizing the mutagenic covariation among traits in three genetic backgrounds of *A. thaliana*. For each component the bars indicate the weight (loading) of each original variable on that eigenvector, and the numbers in the lower left corner of each plot represent the standardized amount of variance explained by the corresponding eigenvector. Notice how the distribution of weights changed dramatically among genetic backgrounds, indicating that the **M**-matrix may have significantly different structure in these three populations. Data from Camara and Pigliucci (1999).

reproductive output—was negative in Landsberg, positive in Dijon, and null in Wassilewskija. While a formal statistical test using common principal components analysis (see Steppan, Chapter 15, this volume) did suggest commonality among the principal components of our three backgrounds, it also rejected both the hypotheses of equality and proportionality among the full trait matrices. It is instructive to compare this result with the common finding of similar or identical **G**- (or **P**, phenotypic) matrices among closely related populations of the same species sampled from the field (see Roff and Mousseau 1999 for a review). Even if **G**-matrices do not diverge significantly among closely related populations, they seem to have the potential to do so once enough mutations have accumulated. If the latter conclusion should be confirmed for natural and mutagenic populations of the same species, they will be an important piece of the puzzle in locating the balance between genetic constraints and selective pressures (see Hansen and Houle, Chapter 6, this volume).

In a different study, Camara, Ancell, and I (2000) also examined another complicating factor of studying **M**-matrices via induced mutagenesis: the effect of different numbers of mutations. The obvious advantage of generating mutations artificially is that one can compress evolutionary time and study attributes of **M** in the laboratory, but an equally clear limit to this is that if one uses high doses of mutagen, the number of mutations that result are likely to greatly diminish the organism's fitness through a panoply of pleiotropic and epistatic effects. In other words, this is another incarnation of the well-known tradeoff between experimental convenience and empirical realism: we do not want to give up the first one, but cannot afford to push it as far as to make our results completely irrelevant to what happens in nature (recently there has been a flurry of studies on the limits and advantages of laboratory experiments in evolutionary ecology: Matos et al. 2000, 2002; Sgrò and Partridge 2000, 2001; Hoffmann et al. 2001; Matos and Avelar 2001).

We then used four different levels of EMS mutagen (plus the nonmutagenic control) on a single genetic background (the Kendalville natural population) to assess the effects on characters' means, variances, and covariances of various numbers of induced mutations. The results were sharply different for the distinct measures: trait means were unaffected by the number of mutations (Fig. 11.4), even at high levels of mutagen. Trait variances, on the other hand, were markedly different, with a general tendency toward an increase in variance with higher mutagenic effects (Fig. 11.5), though several morphological traits actually peaked at intermediate dosages for as yet unclear reasons. When we used common principal components to compare the genetic covariance matrices among mutagenic treatments (details in Camara et al. 2000), we found that similar doses produced matrices with a few common principal components, while more diverging dosages yielded matrices with no common structure at all. Overall, these results suggest that mutations have very different effects on means and variance/covariances, and that—somewhat surprisingly—one can subject *A. thaliana* to considerably high levels of mutagenesis without causing any apparent decrease in the mean fitness of the resulting plants.

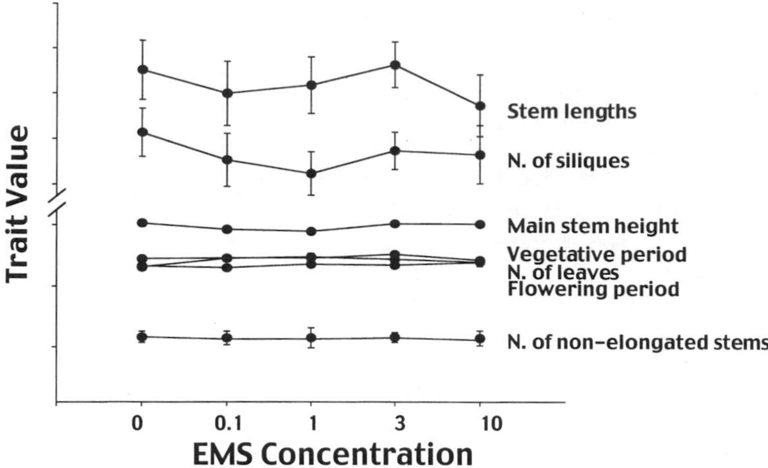


Figure 11.4 The effect of different concentration of mutagen (and increasing number of actual mutations) on the means of quantitative characters in *A. thaliana*. In fact, trait means remained remarkably stable regardless of the number of hits. Data from Camara et al. (2000).

## Selection Along and Away from the Line of Least Resistance

The first two experiments described above found that a well-known genetic constraint linking flowering time and leaf production in *A. thaliana* was present in the mutagenized progeny of the Landsberg genetic background, and the first experiment also found that it channeled the response to selection along the area of phenotypic space identified by the genetic correlation. However, the second experiment also revealed that other genetic backgrounds might not be subject to as strong a constraint as the one found in Landsberg. So, Camara and I (in prep.) set out to repeat the original mutation-selection experiment with a larger sample and using a different genetic background, the Kendalville population. We first established that this population does in fact show natural genetic variation for both traits (i.e., it is not an inbred line), and that the characters in question are linked to each other by a measurable, strong genetic correlation. We then applied the mutagen, thereby increasing the available range of phenotypic variation for both traits, and conducted three generations of selection to shift the population mean of both characters in predetermined directions.

As a control, we grew replicates of both mutagenized and base populations simply propagated by selfing, and verified that these did not in fact wander far from the original population centroid in the phenotypic space identified by flowering time and leaf production (Fig. 11.6a). We also applied selection along the observed genetic correlation, that is, along what should be the line of "least resistance" from an evolutionary standpoint. Both mutagenized and non-muta-

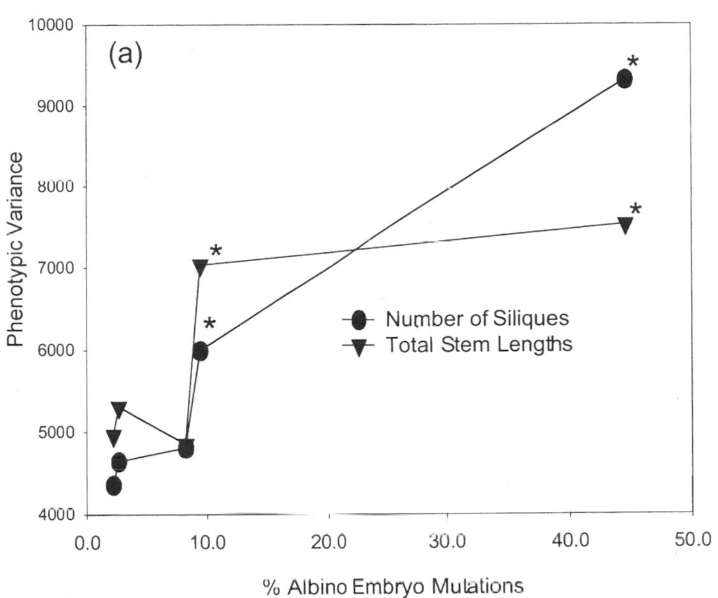


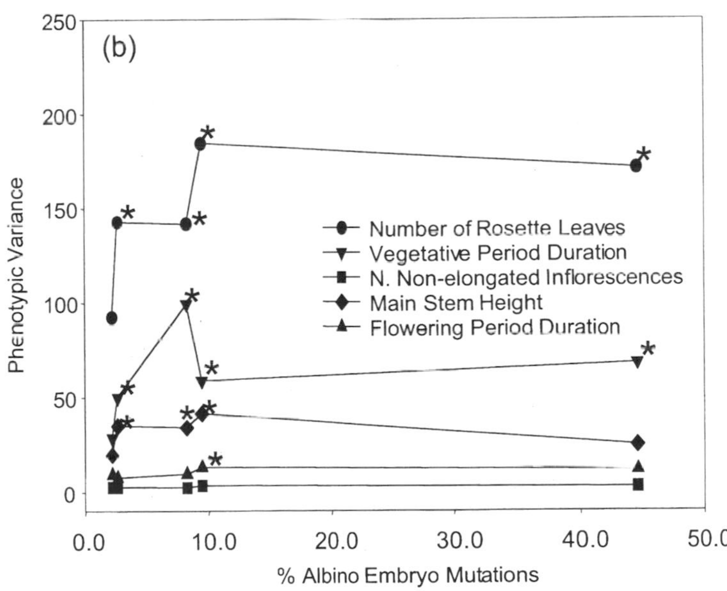


Figure 11.5 The effect of different concentration of mutagen (measured as the increasing number of actual albino mutations) on the variances of quantitative characters in *A. thaliana*. Contrary to the trait means (Fig. 11.4), variances were greatly affected by the number of hits, though not always in a monotonic fashion. Data from Camara et al. (2000).

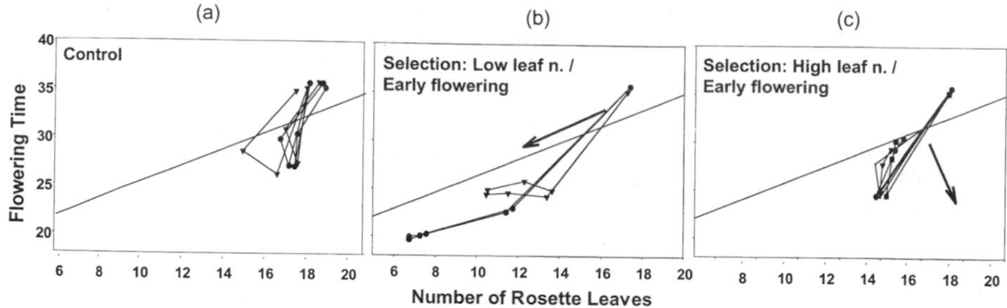


Figure 11.6 Results of a mutation-selection experiment to break a genetic constraint between leaf number and flowering time in *A. thaliana*. Control populations (a) did not move appreciably from their initial position in phenotypic space. Populations selected to move along the diagonal identified by the constraint (b) responded quickly to selection (arrows indicate direction of selection), the mutagenized ones (circles) more so. However, when selection was attempted in a direction orthogonal to the constraint (c) the results were no different from the unselected control. Data from Camara and Pigliucci (in prep.).

genized populations responded quickly to selection (Fig. 11.6b), though the former showed a much prompter response and extended the original phenotypic space significantly more. Indeed, in only three generations we had essentially produced an *A. thaliana* that looked like the Landsberg "lab rat," with only a few leaves and a very early flowering schedule. However, when we attempted to select against the diagonal representing the genetic correlation, we hit a wall with both the mutagenized and nonmutagenized populations (Fig. 11.6c): the selected populations simply did not move from the original region of phenotypic space, as if they had not been selected at all.

It seems clear, therefore, that certain constraints in *A. thaliana* are fairly common, and furthermore, that they are not just a reflection of the observable **G**-matrix (i.e., a matter of variation), but extend to the underlying **M**-matrix (i.e., they are a problem of limited variability). This, of course, cannot be considered a general conclusion (the situation may be different not only for other species, but even for other constraints in *A. thaliana* itself), but it shows the way to an empirical investigation of the **M**-matrix by mutagenesis analyses, thereby significantly extending our experimental grasp of the question of genetic variability.

## It May Be Called M, But Recombination and the Environment Play a Role, Too

Following Wagner's introduction of the idea, the matrix of genetic variability (as opposed to **G**, the matrix of variation) has been referred to popularly as **M** for mutation, although it is obvious that plenty of other evolutionary phenomena actually influence the structure of **M**, for example recombination. It is therefore important to explore the role of these additional factors in shaping and limiting the amount and type of genetic variability that can be available over the short and medium term for a given population. One attempt in this direction is exemplified by my collaboration with Karen Hayden (Pigliucci and Hayden 2001) exploring variability in phenotypic integration in *Arabidopsis thaliana* linked to both genetic recombination and environmental variation.

We used five genotypes of *A. thaliana*, one of which was the standard laboratory line Landsberg, while the remaining four were samples from natural populations from Germany (two accessions), Norway, and Russia. All of these were so-called "early flowering" populations, which probably means that under natural conditions they behave as spring annuals (i.e., they germinate in the spring and go directly to flower, without vernalization). We began by crossing eight individuals of each of the natural accessions with Landsberg, using the latter always as the maternal parent. We planted the seeds obtained from the crosses and grew three generations of progeny, each time propagating the lines by single-plant descent (via selfing). We then used the $F_4$ generation as the material for the actual experiment, having allowed sufficient time for genetic reshuffling to generate new genetic variation (i.e., to translate underlying variability into variation). Given the crossing design, each $F_4$ population was genetically heterogeneous both as a result of recombination and because it originated from eight distinct founder lines used in each of the original crosses. The $F_4$ was grown with the respective parental genotypes under two environmental conditions: 2 ml of standard Hoagland solution administered twice weekly, or the same treatment but with only 10% of the nitrogen concentration.

The results were surprising in that environmentally induced changes in the observed patterns of phenotypic integration were much more marked than differences due to the degree of genetic differentiation among the recombinant lines. For example, population Eil-0 from Germany was characterized by a different arrangement of its first two principal components when the measurements were conducted under low and high nutrients. In particular, the two crucial life-history characters of bolting (flowering) time and length of the reproductive period were independent of each other under high nutrients (the corresponding vectors in PC-space were orthogonal), but negatively correlated to each other (diametrically opposite vectors) under low nutrients. This suggests that the more stressful environment (low nutrients) generated a tradeoff between vegetative and reproductive phases in this population of *A. thaliana*. Even more interestingly, this tradeoff was present *regardless* of the environmental conditions in all three remaining recombinant lines.

When we compared the patterns of phenotypic integration among all line/environment combinations, the emerging picture was remarkably clear (Fig. 11.7): both when measured by overall degree of matrix similarity and by the formal tests provided by common principal components analysis, matrices split first along environmental lines (with most of the low nutrients on one branch and all the high nutrients in the other), and only later by genetic dissimilarity. Therefore, even though the reshuffling of genetic material had contributed to distinct realized patterns of integration, the plasticity of character correlations (Schlichting 1989) was the overwhelming determinant of the observed differences.

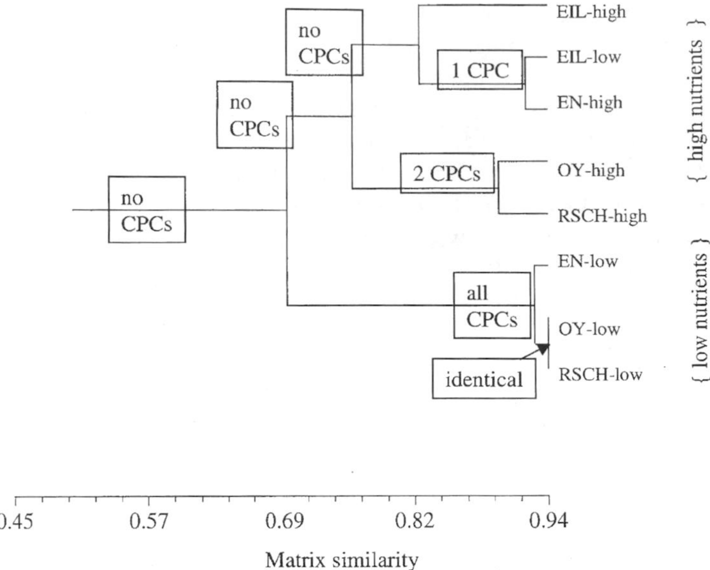


Figure 11.7 Dendrogram depicting the similarity of covariance matrices among four recombinant lines of *A. thaliana* exposed to two different nutrient environments. The boxes at each node indicate the results of formal statistical tests using common principal components (CPCs) analyses to compare the degree of similarity of the matrices on the two branches of a given node. Notice how the two major branches do not share any CPC, and that the major differences in the pattern of phenotypic integration are induced by environmental, not genetic, changes. Data from Pigliucci and Hayden (2001).

## Are Constraints Positive "Forces" in Evolution?

The experiments briefly discussed above begin to address the fundamental question of how it is possible to study the relationship between variation and variability from an experimental point of view. As I have tried to emphasize, they also show the many perils and pitfalls of any empirical research on these matters, which will offer a set of challenges to researchers interested in this topic for years to come. As we have seen, these include the logistics of large experiments, the difficulty of estimating components of **M** given the often overwhelming effect of the environment, as well as the problem of the repeatability of mutagenic and mutation-selection experiments.

All of the above notwithstanding, I should like to conclude this chapter by going back to the big picture of the relationship between (genetic) constraints and natural selection, and in particular to a debate that has been articulated now for some time: should we think of constraints as limitations on natural selection, or as "opportunities" channeling selection in particular directions? I shall argue that recent work in philosophy of science may offer a third alternative: neither selection nor constraints should be thought of as forces at all, contrary to the popular imagery as evoked by Fig. 11.1.

Traditionally, and perhaps most widely, constraints are conceptualized by evolutionary biologists pretty much in line with what the word implies: as *limitations* on the possible evolutionary outcomes for populations subjected to natural selection (Maynard-Smith et al. 1985). Gould, however, has repeatedly argued (most recently in Gould 2002) that constraints should be thought of as *opportunities* for evolutionary change, albeit opportunities that tend to be channeled along preferential routes. The literature on phenotypic integration engages this debate with work showing that evolution does (Schluter 1996) or does not (Merilä and Björklund 1999) in fact occur mostly along lines of "least resistance" as defined by trait correlations.

To some extent this may seem a matter of "semantics," an appellative to which biologists tend to be allergic and that philosophers of science relish. However, at its core the debate is about limits and opportunities for natural selection to shape evolution and to yield adaptation, hardly a secondary matter for evolutionary biologists. It seems to me that the experiments briefly summarized above, as well as anything else we can learn from the ample published literature on the study of genetic correlations, clearly point to the conclusion that constraints (thought of in this case as intercharacter correlations) do both *limit* and *channel* evolution along certain pathways. Just consider the genetic correlation between leaf production and flowering time in *A. thaliana*: on the one hand, it precluded short-term evolution of the population in directions of phenotypic space away from the diagonal represented by the correlation itself. On the other hand, it accelerated the response to selection along the diagonal of those populations characterized by a higher degree of expressed genetic variation for either trait.

This said, it is still not clear at all to me how one can consider, as Gould does, such preferential directions of evolutionary change as "creative" in any sense comparable to the way we understand "unconstrained" natural selection to be. It seems that the problematic idea of constraint as creative force may have arisen from confusion between the *number of directions* available to evolution in a given phenotypic space and the *speed* with which evolutionary change occurs along any one of these directions. If there is no genetic correlation between two traits, then all directions of phenotypic space are available for the population to move toward, with a speed proportional only to the available genetic variation for both traits. If, on the other hand, a genetic correlation exists, the number of directions toward which selection can push the population is greatly diminished (proportionally to the strength of the correlation itself), but the speed of change along the remaining directions may in fact increase because such change is now fueled also by the covariation between the two traits. Another way to think of the latter case is this: the genetic covariation between characters, which is what acts as a brake against evolution away from the "line of least resistance," also acts as an accelerator *along* the same line. But the rate and direction(s) of evolution are distinct concepts, and only confusion between the two may lead somebody to claim that an increased speed in one direction is a "creative" element in the evolutionary process: no matter how one slices it, a genetic correlation is still a (more or less strong) *limit* to what selection can do with the phenotype of a population.

Perhaps a more fundamental challenge to the whole standard way of thinking about selection and constraints has recently come from an external source:

philosophy of science. Throughout the above discussion and presentation of empirical results I have tacitly assumed the validity of the metaphor that casts selection and constraints as "forces" that act independently on populations, sometimes in concert, other times in opposition. This metaphor is in fact very common in technical papers written on selection and constraints, and has been formalized most completely by Sober (1984). For Sober, evolutionary theory can be cast as a theory of forces analogous to Newtonian mechanics, complete with zero-force states, which can be studied through the Hardy-Weinberg equilibrium (the biological equivalent of Newton's statement about bodies on which no forces are acting).

More recently, however, the conception of natural selection as a force has gotten into some trouble, for example when the time comes to distinguish it from the action of other alleged "forces" such as genetic drift (see, for example, a discussion of this in Millstein 2002). The problem has been directly tackled by Matthen and Ariew (2002), who have made the radical proposal that evolutionary "forces," including selection, are not forces at all. Indeed, they claim that the picture of interacting forces is misleading and encourages wrongheaded thinking in both theoretical and empirical evolutionary research. A complete discussion of Matthen and Ariew's argument would lead us too far from the topic of this chapter, but it is instructive to consider their main points briefly and to begin at least to think about the sort of impact they may have on our understanding of the evolution of phenotypic integration.

The first wedge into the "force" conception of evolution is that drift in particular certainly cannot be regarded as a force. For one thing, it does not have predictable and constant direction, as a force is supposed to. Second, and more subtly, since drift is a causative agent that is probabilistic in nature, evolutionary change may have the same cause as no change at all, a notion that would be nonsensical in a framework analogous to Newtonian mechanics.

A second problem for the idea of forces acting during evolution is that forces are, by definition, expressed in the same units of measurement (think of the various laws of Newtonian mechanics). That is obviously not the case for, say, natural selection and genetic constraints, which are literally incommensurable causative agents. They can, of course, both affect gene frequencies over time, but such change is exactly that, an *effect*, not a measure of the cause itself.

Perhaps the most convincing objection to conceiving natural selection and genetic constraints in particular as forces, however, comes from Matthen and Ariew's discussion of what they call the "substrates" of selection. Darwin's original formulation of the theory of evolution by natural selection famously ignored, for lack of knowledge, the genetic bases of the inheritance necessary for selection to produce cumulative change over generations. We now know not only what such substrate is (Mendelian genetics); we also know that evolution *would not work* if different substrates were in place. For example, Fisher (1930) showed that the "blending inheritance" imagined by Darwin would in fact lead to a quick exhaustion of genetic variation, which means that selection could not operate in the long run. As Matthen and Ariew put it, then, it is not natural selection that is responsible for adaptive evolution, but selection within a Mendelian substrate, where the latter is *integral* to the process of evolution, not an obstacle to it. This brings us full circle to the question of the relationship between selection and constraints, which is central to much of this book. To quote Matthen and Ariew (2002, p. 66): "What would happen if selection were to act *by itself* [i.e., without 'constraints']? . . . It is clear that natural selection acts in certain underlying causal media, and the so-called constraints are features of these media. Since natural selection cannot act without such a medium, it is not at all clear what sense can be made of the idea of natural selection acting 'without the intrusion of constraints.' "

If taken seriously, this analysis implies much more than even Gould dared to put forth: not only are constraints not necessarily an obstacle to natural selection, but in general selection cannot work at all *unless* there is a particular genetic substrate that also provides the "constraint." This idea may take a while to sink in, but once one has processed it, one quickly sees all sorts of implications for the way we think about phenotypic evolution. If we abandon the "force" metaphor, then the genetic architecture underlying phenotypic traits, which we can quantify as **G** or **M**, is an integral causal agent of evolutionary change, one that interacts with natural selection in an intimate fashion to yield whatever evolutionary trajectory a particular population takes.

This new paradigm for thinking about selection and constraints also casts the breeders' equation and its multivariate extension with which I started this chapter in a completely different light. While it may be mathematically convenient to think of selection as the vector **S** and of the genetic architecture as the matrix **G**, in reality the two are not that easy to separate. Perhaps a more familiar way to put this is in terms of the uneasiness that at least some biologists (this author, for example) feel when thinking alternatively of **G** as either the result of past selection or as the constraining force limiting future response to selection, depending on the objective of a given study. In reality, of course, **G** and its changes measure the continuous dialectic between past and future evolutionary events, without any clear breaking point, despite the fact that the researcher happens to observe **G** at a particular point in time. Just as there is no clear-cut partitioning of phenotypic variance into genetic and environmental (Pigliucci 2001), so there is no meaningful separation between selection and constraints in explaining changes in phenotypic means.

If abandoning our thinking of selection and constraints (and of other evolutionary causative agents) in terms of forces has the potential to reshape our thinking about how we model long-term phenotypic evolution, then it surely has implications for the way we conceive, carry out, and interpret our experiments too. In particular, it seems to me that this means that we should no longer see **G** as a static attribute of populations or species, but rather we should focus on its *dynamic* as an indispensable partner to natural selection. The coupling of mutational studies of **M** and comparative phylogenetic analyses of **G**, then, becomes a promising venue for empirical research into what we can begin to think of as the continuous coevolution of selective pressures and genetic architectures.

Of course, all of the above will become relevant only if evolutionary biologists are able to work these conceptual suggestions into their empirical research and theoretical models. The proof, so to speak, is in the pudding. However, my goal here has been simply to bring the philosophical discussion to the attention of the evolutionary community: without awareness of the contribution of others, without food for thought as it were, one cannot even begin to think about how to change the direction of one's research program.

*Acknowledgments* I should like to thank Mark Camara for most of the actual work that went into the mutational studies summarized in this chapter, as well as Jonathan Kaplan for many discussions on the potential impact of philosophical thinking on the practice of science, and Katherine Preston for critical reading of the chapter. This research was partly supported by NSF grants BIR-9627564, DEB 1-957551, and DEB-9220593.

## *Literature Cited*

Alonso-Blanco, C., and M. Koornneef. 2000. Naturally occurring variation in *Arabidopsis*: an underexploited resource for plant genetics. Trends in Plant Science 5:22–29.

Antonovics, J. 1976. The nature of limits to natural selection. Annals of the Missouri Botanical Gardens 63:224–247.

Barton, N., and L. Partridge. 2000. Limits to natural selection. BioEssays 22:1075–1084.

Camara, M., and M. Pigliucci. 1999. Mutational contributions to genetic variance/covariance matrices: an experimental approach using induced mutations in *Arabidopsis thaliana*. Evolution 53:1692–1703.

Camara, M. D., C. A. Ancell, and M. Pigliucci. 2000. Induced mutations: a novel tool to study phenotypic integration and evolutionary constraints in *Arabidopsis thaliana*. Evolutionary Ecology Research 2:1009–1029.

Cheverud, J. M. 1984. Quantitative genetics and developmental constraints on evolution by selection. Journal of Theoretical Biology 110:155–171.

Donohue, K. 2002. Germination timing influences natural selection on life-history characters in *Arabidopsis thaliana*. Ecology 83:1006–1016.

Fernandez, J., and C. Lopez-Fanjul. 1997. Spontaneous mutational genotype-environment interaction for fitness-related traits in *Drosophila melanogaster*. Evolution 51:856–864.

Fisher, R. A. 1930. The Genetical Theory of Natural Selection. Clarendon, Oxford.

Fry, J. D., K. A. de Ronde, and T. F. C. Mackay. 1995. Polygenic mutation in *Drosophila melanogaster*: genetic analysis of selection lines. Genetics 139:1293–1307.

Fry, J. D., S. L. Heinsohn, and T. F. C. Mackay. 1996. The contribution of new mutations to genotype-environment interaction for fitness in *Drosophila melanogaster*. Evolution 50:2316–2327.

Gould, S. J. 2002. The Structure of Evolutionary Theory. Harvard University Press, Cambridge, MA.

Hoffmann, A. A., R. Hallas, C. Sinclair, and L. Patridge. 2001. Rapid loss of stress resistance in *Drosophila melanogaster* under adaptation to laboratory culture. Evolution 55:436–438.

Houle, D., K. A. Hughes, D. K. Hoffmaster, J. Ihara, S. Assimacopoulos, D. Canada, and B. Charlesworth. 1994. The effects of spontaneous mutation on quantitative traits. I. Variances and covariances of life history traits. Genetics 138:773–785.

Kauffman, S. A. 1993. The Origins of Order. Oxford University Press, New York.

Kauffman, S. A., and S. Levin. 1987. Towards a general theory of adaptive walks on rugged landscapes. Journal of Theoretical Biology 128:11–45.

Keightley, P. D., and O. Ohnishi. 1998. EMS-induced polygenic mutation rates for nine quantitative characters in *Drosophila melanogaster*. Genetics 148:753–766.

Kirkpatrick, M., and D. Lofsvold. 1992. Measuring selection and constraint in the evolution of growth. Evolution 46:954–971.

Lande, R., and S. J. Arnold. 1983. The measurement of selection on correlated characters. Evolution 37:1210–1226.

Mackay, T. F. C., J. D. Fry, R. F. Lyman, and S. V. Nuzhdin. 1994. Polygenic mutation in *Drosophila melanogaster*: estimates from response to selection of inbred strains. Genetics 136:937–951.

Matos, M., and T. Avelar. 2001. Adaptation to the laboratory: comments on Sgrò and Partridge. American Naturalist 158:655–656.

Matos, M., M. R. Rose, M. T. R. Pitè, C. Rego, and T. Avelar. 2000. Adaptation to the laboratory environment in *Drosophila subobscura*. Journal of Evolutionary Biology 13:9–19.

Matos, M., T. Avelar, and M. R. Rose. 2002. Variation in the rate of convergent evolution: adaptation to a laboratory environment in *Drosophila subobscura*. Journal of Evolutionary Biology 15:673–682.

Matthen, M., and A. Ariew. 2002. Two ways of thinking about fitness and natural selection. Journal of Philosophy 49:55–83.

Maynard-Smith, J., R. Burian, S. Kauffman, P. Alberch, J. Campbell, B. Goodwin, R. Lande, D. Raup, and L. Wolpert. 1985. Developmental constraints and evolution. Quarterly Review of Biology 60:265–287.

Merilä, J., and M. Björklund. 1999. Population divergence and morphometric integration in the Greenfinch (*Carduelis chloris*)—evolution against the trajectory of least resistance? Journal of Evolutionary Biology 12:103–112.

Millstein, R. L. 2002. Are random drift and natural selection conceptually distinct? Biology and Philosophy 17:33–53.

Mitchell-Olds, T. 1996. Genetic constraints on life-history evolution: quantitative-trait loci influencing growth and flowering in *Arabidopsis thaliana*. Evolution 50:140–145.

Mitchell-Olds, T. 2001. *Arabidopsis thaliana* and its wild relatives: a model system for ecology and evolution. Trends in Ecology and Evolution 16:693–699.

Pigliucci, M. 1998. Ecological and evolutionary genetics of *Arabidopsis*. Trends in Plant Science 3:485–489.

Pigliucci, M. 2001. Phenotypic Plasticity: Beyond Nature and Nurture. Johns Hopkins University Press, Baltimore, MD.

Pigliucci, M., and K. Hayden. 2001. Effects of inter-population crossing on phenotypic integration in *Arabidopsis thaliana*. New Phytologist 152:419–430.

Pigliucci, M., G. A. Tyler-III, and C. D. Schlichting. 1998. Mutational effects on constraints on character evolution and phenotypic plasticity in *Arabidopsis thaliana*. Journal of Genetics 77:95–103.

Rausher, M. D. 1992. The measurement of selection on quantitative traits: biases due to environmental covariances between traits and fitness. Evolution 46:616–626.

Redei, G. P. 1992. A heuristic glance at the past of *Arabidopsis* genetics. Pp. 1–15 in N.-H. C. C. Koncz, and J. Schell (eds.). Methods in *Arabidopsis* Research. World Scientific, Singapore.

Roff, D. A., and T. A. Mousseau. 1999. Does natural selection alter genetic architecture? An evaluation of quantitative genetic variation among populations of *Allenomobius socius* and *A. fasciatus*. Journal of Evolutionary Biology 12:361–369.

Schlichting, C. D. 1989. Phenotypic plasticity in *Phlox*. II. Plasticity of character correlations. Oecologia 78:496–501.

Schlichting, C. D., and M. Pigliucci. 1998. Phenotypic Evolution: A Reaction Norm Perspective. Sinauer, Sunderland, MA.

Schluter, D. 1996. Adaptive radiation along genetic lines of least resistance. Evolution 50:1766–1774.

Sgrò, C. M., and L. Partridge. 2000. Evolutionary responses of the life history of wild-caught *Drosophila melanogaster* to two standard methods of laboratory culture. American Naturalist 156:341–353.

Sgrò, C. M., and L. Partridge. 2001. Laboratory adaptation of life history in *Drosophila*. American Naturalist 158:657–658.

Shaw, F. H., R. G. Shaw, G. S. Wilkinson, and M. Turelli. 1995. Changes in genetic variances and covariances: **G** whiz! Evolution 49:1260–1267.

Simpson, G. G., and C. Dean. 2002. *Arabidopsis*, the Rosetta Stone of flowering time? Science 296:285–289.

Sober, E. 1984. The Nature of Selection: Evolutionary Theory in Philosophical Focus. MIT Press, Cambridge, MA.

Turelli, M. 1988. Phenotypic evolution, constant covariances, and the maintenance of additive variance. Evolution 42:1342–1347.

van Tienderen, P. H., and H. P. Koelewijn. 1994. Selection on reaction norms, genetic correlations and constraints. Genetical Research 64:115–125.

Via, S. 1987. Genetic constraints on the evolution of phenotypic plasticity. Pp. 47–71 in V. Loeschcke (ed.), Genetic Constraints on Adaptive Evolution. Springer-Verlag, Berlin.

Wagner, G. P., and L. Altenberg. 1996. Complex adaptations and the evolution of evolvability. Evolution 50:967–976.

## PART IV

# MACROEVOLUTIONARY PATTERNS IN PHENOTYPIC INTEGRATION

In recent years, the field of microevolutionary studies has seen a dramatic increase in comparative approaches applied to questions that have traditionally been the realm of detailed case studies of single species or communities. The accumulation of data from specific studies, from developmental genetics to ecophysiology and community ecology, has helped to underwrite more general hypotheses about evolutionary transitions, organismal function, or community assembly that can be tested by comparing taxa or communities (e.g., Armbruster 1988; Reich et al. 1997; Cunningham et al. 1999; Cavender-Bares and Holbrook 2001; Webb et al. 2002). As a result, the traditional distinction between evolutionary ecology and macroevolutionary studies is starting to look more like a continuum.

Yet, from the perspective of working paleontologists and macroevolutionary biologists, there remain genuine theoretical and methodological problems emerging only at some scales of analysis. Moreover, some central questions in macroevolution and paleontology simply do not pertain to lower levels. These idiomatic lines of research address such issues as the phylogenetic distribution of long-term stasis or evolutionary lability in certain phenotypic traits, broad-scale trends in complexity, or the factors underlying rapid evolutionary radiation of a clade. The three chapters in Part IV address evolution in complex phenotypes from the point of view of macroevolutionary biologists, and they explore the relationship between micro- and macroevolutionary studies.

At virtually all levels of resolution, a salient feature of organisms is that their traits are interdependent. Still, trait integration has a peculiar place in evolutionary biology. Inasmuch as integration affects organismal function, it is subject to experimental or comparative study just as any other character (e.g., Ackerly and Donoghue 1998; Callahan and Waller 2000; Murren et al. 2002). As with other characters, patterns of trait interdependence may be more similar in closely related taxa and fade with phylogenetic distance (Schluter 1996). But phenotypic integration differs from most other organismal properties in that integration itself directly affects the evolutionary trajectories of the integrated traits, at least in the short run. This dual role underlies the contrast between integration as constraint or adaptation, discussed throughout this volume. It also generates some of the uniquely macroevolutionary questions prominent in the following chapters. For example, many microevolutionary comparative studies are concerned with the effect of trait correlations on the evolution of traits or suites of traits, which are the real subjects of interest. By contrast, macroevolutionary studies are more likely to focus on the evolution of integration itself.

In Chapter 12, Eble takes up the topic of integration as an organismal property, in which specific trait relationships are usually less important than the overall level of phenotypic integration. By mapping degrees of integration onto a phylogenetic tree, one can try to reconstruct its effects on clade dynamics. Under this approach, integration is seen as a higher-order property of lineages, similar to evolutionary innovation and exaptation in its potential to influence evolvability or morphological reorganization. In cases where specific trait combinations are considered at a macroevolutionary level, they can be used to investigate the evolution of developmental trajectories and module identity, or they can demonstrate long-term evolutionary stasis in some characters.

Complex changes in shape are often a prominent part of macroevolutionary analysis, which automatically raises several very difficult conceptual and practical issues. A theme common to all three chapters in this section is the problem of characterizing complex phenotypes and ensuring their commensurability across taxa. In other words, the authors question the similarity of the properties chosen for comparison, and ask whether they are the right properties and how they can be measured in the first place. This theme is explored most fully in the two following chapters.

Zelditch and Moscarella (Chapter 13) develop the idea that phenotypic integration is central to life-history evolution because it generates the tradeoffs and therefore the fitness functions underlying various allocation strategies. Across taxa, life history and morphology may be correlated because of their common relationship to growth and developmental rate. The theory of heterochrony formally relates morphogenesis to the timing and rate of growth and development, and thus allows developmental trajectories to be compared across taxa. Although heterochrony has been a powerful tool, its theoretical framework currently does not accommodate a spatial axis; and Zelditch and Moscarella demonstrate that ignoring changes in the location of growth and development can lead to misinterpretations of the differences between taxa in their developmental trajectories. They argue convincingly that incorporating spatial patterns into a broader conception of heterochrony would more accurately describe the complex nature of morphological evolution and would reveal the links between morphogenesis and life-history evolution.

Ackermann and Cheverud (Chapter 14) are similarly concerned about morphological comparisons between taxa, but in their case they are interested not only in phylogenetic trends in size and shape, but also in the evolution of patterns of trait covariation. Here the authors review their work demonstrating that certain features of the primate cranium show a common pattern of integration across the primates, even though changes in absolute size and shape have been dramatic. They use this finding to tackle a common problem in paleontology, namely the limitations imposed by a fragmentary fossil record and fragmented fossils. Along with colleagues Rasmussen and Rehg, Ackermann and Cheverud used the variance-covariance matrix of skull measurements in an extant species to reconstruct the three-dimensional shape of a species known only from a set of poorly preserved fossils. They go on to suggest a number of intriguing questions that may be pursued by using what we know about trait relationships and the developmental biology of extant taxa to interpret evolutionary patterns indicated by the fossil record.

In all three chapters in this section, the authors call for enriching traditional macroevolutionary approaches and drawing on concepts developed in other areas of biology. It seems likely that such a synthesis would also benefit microevolutionary biologists, especially as comparative studies become more sophisticated and more ambitious.

### *Literature Cited*


Ackerly, D. D. and M. J. Donoghue. 1998. Leaf size, sapling allometry, and Corner's rules: a phylogenetic study of correlated evolution in maples (*Acer*). American Naturalist 152:767–791.

Armbruster, W. S. 1988. Multilevel comparative analysis of morphology, function, and evolution of *Dalechampia* blossoms. Ecology 69:1746–1761.

Callahan, H. S. and D. M. Waller. 2000. Phenotypic integration and the plasticity of integration in an amphicarpic annual. International Journal of Plant Sciences. 161: 89–98.

Cavender-Bares, J. and N. Holbrook. 2001. Hydraulic properties and freezing-induced cavitation in sympatric evergreen and deciduous oaks with contrasting habitats. Plant Cell and Environment 24:1243–1256.

Cunningham, S., B. Summerhayes, and M. Westoby. 1999. Evolutionary divergences in leaf structure and chemistry, comparing rainfall and soil nutrient gradients. Ecological Monographs 69:569–588.

Murren, C. J., N. Pendleton, and M. Pigliucci. 2002. Evolution of phenotypic integration in *Brassica* (Brassicaceae). American Journal of Botany 89:655–663.

Reich, P. R., M. B. Walters, and D. S. Ellsworth. 1997. From tropics to tundra: global convergence in plant functioning. Proceedings of the National Academy of Sciences USA 94:13730–13734.

Schluter, D. 1996. Adaptive radiation along genetic lines of least resistance. Evolution 50:1766–1774.

Webb, C. O., D. D. Ackerly, M. A. McPeek, and M. J. Donoghue. 2002. Phylogenies and community ecology. Annual Review of Ecology and Systematics. 33:475–505.

# 12

# The Macroevolution of Phenotypic Integration

GUNTHER J. EBLE

Phenotypic integration is a central aspect of macroevolution. It is also an important concept in macroevolutionary theory. A number of empirical research questions and theoretical debates concerning evolution at and above the species level revolve around the issue of differential phenotypic integration through time. However, macroevolutionary studies in general and evolutionary paleobiology in particular have often conceptualized and documented phenotypic integration in ways that are not always comparable with standard, quantitative-genetic microevolutionary accounts. Thus, subjects such as evolutionary radiations, constraints, morphospace structure and occupation, disparity, allometry, heterochrony and heterotopy, the origin and proliferation of novelties, and clade dynamics, to name but a few, have been interpreted in terms of phenotypic integration (e.g., Valentine and Campbell 1975; Alberch et al. 1979; Maynard Smith et al. 1985; Gould 1989a, 1989b; Jablonski and Bottjer 1990; Erwin 1993; Zelditch and Fink 1996; Foote 1997; McGhee 1999), but with methodological protocols and theoretical motivations often distinct from those of microevolutionary research.

By extension, appreciating the status of phenotypic integration in macroevolution and its relationship with various genetic, developmental, and ecological approaches may help dissolve residual tensions between macroevolutionists and microevolutionists (notwithstanding the fact that some scholars may belong to both communities without apparent loss of sanity!). The essential tension has concerned the issue of reducibility. It is bound to disappear, though, upon recognition that macroevolution is not completely reducible to microevolution because of differences in scale, in the hierarchical manifestation of evolutionary processes, in the extent and sources of contingency and constraint, and in the nature of the focal biological entities (Gould 1982, 2002; Eldredge 1989; Williams 1992; Erwin 2000; Jablonski 2000). None of these justifications for a relative autonomy of macroevolution is necessarily contentious nowadays, as macroevolutionary research has been rendered more rigorous conceptually, empirically, and theoretically, with operational definitions and methodologically sophisticated data collection, analysis, and modeling routinely guiding statistically testable inferences about large-scale evolutionary patterns and processes (Schopf 1972; Raup and Jablonski 1986; Gilinsky and Signor 1991; Erwin and Anstey 1995; Jablonski et al. 1996; Foote 1997; McKinney and Drake 1998; Eble 2000a, 2003; Jablonski 2000; Jackson et al. 2001; Gould 2002). Macroevolution and microevolution are at times factually and theoretically distinct, but material irreducibility is only partial, and complete decouplement is logically impossible, because actual organisms must figure in both domains. There is no fundamental incompatibility between macroevolution and microevolution in a general evolutionary theory, if such theory incorporates biological hierarchies as matters of fact or at least as heuristic representations. Cross-hierarchical organization and dynamics are complex, but not intractable (Simon 1962; Lewontin 1970; Weiss 1971; Eldredge 1985; Vrba and Gould 1986; Vrba 1989; Eble 1999a; McShea 2001; Gould 2002), and the existence of biological hierarchies per se does not necessarily demand level-specific theories of process (e.g., Williams 1992; Maynard Smith and Szathmáry 1995).

At issue, then, is precisely which phenomena, explanations, and theoretical constructs associated with macroevolution are equivalent to those in the microevolutionary realm, which are only partially equivalent, and which—if any—are incommensurable. The study of phenotypic integration is of special relevance in this regard, since the phenotype is important in empirical and theoretical research in population biology, in systematics, in evolutionary developmental biology, and in paleobiology. Clarifying how phenotypic integration is manifested and studied in macroevolution, and understanding the nature of macroevolutionary phenotypic integration, may help in moving beyond the perhaps overly modular multidisciplinarity of current integration studies and toward a more unifying interdisciplinarity (Zelditch 1996; Schlichting and Pigliucci 1998). This in turn may open the way for a more synthetic appreciation of the role of the phenotype in evolutionary theory at large.

In this chapter, I shall examine this state of affairs and discuss macroevolutionary issues that have been or could be analyzed in the context of phenotypic integration, either explicitly or implicitly. In doing so, I shall suggest that research protocols for studying the macroevolution of phenotypic integration are appropriate for certain questions but must be further refined to tackle a wider array of macroevolutionary problems and to allow new models to be advanced, and that certain conceptual and analytical approaches in macroevolution may prove useful, or at least illuminating, in microevolution as well. The intent, as is unavoidable in a subject still in need of a synthesis, is not to be exhaustive or prescriptive, but simply to highlight the unique contexts of macroevolutionary phenotypic integration, and the phenomenology of causes it implies. This can be viewed as a step toward developing a more solid account of the meaning of pattern and process in macroevolution, in a manner that is at the same time compatible with and enriching of microevolutionary theory.

## Phenotypic Integration in Macroevolutionary Studies

### From Comparative Anatomy to Morphospaces

A concern for phenotypic integration and its relationship with interspecific patterns of variation accompanied the history of comparative anatomy, and was particularly evident in the nineteenth century, with the establishment of such notions as the unity of type, functional correlation, structural correlation, the principle of connections, homology, and developmental laws (Russell 1916). Darwin and morphologists after him recast all such notions in evolutionary terms while keeping their heuristic aspects intact. Yet with the new emphasis on phyletic transformation, the growing mechanicism of embryology, and the atomism of Mendelian genetics, organismal integration as a research theme soon faded into the background. In the search for the causes and consequences of variation, a genuine "disintegrative biology" rose to prominence, producing important knowledge but at the same time biasing knowledge production.

Integration thus stood as a secondary problem in evolutionary biology for most of the last century (Schlichting and Pigliucci 1998). In macroevolutionary biology, however, and in particular in paleobiology, the centrality of morphology as data and of taxa as the entities of theoretical discourse ensured that organismal integration was not lost from sight. Simpson (1953), for example, while often emphasizing individual traits in his attempt to render paleontology fully compatible with and explainable within the population-genetic framework of the Modern Synthesis, also asserted the importance of correlated phenotypic evolution and of character complexes. As Olson and Miller (1958) pointed out, Simpson's use of genera as the basic taxonomic units in his studies of evolutionary rate reflected his assessment that differences among genera were proportional to total morphological difference, and that taxa in general referred to whole organisms, even if by proxy.

Olson and Miller's 1958 landmark book attempted to establish the importance of morphological integration in evolutionary biology, and to devise rigorous methods for its study. Their work echoes that of Clausen and colleagues (Clausen et al. 1940; Clausen and Hiesey 1960), but while the latter directly addressed the genetic basis of phenotypic integration with crossing experiments in plants, Olson and Miller's focus was mostly paleobiological. They recognized, with genuine appreciation, the importance of genetic integration, but addressed causality in more exploratory fashion, in terms of trait associations reflecting known function or development, and also potentially referring to size, spatial location within the organism, or "some unifying biological factor." More generally, part of the theoretical motivation was to show that many important questions in paleontology (trends, rates, speciation, hybridization, ontogenetic biases in evolution, and convergence) could and should be fruitfully addressed in terms of morphological integration.

Olson and Miller's work is symptomatic of a time of increasing quantitative sophistication in paleontology, and of more explicit interest in character correlation, allometry, and multivariate evolution (Kermack and Haldane 1950; Kermack 1954; Imbrie 1956; Raup 1956). The trend continued into the 1960s, when quantitative paleontology was relatively well established as an approach, and multivariate morphometrics routine in many studies of invertebrate groups. Morphological integration was often inferred statistically and interpreted in post hoc fashion, rather than tested against a priori predictions. But the premise that character correlation and organismal integration were important and needed to be causally understood was nearly universally agreed upon among students of organismal form.

In this context, the field of theoretical morphology (Raup and Michelson 1965; Raup 1966; McGhee 1999; Eble 2000b) arose in part as an attempt to distill a small number of constructional parameters in terms of which whole organisms could be graphically simulated with mathematical models of form generation. In theoretical morphology, morphospaces define the universe of possibilities, and clusters of *taxa* stand as evidence of phenotypic integration. Emphasis is placed on inter-object relationships within a clade treated as a statistical population, and integration is expressed as co-occurrence of taxa in certain regions of morphospace, regardless of whether the axes of the coordinate system are correlated or not. Theoretical morphology does not prescribe the nature of the biological entities (which can also be individuals within populations, rather than taxa within clades), but it seems clear that, with the explicit construction of morphospaces, a conceptual and analytical framework becomes available to view integration as a property of *clades*, to represent it in relation to the geometry of evolutionary realizations, and to quantify it in terms of inter-object relationships. The framework applies to empirical morphospace studies as well, although here a large number of variables is usually considered, inviting the use of multivariate ordination techniques (such as principal components analysis) to assess both inter-object and inter-variable relationships. The role of morphospaces in inferences about phenotypic evolution needs to be further refined (Foote 1997; Schlichting and Pigliucci 1998; Eble 1998, 2000b, 2003), but morphospace thinking does constitute a distinct contribution of macroevolutionary studies to the topic of integration, by providing a cross-hierarchical platform for understanding constraints on and opportunities for the origin and maintenance of phenotypic variation and covariation.

## Phenotypic Integration as Cohesion and as a Property

However inferred and conceptualized, phenotypic integration in macroevolution relates to organismal cohesion, and as such refers both to constrained variational patterns, the foci of morphometrics and quantitative genetics, as to organizational patterns, where structural, functional, and developmental stability or even invariance must be taken into account. Ultimately, the ontology of macroevolution presupposes phenotypic integration, even if character complexes and the strength of their integration may change in ontogeny and phylogeny (Arnold et al. 1989; Wagner and Misof 1993; Fink and Zelditch 1996;

Shubin and Wake 1996; Chaline et al. 1998; Eble 2000a, 2003). Species and clades exist as units of evolution if at least some of their diagnostic traits are consistently integrated during their evolutionary history, thus supporting statements of homology, relationship, identity, and stasis (Rieppel 1986, 1988; Nelson 1989; Jablonski and Bottjer 1990; Valentine 1990; Gould 2002). Taxa, whether conceived of as classes or individuals, are subject to a dynamics of differential representation, which is conditioned on sustained organismal cohesiveness and stability across generations. Different groups may exhibit different degrees of cohesiveness and stability across characters, different durations of stasis in individual traits or character complexes, and different rates of mosaic evolution, but strong integration of taxon-specific characters is an existence requirement until extinction or pseudoextinction marks the end of a lineage. Homology, in particular, is a key aspect of phenotypic integration in macroevolution, but it is hardly acknowledged in the formal structure of quantitative-genetic models of integration. These models as a rule focus on integration as manifested in correlated patterns of readily changeable quantitative traits, with lack of change and discrete traits that vary only on phylogenetic time scales standing as a nonissue.

In microevolutionary research the default loci of explanation are traits or groups of traits, and their quasi-independence is as much a target of study as their potential integration. Quasi-independence facilitates adaptation and response to selection. As the latter are assumed to have a preeminent role in population differentiation (notwithstanding the fact that established functional interdependences may also promote evolution), one important goal of microevolutionary studies of phenotypic integration is to distill modular units of selection from the complex patterns of trait interactions constituting the organism—in other words, to identify *local* phenotypic integration, and correlated responses to selection (e.g., Lande and Arnold 1983). In macroevolution, such question-oriented disintegration of the organism is invoked routinely in the study of individual lineages through time. Phenotypic disintegration is either accepted *faute de mieux*, when taphonomic bias forces the investigator to focus only on those characters with high preservation potential (e.g., molars in mammals), or because one may be concerned with characters of taxonomic importance or else presumed evolutionary importance (e.g., key innovations).

Yet for most large-scale studies of phenotypic integration, the focus is on integration as a whole-organism property of individuals and the clades they represent. Traits in isolation and trait interactions may be of less immediate interest because overall phenotypic integration is seen as a meaningful macroevolutionary phenomenon, expressed in summary statistics of evolutionary flexibility (e.g., various metrics of disparity) that are comparable across groups and time, regardless of trait identity and the details of trait interaction (Olson and Miller 1958; Hughes 1990; Wagner 1995; Eble 1998, 2000a). Such statistics codify, if not an emergent trait, at least the potential for emergent fitnesses at the species or clade level (much as with variability: Lloyd and Gould 1993). Far from assuming tight organismal integration by default and disregarding dissociability (for a critique of this position, see Gould 1977; Fink and Zelditch 1996), emphasis is placed on an operational "quasi-dependence" of traits in the context of clade definition (homology) and clade persistence (stasis) in geological time. More than adaptation, the interest is on adaptability and, more generally, evolvability (Kauffman 1993; Wagner and Altenberg 1996; Depew 1998). Recent studies of morphological disparity through time (e.g., Foote 1997) reflect this perspective; although, as will be seen below, disparity relates to integration in some respects but not in others (see Cheverud et al. 1989; Eble 2004). And, in macroevolutionary debates, phenotypic integration as a global trait or state is often invoked, and granted empirical reality by the very logic of tests of hypotheses usually proposed. This is the case for the dichotomy between developmental and ecological explanations of evolutionary radiations, both across body plans (Gould 1989a, 1993; Erwin 1993, 1999; McShea 1993; Ridley 1993; Valentine 1995; Eble 1998; Conway Morris 1998; Jablonski 2000) and within body plans (e.g., Wagner 1995; Foote 1999; Hughes et al. 1999; Eble 2000a). There is of course much room for character-by-character analyses of correlations in the context of integration, despite limitations of sampling and temporal resolution. But in the practice of macroevolution, the dynamics of clades is the reference phenomenological domain and the main source of patterns to be explained. While mosaic evolution and dissociability are acknowledged and documented, phenotypic integration in macroevolution is more than a trait-by-trait business. It is also a macroevolutionary trait in its own right.

## Proxies for Phenotypic Integration in Macroevolution

Even though phenotypic integration has been quantified with morphometric data in a number of paleontological studies (e.g., Foote 1991, 1999; Wagner 1995; Hughes et al. 1999; Eble 2000a), changing patterns of taxonomic diversity, origination, and extinction are often used as proxy evidence for differential clade-level phenotypic integration. The morphological disparity among phyla, for example, is still difficult to quantify because of extensive geometric incommensurability across body plans. Higher taxonomic status (phyla, classes, and orders) has thus been used as a proxy for morphological distinctness, and asymmetries (e.g., decline) in the frequency of higher taxa through time have been hypothesized to reflect long-term changes in overall phenotypic integration (Valentine 1986, 1995; Erwin et al. 1987; Kauffman 1989, 1993; Jablonski and Bottjer 1990, 1991; Jacobs 1990; Valentine et al. 1991b; Eble 1998, 1999b).

Further, at lower taxonomic levels, models of taxonomic diversification parameterized with (stochastically) constant per-taxon origination and extinction rates have been reasonably successful at describing macroevolutionary dynamics in a number of groups (e.g., Sepkoski 1978, 1979, 1984; Stanley 1979; Benton 1997). Constant rates of diversification, origination, and extinction in such models have been treated as intrinsic clade properties related to organismal construction and its correlates (e.g., Schopf et al. 1975; Vrba 1983; Eldredge 1989; Stanley 1990; Valentine et al. 1991a). High origination and extinction rates (e.g., trilobites) would suggest more labile, less integrated body plans, while low origination and extinction rates (e.g., bivalves) suggest resistance to change and more integrated phenotypes. In this context, differential per-taxon rates are hypotheses of differential phenotypic integration.

## Approaches to Interspecific Data

Most neontological studies of phenotypic integration center on the quantification of intraspecific variation, which may then be compared among species (e.g., Olson and Miller 1958; Cheverud 1982, 1990; Bookstein et al. 1985; Cheverud et al. 1989). In a phylogenetic context, interesting questions arise, such as whether cladogenesis tends to be associated with more substantial change in correlation structure than is population differentiation (Olson and Miller 1958), or whether niche breadth correlates with degree of integration. Issues of phenotypic integration and dissociability based on species-by-species analyses have recently also been pursued in comparative neontological studies of ontogenetic variation (e.g., Alberch 1983; Shubin and Wake 1996; Smith 1996; Fink and Zelditch 1996), and in applications of phylogeny-based comparative methods (e.g., Harvey and Pagel 1991). The attempt to understand the macroevolution of phenotypic integration by comparing intraspecific patterns of variation across species may be most useful for closely related species, which would be expected to display smoothly gradational patterns of variation and covariation. However, if taxon sampling is sparse, species status or phylogenetic relationships poorly known, or if taxa are very distantly related, the assumption that intra- and interspecific patterns intergrade and are responsive to similar evolutionary processes becomes more difficult to assess, and biological interpretations accordingly become potentially less robust. Further, analysis of large numbers of species over long periods of time on the basis of intraspecific patterns of continuous variation and covariation is complicated by the need to take into account typical macroevolutionary phenomena such as innovations, character gains and losses, and character invariance. This complication is not merely analytical and representational. Aspects of phenotypic variation and integration in macroevolution may simply not be addressable (or modeled in canonical form) with sequential species-by-species comparisons of intraspecific variation, because of potential long-term changes in organismal dimensionality during the history of clades and of potential inhomogeneity of variational properties among species, especially in large clades broadly distributed in time and space. Whether and how temporal extension and geographic distribution may bias inferences about integration in particular groups is an open issue. At any rate, small, young clades are less likely to present such problems, and indeed most comparative analyses of intraspecific variation have appropriately focused on such "intermediate" scales of macroevolution.

Macroevolutionists are often more interested in patterns of integration at the level of more inclusive, species-rich clades, whose usually long histories allow for more robust quantification of patterns of disparity and morphospace occupation. Accordingly, phenotypic integration has been studied by directly considering species as individual data points (much as in evolutionary allometry studies). Thus, variance-covariance, correlation, distance, and other similarity matrices may capture how traits coevolve or co-occur across species in a clade. Species are inferred or assumed to be relatively stable entities on the scale of the whole clade, and randomly selected specimens are treated as representative of each species' morphology. The presence of stasis, or alternatively of some consistent basis for species identity through time, justifies this approach. At least for morphospace studies, the interest is mostly in the large-scale partitioning of morphological variation and in the evolutionary origins and consequences of phenotypic integration of a clade. Specimens are thus seen more as statistical samplers of morphospace regions than as typical members of evolutionary units (although they may figure as such in evolutionary interpretations of pattern).

An additional assumption, often defensible on geological time scales but not always tested, is that intraspecific variation rarely confounds interspecific patterns of variation, such that species (genera, families, etc.) would on average stand as relatively discrete entities in (or samplers of) morphospace (Raup and Boyajian 1988; Sepkoski and Kendrick 1993). Operational issues aside, explicit quantification of interspecific variation and covariation patterns in whole clades complements species-by-species comparisons of intraspecific integration. By identifying other scales and genealogical levels of phenotypic integration, the clade-wide, large scale interspecific approach indicates that an empirical domain for a genuinely macroevolutionary notion of phenotypic integration exists. This naturally allows for hierarchical sorting and selection processes, homology, innovation, and changes in dimensionality to be considered rigorously as part of a broader explanatory framework for long-term patterns of phenotypic integration.

## On the Nature, Representation, and Meaning of Phenotypic Integration in Macroevolution: Beyond the Matrix?

Movie plots are sometimes hard to understand or even meaningless, especially when multiple scales of time and space are considered simultaneously. In the movie *The Matrix*, our anti-hero "Neo" attempts to understand and eventually undermine the inner workings of the virtual reality matrix in which he is enslaved. By analogy with evolution, how much of our current understanding of phenotypic integration and of our confidence in the means to study it are not invariably tied to the formalism of the variance-covariance (or correlation) matrix? What else can be learned, or is there anything else to learn, if one opts to study phenotypic integration with less conventional distillations of the data matrix, or even with other data representations?

The issue is important in light of the multiple kinds, sources, and contexts of integration (Olson and Miller 1958; Cheverud 1996; Roth 1996; Zelditch 1996; Schlichting and Pigliucci 1998; McShea 2001), and of potentially different manifestations of integration in microevolution and macroevolution. Should models for all these variations on the theme of phenotypic integration always refer to pairwise correlations or covariances, or can they also be framed *ab initio* in terms of global measures such as the total variance or total correlation (Van Valen 1974, 1978; Cheverud et al. 1989), which allow but do not demand an explicit matricial representation?

Assessments of phenotypic integration in macroevolution are distinct not only for their emphasis on taxa over variables and by a preference for global metrics, but also because data may be biased by smaller sample sizes and incomplete preservation. As a result, often the reference data matrix is one of nonparametric presence-absence data, inter-object distances, or coefficients of association. Modeling such data could benefit from generalizations of existing models of correlation and variance-covariance structure, but little work has been done in this area. In the absence of models, stochastic simulation (Foote 1996) and resampling methods (Cheverud et al. 1989) may help provide a set of null expectations.

Biological objects, and in particular parts or modular units (McShea and Venit 2001), can vary, covary, or not vary in a panoply of ways. Variation and covariation of discrete characters on macroevolutionary time scales may mean no more than the discrete coming into and going out of existence and coexistence in the phylogenetic tree (Darden 1992; Eble 2004). Organizational and variational properties of the phenotype may be expected to change in concert (Cheverud 1996; Wagner and Altenberg 1996), but patterns of historical constraint, stasis, and homology in macroevolution are often reflected in highly entrenched characters for which only discrete shifts in organization are possible.

The solution to the problem of representation and analysis of integration is not pluralism per se and an "everything goes" methodological attitude, but context-specific operationalism and heuristic methodological diversification. Granted, evolutionary biology is far from exhausting the analytical power of linear algebra, and granted that placing life in "the matrix" has only advanced our understanding of integration, one may look beyond the brackets and consider the possibility that certain kinds of integration, particularly in macroevolution, may invite different formalisms for its analysis and representation. For example, homology and homoplasy have at times been framed in terms of variation and covariation (e.g., Harvey and Pagel 1991; Grafen and Ridley 1996; Maddison 2000), but their representation and interpretation on phylogenies transcend the standard variational framework of variance-covariance matrices. Stasis, in turn, can be studied within lineages and its existence assessed in terms of the extent of variation and covariation of traits (Lande 1986), but many examples of stasis concern single discrete traits, a few taxon-specific traits whose covariation is of less interest than their persistence (i.e., relative invariance) through time, or homologies and their ontogenetic and phylogenetic entrenchment.

On macroevolutionary time scales, traits may be seen as highly integrated as much because of tight correlations with other traits as because of their internal resistance to change (see Wagner and Schwenk 2000) and their persistence through time (which may relate to various causes, both internal and external to the organism). In this sense, a distinction between a state of integration (in a proximal sense) and a dynamic of integration maintenance (in an evolutionary sense, by selection for example) becomes important. Further, while developmental constraints and constraints in general have been interpreted and formalized in terms of covariance structure (Maynard Smith et al. 1985; Lande 1986; Wagner 1988; Arnold 1992; Wagner and Altenberg 1996), many accounts of stasis as an outcome of developmental constraint involve *organizational* criteria cast in terms of either epigenetics or geometric structure in morphospace (Alberch 1982; Maynard Smith et al. 1985; Gould 1989b; Eble 1998; Müller and Newman 1999). While it is tempting to refer to these different styles of study and interpretation of phenotypic integration as the distinction between qualitative ("unrigorous") and quantitative ("rigorous") styles of biological research, qualitative statements convey data that can often be rendered rigorous, objective, testable, and precise.

What the diversity and complexity of macroevolutionary phenomena and the state of macroevolutionary theory suggest is that phenotypic integration, expressing multiple scales of time and space, multiple causes, and different levels in the organizational, genealogical, and ecological hierarchies, may well justify either modifications or alternatives to the "matrix-thinking" that stems from standard quantitative genetic approaches on the one hand, and morphometrics on the other (see also Pigliucci, Chapter 10, this volume). Far from denying the centrality of heritability as a prerequisite for systematic sorting and selection processes, one may consider alternative formalisms that do not reduce to pairwise associations of characters or organisms. Additional sources of pattern are possible to the extent that some aspects of integration may not be contained in the representational structure of genetic or phenotypic variance-covariance matrices. For example, while the genotype-phenotype map can be explored by comparisons of genetic and phenotypic matrices, the mapping itself should ideally be represented in terms of developmental trajectories, patterns of gene expression, regulatory genetic and epigenetic networks, changing identities and roles of modules, and self-organization as a contributor to the emergence of form in ontogeny (Eble 2003).

Novel biological insights on multiscale and multihierarchical organismal integration may invite more extensive use of representations such as graphs (including ontogenetic and phylogenetic trees and networks), contour diagrams, and non-Euclidean or nonmetric topological renditions of the state spaces in which characters and organisms are embedded (e.g., Olson and Miller 1958; Sneath and Sokal 1973; Schlichting and Pigliucci 1998; Newman 2000; Sporns et al. 2000; Magwene 2001; Stadler et al. 2001; Wagner and Stadler 2003).

Further, even when matrices are retained as the default operational tool, the conclusions we standardly draw from current quantitative-genetic models for macroevolution are in need of revision. Rather than explanatorily reducing **P**-matrices and vectors of putatively modular phenotypic units to arbitrarily invoked selection coefficients and a monistic arena of **G**-matrices, one might well consider **P** as a locus of constraint and explanation in its own right (Burger 1986; Wagner 1989, 1994, 1996; Müller 1990; Goodwin 1994; Cheverud 1996; Müller and Newman 1999; Erwin 2000; Jablonski 2000; Gould 2002; Eble 2003, 2004). Other sources of pattern not directly revealed in the structure of genetic variance-covariance matrices, and which may imply causal roles for the phenotype, are patterns of epigenetic covariation, ecophenotypic covariation, phylogenetic covariation, and temporal (stratigraphic) covariation.

In macroevolutionary theory, phenomenological generalization is justifiable because the goal is not to understand characters and their differential integration only, but characters and integration as they relate to a dynamics of taxon sorting that may often depend not only on function and development, but also on external factors (e.g., drastic environmental change, periodic forcing factors such as bolide impacts), on context (biogeographic phenomena, norms of reaction, innovation, morphospace), and on incidental causal consequences of the characters themselves when they lead to emergent fitnesses or organismal exaptation. If phenotypic integration is to be about population-genetic evolutionary processes, the current uniformitarian approach cast in terms of microevolution is appropriate. Yet in macroevolution, uniformitarianism is hardly justifiable; evolutionary processes conditioned on heritability must be generalized across levels and scales, and complemented by biological and nonbiological factors that can change the course of evolution, reset or eliminate evolutionary trends, and modify evolvability and attainability even if they cannot count as "heritable." Can such additional factors be modeled? Perhaps not in detail, but their role is an integral part of the orderliness (or lack thereof) of macroevolutionary phenotypic integration.

## Is Integration the Converse of Modularity?

The simple answer is no, because the whole is more than the sum of the parts. But the answer might also be yes, depending on how one defines and quantifies integration and modularity. Theoretically (Wagner 1996; Wagner and Altenberg 1996), one might suppose that every hypothesis of the macroevolution of phenotypic integration could be expressed in terms of modularity, and vice versa. All other things being equal, this appears reasonable. Indeed, the notions of parcellation (increase in modularity) and integration (decrease in modularity) figure together as opposites in conceptual discussions (Wagner and Altenberg 1996; Eble 2004), in statistical operationalizations (Mezey et al. 2000; Magwene 2001), and in morphometric studies (Klingenberg et al. 2001). But is the opposition *symmetric*? Modularity and integration may actually not be strictly inversely proportional to each other because of factors such as the geometry of organisms, the topology of morphospace, and historical contingency.

### The Geometry of Organisms

Different modules may be differentially integrated and differentially cohesive, and this may depend to a large extent on organismal geometry. Character or part counts can serve as estimates of modularity (McShea and Venit 2001; Eble 2004), but not of integration, for example. The shape of parts itself affects the geometry of connectivity, but if they are homogeneous (e.g., serial homologs), modularity (measured as number of parts) may increase or decrease in evolution without a necessary change in the strength of interactions (within or between those parts), and hence no necessary change in the degree of integration.

### The Topology of Morphospace

Accessibility in morphospace may not be isotropic, because different regions may be accessible to different degrees, such that heterogeneities in the structure of morphospace may determine whether module creation, or increase in modularity, is reversible (see Stadler et al. 2001); if it is not reversible, or if it is but with a lower transition probability, then integration is not the converse of modularization. For example, a narrow route in morphospace may link a primitive taxon restricted in a small, homogeneous region of parameter space to a derived taxon in a broader, heterogeneous region. This configuration of morphospace would by itself promote exploration of the broader region by the derived taxon (assuming no change in intrinsic rates of origination and extinction, and stochastic diffusion in morphospace), with a possible increase in modularity; and it would bias against convergence back to the primitive restricted region because of the narrowness of the route connecting the regions. How to document and model inherent heterogeneities in morphospace, however, is an open issue.

### Historical Contingency

How the geometry of organisms and the topology of morphospace affect modularity and integration may to some extent be a matter of contingency, depending on which phenotypes happen to have evolved and where they are located in morphospace. Contingency plays a role in determining which modules become entrenched or lost in evolution. Some modules, such as limbs, may be lost, and modularity arguably decreases, while the integration of remaining modules also decreases, as former connections with limbs are lost. Thus, a reduction in modularity need not correspond to an increase in integration for historical reasons.

## Temporal Resolution and Scales of Integration

How coeval do organisms need to be for one to quantify integration? Integration is often considered to be an instantaneous property. But some relativization is necessary, because average integration is also biologically interpretable (as a tendency or potentiality, for instance), and temporal resolution is almost never equivalent in different taxa. For example, in an ant colony individuals at any one time are behaviorally integrated, and connectivity is realized by chemical signaling among individuals. At the same time, if one broadens the interval of time over which one observes and infers integration, some individuals that contribute to integration at time $t$ may not have been born yet at time $t - 1$. If one cannot, or does not want to, distinguish between $t$ and $t - 1$, one will assume time-averaged integration to be constant. If birth and death rates are stochastically constant, or if general statements about average integration are useful in a particular theoretical context, then time averaging is acceptable and inconsequential.

In paleobiological studies time averaging is inescapable, and virtually all macroevolutionary research with fossils is pursued at a level of generality where time averaging does not compromise statistical inference or theoretical interpretation. Typically, samples of species are binned in intervals of time that may correspond to millions of years. In a morphometric study, a matrix of correlations might be computed, and morphological integration might be inferred as a property of the group of species on that scale. Characters $a$ and $b$ may be inferred to correlate in general across species even if late in the interval they did not. One may assume that species are randomly distributed over the interval, but if a disproportionately large number of species occurs early on and provides evidence for strong integration, and a smaller number of species occurs later on but provides no evidence for strong integration, a real change in degree of integration would not be identified. Macroevolutionary inference necessarily proceeds by treating time-averaged integration estimates as average properties for the clade, and by comparing such averages across intervals. For long-ranging clades, enough intervals may be included to render temporal patterns of time-averaged integration interpretable as average tendencies reflecting evolutionary potential and evolvability. Debates concerning changes in differential flexibility in macroevolutionary time, such as after the Cambrian explosion or during evolutionary radiations (e.g., Gould 1989a, 1991; Valentine et al. 1991b; Foote 1997; Eble 1998, 1999b, 2000a), are addressed empirically with precisely such a rationale, with degree of integration standing as an average propensity across clades, exceptions notwithstanding (but see Hughes et al. 1999).

## Integration by Chance

For some groups of organisms, a relatively small number of modules at a particular level of organization may compose the organism. In this case, drift may become important in coupling or uncoupling characters by chance, because of the small population of modules. Null models of integration and parcellation need to be devised that take into account the number of modules in an organism. When modules are not numerous and their functionality is not obvious, drift may well be more important than selection or development in reinforcing integration. Further, in macroevolutionary studies a few broad modular regions may be more tractable and of more theoretical interest than the many submodules composing them (e.g., David and Mooi 1996; Mooi and David 1998; Eble 2000a, 2004).

It is encouraging that recent statistics of integration have been proposed that explicitly rely on appropriate null hypotheses (Mezey et al. 2000). Thus, from a range of possible boundary conditions and possible genotype-phenotype interactions, inferences about integration, or lack thereof, are conditioned on it being higher or lower than expected (the expectation potentially varying depending on the situation or the question of interest). While null hypotheses of integration for QTL data take into account gene number, gene effects, and pleiotropic interactions, in macroevolution purely phenotypic considerations are also reasonable. Null hypotheses then can be cast in terms of universes of possible trait interactions at the organismal level, of probabilities of presence or absence of traits with high potential for entrenchment, and of expected trait variances and covariances. The range of possible patterns might also be simulated nonparametrically, with randomization procedures (Cheverud et al. 1989), or mechanistically, with theoretical morphology models.

## Is Disparity a Proxy for Phenotypic Integration?

Disparity refers to the phenotypic distinctness of forms in morphospace. The quantification of disparity has proved to be an important approach to help resolve, in more focally morphological terms, debates about temporal asymmetries in clade histories, the nature of evolutionary radiations, and the relationship between evolvability and innovation (Gould 1991; Foote 1993, 1997; Wills et al. 1994; Wagner 1995; Conway Morris 1998; Eble 1998, 2000a). It also holds promise in evolutionary developmental biology, as a means of assessing the geometry of global phenotypic variation in ontogenetic time and of addressing and testing in more rigorous terms ontogenetic regularities across taxa (Eble 2002, 2003).

Disparity can be quantified in a number of ways (e.g., Foote 1991; Wills et al. 1994), most often as the total variance or the total range of *n* variables in a sample. As a general measure of variation in a clade, disparity can be expected to be promoted by variational modularity, because modularity allows opportunities for semi-independent variation; disparity might also itself be an operational proxy for modularity when common mechanistic sources (e.g., function, development) or causal roles (e.g., in evolvability, in clade selection and sorting, in character evolution) can be identified or hypothesized (Eble 2004).

As discussed previously, modularity and integration need not be strictly the converse of each other, and thus disparity as a proxy for modularity may not be a proxy for integration in the same way. But on macroevolutionary scales and at the level of clades or body plans, change or maintenance of relative phenotypic integration is generally viewed as theoretically more important and empirically more tractable than quantification of absolute degrees of integration, or for that matter of modularity (Gould 1989a, 1991; Foote 1993, 1996, 1997; Wills et al. 1994). And indeed, documentation of disparity patterns through time is usually motivated by an interest in assessing whether integration has increased, decreased, or remained the same in the history of various clades. Integration is treated as a global, average property of clades in any given time interval, and tests of macroevolutionary hypotheses are devised accordingly, using disparity as a rough measure of integration.

Granted that disparity refers to variation and integration to covariation, being thus not formally equivalent (but see Van Valen 1974, 1978; Wagner 1984; Cheverud et al. 1989), the general logic of the evolutionary connection is the same as outlined above for modularity: strong phenotypic integration may constrain the production of novelties and hence disparity, and conversely loose integration may allow novelties to be generated and incorporated, with an associated increase in disparity. Functional and developmental constraints are expressed in patterns of integration, and disparity will reflect remaining possibilities for functional or developmental differentiation of integrated phenotypes. Furthermore, integration as well as disparity can influence evolvability, clade success, and character evolution.

Ultimately, disparity, modularity, and integration in macroevolutionary theory reflect the same concept—constraint. Constraint is a key subject in macroevolution and surfaces in theoretical and analytical morphology (Raup 1972; Foote 1997), constructional morphology (Seilacher 1970), punctuated equilibrium (Eldredge and Gould 1972), heterochrony (Alberch et al. 1979), critiques of adaptationism (Gould and Lewontin 1979; Gould 1980), discussions about the role of chance in macroevolution (Eble 1999a; Gould 2002), and interpretations of evolutionary radiations and the origin of novelties (Valentine and Campbell 1975; Gould 1989a; Jablonski and Bottjer 1990; Erwin 1993; Valentine 1995; Conway Morris 1998; Eble 1998). Disparity appears to be useful as a rough measure of macroevolutionary phenotypic integration, but refining the conceptual and analytical intersections among constraint, integration, modularity, and disparity is much needed to determine the place of disparity in models of phenotypic integration.

## Conclusions

Phenotypic integration is almost universally modeled and interpreted in terms of genetic variance-covariance. Even when matches between phenotypic and genetic correlations hold, analysis and modeling of phenotypic integration proceeds as if it corresponded to genetic integration. Causality is forced to reside in genotypic space, and phenotypic integration stands as a passive object of explanation, devoid of theoretical role.

In macroevolution, the phenotype and phenotypic integration are both observational objects and causal subjects. Phenotypic integration can be modeled in its own terms, taking into account a variety of evolutionary forces and scales of time. This constitutes a challenge for macroevolutionists. A way forward is to conceive of and to model relationships among characters not only in genetic terms, but also in terms of function (degree of functional correlation), epigenetics (degree of developmental correlation, say, by sharing of cell and tissue types or constructional elements), space (degree of contiguity in the whole organism), environment (plasticity integration), biomechanics (degree of shared structural stability, which need not imply shared function), fitness (degree of contribution to overall fitness, even if functions are not shared), and time (characters may be associated or correlated because of shared temporal origins, or else sequential origins, with one character conditioned on the appearance of another). No single model is likely to incorporate successfully all such sources of integration, but a diversity of models may actually be desirable to consider different domains, scales, and levels of integration. Domains, scales, and levels overlap and intersect, and so can models of phenotypic integration. Phenotypic organization, variation, and causal roles differ in microevolution and macroevolution, but are all manifested in the same biological hierarchies, which integrate different kinds of phenotypic integration.

In the vernacular, one meaning of "integration" is "to end segregation of and bring into equal membership in society or in an organization." This has been an aspiration of macroevolutionists for a long time. Phenotypic integration in macroevolution is sufficiently distinct and sufficiently important to justify a more effective incorporation of macroevolutionary data and theory into accounts of phenotypic integration in particular, and into evolutionary theory in general.

*Acknowledgments* I thank W. Callebaut, B. David, D. Erwin, N. Hughes, D. Jablonski, L. McCall, D. McShea, G. Müller, M. Pigliucci, and K. Preston for comments, suggestions, and/or discussion. Research was supported in part by postdoctoral fellowships from the Santa Fe Institute, the CNRS (Dijon, France, through the Conséil Régional de Bourgogne), and the Konrad Lorenz Institute (Vienna, Austria).

*Literature Cited*

Alberch, P. 1982. Developmental constraints in evolutionary processes. Pp. 313–332 in J. T. Bonner, ed. Evolution and Development. Springer, Berlin.

Alberch, P. 1983. Morphological variation in the neotropical salamander genus *Bolitoglossa*. Evolution 37:906–919.

Alberch, P., Gould, S. J., Oster, G. F., and Wake, D. B. 1979. Size and shape in ontogeny and phylogeny. Paleobiology 5:296–317.

Arnold, S. J. 1992. Constraints on phenotypic evolution. American Naturalist 140:S85-S107.

Arnold, S. J., Alberch, P., Csányi, V., Dawkins, R. C., Emerson, S. B., Fritzsch, B., Horder, T. J., Maynard Smith, J., Starck, M. J., Vrba, E. S., Wagner, G. P., and Wake, D. B. 1989. How do complex organisms evolve ? Pp. 403–433 in D. B. Wake and G. Roth, eds. Complex Organismal Functions: Integration and Evolution in Vertebrates. Wiley, New York.

Benton, M. J. 1997. Models for the diversification of life. Trends in Ecology and Evolution 12:490–495.

Bookstein, F., Chernoff, B., Elder, R., Humphries, J., Smith, G., and Strauss, R. 1985. Morphometrics in Evolutionary Biology. Special Publication 15, The Academy of Natural Sciences of Philadelphia.

Burger, R. 1986. Constraints for the evolution of functionally coupled characters: a non-linear analysis of a phenotypic model. Evolution 40:182–193.

Chaline, J., David, B., Magniez-Jannin, F., Malassé, A. D., Marchand, D., Courant, F., and Millet, J.-J. 1998. Quantification de l'évolution morphologique du crâne des Hominidés et hétérochronies. C. R. Acad. Sci. Paris, Sciences de la Terre et des Planètes 326:291–298.

Cheverud, J. M. 1982. Phenotypic, genetic, and environmental integration in the cranium. Evolution 36:499–516.

Cheverud, J. M. 1990. The evolution of morphological variation patterns. Pp. 133–145 in M. Nitecki, ed. Evolutionary Innovations. University of Chicago Press, Chicago.

Cheverud, J. M. 1996. Developmental integration and the evolution of pleiotropy. American Zoologist 36:44–50.

Cheverud, J. M., Wagner, G. P., and Dow, M. M. 1989. Methods for the comparative analysis of variation patterns. Systematic Zoology 38:201–213.

Clausen, J. and Hiesey, W. M. 1960. The balance between coherence and variation in evolution. Proceedings of the National Academy of Sciences USA 46:494–506.

Clausen, J., Keck, D., and Hiesey, W. M. 1940. Experimental studies on the nature of plant species. I. Effect of varied environment on Western North American plants. Publication 564, Carnegie Institution of Washington, Washington, DC.

Conway Morris, S. 1998. The Crucible of Creation. Oxford University Press, Oxford.

Darden, L. 1992. Character: historical perspectives. Pp. 41–44 in E. F. Keller and E. A. Lloyd, eds. Keywords in Evolutionary Biology. Harvard University Press, Cambridge, MA.

David, B. and Mooi, R. 1996. Embryology supports a new theory of skeletal homologies for the phylum Echinodermata. C. R. Acad. Sci. Paris, Sciences de la Vie/Life Sciences 319:577–584.

Depew, D. J. 1998. Darwinism and developmentalism: prospects for convergence. Pp. 21–32 in G. Van de Vijver, S. N. Salthe, and M. Delpos, eds. Evolutionary Systems: Biological and Epistemological Perspectives on Selection and Self-organization. Kluwer, Dordrecht.

Eble, G. J. 1998. The role of development in evolutionary radiations. Pp. 132–161 in M. L. McKinney and J. A. Drake, eds. Biodiversity Dynamics: Turnover of Populations, Taxa, and Communities. Columbia University Press, New York.

Eble, G. J. 1999a. On the dual nature of chance in evolutionary biology and paleobiology. Paleobiology 25:75–87.

Eble, G. J. 1999b. Originations: land and sea compared. Geobios 32:223–234.

Eble, G. J. 2000a. Contrasting evolutionary flexibility in sister groups: disparity and diversity in Mesozoic atelostomate echinoids. Paleobiology 26:56–79.

Eble, G. J. 2000b. Theoretical morphology: state of the art. Paleobiology 26:520–528.

Eble, G. J. 2002. Multivariate approaches to development and evolution. Pp. 51–78 in N. Minugh-Purvis and K. McNamara, eds. Human Evolution Through Developmental Change. Johns Hopkins University Press, Baltimore, MD.

Eble, G. J. 2003. Developmental morphospaces and evolution. Pp. 35–65 in J. P. Crutchfield and P. Schuster, eds. Evolutionary Dynamics: Exploring the Interplay of Selection, Neutrality, Accident, and Function. Oxford University Press, Oxford.

Eble, G. J. 2004. Morphological modularity and macroevolution: conceptual and empirical aspects. In W. Callebaut and D. Rasskin-Gutman, eds. Modularity: Understanding the Development and Evolution of Complex Natural Systems. MIT Press, Cambridge, MA.

Eldredge, N. 1985. Unfinished Synthesis: Biological Hierarchies and Modern Evolutionary Thought. Oxford University Press, New York.

Eldredge, N. 1989. Macroevolutionary Dynamics. McGraw-Hill, New York.

Eldredge, N. and Gould, S. J. 1972. Punctuated equilibria: an alternative to phyletic gradualism. Pp. 82–115 in T. J. M. Schopf, ed. Models in Paleobiology. Freeman, San Francisco.

Erwin, D. H. 1993. The origin of metazoan development: a paleobiological perspective. Biological Journal of the Linnean Society 50:255–274.

Erwin, D. H. 1999. The origin of bodyplans. American Zoologist 39:617–629.

Erwin, D. H. 2000. Macroevolution is more than repeated rounds of microevolution. Evolution and Development 2:78–84.

Erwin, D. H. and R. L. Anstey, eds. 1995. New Approaches to Speciation in the Fossil Record. Columbia University Press, New York.

Erwin, D. H., Valentine, J. W., and Sepkoski, J. J., Jr. 1987. A comparative study of diversification events: the early Paleozoic versus the Mesozoic. Evolution 41:1177–1186.

Fink, W. L. and Zelditch, M. L. 1996. Historical patterns of developmental integration in piranhas. American Zoologist 36:61–69.

Foote, M. 1991. Analysis of morphological data. Pp. 59–86 in N. L. Gilinsky and P. W. Signor, eds. Analytical Paleobiology. The Paleontological Society, University of Tennessee, Knoxville.

Foote, M. 1993. Discordance and concordance between morphological and taxonomic diversity. Paleobiology 19:185–204.

Foote, M. 1996. Models of morphological diversification. Pp. 62–86 in D. Jablonski, D. Erwin, and J. Lipps, eds. Evolutionary Paleobiology. University of Chicago Press, Chicago.

Foote, M. 1997. The evolution of morphological diversity. Annual Review of Ecology and Systematics 28:129–152.

Foote, M. 1999. Morphological diversity in the evolutionary radiation of Paleozoic and post-Paleozoic crinoids. Paleobiology Memoirs, Supplement to Paleobiology 25(2):1–115.

Gilinsky, N. L. and Signor, P. W., eds. 1991. Analytical Paleobiology. The Paleontological Society, University of Tennessee, Knoxville.

Goodwin, B. C. 1994. How the Leopard Changed its Spots. Charles Scribner's Sons, New York.

Gould, S. J. 1977. Ontogeny and Phylogeny. Harvard University Press, Cambridge, MA.

Gould, S. J. 1980. Is a new and general theory of evolution emerging? Paleobiology 6:119–130.

Gould, S. J. 1982. Darwinism and the expansion of evolutionary theory. Science 16:380–387.

Gould, S. J. 1989a. Wonderful Life. Norton, New York.

Gould, S. J. 1989b. A developmental constraint in *Cerion*, with comments on the definition and interpretation of constraint in evolution. Evolution 43:516–539.

Gould, S. J. 1991. The disparity of the Burgess Shale arthropod fauna and the limits of cladistic analysis: why we must strive to quantify morphospace. Paleobiology 17:411–423.

Gould, S. J. 1993. How to analyze Burgess Shale disparity—a reply to Ridley. Paleobiology 19:522–523.

Gould, S. J. 2002. The Structure of Evolutionary Theory. Harvard University Press, Cambridge, MA.

Gould, S. J. and Lewontin, R. C. 1979. The spandrels of San Marco and the Panglossian paradigm: a critique of the adaptationist programme. Proceedings of the Royal Society of London B 205:581–598.

Grafen, A. and Ridley, M. 1996. Statistical tests for discrete cross-species data. Journal of Theoretical Biology 183:255–267.

Harvey, P. H. and Pagel, M. D. 1991. The Comparative Method in Evolutionary Biology. Oxford University Press, Oxford.

Hughes, N. C. 1990. Morphological plasticity and genetic flexibility in a Cambrian trilobite. Geology 19:913–916.

Hughes, N. C., Chapman, R. E., and Adrain, J. M. 1999. The stability of thoracic segmentation in trilobites: a case study in developmental and ecological constraints. Evolution and Development 1:24–35.

Imbrie, J. 1956. Biometrical methods in the study of invertebrate fossils. Bulletin of the American Museum of Natural History 108:211–252.

Jablonski, D. 2000. Micro- and macroevolution: scale and hierarchy in evolutionary biology and paleobiology. Paleobiology 26 (Supplement):15–52.

Jablonski, D., and Bottjer, D. J. 1990. The ecology of evolutionary innovation: the fossil record. Pp. 253–288 in M. Nitecki, ed., Evolutionary Innovations. University of Chicago Press, Chicago.

Jablonski, D., and Bottjer, D. J. 1991. Environmental patterns in the origin of higher taxa: the post-Paleozoic fossil record. Science 252:1831–1833.

Jablonski, D., Erwin, D., and Lipps, J., eds. 1996. Evolutionary Paleobiology. University of Chicago Press, Chicago.

Jackson, J. B. C., Lidgard, S., and McKinney, F. K., eds. 2001. Evolutionary Patterns: Growth, Form, and Tempo in the Fossil Record. University of Chicago Press, Chicago.

Jacobs, D. K. 1990. Selector genes and the Cambrian radiation of the Bilateria. Proceedings of the National Academy of Sciences USA 87:4406–4410.

Kauffman, S. A. 1989. Cambrian explosion and Permian quiescence: implications of rugged fitness landscapes. Evolutionary Ecology 3:274–281.

Kauffman, S. A. 1993. The Origins of Order: Self-Organization and Selection in Evolution. Oxford University Press, Oxford.

Kermack, K. A. 1954. A biometrical study of *Micraster coranguinum* and *M. (Isomicraster) senonensis*. Philosophical Transactions of the Royal Society of London B 237:375–428.

Kermack, K. A. and Haldane, J. B. S. 1950. Organic correlation and allometry. Biometrika 37:30–41.

Klingenberg, C. P., Badyaev, A. V., Sowry, S. M., and Beckwith, N. J. 2001. Inferring developmental modularity from morphological integration: analysis of individual variation and asymmetry in bumblebee wings. American Naturalist 157:11–23.

Lande, R. 1986. The dynamics of peak shifts and the pattern of morphological evolution. Paleobiology 12:343–354.

Lande, R. and Arnold. S. J. 1983. The measurement of selection on correlated characters. Evolution 37:1210–1226.

Lewontin, R. C. 1970. The units of selection. Annual Review of Ecology and Systematics. 1:1–14.

Lloyd, E. A. and Gould, S. J. 1993. Species selection on variability. Proceedings of the National Academy of Sciences USA 90:595–599.

Maddison, W. P. 2000. Testing character correlation using pairwise comparisons on a phylogeny. Journal of Theoretical Biology 202:195–204.

Magwene, P. M. 2001. New tools for studying integration and modularity. Evolution 55:1734–1745.

Maynard Smith, J. and Szathmáry, E. 1995. The Major Transitions in Evolution. Freeman, Oxford.

Maynard Smith, J., Burian, R., Kauffman, S., Alberch, P., Campbell, J., Goodwin, B., Lande, R., Raup, D., and Wolpert, L. 1985. Developmental constraints and evolution. Quarterly Review of Biology 60:265–287.

McGhee, G. R, Jr. 1999. Theoretical Morphology. Columbia University Press, New York.

McKinney, M. L. and Drake, J. A., eds. 1998. Biodiversity Dynamics: Turnover of Populations, Taxa, and Communities. Columbia University Press, New York.

McShea, D. W. 1993. Arguments, tests, and the Burgess Shale—a commentary on the debate. Paleobiology 19:399–402.

McShea, D. W. 2001. Parts and integration: consequences of hierarchy. Pp. 27–60 in J. B. C. Jackson, S. Lidgard, and F. K. McKinney, eds. Evolutionary Patterns: Growth, Form, and Tempo in the Fossil Record. University of Chicago Press, Chicago.

McShea, D. W. and Venit, E. P. 2001. What is a part? Pp. 259–284 in G. P. Wagner, ed. The Character Concept in Evolutionary Biology. Academic Press, San Diego.

Mezey, J. G., Cheverud, J. M., and Wagner, G. P. 2000. Is the genotype-phenotype map modular? A statistical approach using mouse quantitative trait loci data. Genetics 156:305–311.

Mooi, R. and David, B. 1998. Evolution within a bizarre phylum: homologies of the first echinoderms. American Zoologist 38:965–974.

Müller, G. 1990. Developmental mechanisms at the origin of morphological novelty: a side-effect hypothesis. Pp. 99–130 in M. Nitecki, ed. Evolutionary Innovations. University of Chicago Press, Chicago.

Müller, G. and Newman, S. A. 1999. Generation, integration, autonomy: three steps in the evolution of homology. Pp. 65–73 in B. Hall, ed. Homology. Wiley, New York.

Nelson, G. 1989. Species and taxa: speciation and evolution. Pp. 60–81 in D. Otte and J. Endler, eds. Speciation and its Consequences. Sinauer, Sunderland, MA.

Newman, M. E. J. 2000. Models of the small world. Journal of Statistical Physics 101:819–841.

Olson, E. C. and Miller, R. L. 1958. Morphological Integration. University of Chicago Press, Chicago.

Raup, D. M. 1956. *Dendraster*: a problem in echinoid taxonomy. Journal of Paleontology 30:685–694.

Raup, D. M. 1966. Geometric analysis of shell coiling: general problems. Journal of Paleontology 40:1178–1190.

Raup, D. M. 1972. Approaches to morphologic analysis. Pp. 28–44 in T. J. M. Schopf, ed. Models in Paleobiology. Freeman, San Francisco.

Raup, D. M. and Boyajian, G. E. 1988. Patterns of generic extinction in the fossil record. Paleobiology 14:109–125.

Raup, D. M. and Jablonski, D., eds. 1986. Patterns and Processes in the History of Life. Springer, Berlin.

Raup, D. M. and Michelson, A. 1965. Theoretical morphology of the coiled shell. Science 147:1294–1295.

Ridley, M. 1993. Analysis of the Burgess Shale. Paleobiology 19:519–521.

Rieppel, O. 1986. Species are individuals: a review and critique of the argument. Evolutionary Biology 20:283–317.

Rieppel, O. 1988. Fundamentals of Comparative Biology. Birkhäuser, Basel.

Roth, V. L. 1996. Cranial integration in the Sciuridae. American Zoologist 36:14–23

Russell, E. S. 1916. Form and Function. 1982 reprint, University of Chicago Press, Chicago.

Schlichting, C. D. and Pigliucci, M. 1998. Phenotypic Evolution: A Reaction Norm Perspective. Sinauer, Sunderland, MA.

Schopf, T. J. M., ed. 1972. Models in Paleobiology. Freeman, San Francisco.

Schopf, T. J. M., Raup, D. M., Gould, S. J., and Simberloff, D. S. 1975. Genomic versus morphologic rates of evolution: influence of morphologic complexity. Paleobiology 1:63–70.

Seilacher, A. 1970. Arbeitskonzept zur Konstruktions-Morphologie. Lethaia 3:393–396.

Sepkoski, J. J., Jr. 1978. A kinetic model of Phanerozoic taxonomic diversity. I. Analysis of marine orders. Paleobiology 4:223–251.

Sepkoski, J. J., Jr. 1979. A kinetic model of Phanerozoic taxonomic diversity. II. Early Phanerozoic families and multiple equilibria. Paleobiology 5:222–251.

Sepkoski, J. J., Jr. 1984. A kinetic model of Phanerozoic taxonomic diversity. III. Post-Paleozoic families and mass extinctions. Paleobiology 10:246–267.

Sepkoski, J. J., Jr. and Kendrick, D. C. 1993. Numerical experiments with model monophyletic and paraphyletic taxa. Paleobiology 19:168–184.

Shubin, N. and Wake, D. 1996. Phylogeny, variation, and morphological integration. American Zoologist 36:51–60.

Simon, H. A. 1962. The architecture of complexity. Proceedings of the American Philosophical Society 106:467–482.

Simpson, G. G. 1953. The Major Features of Evolution. Columbia University Press, New York.

Smith, K. K. 1996. Integration of craniofacial structures during development in mammals. American Zoologist 36:70–79.

Sneath, P. H. A. and Sokal, R. R. 1973. Numerical Taxonomy. Freeman, San Francisco.

Sporns, O., Toroni, G., and Edelman, G. M. 2000. Theoretical neuroanatomy: relating anatomical and functional connectivity in graphs and cortical connection matrices. Cerebral Cortex 10:127–141.

Stadler, B. M. R., Stadler, P. F., Wagner, G. P., and Fontana, W. 2001. The topology of the possible: formal spaces underlying patterns of evolutionary change. Journal of Theoretical Biology 213:241–274.

Stanley, S. M. 1979. Macroevolution: Pattern and Process. Freeman, San Francisco.

Stanley, S. M. 1990. The general correlation between rate of speciation and rate of extinction: fortuitous causal linkages. Pp. 103–127 in R. M. Ross and W. D. Allmon, eds. Causes of Evolution: A Paleontological Perspective. University of Chicago Press, Chicago.

Valentine, J. W. 1986. Fossil record of the origin of Baupläne and its implications. Pp. 209–222 in D. M. Raup and D. Jablonski, eds. Patterns and Processes in the History of Life. Springer, Berlin and Heidelberg.

Valentine, J. W. 1990. The macroevolution of clade shape. Pp. 128–150 in R. M. Ross and W. D. Allmon, eds. Causes of Evolution: A Paleontological Perspective. University of Chicago Press, Chicago.

Valentine, J. W. 1995. Why no new phyla after the Cambrian? Genome and ecospace hypotheses revisited. Palaios 10:190–194.

Valentine, J. W. and Campbell, C. A. 1975. Genetic regulation and the fossil record. American Scientist 63:673–680.

Valentine, J. W., Tiffney, B. H., and Sepkoski, J. J., Jr. 1991a. Evolutionary dynamics of plants and animals: a comparative approach. Palaios 6:81–88.

Valentine, J. W., Awramik, S. M., Signor, P. W., and Sadler, P. M. 1991b. The biological explosion at the Precambrian-Cambrian boundary. Evolutionary Biology 25:279–356.

Van Valen, L. 1974. Multivariate structural statistics in natural history. Journal of Theoretical Biology 45:235–247.

Van Valen, L. 1978. The statistics of variation. Evolutionary Theory 4:33–43.

Vrba, E. S. 1983. Macroevolutionary trends: new perspectives on the roles of adaptation and incidental effect. Science 221:387–389.

Vrba, E. S. 1989. Levels of selection and sorting with special reference to the species level. Oxford Surveys in Evolutionary Biology 6:111–168.

Vrba, E. S. and Gould, S. J. 1986. The hierarchical expansion of sorting and selection: sorting and selection cannot be equated. Paleobiology 12:217–228.

Wagner, G. P. 1984. On the eigenvalue distribution of genetic and phenotypic dispersion matrices: evidence for a nonrandom organization of quantitative character variation. Journal of Mathematical Biology 21:77–95.

Wagner, G. P. 1988. The significance of developmental constraints for phenotypic evolution by natural selection. Pp. 222–229 in G. de Jong, ed. Population Genetics and Evolution. Springer, Berlin.

Wagner, G. P. 1989. The biological homology concept. Annual Review of Ecology and Systematics 20:51–69.

Wagner, G. P. 1994. Homology and the mechanisms of development. Pp. 273–299 in B. K. Hall, ed. Homology: The Hierarchical Basis of Comparative Biology. Academic Press, San Diego.

Wagner, P. J. 1995. Testing evolutionary constraint hypotheses with early Paleozoic gastropods. Paleobiology 21:248–272.

Wagner, G. P. 1996. Homologues, natural kinds and the evolution of modularity. American Zoologist 36:36–43.

Wagner, G. P. and Altenberg, L. 1996. Complex adaptations and the evolution of evolvability. Evolution 50:967–976.

Wagner, G. P. and Misof, B. Y. 1993. How can a character be developmentally constrained despite variation in developmental pathways? Journal of Evolutionary Biology 6:449–455.

Wagner, G. P. and Schwenk, K. 2000. Evolutionarily stable configurations: functional integration and the evolution of phenotypic stability. Evolutionary Biology 31:155–217.

Wagner, G. P. and Stadler, P. F. 2003. Quasi-independence, homology and the unity of type: a topological theory of characters. Journal of Theoretical Biology 220:505–527.

Weiss, P. A. 1971. The basic concept of hierarchic systems. Pp. 1–43 in P. A. Weiss, ed. Hierarchically Organized Systems. Hafner, New York.

Williams, G. C. 1992. Natural selection: Domains, Levels, and Challenges. Oxford University Press, Oxford.

Wills, M. A., Briggs, D. E. G., and Fortey, R. A. 1994. Disparity as an evolutionary index: a comparison of Cambrian and Recent arthropods. Paleobiology 20:93–130.

Zelditch, M. L. 1996. Introduction to the symposium: historical patterns of developmental integration. American Zoologist 36:1–3.

Zelditch, M. L. and Fink, W. L. 1996. Heterochrony and heterotopy: stability and innovation in the evolution of form. Paleobiology 22:241–254.

# 13

# Form, Function, and Life History

## *Spatial and Temporal Dynamics of Integration*

MIRIAM LEAH ZELDITCH
ROSA A. MOSCARELLA

Developmental integration is usually viewed as a distinctively morphological phenomenon, but organisms are not just morphological systems and developmental integration involves more than form. Over the course of mammalian ontogeny, blind, deaf, toothless, nearly transparent, and virtually immobile infants develop into sighted, hearing, chewing, furred, and active adults. That transformation involves coordinated changes among form and function, and the timing of that transformation determines life-history schedules. In some groups, that transformation occurs very rapidly; in fact, in some clades it is nearly complete at birth, although in others it may take days, weeks, or even months to complete. Both the pace at which the sequence is completed, and some of the details of the individual steps, depend on how limited resources of time and energy are allocated to growth and development. But regardless of how resources are allocated, of how long those transitions take, and of the details of morphology, form and function must be coordinated. This coordination must characterize even the earliest stages of postnatal development, or else infants will not survive the peculiarly vulnerable nestling stage. Thus, integration is a dynamic feature of organisms, and its dynamics are partly a function of life-history strategies.

The connection between development, morphology, function, and life history has been most fully articulated in the theory of heterochrony (discussed below), which postulates that selection on life-history parameters has predictable consequences for both form and function. For example, selection for later age at maturity is expected to result in increased body size and a descendant resembling a scaled-up version of the ancestor. In a remarkably ambitious synthesis of morphological theory, developmental theory, and life-history theory, Gould (1977) undertook to predict not only the conditions under which certain life-history parameters would be altered in specific ways, but also their resultant impact on form and function. A major assumption of that theory is that organisms are highly integrated systems, an assumption critical for predicting morphology because without that assumption we would have no reason to expect that selection on some parameter such as age at sexual maturity would predictably transform size and shape. Unfortunately, the predictions are often inaccurate. For example, we cannot generally predict the morphology of a precocial mammal (i.e., one that is relatively mature at birth) simply by knowing the ancestral ontogeny of form and the age at which it opens its eyes.

The embedded assumptions about organismal integration may explain why the theory so often fails, and, not surprisingly, many workers have abandoned the extreme view of integration at the heart of Gould's theory. But an equally extreme, if contrary, view has replaced it, one in which every morphological feature is viewed as susceptible to its own independent heterochronic perturbation. Each feature that cannot be subsumed under a hypothesis of a single global heterochrony is interpreted as an independent (dissociated) heterochrony in its own right. In some cases, there are nearly as many heterochronic perturbations inferred from the data as there are morphological features to explain (e.g., McKinney 1986). Many of the dissociations reflect transformations in relative growth rates, which are not simply matters of time. Growth rates have a spatial distribution and what might appear to be dissociations in time might instead be global transformations in spatial patterning. Thus, rather than being dissociated heterochronies, these apparently independent changes may instead be due to a coherent transformation in the spatial distribution of growth rates.

Neither spatial repatterning nor changes in spatial integration are recognized by the theory of heterochrony, so in emphasizing them we are outside the domain of the theory. But acknowledging them can serve two major aims of that theory: (1) it restores the emphasis on the importance of integration in organismal evolution, and (2) it reconnects morphogenesis to life-history theory. Herein we begin by briefly reviewing the general theory of heterochrony to lay the foundation for exploring morphogenetic integration and its evolution. After discussing integration in a purely morphological context, we turn to the subject of life-history traits, arguing that these traits are also integrated. Our point is that both morphogenesis and life history are integrated, and so intimately tied to each other that one cannot be understood without comprehending the other. But we cannot understand either solely in terms of heterochrony because integration is not simply a matter of timing. For that reason, we need to look beyond heterochrony to connect morphogenesis and life history.

## A Brief Overview of the Theory and Analysis of Heterochrony

The core of the theory of heterochrony is that selection on developmental life-history parameters, such as developmental rate or age at sexual maturity, has predictable effects on morphology. Morphology evolves as an indirect effect of evolving life histories; for example, selection for increased developmental rate produces a descendant that develops past the point at which the ancestor ceases. The descendant adult will look like an ancestor whose ontogeny has been extended. This extension yields the pattern termed recapitulation; we can see the ancestral morphology in the descendant's ontogeny. Conversely, selection for a lower developmental rate results in a descendant that fails to complete the ancestral ontogeny, so the adult descendant resembles the subadult ancestor (i.e., it is paedomorphic or "childlike"). By linking fundamental parameters of life-history theory to morphology, Gould provided an explanation for morphology that both denied and emphasized adaptation. It denies adaptation because the morphological novelties are not adaptive in their own right, but it emphasizes the adaptive value of life-history parameters.

These simple but radical ideas stimulated a resurgence of interest in heterochrony, and may have prompted much of the current interest regarding the role of development in evolution. The ideas that morphological evolution can be explained as an indirect effect of selection on life history, and that ecological theory allows us to predict which life-history parameters are likely to evolve under particular conditions, combine to provide a predictive theory of morphological evolution. Moreover, the theory predicts that whole morphologies, not just parts of them, will evolve in a correlated fashion; for example, if an ontogeny is truncated to the point that sexual maturity occurs before metamorphosis, the organism (not just one part) will look larval. And such transformations, even if due to minor quantitative changes in ontogeny, can have dramatic effects on form. On their own, it is likely that these ideas would have stimulated research, but they were not on their own. The seminal paper by Alberch et al. (1979, discussed in more detail below) clarified some ambiguities in Gould's presentation, refined terminology, and provided a formalism for applying the concepts to data. Although the scheme could be used to predict morphological evolution given a hypothesized change in life history, it is typically used to infer the change in life history from morphological data.

The scheme (Fig. 13.1A) comprises three axes, size ($S$), shape ($\sigma$), and age ($A$). The ontogenetic trajectory of the ancestor (A) describes a dynamic morphology, that is, one that changes over age. The trajectory is parameterized by $\alpha$, the age at which development begins, $\beta$, the age at which it ends, $k_\sigma$, developmental rate, and $k_s$, growth rate. There are thus six possible pure perturbations of developmental timings and rates, $\forall\delta\alpha$, $\forall\delta\beta$, $\forall\delta k_\sigma$, as well as two for growth rate, $\forall k_s$. Half of them result in a paedomorphic (childlike) morphology; among these are (1) a decrease in developmental rate ($-\delta k_\sigma$; Fig. 13.1B), (2) a decrease in age at the end of development ($-\delta\beta$; Fig. 13.1C), and (3) an increase in age at onset of development ($+\delta\alpha$; Fig. 13.1D). The other three pure perturbations produce a peramorphic morphology, meaning that the adult descendant looks "hypermature" compared to the adult ancestor.

An important feature of the Alberch et al. analytic scheme, obscured by this discussion of adults, is that morphology is viewed as a continuum of shapes. That continuum is represented by the shape axis $\sigma$. At each point on $\sigma$, from the age at which development begins ($\alpha$) to when it ends ($\beta$), the organism has a shape, $\sigma$. For example, we can represent two stages in skull ontogeny, (1) the youngest age

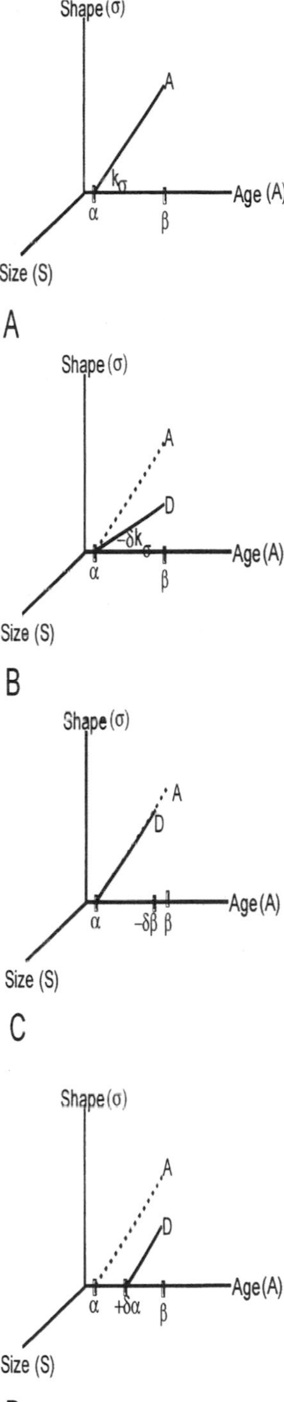


Figure 13.1 The scheme devised by Alberch et al. (1979) for analyzing heterochrony. A. Formalizing the ancestral ontogeny as a trajectory (A) as a three-dimensional vector describing the change in shape (and size) from the point at which development begins, $\alpha$, to when it ends $\beta$, at the rate $k_\sigma$ (for shape) and $k_s$ (for size; ignored throughout this brief summary of the scheme). B. Decrease in developmental rate; the rate of development, $k_\sigma$, is lower for the descendant trajectory (D), a modification formalized by $-\delta k_\sigma$ and termed "neoteny." C. Decrease in the age at which development ends; $\beta$ occurs at a younger age in the descendant, a modification formalized by $\delta\beta$ and termed "progenesis." D. Increase in the age at which development begins; $\alpha$ occurs at an older age in the descendant, a modification formalized by $+\delta\alpha$ and termed "postdisplacement."

at which it is sufficiently ossified to measure (Fig. 13.2A) and (b) sexual maturity (Fig. 13.2B), and visualize the transformation between those two shapes (Fig. 13.2C). Given the formalism, we can predict the impact of any change in $\alpha$, $\beta$ or $k_\sigma$ on shape.

The scheme is far easier to apply in principle than in practice, and it might be regarded more as a heuristic device than an operationally useful formalism. That is partly because of the difficulties of measuring shape, but it is also partly because important assumptions are embedded in the theory and therefore incorporated in the formalism. When data fail to meet those assumptions, the inferences based on the scheme can be actively misleading, misconstruing both the process by which morphogenesis evolves and its relationship to evolving life histories.

One of the most important theoretical claims, the one on which the entire scheme is based, is that development evolves *solely* in rate or timing. If organisms differ in morphogenesis as well as in rate or timing, the scheme is not truly applicable. In the next section we discuss the rationale for that expectation, which comes from a theory of developmental integration, and why that expecta-

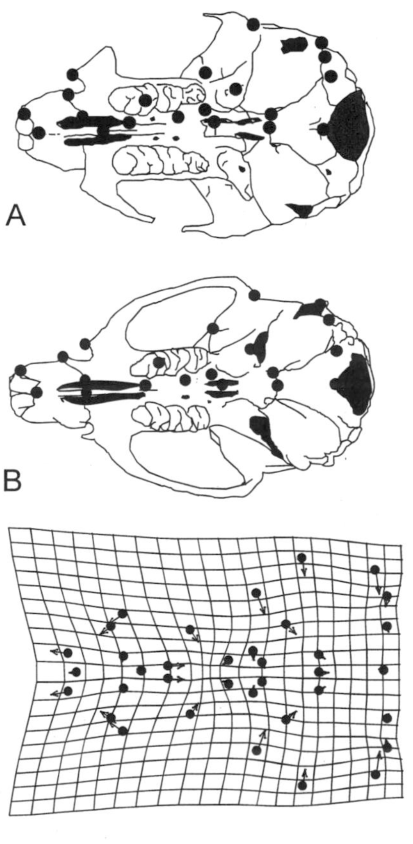


Figure 13.2 The ontogenetic transformation in ventral skull shape, sampled at homologous anatomical loci (landmarks), from (A) the age at which the skull is sufficiently ossified to measure, to (B) sexual maturity. C. The shape transformation shown both by displacements of landmarks relative to others (indicated by vectors at landmarks) and by a deformed grid that interpolates the change between landmarks (the interpolation function is the thin-plate spline). This figure is based on data from the cotton rat *Sigmodon fulviventer*, measured on the day of birth and at the age when the first litter is conceived. The most visually striking changes are the elongation of the skull (especially anteriorly), and the lengthening of the incisive foramen and anterior cranial base relative to the palate.

tion of change solely in rate or timing might be unrealistic even if development is indeed integrated. We then consider how the Alberch et al. scheme can be actively misleading when the data do not meet that important condition.

### Developmental Integration

As formulated by Gould (1977), the theory of heterochrony presumes that organisms are highly integrated systems and, consequently, only three components of development can evolve independently: (1) growth, (2) morphogenesis, and (3) sexual maturation. These are the only lines of dissociation admitted by the theory, which is why the Alberch et al. scheme concentrates on changes in the timing and rate of these components. Even though Gould rejects the Cuvierian vision of perfect integration, he does present heterochrony as a global phenomenon, one that affects both the whole organism and the whole ontogenetic trajectory. An underlying assumption is that morphogenesis evolves as a single integrated unit, dissociable from growth and sexual maturation but not internally dissociable. That emphasis on heterochrony as a global phenomenon makes sense in light of life-history theory because the basic parameters of the theory of heterochrony, that is, developmental rate, growth rate, and age at sexual maturity, are properties of whole organisms in life-history theory. But making sense from that perspective is not enough; it must also make sense at the developmental, mechanistic level. And the developmental argument for global heterochrony is less compelling, largely because it seems to be based on a rather vaguely articulated theory of developmental correlations.

The correlations relevant to the theory of heterochrony are correlations over ontogenetic time. Features that change over time are correlated with each other by virtue of their joint relationship to time—even if they are merely temporally coincident. Those temporal correlations do not imply that any common process affects the whole organism as an integrated unit; just because two features change over the same ontogeny does not mean that they covary, nor does it imply that their covariance has a genetic basis. Thus, to explain global heterochrony in terms of developmental correlations, the correlations must be due to more than just temporal coincidence. We need a rationale for expecting that a whole organism comprises a globally integrated morphogenetic unit.

Gould (1997, p. 395), turning to D'Arcy Thompson, offered an explanation for the expectation of global heterochrony, which is that morphogenesis might be reducible to a few generating parameters: "The extreme atomism of 'bean-bag' genetics might seek an independent efficient cause for each [modification in form], tying their coordinated appearance only to adaptive requirements. But I share D'Arcy Thompson's conviction that complex organic pattern can usually be reduced to fewer and simpler generating factors." But as Gould terms it, this view of development is a "conviction" rather than an empirically well-supported conclusion, and no mechanistic theory suggests or supports it. D'Arcy Thompson (1942) views integration as spatial and geometric—it is a matter of spatial organization and coherence. We can depict hypothetical examples consistent with D'Arcy Thompson's expectations. For example, perhaps the most highly integrated transformation possible is an elongation of the skull at a uniform rate over the entire skull (Fig. 13.3A); another example postulates a moderately graded decrease in growth rates from the anterior to posterior ends of the skull (Fig. 13.3B), and a third conceivable case is a global anteroposterior growth gradient (Fig. 13.3C). Such spatially coherent transformations can be described by one or

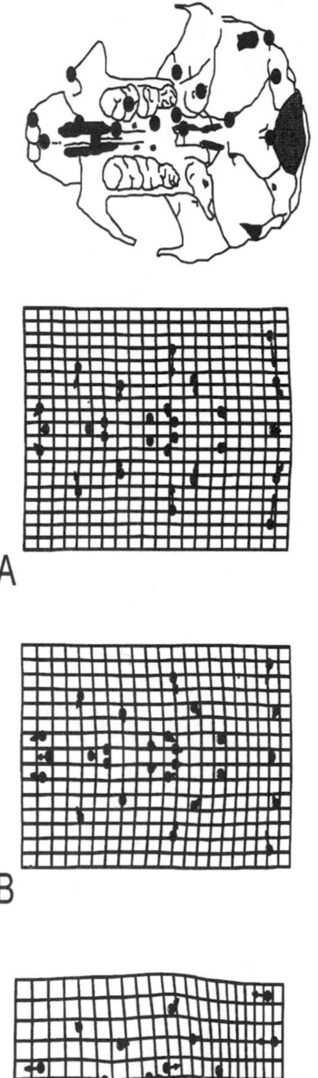


Figure 13.3 Three hypothetical cases of spatially integrated mammalian skull ontogenies. A. Skull growth by uniform elongation. B. Skull growth deviating from uniform in that rates of elongation decrease along the anteroposterior axis. C. Skull growth along an anteroposterior growth gradient.

two parameter(s), meaning that the whole of skull development could be reduced to a single or two generating factors. Importantly, any modification in developmental rate would affect the entire skull, not just a part.

A vision of spatially coherent development may appeal on aesthetic grounds, but there is little empirical evidence in its favor. On the other hand, such lack of evidence is not as strong an argument as it might appear to be since few studies have examined spatial integration of development. One of the few studies that has done so concludes that postnatal skull development of the cotton rat *Sigmodon fulviventer* is indeed integrated, and that both the intensity of integration and its spatial structure, are ontogenetically dynamic (Zelditch et al. 1992). Two ontogenetic stages are shown in Fig. 13.4, the first describing the change from 1 to 10 days postnatal, the second from 10 to 20 days postnatal (the interval in which weaning occurs). Sometimes, as in the second stage (Fig. 13.4B) the change in skull form seems particularly well integrated, if not to the extraordinary degree of the hypothetical cases. The two components figured above, that is, uniform elongation and a global growth gradient, account for only 67% of the total change in skull form over the earliest phase (Fig. 13.4A), but for as much as 82% of that over the next developmental phase (Fig. 13.4B).

Should spatial integration be a general phenomenon, we might anticipate that development ought to evolve in an integrated, coherent fashion. But numerous empirical studies challenge that expectation, or, more precisely, they challenge the hypothesis that *heterochrony* is a global phenomenon. That is because many studies, perhaps even most, conclude that different parts of the organism are affected by different heterochronic perturbations, a pattern termed "dissociated heterochrony" which, according to some workers, is the general rule (e.g., McNamara 1988). Dissociated heterochrony suggests that morphogenesis evolves in a mosaic fashion, which seems to imply that development is weakly integrated; were parts (or processes) integrated, dissociating them would involve more than a simple change in rate or timing, it would also involve uncoupling them (Raff and Wray 1989).

Because dissociated heterochrony seems to be such a common phenomenon, it must be taken seriously. The empirical evidence needs to be critically reviewed and a plausible mechanistic explanation is also required. One explanation for dissociation is that different heterochronic perturbations affect individual growth fields independently (McKinney and McNamara 1991). If this is indeed the explanation for dissociated heterochrony, then many organisms must be highly dissociable because so many dissociations are required to explain the data. But apparent dissociations could also result from methodological artifacts, and from mistaking the predictions of a hypothesis of global heterochrony. In some notable cases, such as that of human neoteny, some of the features interpreted as outcomes of dissociated heterochrony are actually expected outcomes of global neoteny. In particular, we expect an adult mammalian neotene to have a large bulbous braincase, small face, and flexed cranial base, and it is this combination of features that has been taken as evidence that the brain evolves by hypermorphosis (increased developmental time) whereas the face evolves by neoteny (decreased developmental rate, e.g., McKinney and McNamara 1991; for more on the debate, see McKinney 1998; Godfrey and Sutherland 1996; McNamara 2002). It is unlikely

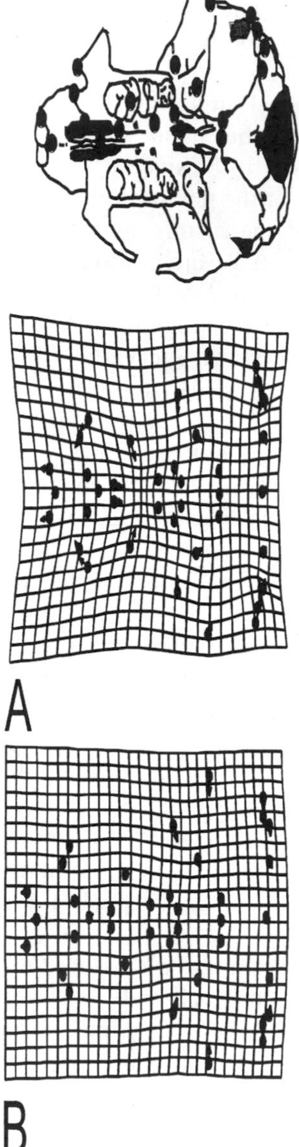


Figure 13.4 Two sequential phases of shape change for the cotton rat *S. fulviventer* that differ significantly in spatial patterning and also in degree of integration. A. From 1 to 10 days postnatal age. B. From 10 to 20 days postnatal age.

that humans do evolve by either kind of global heterochrony (Williams et al. 2002), but the conflict among interpretations results partly from inconsistent views of the predictions. One major distinction is that some workers interpret these concepts in terms of growth rates while others interpret them in terms of developmental rates. As a result, some workers expect neoteny to reduce *growth rates* (both rates of body growth and relative growth rates: see McNamara 2002) whereas others expect reductions in *developmental rates* (Gould 1977). Additionally, analyses can be complicated by dynamics of mammalian skull growth, especially when comparisons are made among shapes at particular stages rather than whole ontogenies (see Alba 2002). However, the difference between interpretations may depend most heavily on analytic methods. When the brain and face are analyzed separately, their changes appear to be dissociated. By replacing the original multivariate view of neoteny by a univariate one, global heterochrony can be misconstrued as dissociated heterochrony. That is because global heterochrony predicts that some allometric coefficients increase when others decrease, but such inconsistent changes are sometimes taken as the primary empirical evidence for dissociation. Viewed in multivariate context, such conflicting changes are exactly what we expect from global heterochrony. In effect, the algebraic dissociation of a complex morphology into individual measurements accounts for the inference of biological dissociation.

An additional and more subtle complication is that global changes in spatial patterning can appear to be dissociated heterochrony even when analyzed multivariately. To show the impact of the analytic methods, especially that global spatial repatterning can indeed look like dissociated heterochrony, we first examine a case of global acceleration (Fig. 13.5), then a more complex evolutionary scenario in which the skull ontogeny is highly integrated spatially through the evolutionary transformations of ontogeny, the case already discussed in the context of our hypothetical examples, namely, from uniform elongation (Fig. 13.3A) through a moderately graded decrease in rates along the anteroposterior skull axis (Fig. 13.3B) to a pure skull-wide gradient (Fig. 13.3C).

### Algebraic Dissociation: Univariate Analysis of Heterochrony

Empirical studies of heterochrony typically begin by applying the Alberch et al. (1979) scheme to the data. In principle, we must plot age on the $x$-axis, shape on the $y$-axis, and size on the $z$-axis. This is easier said than done when the data are as complex as the mammalian skull because there are so many dimensions of form. Not surprisingly, many studies examine these dimensions separately, as criticized below. There is a multivariate alternative because the ontogeny of the most complex form can be represented by a single line, the one describing the covariance between shape and age (the shape axis $\sigma$). The coefficients of that line can be estimated by multivariate regression of shape on age, thus there is no technical barrier to constructing a multivariate shape axis $\sigma$. But doing so constrains the analysis because there is only one shape axis in the Alberch et al. scheme; it is therefore necessary that species share the same ontogenetic transformation of shape. Determining whether they do is straightforward because the similarity between the vectors of multivariate regression coefficients can be assessed by estimating the angle between them; the cosine of that angle is the vector correlation. When two vectors point in the same direction, the angle between them is 0.0 and the vector correlation ($R_v$) is 1.0, as it is in the case of global acceleration shown in Fig. 13.5. Of course, we would not realistically expect to find truly identical ontogenies in empirical cases, not even if we are comparing individuals from the same species. We instead ask whether the observed angle is larger than expected by chance (see Zelditch et al. 2000, 2003, for a resampling-based approach to testing these angles statistically).

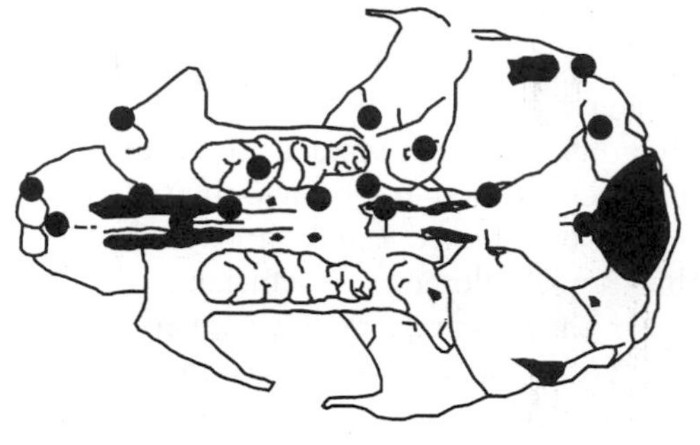

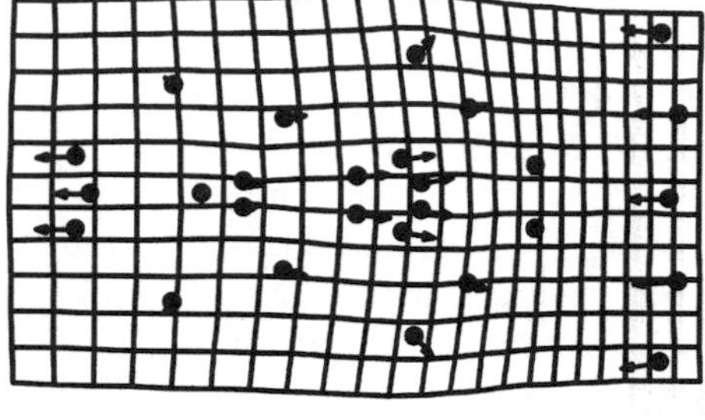

A

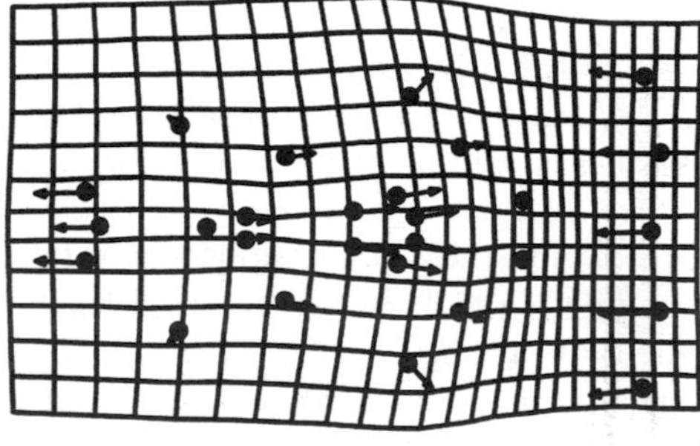

B

Figure 13.5 Ontogeny and phylogeny of skull development for a hypothetical case of accelerated developmental rate ($+\delta k_{\sigma}$). A. Ancestral ontogeny. B. Descendant ontogeny. The direction of evolutionary change in adult skull shape is the same as the direction of the (shared) ontogenetic trajectory.

Once we have shown that species do have the same ontogeny, we can progress to comparing their rates and timings. At least in principle, we could score each specimen for its position along $\sigma$ as a function of age, or we could measuring the morphometric distance between each specimen and that of the youngest age-class which gives an estimate of the amount of change accomplished over time; regressing that distance on age gives a measure of the rate of shape differentiation. As expected, when the rate of differentiation is analyzed in that way for the hypothetical case, the descendant has a higher rate of development (Fig. 13.6).

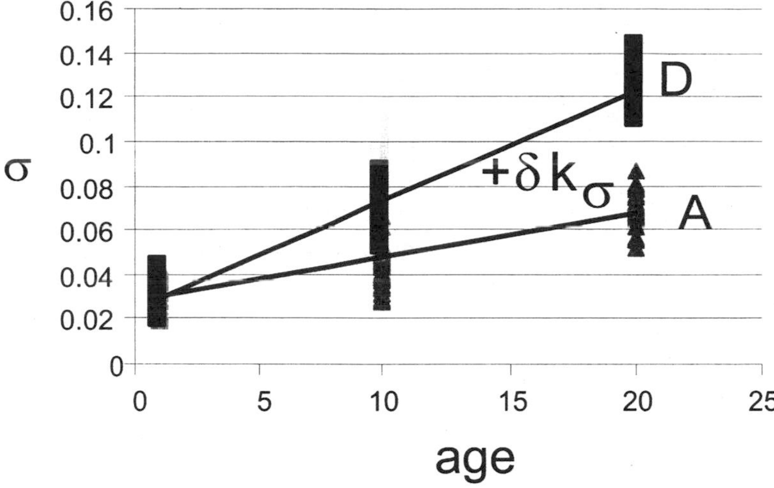


Figure 13.6 A comparison of developmental rates, measured as the rate of differentiation away from the shape of the youngest form relative to age (measured as the Procrustes distance between each shape and the mean of the youngest age-class); the comparison is made between the two ontogenies shown in Fig. 13.5. The rate is higher in the descendant (D) than in the ancestor (A), hence this is a case of acceleration, $+\delta k_{\sigma}$.

Instead of taking a multivariate approach to shape, workers engaged in studies of heterochrony often use simple measures of size, such as length and width. These are then regressed on a measure of body size, producing familiar bivariate allometry plots. But we cannot apply the Alberch et al. scheme to bivariate allometric plots without first translating the predictions, which are written in multivariate, geometric terms. Without that translation, even global acceleration will be misconstrued as dissociated heterochrony. This can be seen by reanalyzing the data of Fig. 13.3 (rescaled for the sake of realism, Fig. 13.7) in the classical terms of bivariate allometry. Doing so requires first measuring the distances between landmarks (to obtain the traditional measurements of length and width), then regressing those size measurements (log-transformed) on a measure of overall size (here skull length, also log-transformed). As is evident in the allometric coefficients plotted on the skull, some ancestral allometric coefficients are lower than those of the descendant, whereas others are higher. In the case of global acceleration, we expect all coefficients to become "less isometric" (because isometry means that no shape change occurs, and increasing the developmental rate has the effect of increasing the amount of shape change that occurs in the given time. Thus, positively allometric coefficients should increase whereas negatively allometric ones should decrease. That pattern of contrasting changes is not enough to document acceleration because allometric coefficients might be altered in the expected direction but not by the expected amounts and, according to the hypothesis of global acceleration, ratios among allometric coefficients are not altered. If they are, the ontogenetic trajectory of shape has evolved, not just its rate or timing.

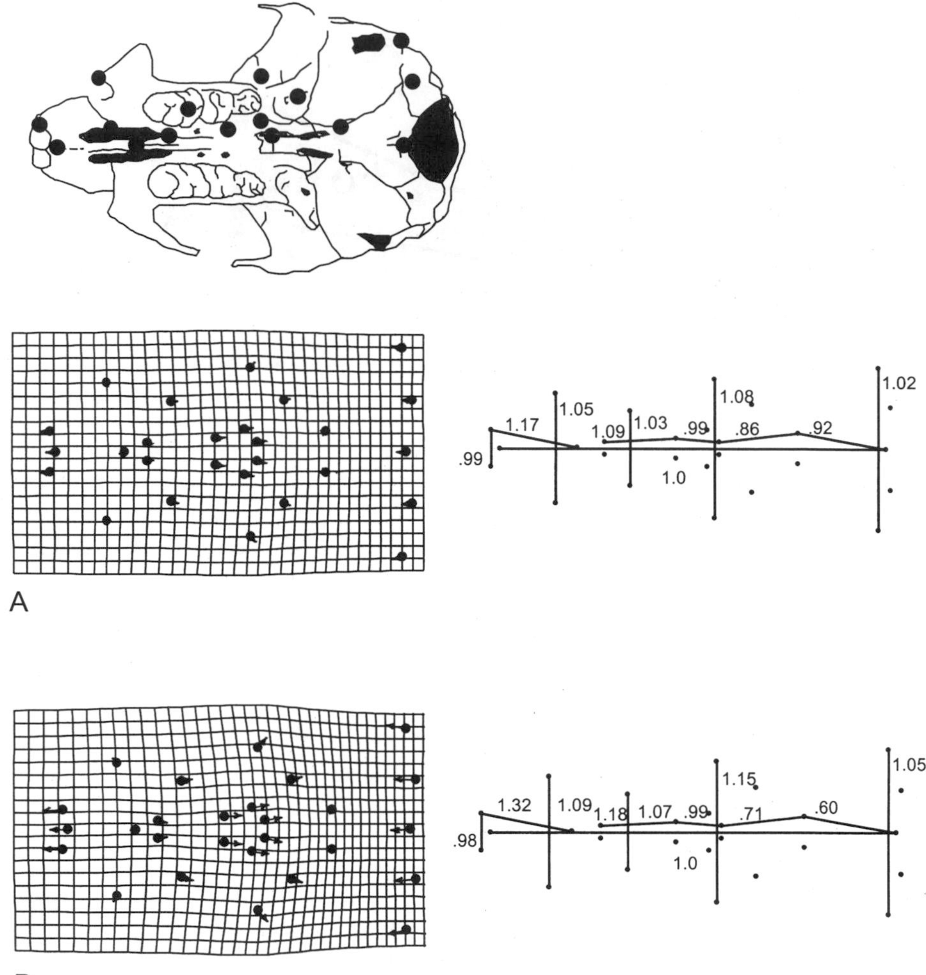


Figure 13.7 Acceleration of developmental rate shown both by a deformation of skull geometric shape and by allometric coefficients of traditional measurements obtained by measuring the distance between landmarks, then regressing the (log-transformed) measures of length on overall skull length (calculations are based on the landmark coordinates obtained by applying the hypothetical model of skull shape ontogeny to the coordinates of the neonatal *S. fulviventer* landmarks). A. Ancestral ontogeny. B. Descendant ontogeny.

If the hypothesis of global acceleration is misinterpreted and read to predict increases in all allometric coefficients, we would infer acceleration of skull width and anterior lengths, but neoteny of posterior lengths. This case of global acceleration would thus be interpreted as dissociated heterochrony. Many workers have addressed this issue, pointing out that the sign of the allometric coefficient is critically important for predicting its change under any hypothesis of heterochrony (e.g., Gould 1966; Shea 1985; Godfrey and Sutherland 1996; Klingenberg

1998). But the univariate view of heterochrony still occasionally intrudes, such as when considering the relative likelihood of different kinds of heterochrony. Because they involve contrasting changes in allometric coefficients, neoteny ($-\delta k_\sigma$) and acceleration ($+\delta k_\sigma$) have been seen as more complex than modifications requiring no such contrasting changes, such as progenesis ($-\delta\beta$) or hypermorphosis ($+\delta\beta$). But biologically, each and every one of these involves merely one dissociation even if it might seem that more are required to produce contrasting changes of allometric coefficients. Whereas progenesis and hypermorphosis dissociate maturation from growth + development, neoteny and acceleration decouple development from growth + maturation. Neoteny and acceleration are no more complex than progenesis and hypermorphosis.

### Implications of Spatial Repatterning for Inferences of Dissociated Heterochrony

Global spatial repatterning alters the ontogenetic trajectory of form, not just its rate or duration. This can be misinterpreted as dissociated heterochrony for more than purely methodological reasons. The other reason is the broadened definition of heterochrony, which expands the meaning of heterochrony to the point that every single change in any parameter of any regression, of any measurement on any other, is interpreted as heterochrony (see, e.g., McKinney and McNamara 1991). In light of this broad definition, changes in spatial patterning can also count as heterochrony because every change in the spatial distribution of growth rates counts as a change in growth rates. If heterochrony is assumed to be the cause of these changes in spatial patterning, global repatterning will be construed as dissociated heterochronies, even though they are neither dissociated nor heterochrony.

It may seem reasonable to view any change in relative growth rates as heterochrony because they are changes in rates. But there are at least three reasons to avoid doing so. First, growth rates are not simply matters of time any more than development is simply a matter of time—growth rates have locations and spatial organization as well. The broadened definition of heterochrony reduces the evolution of development to changes in rate or timing, thereby reducing all of development and all of space to mere time. Second, the theory of heterochrony (which stimulated the interest in this phenomenon) is predicated on the original, restrictive meaning of the concept. And third, the broadened definition is scientifically meaningless because it encompasses all conceivable possibilities. But even if we retain Gould's strict definition, changes in spatial patterning can still be misinterpreted as dissociated heterochronies, even when analyses are done multivariately.

To demonstrate how a globally integrated change in spatial pattern could be misinterpreted as dissociated heterochrony, we use another hypothetical example, a phylogenetic sequence of ontogenetic trajectories (Fig. 13.8). It is obvious by visual inspection that these ontogenies are different, an impression corroborated by statistical analysis: the vector correlations between sequential pairs of age-classes is only 0.64, which is both statistically significantly less than 1.0 and indicative of substantial change. Nevertheless, suppose we omit statistically testing the vectors for similarity and proceed directly to comparing the

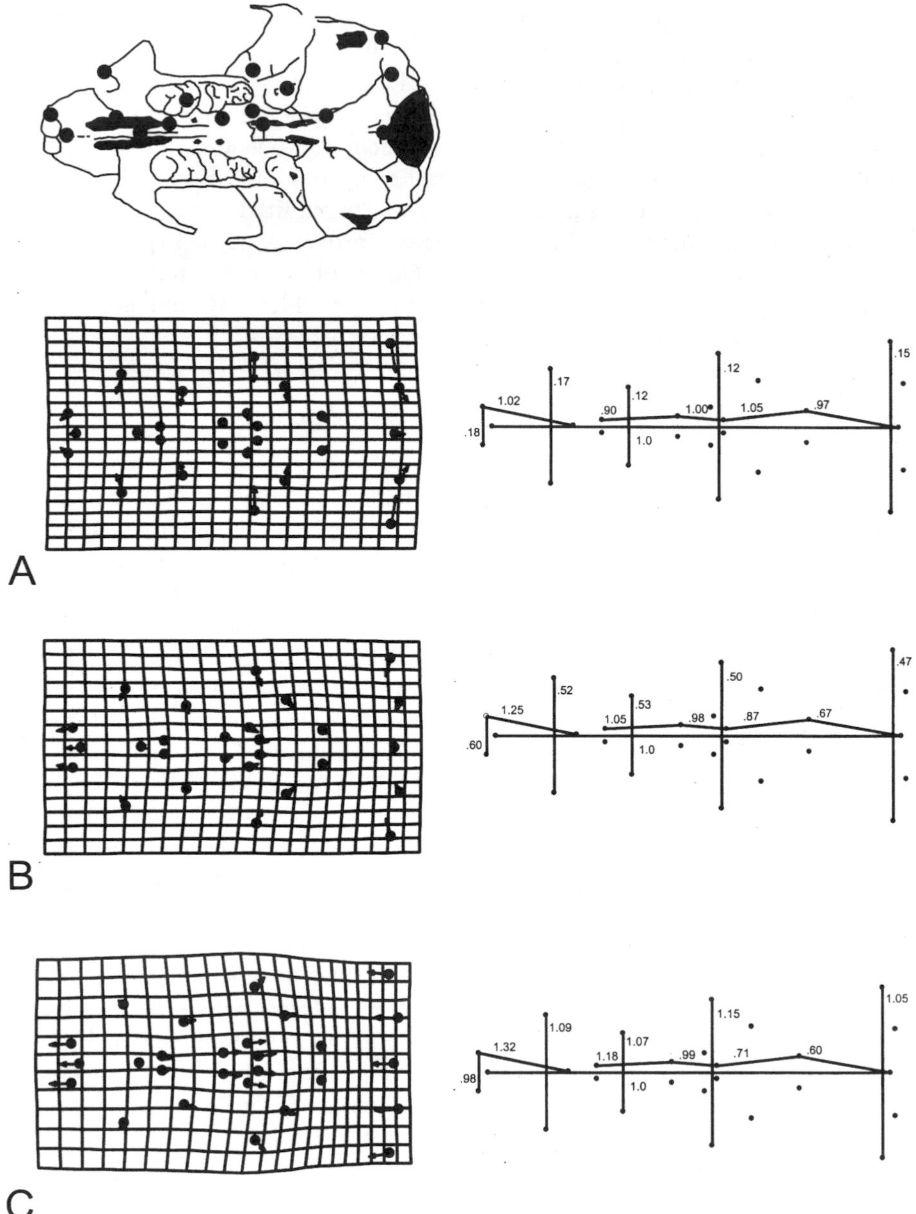


Figure 13.8 An evolutionary sequence of ontogenies (from species A to B to C) by global spatial repatterning, represented both as a geometric deformation of skull shape and by allometric coefficients of traditional measurements, obtained as above. This transformation proceeds from pure elongation of the skull (A) to a departure from uniform elongation due to (B) moderately graded rates that decrease along the anteroposterior skull axis, to (C) a global anteroposterior skull growth gradient.

ontogenies, as is often done. Comparing the ontogeny shown in Fig. 13.8A to that of Fig. 13.8B reveals a change from uniform elongation of the skull to moderately graded growth rates (along the anteroposterior axis). Given the allometric coefficients shown on the skull, we might infer that the first step involves primarily neoteny ($-\delta k_\sigma$) because negatively allometric coefficients are generally increased while the sole positively allometric coefficient is decreased. But two measurements deviate from the expectations of a hypothesis of global neoteny: the most posterior length measurement, which is negatively allometric but decreases, and the isometric coefficient of the most anterior length measurement, which increases. Thus, we might infer a nearly global neoteny plus two dissociated heterochronies (of the most anterior and most posterior regions of the skull). Over the second step, from Fig. 13.8B to Fig. 13.8C, we can see an intensification of the gradient; several negatively allometric coefficients increase, but so do two positively allometric coefficients. One possible (heterochronic) interpretation of these results is that the process of skull widening undergoes neoteny whereas skull lengthening undergoes acceleration.

Thus, simple global change in spatial patterning may seem to imply dissociated heterochrony, even if a multivariate approach is taken to the data. Not surprisingly, a univariate approach also yields an inference of dissociation. Taking the univariate approach to these data, the first evolutionary transition seems to involve an acceleration of skull growth (most coefficients increase) with the exception of the posterior skull length measurements, which evolve by neoteny (those coefficients decrease). The second transition also implies acceleration of most of the skull and neoteny of the posterior skull (except for its most posterior width). Given this scheme, it is not the processes of lengthening and widening that dissociate; rather, it is the anterior and posterior parts of the skull that evolve independently.

The disturbing conclusion is that spatial repatterning, which need not involve any change in an organismal developmental rate, can be interpreted in terms of multiple changes in developmental rate. Whether the data of global spatial repatterning are analyzed multivariately or univariately, they suggest dissociated heterochrony. In the case of our hypothetical scenario, both interpretive schemes inferred a combination of neoteny and acceleration even though none of the species differ in developmental rate. We do not mean to say that global spatial repatterning actually accounts for most empirically detected cases of dissociated heterochrony, nor that most cases of spatial repatterning are global. Some cases of spatial repatterning clearly are not global (Zelditch et al. 2001), but too few studies have examined the scale of spatial repatterning to draw any generalizations from the empirical literature.

At its worst, the tendency to interpret the evolution of ontogeny solely in terms of heterochrony oversimplifies development by reducing it to simply a matter of time, and artificially modularizes it. Furthermore, it can misrepresent the evolution of life history in two important ways. One is that developmental life-history parameters may seem to evolve even when they do not. The second is that it may suggest that there is no such thing as an organismal developmental rate because different parts of an organism have independent developmental rates. To the extent that the concept of developmental rate matters in life history theory, denying its existence has profound theoretical consequences. And by emphasizing modularity over integration, the idea that life history and morphology comprise an integrated unit loses its theoretical significance. We now turn to the integration of life history, emphasizing the importance of developmental rates and timings, and their integration, to show that they matter both to theory and organisms.

## Integration and Life-History Theory

Neither development nor morphology has figured prominently in life-history theory, but integration has. Life-history theory is founded on the premise that time and energy are limited and that resources must be divided up among (1) growth and development, (2) reproduction, and (3) survivorship. Because resources allocated to one activity are unavailable to others, tradeoffs play a major role in the theory (Stearns 1976; Roff 1992). Tradeoffs are one consequence of integration. Thus, integration is at the heart of life-history theory. But integration is even more deeply embedded in the theory because life history parameters are complex dimensions, not simple one-dimensional variables. Moreover, morphology has a pervasive effect on life-history variables, if only because many scale with body size; Millar (1977) goes so far as to say that virtually all reproductive traits in mammals scale with body size except for litter size and age at weaning, which do scale with body size in some groups (e.g., Harvey and Clutton-Brock 1985; Gittleman 1986). In effect, life history theory can be viewed as predicting covariance structures among a number of traits, some morphological, that determine the dimensions of life-history strategies.

We can diagram components of a life-history strategy (Fig. 13.9), which allows us to specify its constituent parts as well as the interactions among components that compete for resources. However, we cannot fully depict all the interactions

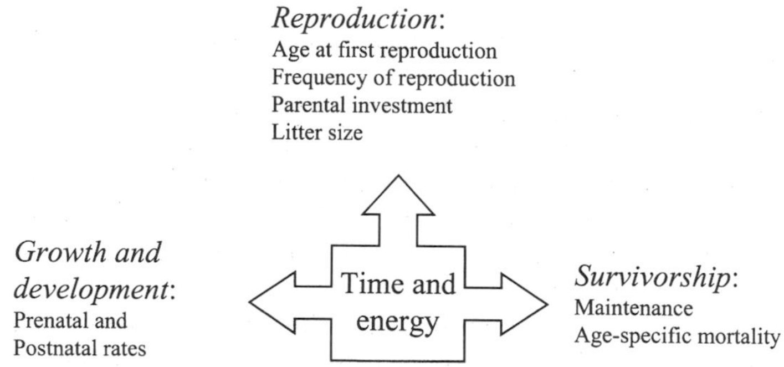


Figure 13.9 The components of life history.

because of their complexity. For example, Case (1978) suggests that species with nearly equal rates of nestling and adult mortality ought to grow at moderate rates, in contrast to those that experience exceptionally high rates of infant mortality relative to adult mortality (which are expected to grow rapidly). Thus, growth rates might depend on the balance between mortality rates at two stages. However, Arendt (1997) points out that infants at a relatively high risk should *develop* rapidly, not necessarily *grow* rapidly. Risks are reduced by being mature rather than just large. In theory, it might be optimal to grow and also to develop rapidly, but the possibility of doing so depends on the relationship between growth and development, specifically, on whether they are antagonistic processes (as suggested by Ricklefs and colleagues: Ricklefs 1979; Ricklefs et al. 1994).

Even these preliminary considerations of theory reveal the importance of integration, both that implied by the idea of tradeoffs and that induced by developmental interactions (such as between growth and development). Growth and development may be among the major dimensions of life histories. Growth has been judged a dominant component of variation in mammals, and developmental timing sometimes emerges as another (Stearns 1983; Swihart 1984). In addition, because time is one of the limiting resources, tradeoffs have a temporal component, as do some of the important parameters of life histories that are temporally structured, such as age-specific mortality rates. Moreover, some features conventionally viewed as life-history traits, such as age at weaning, may depend on morphology and thus on rates of morphogenesis. As a result, interactions among morphogenetic processes may influence life-history schedules.

Because much of life-history theory is concerned with optimal strategies, correlations often enter into the theory as "constraints." From this perspective, constraints are interactions among variables that preclude simultaneously optimizing two or more parameters. From the perspective of developmental theory, one of the most important questions is whether growth rates and developmental timings are mutually constraining or can be jointly optimized. Growth and development have been viewed as antagonistic at the cellular level, an argument that extends the allocation principle to cellular physiology: cells can either divide or differentiate, but not both (Ricklefs 1979; Ricklefs et al. 1994). Assuming that cells are a limited resource, allocating cells to one process would deprive the other, hence high rates of development could depress growth. Whether cells are a limiting resource is unclear because development, particularly at later stages, cannot be easily dichotomized into morphogenesis versus growth. Differential growth can be viewed as morphogenetic because it shapes organismal structures, yet it involves cell proliferation (even if the process includes far more than just cell division). Indeed, some authors view growth and differentiation as positively correlated processes, a relationship evident in the tendency toward allometric scaling (Klingenberg 1998; Shea 2002), and others view them as dissociable (e.g., Gould 1977; Alberch et al. 1979).

Given the importance of body size in mammalian life-history theory, a theory of optimal growth rates is of considerable interest. But growth may be less relevant than developmental timing when it comes to factors affecting age-specific mortality rates. The rate of passage through the vulnerable nestling period is not so much a function of growth rates as it is a function of developmental rates. A large but blind, deaf, and immobile nestling is likely to be at higher risk from predators (and from environmental vicissitudes that cannot be controlled by the mother) than is a small but sighted, hearing, and mobile infant. When we consider that vision, hearing, and mobility are functions dependent on morphology, it is clear that morphology and morphogenetic processes are vital to structuring life histories.

One basic question regarding that structure is whether stages of the life cycle are integrated or can be decoupled. Is there, for example, an optimal age at weaning independent of the optimal age at eye-opening, or is the age at weaning a function of the age at eye-opening? One general way to frame this question is to ask whether organisms have "a" developmental rate, meaning that the rate at which they progress through their life cycle is governed by a single set of parameters, or if instead there are several rates, each of which governs an individual phase and can be independently modified, although these phases could still be modified in a correlated fashion even if governed by different sets of parameters. Because the degree of maturity at birth seems to be of particular importance in mammalian life histories, we could ask whether rates of prenatal development can be modified independently of postnatal rates. Similarly, we can ask whether prenatal growth rates can be modified independently of postnatal rates. And finally, returning to the issue of the potential tradeoffs between growth and development, we can ask whether prenatal growth rates can be modified independently of postnatal developmental rates.

Although life-history studies rarely discuss either morphology or developmental processes (but see Pigliucci and Schlichting 1998; Pigliucci and Marlow 2001), we can think of a life-history schedule as an ontogenetic trajectory of functions, intimately related to form. The temporal correlations among phases along the trajectory, and the functional and developmental integration among form and function at any one phase, may both direct the evolution of life-history strategies. All the theories about the impact of covariance structures on rates and directions of evolution that are devised to explain morphological evolution could apply equally well to evolving life-history strategies (e.g., Wagner 1988).

Given the central role that integration may play in the evolution of life-history strategies, it is surprising that so few studies exploit multivariate methods. Instead, comparisons are based on individual variables or pairwise correlations (Millar 1977; Case 1978; Eisenberg 1981; Harvey and Clutton-Brock 1985; Gittleman 1986). To the extent that multiple variables are analyzed jointly, this is usually done by multiple regression (e.g., Sacher and Staffeldt 1974; Case 1978), a procedure that should not be used when the independent variables are mutually collinear. One of the few multivariate analyses of mammalian life-history strategies came to important, if still somewhat debatable, conclusions, namely, that body size is the dominant dimension of mammalian life histories, and that after this factor is removed, developmental timing emerges as the dominant dimension (Stearns 1983)

These conclusions are suspect because comparisons were made using nested analysis of variance, treating taxa as "factors," to determine the role of phylogeny in structuring life histories. But variance "explained" at the family level is not truly explained because the taxonomic rank "family" is not a biological factor, hence cannot explain variation. Analysis by rank does not provide a useful approach either to discovering or controlling for phylogenetic pattern, especially in the case of rodent life histories, because 234 species of precocial hystricomorphs are subdivided into 17 families whereas over 1200 species (and multiple cases of origins of precociality) are contained within one myomorph, Muridae. Rather than analyses of variance based on rank, we need analyses based on phylogenies.

Multivariate analyses, unlike phylogenetic analyses, cannot reveal the evolutionary dynamics of life-history integration but they do reveal covariances among life-history variables. For example, in an analysis of 55 species of rodents for which we could find data on seven variables: (1) neonatal weight, (2) gestation length, (3) litter size, (4) age at eye-opening, (5) age at eruption of the lower and (6) upper incisors, and (7) age at weaning, as well as adult body size, we find two principal components after removing scaling effects by regression on body size (species are listed in the Appendix, along with sources of the data). The first of these components accounts for 48% of the variance and shows an inverse relationship between developmental timing (age at eye-opening, age at eruption of the two incisors, and, much more weakly, lactation period) and gestation period plus neonatal weight (Table 13.1). Species with high scores on this component have short gestation periods and small infants that reach the tabulated milestones relatively late. Not surprisingly, this component discriminates precocial species, which are born sighted, hearing, furred, and mobile, from altricial species, which are born blind, deaf, naked, and incapable of independent locomotion (Fig. 13.10). The second component, which accounts for 29% of the variance, describes parental investment strategy; it is dominated by three variables, gestation period and lactation period (which are positively related), and litter size, which is negatively related to the other two variables. Thus, this component distinguishes between species that devote large quantities of energetic and timing resources to a few offspring (which have high scores on this component), or fewer resources to many.

Table 13.1 Principal component loadings for life-history variables and variance explained by each component

| Life-history traits | PC1 | PC2 |
|---|---|---|
| Gestation period | −0.630 | 0.701 |
| Litter size | 0.565 | −0.677 |
| Noenatal weight | −0.796 | 0.440 |
| Lower incisors | 0.791 | 0.356 |
| Upper incisors | 0.765 | 0.491 |
| Eye-opening | 0.804 | 0.289 |
| Lactation period | 0.427 | 0.667 |
| % Variance explained | 48.41 | 29.16 |

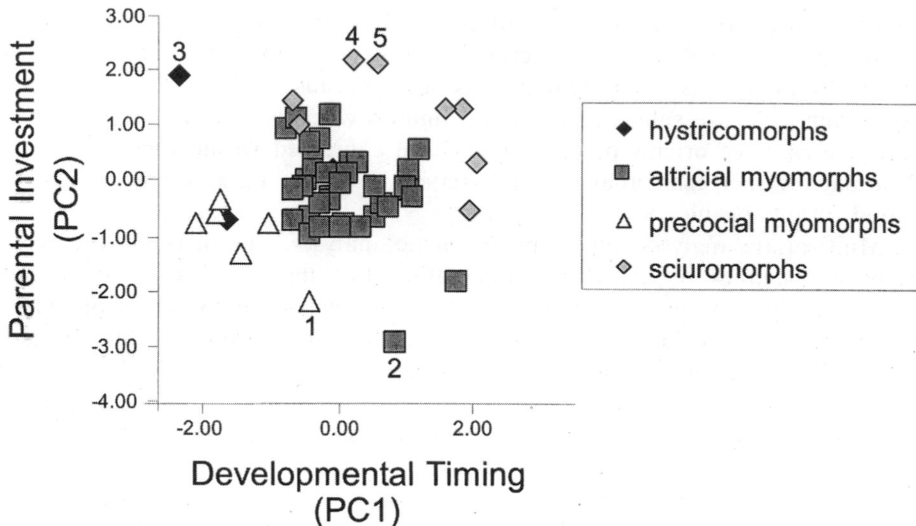


Figure 13.10 The principal components of life-history variation in 55 selected species of rodents (listed in the Appendix); taxa discussed in the text are numbered for ease of reference: (1) *Sigmodon hispidus*, (2) *Mesocricetus auratus*, (3) *Erethizon dorsatum*, (4) *Funisciurus congicus*, and (5) *Paraxerus palliatus ornatus*.

An interesting feature revealed by the PCA (Fig. 13.10) is that altricial and precocial species vary considerably in parental investment strategy. Apparently, having mature neonates is uncorrelated with having many cheap or few costly offspring. For example, the precocial myomorph cotton rat *Sigmodon hispidus*, which has a low score on PC1 for a precocial rodent, also has one of the lowest scores on PC2 (a value exceeded only by that of the altricial hamster *Mesocricetus aureatus*). In contrast, the extremely precocial porcupine *Erethizon dorsatum*, with one of the highest scores on PC1, also has one of the highest scores on PC2 (exceeded only by those of two altricial African squirrels, the arboreal species *Funisciurus congicus* and the terrestrial species *Paraxerus palliatus ornatus*).

To the extent that we can regard this sparse sample as representative of rodents, developmental timing emerges as the dominant component of rodent life-history strategies and parental investment as the second most influential. In one sense, this is not surprising, but some previous studies have questioned whether degree of maturity at birth predicts the timing of life-history events (e.g., Neal 1990). Our results suggest that the timing of some events, particularly age at weaning, may be related both to developmental life history and also to parental investment. Unfortunately, given our lack of information on the timing of later postnatal events, it is not clear whether the timing of early and late events is correlated. As a result, we cannot say whether the whole of ontogeny represents a single integrated component of "developmental timing" or if, instead, there are dissociable phases of the life cycle. Also, in the absence of all morphological data other than body size, we cannot explore the relationship between morphogenesis and life history (which could be examined using partial least squares: Sampson et al. 1989; Rohlf and Corti 2000; see also Bookstein 1982; Joreskog and Wold 1982).

Addressing that last question will prove difficult without measures of developmental rates because one approach to answering it is to fit models to the data and ask whether a single set of rate and timing parameters fits the entire curve. Growth data typically evince such continuity, as shown by numerous studies of mammalian growth (Laird 1965; Zullinger et al. 1984; Fiorello and German 1997). Should those models apply equally well to development, that would suggest (but not prove) that development is also a single continuous process rather than a sequence of individual phases characterized by phase-specific rates.

## Conclusions

Our major theme throughout this chapter, namely, that organisms are developmentally integrated in morphogenesis and life history, is hardly a radical idea. This, after all, is the basic premise underlying the theory of heterochrony formulated by Gould (1977). The connection between life history and heterochrony probably does lie primarily in rates and timings of growth and development, but in pursuing that connection we should not oversimplify development to the point that it is a purely temporal phenomenon. In doing so, we not only miss the important spatial component of integration, we can also misinterpret developmental timing. One of the most striking ironies of the modern literature on heterochrony is the retreat from the view of organisms as integrated systems, the argument so passionately defended by Gould. That perspective has been replaced by the idea that each organismal part (even each individual measurement) has its own ontogenetic trajectory and perhaps even many of them—one for each age. By departing from the theory of heterochrony in recognizing that development may evolve in both its spatial and temporal patterns, we may paradoxically achieve what that theory failed to do: place organismal integration at the heart of the link between morphogenesis and life history.

*Acknowledgments* We thank William L. Fink, Barbara L. Lundrigran, Eladio Marquez, and Donald L Swiderski for their valuable contributions to our understanding of mammalian life histories and morphological integration.

### *Literature Cited*

Alba, D. M. 2002. Shape and stage in heterochronic models. Pp. 28–50 in N. Minugh-Purvis and K. J. McNamara, eds. Human Evolution Through Developmental Change. Johns Hopkins University Press, Baltimore, MD.

Alberch, P., S. J. Gould, G. F. Oster, and D. B. Wake. 1979. Size and shape in ontogeny and phylogeny. Paleobiology 5:296–317.

Arendt, J. D. 1997. Adaptive intrinsic growth rates: an integration across taxa. Quarterly Review of Biology 72:149–177.

Bookstein, F. L. 1982. The geometric meaning of soft modeling, with some generalizations. Pp. 55–74 in Joreskog, K. G. and H. Wold, eds. 1982. Systems under Indirect Observation: Causality–Structure–Prediction, Parts I and II. Elsevier, Amsterdam.

Case, T. J. 1978. On the evolution and adaptive significance of postnatal growth rates in the terrestrial vertebrates. Quarterly Review of Biology 53:243–280.

Eisenberg, J. F. 1981. The Mammalian Radiations. University of Chicago Press, Chicago.

Fiorello, C. V., and R. Z. German. 1997. Heterochrony within species: craniofacial growth in giant, standard and dwarf rabbits. Evolution 51:250–261.

Gittleman, J. L. 1986. Carnivore life history patterns: allometric, phylogenetic, and ecological associations. American Naturalist 127:744–771.

Godfrey, L. R., and M. R. Sutherland. 1996. Paradox of peramorphic paedomorphosis: heterochrony and human evolution. American Journal of Physical Anthropology 99:17–42.

Gould, S. J. 1966. Allometry and size in ontogeny and phylogeny. Biological Reviews 41:587–640.

Gould, S. J. 1977. Ontogeny and Phylogeny. Harvard University Press, Cambridge, MA.

Harvey, P. H., and T. H. Clutton-Brock. 1985. Life history variation in primates. Evolution 39:559–581.

Joreskog, K. G., and H. Wold. 1982. Systems under Indirect Observation: Causality–Structure–Prediction, Parts I and II. Elsevier, Amsterdam.

Klingenberg, C. P. 1998. Heterochrony and allometry: the analysis of evolutionary change in ontogeny. Biological Reviews 73:79–123.

Laird, A. K. 1965. Dynamics of relative growth. Growth 29:249–263.

McKinney, M. L. 1986. Ecological causation of heterochrony: a test and implications for evolutionary theory. Paleobiology 12:282–289.

McKinney, M. L. 1998. The juvenilized ape myth—our "overdeveloped" brain. BioScience 48:109–116.

McKinney, M. L., and K. J. McNamara. 1991. Heterochrony: The Evolution of Ontogeny. Plenum Press, New York.

McNamara, K. J. 1988. The abundance of heterochrony in the fossil record. Pp. 287–325 in M. L. McKinney, ed. Heterochrony in Evolution: A Multidisciplinary Approach. Plenum Press, New York.

McNamara, K. J. 2002. What is heterochrony? Pp. 1–4 in N. Minugh-Purvis and K. J. McNamara, eds. Human Evolution Through Developmental Change. Johns Hopkins University Press, Baltimore, MD.

Millar, J. S. 1977. Adaptive features of mammalian reproduction. Evolution 31:370–386.

Neal, B. R. 1990. Observations on the early post-natal growth and development of *Tatera leucogaster*, *Aethomys chrysophilus* and *A. namaquensis* from Zimbabwe, with a review of the pre- and post-natal growth and development of African murid rodents. Mammalia 54:245–270.

Pigliucci, M., and E. T. Marlow. 2001. Differentiation of flowering time and phenotypic integration in *Arabidopsis thaliana* in response to season length and vernalization. Oecologia 127:501–508.

Pigliucci, M., and C. D. Schlichting. 1998. Reaction norms of *Arabidopsis*. V. Flowering time controls phenotypic architecture in response to nutrient stress. Journal of Evolutionary Biology 11:285–301.

Raff, R. A., and G. A. Wray. 1989. Heterochrony: developmental mechanisms and evolutionary results. Journal of Evolutionary Biology 2:409–434.

Ricklefs, R. E. 1979. Adaptive constraint and compromise in avian postnatal development. Biological Reviews 54:269–290.

Ricklefs, R. E., J. M. Starck, and I. Choi. 1994. Inverse relationship between functional maturity and exponential growth rate of avian skeletal muscle: a constraint on evolutionary response. Evolution 48:1080–1088.

Roff, D. A. 1992. The Evolution of Life Histories: Theory and Analysis. Chapman & Hall, New York.

Rohlf, F. J., and M. Corti. 2000. Use of two-block partial least squares to study covariation in shape. Systematic Biology 49:740–753.

Sacher, G. A., and E. F. Staffeldt. 1974. Relation of gestation time to brain weight for placental mammals: implications for the theory of vertebrate growth. American Naturalist 108:593–615.

Sampson, P. D., A. P. Striessguth, H. M. Barr, and F. L. Bookstein 1989. Neurobehavioral effects of prenatal alcohol. II. Partial least squares analysis. Neurotoxicology and Teratology 11:477–491.

Shea, B. T. 1985. Bivariate and multivariate growth allometry—statistical and biological considerations. Journal of Zoology 206:367–390.

Shea, B. T. 2002. Are some heterochronic transformations likelier than others? Pp. 79–101 in N. Minugh-Purvis and K. J. McNamara, eds. Human Evolution Through Developmental Change. Johns Hopkins University Press, Baltimore, MD.

Stearns, S. C. 1976. Life-history tactics: a review of the ideas. Quarterly Review of Biology 51:3–47.

Stearns, S. C. 1983. The influence of size and phylogeny on patterns of covariation among life-history traits in the mammals. Oikos 41:173–187.

Swihart, R. K. 1984. Body size, breeding season, length, and life history traits of lagomorphs. Oikos 43:282–290.

Thompson, D. A. W. 1942. On Growth and Form. Cambridge University Press, Cambridge.

Wagner, G. P. 1988. The influence of variation and of developmental constraints on the rate of multivariate phenotypic evolution. Journal of Evolutionary Biology 1:45–66.

Williams, F. L., L. Godfrey, and M. R. Sutherland. 2002. Heterochrony and the evolution of Neandertal and modern human craniofacial form. Pp. 405–441 in N. Minugh-Purvis and K. J. McNamara, eds. Human Evolution Through Developmental Change. Johns Hopkins University Press, Baltimore, MD.

Zelditch, M. L., F. L. Bookstein, and B. L. Lundrigan. 1992. Ontogeny of integrated skull growth in the cotton rat *Sigmodon fulviventer*. Evolution 46:1164–1180.

Zelditch, M. L., H. D. Sheets, and W. L. Fink. 2000. Spatiotemporal reorganization of growth rates in the evolution of ontogeny. Evolution 54:1363–1371.

Zelditch, M. L., H. D. Sheets and W. L. Fink. 2001. The spatial complexity and evolutionary dynamics of growth. Pp. 145–194 in M. L. Zelditch, ed. Beyond Heterochrony: The Evolution of Development. John Wiley, New York.

Zelditch, M. L., H. D. Sheets, and W. L. Fink. 2003. The ontogenetic dynamics of disparity. Paleobiology 19:139–156.

Zullinger, E. M., R. R. Ricklefs, K. H. Redford, and G. M. Mace. 1984. Fitting sigmoidal equations to mammalian growth curves. Journal of Mammalogy 65:607–636.

## Appendix: Species Analyzed and Source of Data

| Species | References |
|---|---|
| *Cavia porcellus* | Sacher and Staffeldt (1974), Weir (1974), Ediger (1976), Eisenberg (1981) |
| *Erethizon dorsatum* | Dieterlen (1963), Costello (1966), Sacher and Staffeldt (1974), Eisenberg (1981) |
| *Hydrochoerus hydrochaeris* | Ojasti (1973), Cueto (1999) |
| *Acomys cahirinus* | Dieterlen (1961, 1963) |
| *Aethomys chrysophilus* | Neal (1990), Hayssen et al. (1993) |
| *Aethomys namaquensis* | Neal (1990), Hayssen et al. (1993) |
| *Arvicanthis niloticus* | Delany and Monro (1985) |
| *Bolomys lasiurus* | Mello and Cavalcanti (1982), Hayssen et al. (1993) |
| *Calomys callosus* | Mello (1984), Hayssen et al. (1993) |
| *Calomys hummelincki* | Martino (unpublished) |
| *Dipodomys stephensi* | Lackey (1967), Hayssen et al. (1993) |
| *Holochilus brasiliensis* | Mello (1986), Hayssen et al. (1993) |
| *Holochilus sciureus* | Aguilera (1987) |
| *Mesocricetus aureatus* | Kent (1968), Magalhaes (1968), Sacher and Staffeldt (1974), Eisenberg (1981), Siegel (1985) |
| *Microtus ochrogaster* | Nadeau (1985), Stalling (1990) |
| *Microtus pennsylvanicus* | Nadeau (1985) |
| *Mus minotoides* | Willian and Meester (1978) |
| *Mus musculus* | Grneberg (1952), Sacher and Staffeldt (1974), Eisenberg (1981) |
| *Napaeozapus insignis insignis* | Layne and Hamilton (1954), Hayssen et al. (1993) |
| *Neotoma albigula venusta* | Richardson (1943), Schwartz and Bleich (1975), Hayssen et al. (1993) |
| *Neotoma floridana* | Svihla and Svihla (1933), Pearson (1952), Hamilton (1953), Dieterlen (1963), McClure (1987) |
| *Neotoma lepida intermedia* | Schwartz and Bleich (1975), Hayssen et al. (1993) |
| *Ochrotomys nutalli nutalli* | Linzey and Linzey (1967) |
| *Onychomys longicaudatus* | Horner and Taylor (1968), Taylor (1968) |
| *Oryzomys albigularis* | Moscarella and Aguilera (1999) |
| *Otomys angoniensis* | Phillips et al. (1997) |
| *Otomys irroratus* | Neal (1990), Hayssen et al. (1993), Phillips et al. (1997) |
| *Otomys unisulcatus* | Pillay (2001) |
| *Peromyscus californicus* | Layne (1968) |
| *Peromyscus gossypinus* | Layne (1968), Sacher and Staffeldt (1974), Eisenberg (1981) |
| *Peromyscus leucopus* | Dieterlen (1963), Layne (1968), Sacher and Staffeldt (1974), Eisenberg (1981) |
| *Peromyscus maniculatus bairdii* | Layne 1968), Sacher and Staffeldt (1974), Eisenberg (1981) |
| *Peromyscus maniculatus gambelii* | Layne (1968), Sacher and Staffeldt (1974), Eisenberg (1981) |
| *Peromyscus thomasi* | Layne (1968) |
| *Praomys albipes* | Bekele (1995) |
| *Pseudomys novaehollandiae* | Kemper (1976a, 1976b) |
| *Rattus assimilis* | Taylor (1961) |
| *Rattus lutreolus* | Fox (1979, 1985), Hayssen et al. (1993) |
| *Rattus norvegicus* | Greenman and Duhring (1923), Farris (1949), Schour and Massler (1949), Sacher and Staffeldt (1974), Baker (1979), Eisenberg (1981) |
| *Scotinomys teguina* | Hooper and Carleton (1976) |
| *Scotinomys xerampelinus* | Hooper and Carleton (1976) |
| *Sigmodon hispidus hispidus* | Meyer and Meyer (1944) |

| Species | References |
|---|---|
| *Sigmodon ochrogthus* | Hoffmeister (1963) |
| *Tatera leucogaster* | Neal (1990) |
| *Tylomys nudicaudatus* | Tesh and Cameron (1970) |
| *Zapus hudsonius* | Quimby (1951) |
| *Zygodontomys brevicauda* | Aguilera (1985), Voss et al. (1992) |
| *Ammospermophilus harrisii* | Neal (1965) |
| *Spermophilus tereticaudus neglectus* | Neal (1965) |
| *Funisciurus congicus* | Viljoen and Du Toit (1985) |
| *Paraxerus cepapi cepapi* | Viljoen and Du Toit (1985) |
| *Paraxerus paraxerus ornatus* | Viljoen and Du Toit (1985) |
| *Paraxerus paraxerus ongensis* | Viljoen and Du Toit (1985) |
| *Sciurus vulgaris* | Sacher and Staffeldt (1974), Eisenberg 1981, Viljoen and Du Toit (1985) |
| *Tamiasciurus hudsonicus* | Eisenberg (1981), Viljoen and Du Toit (1985) |

### *Literature Cited in Appendix*

Aguilera, M. 1985. Growth and reproduction in *Zigodontomys microtinus* (Rodentia, Cricetidae) from Venezuela in a laboratory colony. Mammalia 49:75–83.

Aguilera, M. 1987. Ciclo de vida, morfometría craneana y cariología de *Holochilus venezuelae* Allen 1904 (Rodentia, Cricetidae). Trabajo de Ascenso, Universidad Simón Bolívar, Caracas, Venezuela.

Baker, D. E. J. 1979. Reproduction and breeding. Pp. 153–168 in H. J. Baker, J. R. Lindsey, and S. H. Weisbroth, eds. The Laboratory Rat, Vol. 1: Biology and Diseases. Academic Press, Orlando, FL.

Bekele, A. 1995. Post-natal development and reproduction in captive bred *Praomys albipes* (Mammalia: Rodentia) from Ethiopia. Mammalia 59:109–118.

Costello, D. E. 1966. The World of the Porcupine. J. B. Lippincott, Philadelphia.

Cueto, G. R. 1999. Biología reproductiva y crecimiento del carpincho (*Hydrochoerus hydrochaeris*) en cautiverio: una interpretación de las estrategias poblacionales. Tesis Doctoral, Facultad de Ciencias Exactas y Naturales, Universidad de Buenos Aires, Buenos Aires, Argentina.

Delany, M. J., and R. H. Monro. 1985. Growth and development of wild captive Nile rats, *Arvicanthis niloticus* (Rodentia: Muridae). African Journal of Ecology 23:121–131.

Dieterlen, F. 1961. Beiträge zur Biologie der Stachelmaus, *Acomys cahirinus dimidiatus* Cretzschmar. Zeitschrift Säugetierkunde 26:1–13.

Dieterlen, F. 1963. Vergleichende Untersuchungen zur Ontogenese von Stachelmaus (*Acomys*) und Wanderratte (*Rattus norvegicus*). Beiträge zum Nesthocker-Nestflüchter Problem bei Nagetieren. Zeitschrift Säugetierkunde 28:193–227.

Ediger, R. D. 1976. Care and management. Pp. 5–12 in J. E. Wagner and P. J. Manning, eds. The Biology of the Guinea Pig. Academic Press, New York.

Eisenberg, J. F. 1981. The Mammalian Radiations: An Analysis of Trends in Evolution, Adaptation, and Behavior. University of Chicago Press, Chicago.

Farris, E. J. 1949. Breeding of the rat. Pp. 1–18 in E. J. Farris and J. Q. Griffith, eds. The Rat in Laboratory Investigation. J. B. Lippincott, Philadelphia.

Fox, B. J. 1979. Growth and development of *Rattus lutreolus* (Rodentia: Muridae) in the laboratory. Australian Journal of Zoology 27:945–957.

Fox, B. J. 1985. A graphical method for estimating length of gestation and estrous cycle from birth intervals in rondents. Journal of Mammalogy 66:168–173.

Greenman, M. J., and F. L., Duhring. 1923. Breeding and care of the albino rat for research purpose. The Wistar Institute of Anatomy and Biology. Philadelphia.

Grüneberg, H. 1952. The Genetics of the Mouse. Martinus Nijhoff, The Hague.

Hamilton, W. J. 1953. Reproduction and young of the Florida wood rat, *Neotoma floridana floridana* (Ord). Journal of Mammalogy 34:180–189.

Hayssen, V., A. van Tienhoven, and A. van Tienhoven. 1993. Asdell's Patterns of Mammalian Reproduction. A Compendium of Species-Specific Data. Cornell University Press, Ithaca, NY.

Hoffmeister, D. F. 1963. The yellow-nosed cotton rat, *Sigmodon achrognathus*, in Arizona. American Midland Naturalist 70:429–441.

Hooper, E. T., and M. D. Carleton. 1976. Reproduction, growth and development in two contiguously allopatric rodent species, genus *Scotinomys*. Miscellaneous Publications, Museum of Zoology, University of Michigan 151:1–52.

Horner, B. E., and J. M. Taylor. 1968. Growth and reproduction behavior in the southern grasshopper mouse. Journal of Mammalogy 49:644–660.

Kemper, C. M. 1976a. Growth and development of the Australian Murid *Pseudomys novaehollandiae*. Australian Journal of Zoology 24:27–37.

Kemper, C. M. 1976b. Reproduction of *Pseudomys novaehollandiae* (Muridae) in the laboratory. Australian Journal of Zoology 24:159–167.

Kent, G. C., Jr. 1968. Physiology of reproduction. Pp. 119–138 in R. A. Hoffman, P. F. Robinson, and H. Magalhaes, eds. The Golden Hamster: Its Biology and Use in Medical Research. Iowa State University Press, Ames.

Lackey, J. A. 1967. Growth and development of *Dipodomys stephensi*. Journal of Mammalogy 48:624–632.

Layne, J. N. 1968. Ontogeny. Pp. 148–253 in J. A. King, ed. Biology of *Peromyscus* (Rodentia). Special Publication No. 2, American Society of Mammalogists.

Layne, J. N., and W. J. Hamilton, Jr. 1954. The young of the woodland jumping mouse, *Napaeozapus insignis insignis* (Miller). American Midland Naturalist 52:242–247.

Linzey, D. W., and A. V. Linzey. 1967. Growth and development of the golden mouse *Ochrotomys nutalli nutalli*. Journal of Mammalogy 48:445–458.

Magalhaes, H. 1968. Housing, care, and breeding. Pp. 15–23 in R. A. Hoffman, P. F. Robinson, and H. Magalhaes, eds. The Golden Hamster: Its Biology and Use in Medical Research. Iowa State University Press, Ames.

McClure, P. A. 1987. The energetics of reproduction and life histories of cricetine rodents (*Neotoma floridana* and *Sigmodon hispidus*). Pp. 241–258 in A. S. I. Loudon and P. A. Racey, eds. Reproductive Energetics in Mammals. Clarendon Press, Oxford.

Mello, D. A. 1984. *Calomys callosus* Rengger, 1830 (Rodentia, Cricetidae): sua caracterização, ditribuição, criação e manejo de uma cepa em laboratorio. Mem. Inst. Oswaldo Cruz 19:37–44.

Mello, D. A. 1986. Estudios sobre o ciclo biológico de *Holochilus brasiliensis* (Rodentia, Cricetidae) em laboratório. Bol. Mus. Para. Emílio Goeldi 2:181–192.

Mello, D. A., and I. P. Cavalcanti. 1982. Biología de *Zigodontomys lasiurus* (Rodentia, Cricetidae) em condiçôes de laboratório. Brasil Florestal 50:57–64.

Meyer, B. J., and R. K. Meyer. 1944. Growth and reproduction of the cotton rat, *Sigmodon hispidus hispidus*, under laboratory conditions. Journal of Mammalogy 25:107–129.

Moscarella, R. A., and M. Aguilera. 1999. Growth and reproduction of *Oryzomys albigularis* (Rodentia: Sigmodontinae) under laboratory conditions. Mammalia 63:349–362.

Nadeau, J.H. 1985. Ontogeny. Pp. 254–285 in R. H. Tamarin, ed. Biology of New World *Microtus*. Special Publication No. 8, American Society of Mammalogists.

Neal, B. J. 1965. Growth and development of the round-tailed and Harris antelope ground squirrels. American Midland Naturalist 73:479–489.

Neal, B. R. 1990. Observations on the early post-natal growth and development of *Tatera leucogaster*, *Aethomys chrysophilus* and *A. namaquensis* from Zimbabwe, with a review of the pre- and post-natal growth and development of African murid rodents. Mammalia 54:245–270.

Ojasti, J. 1973. Estudio biológico del chigüire o capibara. Fondo Nacional de Investigaciones Agropecuarias, Venezuela.

Pearson, P. G. 1952. Observations concerning the life history and ecology of wood rat, *Neotoma floridana floridana* (Ord). Journal of Mammalogy 33:459–463.

Phillips, J., T. Kearney, N. Pillay, and K. Willan. 1997. Reproduction and postnatal development in the Angoni vlei rat *Otomys angoniensis* (Rodentia, Muridae). Mammalia 61:219–229.

Pillay, N. 2001. Reproduction and postnatal development in the bush Karoo rat *Otomys unisulcatus* (Muridae, Otomynae). Journal of Zoology, London 254:515–520.

Quimby, D. C. 1951. The life history and ecology of the jumping mouse, *Zapus hudsonius*. Ecological Monographs 21:61–95.

Richardson, W. B. 1943. Wood rats (*Neotoma albigula*): their growth and development. Journal of Mammalogy 24:130–143.

Sacher, G. A., and E. F. Staffeldt. 1974. Relation of gestation time to brain weight for placental mammals: implications for the theory of vertebrate growth. American Naturalist 108:593–615.

Schour, I., and M. Massler. 1949. The teeth. Pp. 104–165 in E. J. Farris and J. Q. Griffith, eds. The Rat in Laboratory Investigation. J. B. Lippincott, Philadelphia.

Schwartz, O. A., and V. C. Bleich. 1975. Comparative growth in two species of woodrats, *Neotoma lepida intermedia* and *Neotoma albigula venusta*. Journal of Mammalogy 56:653–666.

Siegel, H. I. 1985. The Hamster: Reproduction and Behavior. Plenum Press, New York.

Stalling, D. T. 1990. *Microtus ochrogaster*. Mammalian Species 355:1–9.

Svihla, A., and R. D. Svihla. 1933. Notes on the life history of the wood rat *Neotoma floridana rubida* Bangs. Journal of Mammalogy 14:73–75.

Taylor, J. M. 1961. Reproductive biology of the Australian bush rat *Rattus assimilis*. University of California Publications in Zoology 60:1–66.

Taylor, J. M. 1968. Reproductive mechanisms of the female southern grasshopper mouse, *Onychomys longicaudatus*. Journal of Mammalogy 49:303–309.

Tesh, R. B., and R. V. Cameron. 1970. Laboratory rearing of the climbing rat, *Tylomys nudicaudatus*. Laboratory Animal Care 20:93–96.

Viljoen, S., and S. H. C. Du Toit. 1985. Postnatal development and growth of southern African tree squirrels in the genera *Funisciurus* and *Paraxerus*. Journal of Mammalogy 66:119–127.

Voss, R. S., P. D. Heideman, V. L. Mayer, and T. M. Donnely. 1992. Husbandry, reproduction and postnatal development of the Neotropical muroid rodent *Zygodontomys brevicauda*. Laboratory Animals 26:38–46.

Weir, B. J. 1974. Reproductive characteristics of hystricomorph rodents. Symposium of the Zoological Society of London 34:265–301.

Willian, K. and J. Meester. 1978. Breeding biology and postnatal development of the African dwarf mouse. Acta Theriologica 23:55–73.

# 14

# Morphological Integration in Primate Evolution

REBECCA ROGERS ACKERMANN
JAMES M. CHEVERUD

Morphological integration plays an important role in directing evolutionary change. As evolutionary morphologists, we are interested in how the relationships among integrated morphological elements constrain or facilitate the evolution of complex phenotypes. Quantitative genetic models and related empirical tests suggest that functional and developmental integration at the level of the individual causes coinheritance of morphological elements, thereby affecting the evolution of heritable variation, and causing coordinate evolution at the population level (Cheverud, 1982, 1984, 1989a, 1995, 1996a, 1996b). Olson and Miller (1958) developed the concept of morphological integration, proposing that traits that interacted in development or function would tend to evolve together. Later theoretical work has supported this intuition. It was proposed that functionally or developmentally interacting traits would come under correlating selection (Sokal 1978; Cheverud 1982, 1984; Lande and Arnold 1983), so that functionally well-integrated phenotypes would be favored by natural selection. Correlating selection is a form of stabilizing selection in which the fitness associated with any given trait depends on the value of functionally interacting traits. Lande (1980a, 1984) showed that correlating selection would cause the evolution of the genetic correlation between traits to approximate the level of correlation specified by correlating selection. Finally, Lande's (1979, 1980b) models of multivariate evolution under genetic drift and natural selection demonstrate that selection on any one trait also causes the evolution of genetically correlated traits. Hence, functional and developmental integration result in correlating selection, which in turn leads to genetic integration and the coevolution of functionally and developmentally related traits.

Because our understanding of evolutionary morphological change primarily comes from the products of evolution, we assess the role of integration in evolutionary change through empirical approaches that interpret patterns of extant covariation in the context of phylogenetic relationships. This approach has been particularly useful for interpreting evolutionary change in primate crania (Cheverud 1982, 1995, 1996a, 1996b; Ackermann and Cheverud 2000, 2002). In this chapter we shall first review comparative morphological integration within Anthropoid primates (e.g., monkeys and apes). This shall include the broad contextualization of these relationships within a phylogenetic framework. We shall then demonstrate two approaches that use morphological integration in extant primates to understand primate evolution. In the first approach, we shall illustrate how the principles of morphological variation can be applied to the task of interpreting the primate fossil record—a record that is scanty by nature and usually does not lend itself to direct analysis of complex integration patterns. In the second approach, we shall demonstrate how morphological integration can be used to test hypotheses of evolutionary diversification.

## Morphological Integration in Primate Crania

Primates traditionally have been divided into two suborders, Prosimii and Anthropoidea. Within Anthropoids—"higher primates"—there are two major groups: the Platyrrhines (New World monkeys) from South and Central America, and the Catarrhines (Old World monkeys and the hominoids, also known as apes and humans) from Africa, Europe, and Asia. Despite considerable disagreement over a number of the finer details of Anthropoid relationships, there is general consensus regarding the basic structure of the phylogenetic tree, shown in Fig. 14.1.

Understanding cranial morphology is particularly important for studying primate evolution, as the suite of morphological characters most often used to delineate evolutionary relationships in primates is found primarily in the skull (Fleagle 1999). Additionally, most of our knowledge of fossil primates is based on cranial remains, and especially teeth, which are extremely hard and therefore plentiful by paleontological standards. But more importantly for studies of morphological integration, the cranium houses a number of developmentally and functionally important regions, especially the brain, special sense organs of sight, smell, and hearing, and the masticatory complex. Much of the diversity in primate cranial morphology is no doubt closely tied to the relative import of these regions as different clades evolve along distinct evolutionary trajectories.

Studies of mammalian cranial development indicate that the skull is composed broadly of two regions distinguished on the basis of growth pattern: the neurocranium and the face (Moore 1981; Cheverud 1982, 1995; Hanken and Hall 1993). Although these regions are connected to each other, and therefore influence each other's size and shape, they also function independently, since they are influenced by the growth of different organs and by the action of distinct hormonal factors. The neurocranium can be further subdivided into three units: the cranial vault, which is closely tied to brain growth; the orbit, which is affected by the growth of

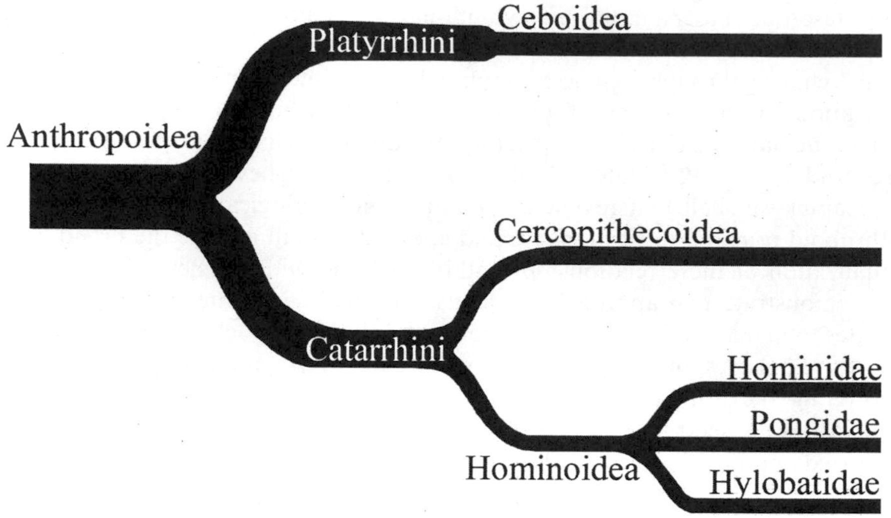


Figure 14.1 The basic structure of the higher primate (suborder Anthropoidea) tree. The highest hierarchical divisions occur (1) within Anthropoids, between the New World monkeys (Platyrrhines) and Old World monkeys and apes (Catarrhines), and (2) within Catarrhines, between the Old World monkeys (Cercopithecoids) and apes (Hominoids).

the eye and its surrounding tissues; and the cranial base, which is tied to brain growth as well as other late-acting somatic factors (Fig. 14.2). The face can be further subdivided into the oral region, associated with growth in the teeth and oral cavity; the nasal region, which responds to the growth of the nasal septum; and the zygomatic region, which is associated with the presence and activity of the muscles of mastication. Using these regional classifications, we can explore the integration within functional/developmental sets, as well as relationships between them, including assessment of total morphological integration.

## New World Monkeys: Infraorder Platyrrhini, Superfamily Ceboidea

The living Anthropoids that inhabit the tropical areas of Central and South America—the Platyrrhines—diverged from the rest of the Anthropoids nearly 30 million years ago. They have been geographically isolated from all other primates since that time, and as a result are uniquely diverse both in behavior and in morphology. Despite a broad range of diversity across this primate taxon, recent analysis of over 5000 New World monkey crania suggests there is a common pattern of morphological integration—resulting from common patterns of skull development—which is shared by all Platyrrhines (Marroig and Cheverud 2001). These Platyrrhine species (see Table 14.1) share a basic pattern of morphological integration across the whole cranium with relatively high correlations among developmentally related traits, including distinct integrated sets of neural and somatic traits, and relatively strong intercorrelation of oral traits, both at the

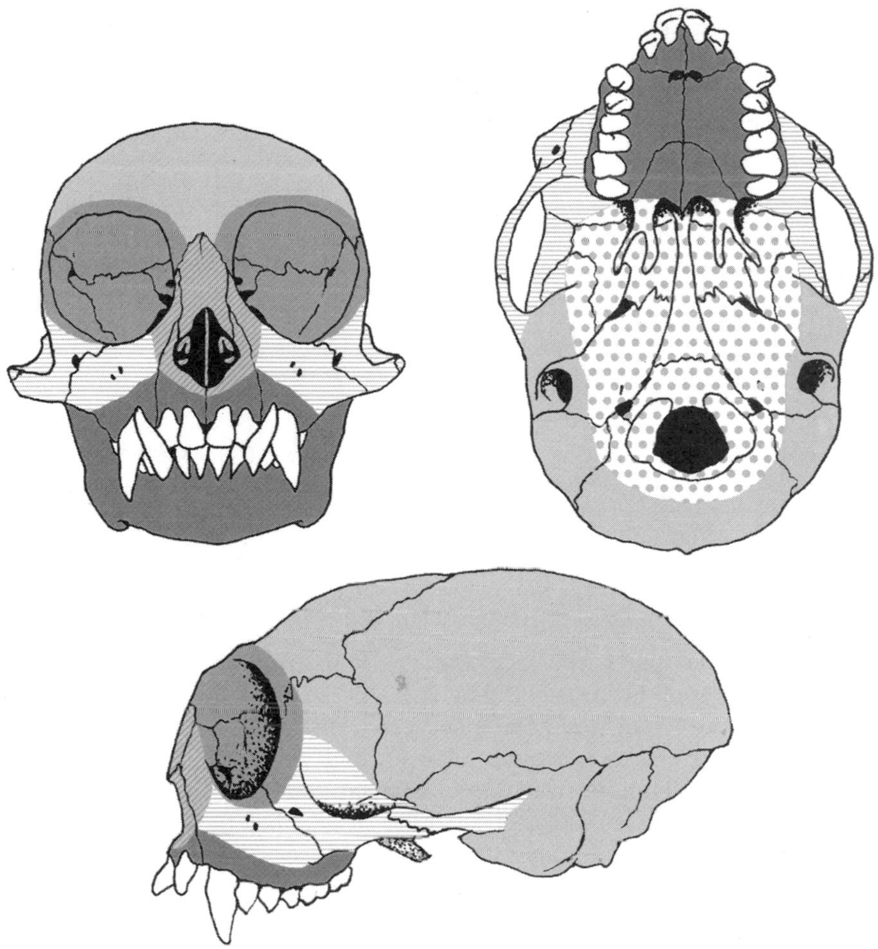

Figure 14.2 The regions of the primate cranium. Light gray = cranial vault; medium gray = orbit; dark gray = oral; light stripes = zygomatic; dark stripes = nasal; dots = cranial base.

species and at the genus level. Across all the genera, correlation among facial traits is generally stronger than among neurocranial traits; however, when significant neural integration exists, it is as strong as average facial integration. Across the taxon, the most highly integrated trait set is the set of oral traits, while the cranial base, nasal region, zygomatic region, and orbital region are not highly integrated.

Interestingly, there is a tight correlation among characters of the face in all South American primates, with one exception. Some callitrichids do not display significant integration in the face, but do in the neural region. Conversely, most taxa that are significantly integrated in the face are not in the neural region. This dichotomy among species displaying neural and facial integration is probably due to the relative magnitude of the correlations. For example, given that the cranium

Table 14.1 The genera included in studies of primate morphological integration, as referred to in the text. More detailed information, including lists of species names, can be found in the referenced sources.

| Superfamily | Genera | References |
|---|---|---|
| Ceboidea | *Ateles* (spider monkey); *Brachyteles* (wooly spider monkey); *Lagothrix* (wooly monkey); *Alouatta* (howler monkey); *Cacajou* (uakarie monkey); *Chiropotes* (bearded saki monkey); *Pithecia* (saki monkey); *Callicebus* (titi monkey); *Cebus* (capuchin monkey); *Saimiri* (squirrel monkey); *Callimico* (Goeldi's monkey); *Callithrix* (marmoset); *Cebuella* (pygmy marmoset); *Leonotopithecus* (lion tamarin); *Saguinus* (tamarin); *Aotus* (night monkey) | Cheverud (1995, 1996b); Marroig and Cheverud (2001); Ackermann and Cheverud (2000, 2002) |
| Cercopithecoidea | *Macaca* (macaque monkey); *Papio* (baboon); *Cercocebus* (mangabey) | Cheverud (1982, 1989b) |
| Hominoidea | *Gorilla* (gorilla); *Pan* (chimpanzee and bonobo); *Homo* (human) | Ackermann (1998, 2002a, 2003) |

is divided into two parts, the face and the neurocranium, if correlations are relatively high among facial traits, they must be relatively low among neurocranial traits and between neurocranial and facial traits. Thus, a finding of relatively high facial intercorrelation precludes relatively high neurocranial intercorrelation and vice versa. In summary, while a neurosomatic integration pattern is shared among all New World monkeys, taxa differ in whether facial or neurocranial traits are (relatively) more highly integrated.

### Old World Monkeys: Infraorder Catarrhini, Superfamily Cercopithecoidea

Does a similar pattern occur in Old World monkeys? Because Old and New World monkeys have evolved separately for nearly 30 million years, it seems quite plausible that their patterns of morphological integration would have diverged. However, this is not what we see. Despite the fact that a long list of cranial traits sets Old World monkeys apart from their New World relatives, earlier studies indicate that papionins (macaques, baboons, and mangabeys; see Table 14.1) have very similar patterns of morphological integration in the cranium; phenotypic correlations are higher among functionally and developmentally related craniofacial traits than among unrelated traits. Analysis of morphological integration in the rhesus macaque cranium using similar traits and developmental classifications found that phenotypic correlations were two to three times higher among functionally and developmentally related traits than among unrelated traits (Cheverud 1982). Once again, integration occurs within the orofacial and neurocranial regions, so that developmentally integrated traits tend to be inherited together, in contrast to developmentally unrelated traits (Cheverud 1982). Similarly, in a study of phenotypic craniofacial integration in a number of papionin species (Cheverud 1989b), oral characters were relatively highly correlated in all seven species.

### Apes and Humans: Infraorder Catarrhini, Superfamily Hominoidea

Since the same general pattern of morphological integration occurs in both New and Old World monkeys, it seems likely that it occurs generally throughout the Anthropoid primates (monkeys, apes, and humans), including our hominid ancestors. Furthermore, because the developmental patterns producing genetic integration in primates are based on physiological phenomena that occur throughout eutherian mammals, it is possible that this pattern of integration is common to the entire Primate Order.

Recent studies comparing living African apes (gorillas, chimps, bonobos) and humans (see Table 14.1) are consistent with this hypothesis (Ackermann 1998, 2002a, 2003). These studies focused on the face, and showed similar patterns of covariation and integration among all four species. The apes and humans fit the model of total morphological integration, with primary contributions from the oral, and to a lesser extent the zygomatic, regions. This varies slightly from what was seen in analyses in the New and Old World monkeys, where the oral region alone was the primary facial contributor to total integration. The results indicate a high degree of connectivity or relationship among those skeletal elements associated most closely with mastication, and suggest that a consistent pattern of total morphological integration within the cranium may exist more broadly among great apes and humans. However, the results also indicate some divergence within this group; all four species were integrated in the oral regions, but only the humans and the gorillas showed significant integration in the zygomatic region. Aside from this one exception, the degree of integration in the oral region follows the pattern displayed by the rest of the Anthropoid primates, suggesting that (1) there is a general pattern of morphological integration in primates, and (2) we may be able to learn about developmental and evolutionary divergence in our hominid ancestors by exploring this pattern in the apes.

## Integration and Evolutionary Change

Together, the results from across the Anthropoid primates might shed light on the question of whether morphological integration evolves. Results point to a general shared pattern of developmentally based morphological integration, with some differences in the details of this integration within some taxa. We know that across eutherian mammals, including primates, the growth of the cranial vault is dependent on the growth of the brain (especially the cerebrum and cerebellum), and occurs early in ontogeny, during the prenatal and neonatal periods in eutherian mammals (including primates). Conversely, most facial growth occurs relatively late in ontogeny, after the brain stops growing. Because the hormonal and genetic factors that underlie early and later mammalian growth differ, early- and late-growing organs generally develop and function independently (Sara et al. 1981;

Atchley et al. 1984; Riska et al. 1984; Riska and Atchley 1985; Cheverud 1996a; Cheverud et al. 1996; Vaughn et al. 1999). For example, the brain and eye complete their growth early, before the influence of the growth hormone, while facial features, especially those influenced by the size of the attaching muscles and the oral cavity, continue to grow under the influence of growth hormone. Therefore, taxa with higher levels of neural integration—such as some Platyrrhines—probably had stronger contributions of early developmental factors to their overall integration, while taxa with higher levels of facial integration had stronger contributions from later-acting developmental factors. These evolutionary changes in facial-neural integration may result from changes in signaling factors (like growth hormone or homeobox genes) that occur during ontogeny (Kim et al. 1998).

It is interesting to consider the lack of strong association among facial characters in *Saguinus*, *Callimico*, and *Aotus*, in relation to some of the unusual developmental and functional adaptations in these genera. All three genera have little or no cranial sexual dimorphism and are monogamous (*Aotus*, and perhaps *Callimico*) or polyandrous (*Saguinus*: Fleagle 1999). A lack of relatively strong intercorrelation among facial measurements could be explained if selection for sexual dimorphism in the face led to the integration of these characters (or in this case, lack of selection equals lack of integration). *Saguinus* and marmoset species typically birth twins from a unicornate uterus, which undoubtedly has extreme implications for prenatal and neonatal growth (see Martin 1992), including the evolution of cooperative group care of neonates and infants. Thus the whole breeding group provides the nurturing environment for neonates in these primates, rather than just the mother as in most mammals. It is possible that the extreme importance of this early developmental period could have increased the degree of neural integration; combined with a lack of facial integration (above) this could result in a unique integration pattern. Additionally, *Aotus* is arguably the most primitive of all Anthropoids, as it is uniquely nocturnal in this taxon and consequently has distinctive cranial morphology associated with night vision; these features could be associated with changes in functional relationships among cranial regions, resulting in a divergent patterns of cranial integration.

Similarly, the divergence of the human and gorilla facial integration patterns from the rest of the Anthropoid primates reflects evolutionary change, potentially linked to specific functional and developmental differences. Both gorillas and humans are integrated in the zygomatic region of the face. Because this is not the general pattern in the rest of the Anthropoid primates, it may be that it evolved separately in the gorilla and human lineages, after the divergence of the gorilla/chimp + human ancestor and the chimp/human ancestor, respectively. It is certainly true that both gorillas and humans have unique facial morphology. In gorillas, certain facial features associated with their extremely herbaceous diet—such as high molar tooth crests and massive chewing muscles—are more robust than in any other ape (Fleagle 1999). The shape of the zygomatic bone is highly correlated with the muscles of mastication, and high levels of zygomatic integration could be the result of selection for the unique gorilla dietary adaptation, and the subsequent functional integration of this region. Changes in the facial integration pattern of gorillas could also result from the evolutionary divergence of developmental patterns in the gorilla lineage. Gorillas grow quickly, reaching extreme sizes and shapes in the same length of time that chimps reach their less extreme morphological sizes and shapes. Gorillas are larger than any other primate and have extreme levels of sexual dimorphism (females weigh 70–90 kg and males up to 200 kg), which is also reflected in overall muscular and associated craniofacial changes. Selection in this lineage for extreme body size and shape could result in the developmental integration of regions of the cranium.

Humans are also facially unique among apes, albeit in different ways. Humans are distinguishable dentally by our relatively small canines (resulting in low levels of sexual dimorphism in the face) and reduced (or absent) third molars. We are also uniquely dependent on tools and fire for our food processing, and therefore rely little on our teeth as tools. Additionally, human cranial development is exceptional among primates, as our brain grows to a relatively larger size than any living primate. Integration in the zygomatic region could reflect a reorganization of facial integration related to any or all of these factors. These possibilities, and particularly their implications for understanding the fossil record of human evolution, are speculative and require further exploration.

## Morphological Integration and the Interpretation of Fossil Remains

As already mentioned, most of our knowledge of fossil primates is based on cranial remains. But the primate fossil record is by nature spotty, and small sample sizes form the greatest barrier to investigating variation directly in fossil primates. Therefore, our interpretation of primate fossils is often grounded in our understanding of variation in extant populations. In such a situation, extant variation is typically used as a "yardstick" of sorts, against which fossil differences are measured. However, because most of the phenotypic variation in cranial traits of interest to paleontologists has complex genetic underpinnings (Rogers et al. 1999), and because these drive the coinheritance of sets of morphological elements (Cheverud 1996a), it is important to incorporate explicitly the concept of morphological integration into studies of fossil morphology. Morphological integration provides a conceptual framework for interpreting primate—and particularly human—evolution that currently exists in theory, if not in practice (see Strait 2001, and references therein). Variation among fossils usually must be judged against a background of patterns of variation and integration found in extant forms because too few fossils exist to give an adequate picture of within-taxon variation. Furthermore, patterns of morphological integration must be taken into account in studies of phylogenetic reconstruction to provide proper weighting of morphological features in cladistic analyses.

However, with few exceptions (Strait 2001; Ackermann 2003a), paleoanthropologists have not explicitly tested hypotheses of morphological integration, and have rarely incorporated the principles of morphological integration directly into their methods. This despite common understanding that variation (and by extension integration) in extant populations is complicated, making extant models imperfect, and that the evolution of trait complexes is an important concern (Skelton et al. 1986; Skelton and McHenry 1992, 1998; Begun 1992;

Strait et al. 1997; Strait and Grine 1998; McCollum 1999; Asfaw et al. 1999; Lovejoy et al. 1999; Strait 2001). A viable explanation for this deficiency might stem from a lack of clarity surrounding the applicability of morphological integration to fossil-bound issues. Here, we shall give two examples drawn from our work that illustrate how morphological integration can be incorporated into paleontological studies, and will endeavor to outline avenues for further research.

## Monkey Analogs and *Catopithecus*

In a study of *Catopithecus browni,* an Eocene anthropoid from North Africa, we used monkey models of craniofacial integration to reconstruct the relative proportions of the skull (Rasmussen et al. 2003). *C. browni* is represented by dozens of dental and jaw specimens, and several crania (Godinot 1994; Simons et al. 1994; Simons and Rasmussen 1996). The six most complete crania all are badly crushed (Simons and Rasmussen 1996). While morphological details of local regions of the skull can be discerned on each of these damaged crania, overall shape and proportions are distorted. However, each skull is crushed in a different plane, and with varying degrees of severity, and as a result each one preserves some unique morphological regions that are relatively undisturbed. This partial preservation of three-dimensional shape in localized areas of the crania provides an opportunity to attempt three-dimensional reconstruction of the entire cranium using patterns of morphological integration from extant species. We did this by combining the intact morphological bits from parts of the crushed skulls with information on morphological integration in small-bodied living Platyrrhines.

Tamarins (*Saguinus oedipus* and *Saguinus fusicollus;* Hershkovitz 1977; Garber and Teaford 1986) were used as models to guide the three-dimensional reconstructions. *Catopithecus* was a small anthropoid, with a cranium comparable to the small-bodied callitrichids (Rasmussen and Simons 1992; Simons and Rasmussen 1996). A series of regression equations developed in these model tamarin species was used to predict the values of the measurements missing in the fossil due to lack of preservation, damage, or distortion, from measurements available in the fossil specimen. With a complete set of these imputed and measured distances it was then possible to produce the three-dimensional coordinates of landmarks used to define the measurements in the fossil form.

What is most important for studies of morphological integration is that the computer reconstructions take as a "model" the intertrait relationships represented in the variance/covariance matrix of skull measurements, not the values of the traits per se in the extant tamarin model species. Because covariation patterns are used to reconstruct missing regions, they should be accurate to the extent that *Catopithecus* craniofacial morphology varied and was integrated in a manner similar to living primates of approximately the same body size. Because of the similarity in integration across primates, this seems a reasonable assumption. The reconstructions based on tamarin morphological integration proved very successful when judged visually, and similar to reconstructions produced by skilled paleontologists (Fig. 14.3).

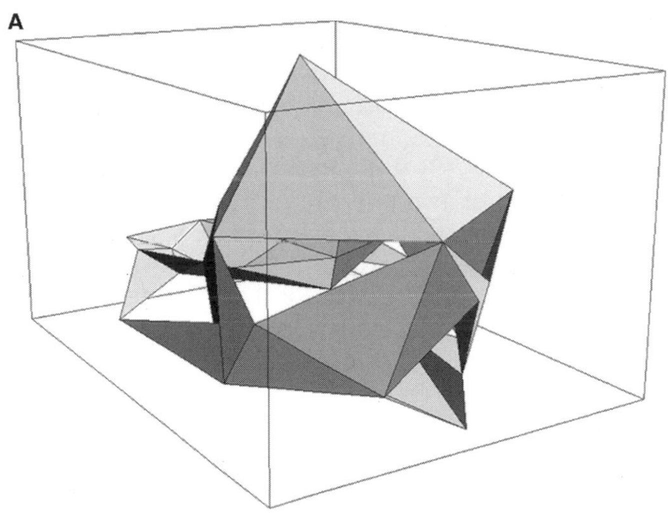


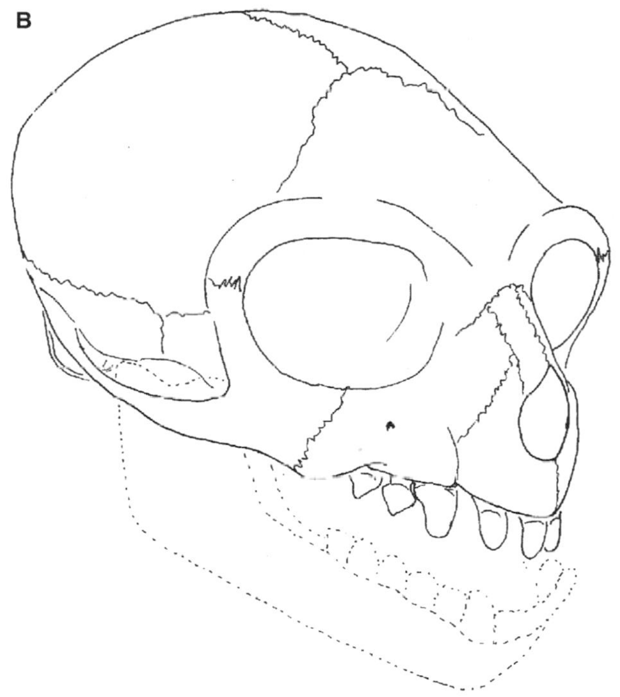


Figure 14.3 Three-dimensional reconstruction of *Catopithecus browni*, based on tamarin (*S. fuscicollis*) models. A. Wire-frame surface rendering, produced using Mathematica. B. Artistic rendering of the same reconstruction. From Rasmussen et al. (2003).

## Ape Analogs and Early *Homo*

The relationships between fossil members of the genus *Homo* from East Africa are uncertain, and discussion revolves around whether variation among the fossil specimens is consistent with separate species rank. The affiliation between the specimens KMN-ER 1470 and KNM-ER 1813—that is, whether they represent one species (*Homo habilis*) or two (*Homo rudolfensis* and *Homo habilis*, respectively)—is a matter of debate, and many of the classic papers splitting or lumping these fossils depend on models of extant variation drawn from living ape and human populations (see Fig. 14.4). KNM-ER 3733 is a third specimen of *Homo* from East Africa, and considered a member of what was traditionally called "African *Homo erectus*" and what some researchers now call *Homo ergaster* (see Collard and Wood 2000; Wood and Richmond 2000).

In a recent study of these three fossils (Ackermann 2002b, 2003b), patterns of morphological integration in living apes and humans served as models for estimating appropriate levels of variability associated with proposed systematic relationships. Again, the assumption that fossils and extant analogs vary in the same way is not a new one, but the incorporation of a complex understanding of variation and integration into methodological approaches is. The models that paleoanthropologists most often use for evaluating fossil hominids are built from great ape and human samples—populations that undoubtedly have different histories and different patterns of dimorphism. Therefore, any similarities and differences perceived among the fossils are influenced by properties inherent in the extant samples, or by artifacts of their sampling.

As with the *Catopithecus* example, what is most important here is that complex patterns of variation from living species stand as explicit surrogates for fossil variation. Fossils were compared using the Mahalanobis distance statistic based on variance/covariance matrices of living species. This calculation is repeated using different variation surrogates drawn from human, chimp, and gorilla populations. Significance is calculated using extant pairs from different species, that is,

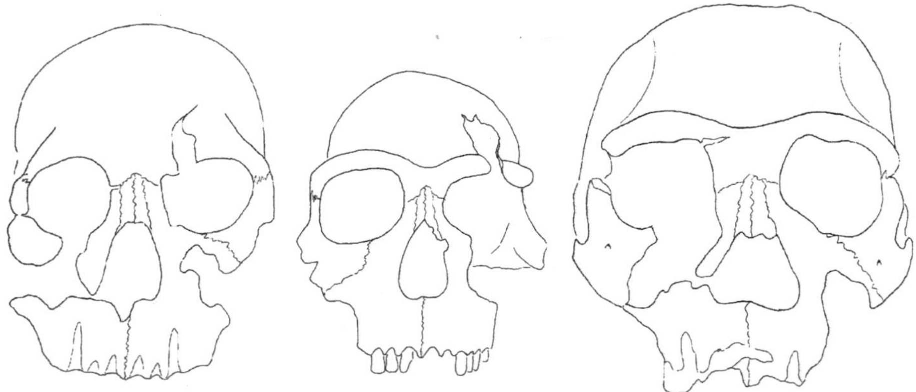

Figure 14.4 Fossil crania of our early *Homo* ancestors. From left to right: KNM-ER 1470 (*H. rudolfensis*?), KNM-ER 1813 (*H. habilis*), KNM-ER 3733 (*H. erectus*).

pairs of chimp skulls or pairs of gorilla skulls compared using patterns of human variation, or human or chimp pairs compared using gorilla variation patterns, and so on, therefore providing a null model against which different-species significance can be judged. This assumes that to a certain extent variation will be different among species, and factors this assumption into the results. The results are surprising, and show a closer relationship between the KNM-ER 1470 and the *H. erectus* specimen than between KNM-ER 1813 and KNM-ER 1470. This indicates that KNM-ER 1470 and KNM-ER 3733 differ from each other along morphological dimensions that tend to be quite variable within species and are more similar along relatively invariant morphological dimensions.

Importantly, building a model that explicitly incorporates the variation and integration of the analog species can give a different interpretation of fossil relationships than traditional methods. Perhaps our complex evolutionary history is not well served by simple models of variation; patterns of covariation and integration can help us to build more complex, nuanced models.

## Further Paleoanthropological Research Questions

These are but two examples of how morphological variation and integration can be used to understand fossils. In addition, there are a number of questions arising from our work on living primates that could shed light on some of the more vexing issues in human evolution, and are promising avenues for future empirical studies. One important avenue for research concerns the role of mosaic morphologies in human evolution. We know that functionally and developmentally related traits are coinherited, while unrelated character complexes are inherited independently of one another. This facilitates the integrated evolution of functionally related characters while permitting the mosaic evolution of unrelated characters (Cheverud 1982; Wagner 1996; Wagner and Altenberg 1996; Magwene 2001).

One conclusion that arose from studies of Platyrrhines—that facial and neurocranial morphology are independently integrated—could explain strange chimeric phenotypes like that seen in *H. habilis sensu lato* (a broad taxon which incorporates fossils split by some into two distinct species, *Homo rudolfensis* and *H. habilis*: see Collard and Wood 2000; Wood and Richmond 2000). This taxon is plagued by a mosaic of dental versus neural versus postcranial morphology that makes taxonomic judgments difficult: *H. rudolfensis* has a large, *Homo*-like braincase and endocranial morphology combined with relatively robust facial and dental features that resemble those found in the robust australopithecines. *H. habilis*, on the other hand, has a smaller braincase, with an essentially *Homo*-like masticatory complex and somewhat australopithecine-like postcranium. It is possible that this mosaic morphology could be caused by the decoupling of separately integrated regions, driven by underlying differences in the relative degree of morphological integration in cranial complexes. Because development of these two regions is also distinct, the differences in integration could be caused by relatively small changes in the timing of developmental processes. That small changes in development could cause substantial evolutionary change is nothing new, but that differences in cranial integration could influence this process is, and might also offer a route for evaluating fossil morphology when direct observation of development is not an option.

But there are many other research questions. The autonomy of integrated modules could allow evolutionary change in some parts of an organism relative to others, and could provide an explanation for unusual morphologies. Could such a scenario explain the evolution of the human brain—a pattern of increased brain relative to reduced face clearly shown in the human fossil record? If integration in the face is stronger than in the brain, is it necessarily more constrained? Could the tendency toward increasing brain size throughout human evolution be due in part to it being the less constrained path? Primates have much larger brain size relative to body size, and extended growth periods compared to other mammals (Martin 1990). It is possible that the modularization of braincase and face in primate evolution could be the consequence of prolonging the early growth period (the time when the brain grows most) to larger brain and body size, resulting in the relatively large primate brain. Perhaps the evolutionary emergence of the relatively huge human brain is tied to the extension of such changes in early growth? These kinds of research directions can facilitate the establishment of a much-needed link between paleontology and developmental biology.

## Using Morphological Integration to Test Hypotheses of Evolutionary Diversification

Evolutionary theory lays a foundation for understanding morphological change that is often grounded in an understanding of patterns of variation/covariation (and hence integration). This is because evolutionary forces rely on intraspecific variation as fuel for population diversification. For example, the expected dispersal of average population phenotypes through random drift over generations is a function of genetic variation/covariation, the effective size of the evolving populations, and the time since divergence (Lande 1979, 1980b; Lofsvold 1988). Similarly, the pattern of differential selection responsible for species differences is a function of the genetic variation/covariation and observed mean differences (Lande 1979). The phenotypic variance/covariance (V/CV) matrix is often proportional to the additive genetic V/CV for morphological traits (Cheverud 1988; Roff 1995, 1996; Koots and Gibson 1996) and may be substituted for it, although this is not universally agreed (Willis et al. 1991). Thus, it is possible to use information on phenotypic morphological integration to test hypotheses of evolutionary diversification.

We recently used this approach to test whether the morphological diversity seen among tamarins could be explained by genetic drift alone (Ackermann and Cheverud 2002). If populations have diversified through random evolutionary processes such as genetic drift, evolutionary theory predicts a proportional relationship between within-group and between-group phenotypic variation (Lande 1979, 1980b; Lofsvold 1988); a nonproportional relationship would indicate nonrandom evolution. This is shown by the equation:

$$\mathbf{B}_t = \mathbf{G}(t/N_e)$$

where $\mathbf{B}_t$ is the dispersion matrix, or between-population variance/covariance matrix, in generation $t$, $\mathbf{G}$ is the additive genetic variance/covariance matrix of the base population from which the group of species is derived, and $N_e$ is the effective population size of the individual taxa (Lande 1979, 1980b; Lofsvold 1988). Substituting the phenotypic within-group V/CV matrix (**W**) for the additive genetic V/CV, we get:

$$\mathbf{B} \propto W(t/N_e)$$

Because $t$ and $N_e$ are constants for any particular comparison, the expected pattern of between-group phenotypic variation should be proportional to the within-group phenotypic variation ($\mathbf{B} \propto \mathbf{W}$), if the populations have diversified by random evolutionary processes. Similarly, if these patterns of variation are not proportional, one can postulate that other modes of evolutionary phenotypic divergence—such as differential selection—were at work.

Patterns of within- and between-taxon variation at various levels in the tamarin hierarchy were compared in order to identify the evolutionary processes involved in the morphological diversification of tamarins. Our analysis indicated that the genus *Saguinus* is too variable in some features for divergence to have occurred through random drift alone. We concluded that a large proportion of this difference was due to divergent size-selection of the two major tamarin clades, although some of it must also have resulted from diversifying selection on nonallometric aspects of cranial shape. Interestingly, our results are consistent with evolutionary hypotheses arguing that selection for smaller size occurred in the small-bodied tamarin clade as part of a specialization for vertical postures (for insect-foraging on large vertical supports: Garber 1992). At lower hierarchical levels, we were able to demonstrate a general pattern of drift, with some evidence for nonrandom divergence, indicating that subgroups within each clade diverged through a combination of selection and drift processes. Our analysis is consistent with biogeographic hypotheses of tamarin dispersal.

In an earlier study, one of us (Cheverud 1996b) applied Lande's (1979) theory—that the pattern of selection responsible for species differences could be reconstructed from observed mean differences—to the task of understanding morphological diversification between saddle-back (*Saguinus fuscicollis*) and cotton-top (*Saguinus oedipus*) tamarins. Lande (1979) suggested that the pattern of selection responsible for species differences could be reconstructed from observed mean differences using the following relationship:

$$\boldsymbol{\beta} = \mathbf{G}^{-1}[z_i - z_j]$$

where $\boldsymbol{\beta}$ is the differential selection gradient summed over the generations (Lande and Arnold 1983), $\mathbf{G}^{-1}$ is the inverse of the pooled within-species genetic V/CV matrix, and $[z_i - z_j]$ is the difference in means between species $i$ and $j$ (Lande 1979; Lofsvold 1988). Again, the phenotypic within-group V/CV matrix is substituted for the genetic V/CV matrix. The results suggested selection for an increased area of attachment for the anterior temporalis muscle in cotton-top tamarins, and associated changes in the anterior neurocranium. The anterior portion of the temporalis muscle is associated with incisive food preparation in primates (Cachel 1979). Relatively increased prognathism (snout length) is also selected for in the cotton-top tamarins, which may serve to widen the gape for preparing relatively large food items. Together, these results suggest that diversifying selection produced increased masticatory efficiency in the anterior mouth in the *S. oedipus* clade.

Both tests of diversification by genetic drift or selection show that knowledge of variation and integration can be applied to the task of understanding morphological evolution in tamarins. Because of the similarity in patterns of variation and integration across primates (as discussed above), it is reasonable to assume that such approaches can be applied more broadly. In fact, using a model-based V/CV matrix, it should be possible to extend these approaches into the fossil record. We could test for the selection required to produce observed differences in fossil skulls, such as those observed during the transition from *Homo habilis* to *Homo erectus* around 1.8 million years ago. For example, when facial morphology is compared in the two *Homo habilis* specimens described above (KNM-ER 1813 and KNM-ER 1470) and one *H. erectus* specimen (KNM-ER 3733), and these three individuals are treated as group means (as *i* and *j* above), and are compared using a human phenotypic V/CV matrix, a similar amount and pattern of selection is necessary to produce the *H. erectus* from the smaller *H. habilis* (KNM-ER 1813) as to produce the bigger *H. habilis* (KNM-ER 1470) from the smaller *H. habilis* (KNM-ER 1813) (Ackermann, unpublished data). This suggests that there may be problems with the taxonomy of this genus as it currently stands. Such approaches may offer viable alternatives to traditional methods for understanding diversity in the fossil record.

## Conclusions

Our survey of the literature showed that there are many common aspects to morphological integration in the primate cranium based on the functional and developmental relationships among cranial features, especially the relatively strong integration within the face, particularly the oral region, and within the neurocranium. Commonality of integration across the order highlights the usefulness of quantitative analyses of patterns of morphological integration within extant species in primate paleontological studies. We have illustrated many aspects of paleontological research that can benefit from a perspective on quantitative evolutionary theory and morphological integration, including the use of morphological integration patterns to do the following: aid in the reconstruction of fossil specimens; aid in judging whether the pattern and extent of morphological differences between fossil specimens is consistent with their taxonomic and phylogenetic placement; determine whether the morphological divergence within a clade is consistent with random evolutionary processes; and measure the differences in selection required to produce observed morphological differences among fossil taxa. Further research into patterns of morphological integration and their evolutionary consequences will be important for the interpretation of the history of these groups.

*Literature Cited*

Ackermann, R. R. 1998. A quantitative assessment of variability in the australopithecine, human, chimpanzee, and gorilla face. Ph.D. dissertation, Washington University, St. Louis.

Ackermann, R. R. 2002a. Patterns of covariation in the hominoid craniofacial skeleton: implications for paleoanthropological models. J. Hum. Evol. 43:167–187.

Ackermann, R. R. 2002b. What can morphological variation tell us about phylogenetic divergence? Am. J. Phys. Anthropol. Suppl. 32:35.

Ackermann, R. R. 2003a. Morphological integration in hominoids: a tool for understanding human evolution. Am. J. Phys. Anthropol. Suppl. 36:55.

Ackermann, R. R. 2003b. Using extant morphological variation to understand fossil relationships: a cautionary tale. S. Afr. J. Sci. 99:255–258.

Ackermann, R. R., and J. M. Cheverud. 2000. Phenotypic covariance structure in tamarins (genus *Saguinus*): a comparison of variation patterns using matrix correlation and common principal component analysis. Am. J. Phys. Anthropol. 111:489–501.

Ackermann, R. R., and J. M. Cheverud. 2002. Discerning evolutionary processes in patterns of tamarin (genus *Saguinus*) craniofacial variation. Am. J. Phys. Anthropol. 117:260–271.

Asfaw, B., T. White, O. Lovejoy, B. Latimer, S. Simpson, and G. Suwa. 1999. *Australopithecus garhi*: a new species of early hominid from Ethiopia. Science 284:629–635.

Atchley, W. R., B. Riska, L. Kohn, A. Plummer, and J. Rutledge. 1984. A quantitative genetic analysis of brain and body size associations, their origin and ontogeny: data from mice. Evolution 38:1165–1179.

Begun, D. R. 1992. Miocene fossil hominoids and the chimp-human clade. Science 257:1927–1932.

Cachel, S. M. 1979. A functional analysis of the primate masticatory system and the origin of the anthropoid post-orbital septum. Am. J. Phys. Anthropol. 50:1–18.

Cheverud, J. M. 1982. Phenotypic, genetic, and environmental morphological integration in the cranium. Evolution 36:499–516.

Cheverud, J. M. 1984. Quantitative genetics and developmental constraints on evolution by selection. J. Theoret. Biol. 110:155–172.

Cheverud, J. M 1988. A comparison of genetic and phenotypic correlations. Evolution 42:958–968.

Cheverud, J. M. 1989a. The evolution of morphological integration. In H. Splechtna and H. Helge (eds.), Trends in Vertebrate Morphology. Proceedings of the 2nd International Symposium of Vertebrate Morphology, Fortschritte der Zoologie 35:196–197. Fisher, New York.

Cheverud, J. M. 1989b. A comparative analysis of morphological variation patterns in the papionins. Evolution 43:1737–1747.

Cheverud, J. M. 1995. Morphological integration in the saddle-back tamarin (*Saguinus fuscicollis*) cranium. Am. Nat. 145:63–89.

Cheverud, J. M. 1996a. Developmental integration and the evolution of pleiotropy. Am. Zool. 36:44–50.

Cheverud, J. M. 1996b. Quantitative genetic analysis of cranial morphology in the cotton-top (*Saguinus oedipus*) and saddle-back (*S. fuscicollis*) tamarins. J. Evol. Biol. 9:5–42.

Cheverud, J., E. Routman, F. M. Duarte, B. van Swinderen, K. Cothran, and C. Perel. 1996. Quantitative trait loci for murine growth. Genetics 142:1305–1319.

Collard, M., and Wood, B. A. 2000. How reliable are human phylogenetic hypotheses? Proc. Natl. Acad. Sci. USA 97:5003–5006.

Fleagle, J. G. 1999. Primate Adaptation and Evolution. Academic Press, San Diego.

Garber, P. A. 1992. Vertical clinging, small body size, and the evolution of feeding adaptations in the callitrichinae. Am. J. Phys. Anthropol. 88:469–482.

Garber, P., and M. Teaford. 1986. Body weights in mixed species troops of *Saguinus mystax mystax* and *Saguinus fuscicollis nigrifrons* in Amazonian Peru. Am. J. Phys. Anthropol. 71:331–336.

Godinot, M. 1994. Early North African primates and their significance for the origin of Simiiformes (= Anthropoidea). Pp. 235–295 in J. G. Fleagle, and R. F. Kay (eds.), Anthropoid Origins. Plenum Press, New York.

Hanken J., and B. K. Hall. 1993. The Skull, Vol. 1: Development. University of Chicago Press, Chicago.

Hershkovitz, P. 1977. Living New World Monkeys (Platyrrhini), with an Introduction to the Primates, Vol 1. University of Chicago Press, Chicago.

Kim, H.-J., D. P. C. Rice, P. J. Kettunen, and I. Thesleff. 1998. FGF, BMP- and Shh-mediated signaling pathways in the regulation of cranial suture morphogenesis and calvarial bone development. Development. 125:1241–1251.

Koots K. R., and J. P. Gibson. 1996. Realized sampling variances of estimates of genetic parameters and the difference between genetic and phenotypic correlations. Genetics 143:1409–1416.

Lande, R. 1979. Quantitative analysis of multivariate evolution, applied to brain:body size allometry. Evolution 33:402–416.

Lande, R. 1980a. The genetic covariance between characters maintained by pleiotropic mutations. Genetics 94:203–215.

Lande R. 1980b. Genetic variation and phenotypic evolution during allopatric speciation. Am. Nat. 116:463–479.

Lande, R. 1984. The genetic correlation between characters maintained by selection, linkage and inbreeding. Genet. Res., Camb. 44: 309–320.

Lande, R., and S. J. Arnold. 1983. The measurement of selection on correlated characters. Evolution 37:1210–1226.

Lofsvold, D. 1988. Quantitative genetics of morphological differentiation in *Peromyscus*. II. Analysis of selection and drift. Evolution 42:54–67.

Lovejoy, C. O., M. J. Cohn, and T. D. White. 1999. Morphological analysis of the mammalian postcranium: a developmental perspective. Proc. Natl. Acad. Sci. USA 96:13247–13252.

Magwene, P. M. 2001. New tools for studying integration and modularity. Evolution 55:1734–1745.

Marroig G., and J. M. Cheverud. 2001. A comparison of phenotypic variation and covariation patterns and the role of phylogeny, ecology and ontogeny during cranial evolution of New World monkeys. Evolution 55:2576–2600.

Martin, R. D. 1990. Primate Origins and Evolution. Princeton University Press, Princeton, NJ.

Martin, R. D. 1992. Goeldi and the dwarfs: the evolutionary biology of the small New World monkeys. J. Hum. Evol. 22:367–393.

McCollum, M. A. 1999. The robust australopithecine face: a morphogenetic perspective. Science 284:301–305.

Moore, W. 1981. The Mammalian Skull. Cambridge University Press, Cambridge.

Olson, E., and R. Miller. 1958. Morphological Integration. University of Chicago Press, Chicago.

Rasmussen, D. T., and E. L. Simons. 1992. Paleobiology of the oligopithecines, the world's earliest anthropoid primates. Int. J. Primatol. 13:477–508.

Rasmussen, D. T., R. R. Ackermann, J. Rehg, J. M. Cheverud, and E. L. Simons. 2003. Visualizing *Catopithecus*: 3-D reconstructions of an early anthropoid. In preparation.

Riska, B., and W. R. Atchley. 1985. Genetics of growth predicts patterns of brain size evolution. Science 229:668–671.

Riska, B., W. R. Atchley, and J. J. Rutledge. 1984. A genetic analysis of targeted growth in mice. Genetics 107:79–101.

Roff, D. A. 1995. The estimation of genetic correlations from phenotypic correlations: a test of Cheverud's conjecture. Heredity 74:481–490.

Roff, D. A. 1996. The evolution of genetic correlations: an analysis of patterns. Evolution 50:1392–1403.

Rogers, J., M. C. Mahaney, L. Almasy, A. G. Comuzzie, and J. Blangero. 1999. Quantitative trait linkage mapping in anthropology. Yrbk. Phys. Anthropol. 42:127–151.

Sara, V., K. Hall, and L. Wetterberg. 1981. Fetal brain growth: a proposed model for regulation by embryonic somatomedin. Pp. 241–252 in M. Ritzen, A. Aperia, K. Hall, A. Larsson, A. Zetterberg, and A. Zetterstrom (eds.), Biology of Normal Human Growth. Raven, New York.

Simons, E. L., and D. T. Rasmussen. 1996. Skull of *Catopithecus browni*, an early Tertiary catarrhine. Am. J. Phys. Anthropol. 100:261–292.

Simons, E. L., D. T. Rasmussen, T. M. Bown, and P. S. Chatrath. 1994. The Eocene origin of anthropoid primates: adaptation, evolution, and diversity. Pp. 179–201 in J. G. Fleagle and R. F. Kay (eds.), Anthropoid Origins. Plenum Press, New York.

Skelton, R. R., and H. M. McHenry. 1992. Evolutionary relationships among early hominids. J. Hum. Evol. 23:309–349.

Skelton, R. R., and H. M. McHenry. 1998. Trait list bias and a reappraisal of early hominid phylogeny. J. Hum. Evol. 34:109–113.

Skelton, R. R., H. M. McHenry, and G. M. Drawhorn. 1986. Phylogenetic analysis of early hominids. Curr. Anthropol. 27:329–340.

Sokal, R. 1978. Population differentiation: something new or more of the same? Pp. 215–239 in P. F. Brussard (ed.), Ecological Genetics: The Interface. Springer-Verlag, Berlin.

Strait, D. S. 2001. Integration, phylogeny, and the hominid cranial base. Am. J. Phys. Anthropol. 114:273–297.

Strait, D. S., and F. E. Grine. 1998. Trait list bias? A reply to Skelton and McHenry. J. Hum. Evol. 34:115–118.

Strait, D. S., F. E. Grine, and M. A. Moniz. 1997. A reappraisal of early hominid phylogeny. J. Hum. Evol. 32:17–82.

Vaughn, T. T., L. S. Pletscher, A. Peripato, K. King-Ellison, E. Adams, C. Erikson, and J. M. Cheverud. 1999. Mapping quantitative trait loci for murine growth: a closer look at genetic architecture. Genet. Res., Camb., 74:313–322.

Wagner, G. P. 1996. Homologues, natural kinds and the evolution of modularity. Am. Zool. 36:36–43.

Wagner, G. P., and L. Altenberg. 1996. Complex adaptations and the evolution of evolvability. Evolution 50:967–976.

Willis, J. H., J. A. Coyne, and M. Kirkpatrick. 1991. Can one predict the evolution of quantitative characters without genetics? Evolution 45:441–444.

Wood, B. A., and B. G. Richmond. 2000. Human evolution: taxonomy and paleobiology. J. Anat. 196:19–60.

## PART V

# THEORY AND ANALYSIS OF PHENOTYPIC INTEGRATION

Part V comprises two distinct, albeit related, issues. On the one hand, there is the practical (operational) problem of what to *do* with data sets concerning phenotypic integration. The statistical challenges posed by complex multivariate data are not trivial, although there is a long tradition of research in this respect (Sneath and Sokal 1973). On the other hand, there is the broader, more conceptual, problem of how to *think* about phenotypic integration, together with the related problem of how to weld such thinking with the more general framework of modern evolutionary theory (Gould 2002). To the extent that we intuitively judge that our theoretical approaches should be related to the way we analyze our data, the two problems are of course related.

Perhaps the biggest question concerning theory and analysis of phenotypic integration is if, and to what extent, we need to alter our classical neo-Darwinian thinking to account for what we are learning about integration. Calls for the radical rewriting of the Modern Synthesis have been made repeatedly throughout the twentieth century (e.g., Goldschmidt 1940; Sultan 1992; Carroll 2000; Johnson and Porter 2001), and equally often dismissed (e.g., Stebbins and Ayala 1981; Mayr 1993). The approach taken in this book is in some sense similar to that advocated by Gould (2002): while the core ideas of Darwinism are still vital and represent part of the fundamental infrastructure of how we understand organic evolution, some major conceptual reassessment and shift in emphasis are increasingly demanded by our more sophisticated understanding of various aspects of phenotypic evolution (Schlichting and Pigliucci 1998). In this book, this middle ground approach emerges particularly clearly in this section on theory and analysis.

The section is structured as a sequence from more analytical to more theoretical chapters, ending with a contribution from philosophy of science, the relevance of which has been emphasized in different chapters of this book. The first chapter of this section, by Scott Steppan (Chapter 15), tackles the difficult problem of not only dealing with complex sets of multivariate data, but of the necessity of doing so within a comparative phylogenetic context. Steppan highlights how exactly multivariate data differ from univariate ones when it comes to phylogenetic analyses, and then examines all the major techniques proposed so far in this field: from element-by-element matrix analyses to matrix correlations, from ordination techniques to the use of indices of integration. It is a tour through the intricacies of the statistics necessary for integration studies, and one that for a long time to come should prove an invaluable reference point for the researcher seriously interested in this field.

Derek Roff (Chapter 16) then shifts the focus to the problems posed by genetic architecture, its evolution, and how we measure it. After discussing what one might mean by "genetic architecture," Roff goes back to the fundamental equations of evolutionary quantitative genetics and to how they can be utilized to describe the long-term evolution of phenotypic integration. Following the theoretical treatment, the reader is presented with actual examples, which nicely bridge the gap with the more empirically oriented sections of this book.

Jason Wolf, Cerisse Allen, and Anthony Frankino (Chapter 17) move further from the analytical toward the theoretical end of the spectrum by exploring the idea of "phenotype landscapes" and how they can help in conceptualizing issues that emerge from research on phenotypic integration. They consider separately the questions of the evolution of phenotypic means and of phenotypic variances, the latter related to the much-discussed idea of canalization (Waddington 1942; Pigliucci and Murren 2003). Finally, they move to a discussion of the evolution of trait covariances on phenotype landscapes.

Kurt Schwenk and Günter Wagner (Chapter 18) tackle another concept that recurs throughout the book, that of constraint. It is a crucial idea, as it relates the mutability and the stability of phenotypes over evolutionary time, arguably among the most pervasive and puzzling characteristics of phenotypic evolution. Schwenk and Wagner put forth the idea that the concept of constraint has theoretical force only if considered in relative terms; hence, an accurate specification of the local context is always necessary to make sense of talk of constraints. They arrive at a definition of constraint as "a mechanism or process that limits the evolutionary response of a character or set of characters to external selection acting during a focal life stage," which immediately highlights the further important distinction between internal and external selection, so often discussed in the literature.

The last contribution to this section (Chapter 19) is by two philosophers of science, Paul Griffiths and Russell Gray, and it deals with developmental systems theory (DST) and how it relates to Darwinism. Essentially, DST proposes that the genetic locus of analysis, while important, is no more causally relevant to our understanding of phenotypic evolution than other phenomena, which are too often relegated to the secondary role of ancillary conditions. Among these are epigenetic interactions, phenotypic plasticity, and even ecological phenomena such as niche construction. DST has been widely discussed by both biologists and philosophers during recent years (Oyama et al. 2001), but it has somehow remained marginal to the mainstream of contemporary evolutionary thought. We think this needs to change, and wish to help bring some of the concepts elaborated by developmental systems theorists to the attention of a wider audience of evolutionary biologists. The goal is to challenge both philosophers and biologists: the former need to show that the concepts they advance can be operationalized by practicing biologists to advance their field of inquiry. On the other hand, biologists can benefit from occasionally stepping back from the nitty-gritty details of their trade in order to question, validate, or perhaps replace, some of the basic assumptions they adopt in order to produce their theoretical and empirical studies.

### *Literature Cited*

Carroll, R. L. 2000. Towards a new evolutionary synthesis. Trends in Ecology and Evolution 15:27–32.

Goldschmidt, R. 1940. The Material Basis of Evolution. Yale University Press, New Haven, CT.

Gould, S. J. 2002. The Structure of Evolutionary Theory. Harvard University Press, Cambridge, MA.

Johnson, N. A., and A. H. Porter. 2001. Toward a new synthesis: population genetics and evolutionary developmental biology. Genetica 112/113:45–58.

Mayr, E. 1993. What was the evolutionary synthesis? Trends in Ecology and Evolution 8:31–33.

Oyama, S., P. E. Griffiths, and R. D. Gray. 2001. Cycles of Contingency: Developmental Systems Theory and Evolution. MIT Press, Cambridge, MA.

Pigliucci, M., and C. Murren. 2003. Genetic assimilation and a possible evolutionary paradox: can macroevolution sometimes be so fast as to pass us by? Evolution 57:1455–1464.

Schlichting, C. D., and M. Pigliucci. 1998. Phenotypic Evolution: A Reaction Norm Perspective. Sinauer, Sunderland, MA.

Sneath, P. H. A., and R. R. Sokal. 1973. Numerical Taxonomy. W.H. Freeman, San Francisco.

Stebbins, G. L., and F. J. Ayala. 1981. Is a new evolutionary synthesis necessary? Science 213:967–971.

Sultan, S. E. 1992. What has survived of Darwin's theory? Phenotypic plasticity and the neo-darwinian legacy. Evolutionary Trends in Plants 6:61–71.

Waddington, C. H. 1942. Canalization of development and the inheritance of acquired characters. Nature 150:563–565.

# 15

# Phylogenetic Comparative Analysis of Multivariate Data

SCOTT J. STEPPAN

## The Problem: Comparative Analysis of Multivariate Data

A central question in fields ranging from quantitative genetics to phenotypic integration and developmental biology is how do patterns of covariation among traits evolve? The fields of inquiry differ in which underlying causes for the covariation they investigate, but they share the challenge of comparing multivariate data sets. In order to understand the evolution of those patterns, as opposed to their maintenance within an organism, it is critical to do so in a phylogenetic context. That is, history must be incorporated into the analysis. A phylogenetic perspective is only recently gaining much application in the pursuit to understand how organisms are built, how their structures are integrated through genetic organization and developmental processes, and how these integrating features both shape and are shaped by evolution. This chapter will discuss the methods available for comparing patterns of integration among taxa and how to conduct those comparisons phylogenetically.

The comparative analysis of multivariate data presents special challenges. In addition to analyzing and visualizing patterns of variation among individuals within a group, one must simultaneously integrate phylogenetic information. Not only must one identify the patterns present within a group (species or population), one must extract the subtle differences among the patterns. These subtle differences will be distributed among groups (taxa) and will create a pattern associated with the phylogenetic history of the taxa. These patterns within patterns must be analyzed and understood simultaneously with the within-group patterns and thus add another dimension to the data that cannot be identified using conventional statistical analyses (see Fig. 15.1, page 334). In addition, phylogenetic comparisons implicitly involve multiple taxa, because just two taxa provide no phylogenetic information. The pattern of differences among taxa thus becomes hierarchical in nature (i.e., reflecting the phylogenetic hierarchy), which extends the inherently hierarchical nature of multivariate data because the developmental program is also hierarchical. These inherent complexities of multivariate data produce the greatest challenges for comparative analyses.

Two sets of issues that must be addressed to analyze multivariate data in a comparative context are how to compare data across taxa and how to incorporate phylogenetic information into those comparisons. In this chapter I shall argue that the phylogenetic approach is critical to identifying shared multivariate features and reconstructing their evolution. I shall then summarize some of the special problems with the comparative analysis of multivariate data, survey the most common set of methods used to compare multivariate data, and highlight approaches that incorporate phylogenetic information. Because most studies of phenotypic integration and related questions summarize multivariate data in matrix form, I shall focus on comparisons between matrices. Additional discussions of these issues and methods can be found in Chernoff and Magwene (1999) and Cheverud et al. (1989).

This chapter looks at multivariate data, whether intended for study of phenotypic integration or comparative quantitative genetics (Steppan et al. 2002), from a phylogenetic perspective. The focus is on understanding how patterns of covariation among traits evolve. This objective requires an independently derived phylogeny, typically estimated from the explosively growing body of DNA sequence data. I shall not discuss the use of morphometric or multivariate data to estimate phylogenies. That is an entirely separate issue with its own set of challenges and controversies (Rohlf 1998; Zelditch et al. 1998).

## Why Phylogeny Is Important

Some form of phylogenetic information is absolutely necessary if we are to reconstruct the evolutionary history of traits. At its simplest, the information could be derived from a pairwise comparison between taxa, but only if the pair consists of ancestor and descendant, wherein the phylogenetic information establishes a temporal polarity. Any other single-pair comparison merely documents a difference, not how that difference evolved. When conducting multitaxon comparisons, additional information is needed to polarize pairwise comparisons or partition the data into a set of phylogenetically independent comparisons (e.g., phylogenetic independent contrasts, Felsenstein 1985). Those partitions are the clades or ancestor-descendant pairs in a phylogeny. The statistical and philosophical justification for the use of phylogenetic information in comparative studies (hereafter referred to as the comparative method) has been discussed in detail elsewhere (Felsenstein 1985; Brooks and McLennan 1991; Harvey and Pagel 1991; Garland and Adolph 1994). Briefly, multitaxon comparisons without a phylogeny inflate the degrees of freedom in statistical tests by treating the values of traits measured for each taxon as independent observations when in fact they are not. Taxa may be similar for a given trait not because they evolved that trait independently but because they inherited it from an ancestor. Statistically significant correlations between traits (e.g., an environmental factor and a putative adaptation or two features hypothesized to share developmental origins) can be due entirely to common ancestry rather than common adaptive or genetic mechanisms (Felsenstein 1985). Multitaxon comparisons without a phylogeny are also unable to determine the temporal sequence of events (Coddington 1988), the appropriate taxon comparisons, or which condition is ancestral and which is derived, and thus cannot determine either the magnitude or the direction of evolution.

Phylogenetic comparisons of DNA sequences use patterns of variation across nucleotides to estimate bias (transformation rates among nucleotides or variation in such rates among nucleotide sites) and identify gene regions under functional constraint or selection. Phylogenetic analysis of phenotypes can uncover the evolutionary lability of traits and of phenotypically integrated features. The remainder of this chapter explores numerous questions that can be addressed effectively only by adopting a phylogenetic perspective within a basic comparative approach, but here I shall mention some of the key questions. How evolutionarily labile are covariance traits? (covariance traits are the elements of phenotypic integration or covariance structure, identifiable as the covariances or sets of covariances among traits). What is the rate of change in covariance traits? And, although it does not necessarily require a phylogeny, which traits are labile and which are conserved?

## How Multivariate Data Differ from Univariate Data in Comparative Studies

Multivariate data are more than just a set of univariate traits; they include the correlations or covariances among the traits. Mathematically, this is the difference between a vector of means and a matrix of variances and covariances. Students of phenotypic integration are interested primarily in the patterns and magnitudes of covariances, not simply the means of traits, but the comparative methods that have been developed to date are designed for univariate data, as will be described below. The most common methods test for an association between evolutionary changes of two traits, specifically in their means or modes. The association tested occurs across time and clades, not among individuals of a taxon. In the phylogenetic comparative method, each trait is typically optimized independently onto a phylogeny (i.e., "mapped"), and then a pairwise comparison or regression is made between the optimizations of two traits. Most optimization involves reconstructing ancestral states so as to minimize some objective function (e.g., number of steps, sum of squares, likelihood) over the entire phylogeny. Optimization is relatively well understood for univariate data (Swofford et al. 1996; Schluter et al. 1997) provided there is no significant directional bias (Cunningham et al. 1998) or if that bias is known a priori. With multivariate data, the variances and covariances of many traits must be simultaneously optimized, and theory is lacking on what model to use for these covariances or how best to do that.

A unique challenge presented by multivariate data is how to simultaneously optimize or reconstruct ancestral matrices when the traits forming them are correlated. Should each correlation and covariance be optimized independently? In addition, the correlations/covariances are themselves correlated (hence the integration that is being studied) and they are correlated to varying degrees. This problem just exacerbates that of accurate ancestral reconstruction (Schluter et al. 1997; Cunningham et al. 1998), where even standard reconstructions of ancestral character means can be prone to a high degree of error. Continuous traits often have such large optimization errors that confidence intervals for the oldest ancestors can exceed the range of variation observed among all extant taxa (Schluter et al. 1997).

Multivariate studies place greater demands on sample sizes than do univariate comparative studies because group parameter estimates (variances and covariances) are reduced to single observations, and because the analyses must be able to detect subtle variations among patterns. The matrices must be well estimated from large sample sizes or the analyses may just be interpreting estimation error as a signal. The greater sampling requirement may be met for some taxa, but comparative studies often include taxa that are uncommon. This problem should be taken into account when designing research projects.

None of the current techniques for the comparative method addresses fully the concerns discussed here and most do not address them at all. In the next section I shall survey available techniques for matrix comparisons and then in the following section I shall discuss applying them in a phylogenetic framework. Although some modifications and future directions will be noted, for the time being we are faced with incomplete analytical solutions to the problem.

## Matrix Comparison Techniques

The array of analytical techniques for comparing matrices can be bewildering. The choice of technique will depend primarily on the question asked and on the nature and structure of the data. For example, element-by-element comparisons are most appropriate when specific correlations or covariances are of interest, but it is excessively cumbersome in searches for patterns of integration. I shall describe several classes of techniques below, highlighting how they can be applied to multiple taxa. Example references and the strengths and weakness of these techniques are briefly summarized elsewhere (Steppan et al. 2002).

### Element-by-Element

The element-by-element approach is the most detailed, as it involves taking each element of a matrix in turn and testing for significant differences. When many elements are being tested, adjustments like the Bonferroni correction (Rice 1989) must be made to the significance level before interpretation. Statistical tests can be as simple as a *t*-test in conjunction with resampling techniques (Brodie 1993; Roff 2000), but for complex situations other tests may be preferred, such as Shaw's restricted maximum likelihood (Shaw 1987, 1991). The element-by-element approach does not take into account the nonindependence of matrix elements (covariances); several may be correlated because of changes in an underlying factor (e.g., pleiotropy). For an extreme example, if two matrices are proportional (as when variance/integration has increased population-wide), all corresponding elements in the two matrices will be significantly different from each other but as a result of only one underlying biological cause. This approach is typically used for taxon pairs, but to extend it to multiple taxa, one could ask whether an element varies significantly across many taxa.

### Matrix Correlation

Matrix correlations are a measure of similarity in pattern that is calculated by testing for correlation between the values of corresponding elements. In effect, the elements are the observations, and the taxa are the variables. Again, elements are considered to be independent of each other, an assumption that is frequently violated. Proportionality cannot be distinguished from matrix equality, and many different changes to a matrix can result in the same correlation. The latter is a greater problem for matrix correlation than most other techniques described here because it reduces all differences between matrices to a single dimension, whereas some other techniques are able to discriminate between different kinds of changes. For some questions, these shortcomings may be unimportant, as when a general measure of similarity is needed. Used in this way, matrix correlation becomes a distance (similarity) metric, but if the patterns that produce changes are of interest, then other methods are preferable (Steppan 1997a). Matrix permutation methods like Mantel's test (Dietz 1983) can test for significant matrix correlation between taxa. The null hypothesis is no correlation, an uncommon condition for morphological traits, such that rejection of the null can be trivial and uninteresting (see Chapter 7, this volume). If the question of interest is whether significant divergence in patterns of covariation (integration) has occurred, then techniques that assume equality as the null (common principal components analysis, maximum likelihood) are preferable. Multiple taxa cannot be compared directly, but comparisons can be structured hierarchically (see Single Pairs section, page 000 below).

### Maximum Likelihood

Maximum-likelihood methods (Shaw 1987, 1991) can be used to test a wide variety of hypotheses about variance-component matrices. Hypotheses of equality can be tested for individual elements, subsets of matrices (see Pleiades section, page 332), and entire matrices. With a well-estimated data set, this method permits statistically precise statements about which parts of the matrix differ and by how much. Although not part of the current implementation of maximum likelihood, the great potential of this method might be in its integration with likelihood estimates for ancestral reconstruction (see below) and perhaps even phylogeny reconstruction (Swofford et al. 1996). Models of the evolution of covariance structure could be tested within a likelihood framework that integrates the likelihood of a given model over the range of possible error in the estimation of covariance elements, uncertainty in ancestral reconstruction, and perhaps even phylogenetic error (Huelsenbeck et al. 2000). This integrative approach with likelihood is computationally demanding but may be within reach in the near future, perhaps by applying a Bayesian statistical framework.

### Ordination

Rather than comparing the many elements in a matrix individually, ordination techniques can be used to transform the data according to some optimality criterion and comparisons can be made among these new variables. An objective is often to reduce the dimensionality of the data so that analysis and interpretation are simplified. Ordination can also be effective at data exploration, and principal components analysis (PCA) is often used for that purpose. Simplification may be the primary objective. The choice of ordination technique should depend on the hypotheses tested. In one example, Voss and colleagues used PCA to identify a size vector so that ontogenies could be compared across populations (Voss et al. 1990), whereas later factor analysis (FA) was used to examine differences in patterns of integration (Voss and Marcus 1992). In PCA or FA, analyses are conducted on each operational taxonomic unit (OTU; whether each OTU is a population or species) separately and then pairwise or other comparisons made between OTU factor loadings. A modification of FA, confirmatory factor analysis (CFA), can be used to compare observed factor structure to a hypothesized factor structure (Zelditch 1987; Zelditch et al. 1990). Although none of these techniques is designed for multiple OTUs, paired comparisons can be structured in groups to test questions in an evolutionary context. An underappreciated concern when applying ordination methods is that the structure in the data being maximized may not match the patterns produced by the biological mechanisms or predicted by the conceptual models being tested. Major causal mechanisms may not result in unrelated, orthogonal vectors of variation (PCA; Houle et al. 2002) or maximized covariance (FA). For example, changes in allometry will likely rotate the orientation of the first principal component in PCA and that may force a rotation in the other eigenvectors even if the localized underlying developmental programs have not changed.

The most notable ordination technique that can incorporate multiple taxa simultaneously is common principal components analysis (CPCA; Flury 1987a, 1987b). In CPCA, a set of matrices is tested for the presence of a common vector or vectors under a PCA model. The number of matrices that can be considered is unlimited. Each level of shared structure is compared to an adjacent level, creating a hierarchical set of tests. The method discriminates among matrices with a wide variety of levels of shared structure, including equality (not significantly different), proportionality (unequal, but the hypothesis of proportional eigenvalues is not rejected), CPC, partial CPC (some but not all eigenvectors in common), and unrelated (no shared structure). There are $N - 2$ levels, where $N$ equals the number of traits. The method has been reviewed repeatedly (Steppan 1997a; Phillips and Arnold 1999; Roff 2000; Steppan et al. 2002) and will not be presented in detail here. The key advantages of CPCA for comparative phenotypic integration studies are its ability to decompose matrix structure into a hierarchical set of structures and its ability to analyze many taxa together under the same model. It can also be applied to developmental data in a variety of contexts (Klingenberg and Zimmermann 1992; Klingenberg et al. 1996). Despite its growing popularity, CPCA also has several deficiencies that should be considered, including the underlying PCA assumption of orthogonality of factors (Houle et al. 2002; Steppan et al. 2002). Additionally, we have a poor understanding of how pleiotropy affects covariance structure, and the detailed and seemingly clear-cut results of CPCA may entice a researcher to be overconfident in the results.

An alternative approach would be to perform ordinations across taxa, where the observations are the matrix elements (variances and covariances) rather than the individual measurements. In this approach, matrices are converted into vectors, creating taxon-by-covariance-element matrices (each column is a taxon, each row is a matrix element) that could themselves be subjected to ordination analyses. Taxa (OTUs) would be grouped by similarity in covariance structure and resultant axes would be determined by the variation in patterns of covariance structure. This method would have three advantages. First, it would allow multi-taxon comparisons. Second, examination of those covariances that are highly correlated with the new factors would identify those suites of covariances that vary most among taxa or help identify correlation pleiades (see below; Berg 1960). By subjecting them to canonical variates analysis or similar techniques, which maximize correlations among sets of both independent and dependent variables, one could identify which suites of covariances are maximally correlated with ecological or behavioral variables and thus test models of adaptive integration. Third, the OTUs could be ordinated or plotted by their similarity in covariance structure ("covariance space"), and OTUs joined graphically according to their phylogenetic relationships. The phylogeny would thus be mapped onto that ordination, in contrast to mapping covariance structure onto the phylogeny, as suggested below.

The taxon-by-covariance-element approach just described is similar in its restructuring of the data to a MANOVA approach advocated by Roff (2002). In Roff's MANOVA method, developed specifically for the difficult statistical problem of comparing genetic variance/covariance matrices, matrix elements become the columns and jackknife pseudoreplicates become the rows (each row is a jackknife run where a single family has been excluded from the analysis). The new matrix follows a MANOVA structure in adding columns that are coded to specify the OTU (all OTUs are combined in a single data set) and any environmental variable of interest. Various experimental designs can be used including nested analyses. The relative contributions to the variances/covariances by the variables of interest (e.g., species, phylogeny, sex, environment) can then be partitioned and the significance level of the effect estimated.

### Integration Indices

Measures of overall integration, rather than the patterns of integration within matrices, can also be employed (Chernoff and Magwene 1999). Examples include the average absolute values of the correlations (Cane 1993) and the variance of the correlation matrix's eigenvalues (Wagner 1984; Cheverud et al. 1989). These are perhaps more difficult to interpret in a phylogenetic context but can be mapped onto a phylogeny as a single trait. This approach can determine whether some clades display more integration then others, but these methods cannot detect homoplasy in integration. That is, higher integration values can be achieved by different mechanisms (e.g., increased canalization of different trait complexes), and thus high or low values of integration may not be homologous. Phylogenetic methods assume homology.

### Distance

Appropriate distance metrics have been little explored. A careful selection of metrics could be used to explore different aspects of integration or covariance structure that may be shared among taxa. The most common metric is disparity (Willis et al. 1991; Steppan 1997b), the summed off-diagonal, element-by-element differences in a covariance matrix. Variants of disparity have been used as well. Podolsky et al. (1997) used a standardized disparity, $GD^2$, in which the differences between covariance elements are squared before summing and then the sum is divided by the number of elements times the mean covariance. This latter step was done to standardize the metric when sets of matrices with different numbers of traits are compared, but it would be unnecessary with matrices of the same form. Roff and colleagues included the diagonal elements (variances) and called this version the T method when it was used with permutation to test for significant disparity (Roff et al. 1999; Begin and Roff 2001). Permutation adds an important statistical test to a general approach usually lacking statistics. Compared to disparity, the T method will be more influenced by changes in population variance than covariance structure, because the absolute values of variances and their differences is often much greater than the covariances.

Matrix correlations could be converted to distances by subtracting them from 1, but these distance measures would obviously still have some of the limitations mentioned above. Other metrics include correlation disparities (i.e., disparity scaled by variances) or other variations (largely unexplored) chosen to emphasize different aspects of matrix structure. The "size" and "shape" of the differences between matrices could be calculated in a manner analogous to removing size in multivariate studies by regressing corresponding matrix elements, and then calculating disparity among the residuals (i.e., disparity corrected for proportional changes in matrices). It might then be informative to subject the derived distances to further analysis, including ordination analyses like multidimensional scaling (MDS, a dimension-reducing method that estimates new dimensions that minimize the distances between OTUs).

### "Pleiades" Approaches

Berg (1960) introduced the concept of correlation pleiades to describe suites of highly correlated traits that are relatively independent of other such suites within an organism. Despite the intuitive appeal of this concept, actually identifying pleiades within a taxon can be difficult since there are few clear criteria for grouping traits either a priori (on the basis of first principles) or a posteriori (based on observed trait correlations). Nevertheless, several different empirical approaches have been developed. Olson and Miller (1958) identified $\rho$-groups as traits bonded by phenotypic correlations greater than some set value or values. They compared the $\rho$-groups to $F$-groups, sets of traits hypothesized to be functionally integrated a priori. Cheverud (1982) extended this framework to comparisons with $G$-sets, analogous to $\rho$-groups but based on genetic correlations. The maximum-likelihood method of Shaw has been extended to determine whether taxa share subsets of a matrix, where subsets are analogous to pleiades. This restricted maximum-likelihood model (REML) has been applied mostly to quantitative genetic parameters (Shaw 1987; 1991; Service 2000). Unlike CPCA, it has not yet been extended to multiple taxa, but it could be quite powerful. Other approaches are discussed in other chapters of this book and elsewhere (Cheverud et al. 1989; Chernoff and Magwene 1999; Magwene 2001). These include common space analysis (Flury 1987b, 1988) and dimensionality (Kirkpatrick et al. 1990). (See Steppan et al. 2002 for a brief review.)

## Phylogenetic Approaches to Multivariate Data

Once the patterns of phenotypic integration have been characterized for a set of OTUs, the focal question becomes how to compare these patterns within a phylogenetic context. Extrapolating studies of phenotypic integration beyond single OTUs to phylogenetic comparisons adds three demands. First, multiple comparisons are required, and second, to be phylogenetic, either those comparisons must be structured according to the phylogeny or special phylogenetic analyses must be conducted. As stated before, the latter are noticeably lacking for multivariate data, so much of this discussion will examine means of adding phylogenetic structure. Third, because the objective is to extract subtle differences from the within-taxon covariance structure, precise estimation of the individual matrices is even more important than is the case for noncomparative studies. I shall not discuss this third demand here but raise it as an underappreciated consequence of comparative studies.

Comparative methods can be subdivided into those for discrete data and those for continuous data (Harvey and Pagel 1991). Discrete data, such as presence/absence (Brooks and McLennan 1991), can be analyzed by character mapping (optimizing the character state transformations onto a phylogeny under a given optimality criterion) or sister-group comparisons. In the latter, sets of sister groups, where one member of each pair possesses a key trait and the other lacks it, are surveyed for conformity to a hypothesized pattern. The number of observations is equal to the number of sister-group pairs. There is a wider variety of methods for continuous data and this type of data is generally of more interest in the field of integration. Continuous methods include minimum evolution, phylogenetic autocorrelation, and independent contrasts and its variants. I shall discuss the application of these methods in more detail below.

I have organized the various approaches into five conceptual groups that differ by the manner in which the data are integrated with phylogenetic information: character mapping, single pairs, hierarchical group, ancestral reconstruction, and phylogenetic correlation. Graphical examples of three of these approaches are shown in Fig. 15.2.

## Character Mapping

Character mapping is not a multivariate technique but one where discrete or continuous characters are extracted from a matrix and then mapped (optimized) individually onto a phylogeny. An example is shown in Fig. 15.1, where one can recognize a change in $\rho$-group membership coinciding with the divergence of the species pair on the left. One would reconstruct the loss of a strong correlation and the gain of another on the branch leading to this clade. If one had hypothesized such a change on the basis of a shift in function of a structure made up by those

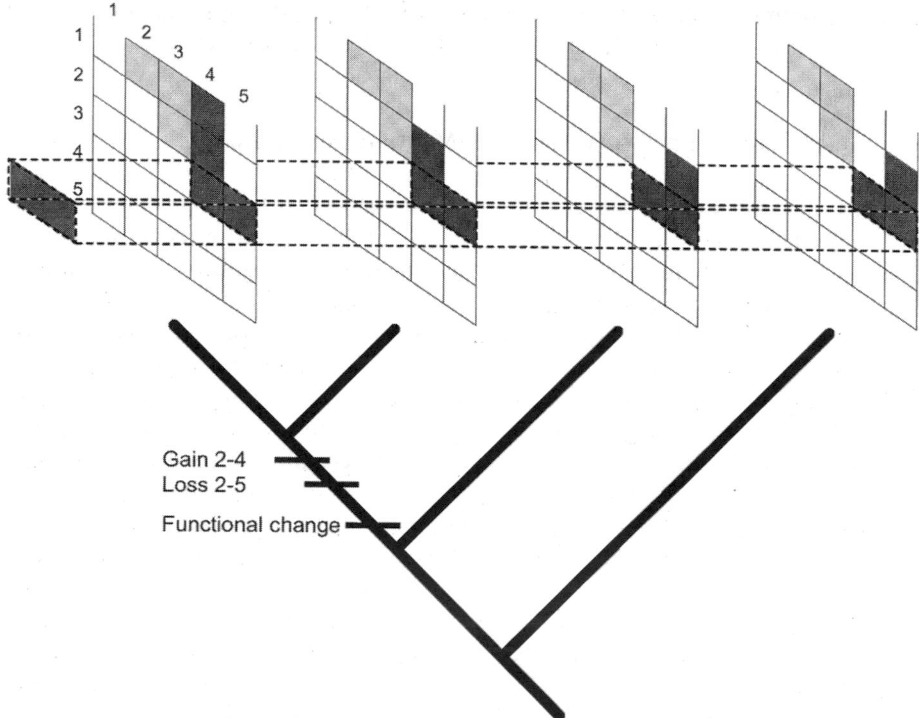


Figure 15.1 Hypothetical phylogeny for four species with each species represented by a correlation or variance/covariance matrix for five quantitative traits (rectangles). There are two sets of correlations (covariances) with particularly high values that could be considered correlation pleiades (*sensu* Berg 1960). These two sets are indicated by the shaded elements. Only sets above the diagonal are shown. Assuming the sets are homologous across all four species, mapping correlations as characters onto the phylogeny results in the conclusion that there have been changes in the membership in the dark set. On the branch leading to the left-hand clade, there has been a loss of the correlation 2-with-5 (2–5) and the gain of the correlation 2–4. Subsequently on the branch leading to the leftmost species, the correlation 1–4 was added to the dark set. Across all four species there is a conserved pair of correlations (3–4, 3–5) that is indicated by the prism transcending the four matrices. See text for discussion of the functional change and its interpretation in a phylogenetic context. The figure illustrates some of the principles involved in identifying both shared patterns of integration and modifications across the evolution of a diversifying lineage.

traits, then the data would support the functional hypothesis, *sensu* Coddington (1988). Repeated examples from other clades or subclades would strengthen the conclusion. In the example in Fig. 15.1, the change in group membership represents a single observation, and many more clades must be sampled for statistical power. Of the various ways of characterizing integration, element-by-element and pleiades approaches are most amenable to extraction of single characters for mapping, but gain or loss of factors from FA could also be mapped (B. Chernoff, pers. comm.). One could also treat covariances or correlations as individual continuous characters and map them, but estimation error could be significant, and minor changes in correlations could be overinterpreted when mapped onto a phylogeny. The following approaches are explicitly multivariate.

## Single Pairs

The single pairs approach simply makes all possible pairwise comparisons between taxon matrices (Fig. 15.2a). Statistically this is the least justifiable approach because it involves repeated comparisons (each of $N$ OTUs is involved in $N - 1$ comparisons), but it can provide a useful overview of patterns or range of variation for a metric (Steppan 1997a; Ackermann and Cheverud 2000; Marroig and Cheverud 2001). This approach has been used when others are difficult to apply because of the nature of the question being asked or the matrix comparison method. For example, in leaf-eared mice *Phyllotis*, all possible matrix correlations were obtained, then grouped by the taxonomic level at which each was calculated (Steppan 1997a). The comparison of interest was the matrix correlation, a simple measure of overall similarity, but which cannot be used with the hierarchical group analysis (a multigroup method) or autocorrelation and is less appropriate than character-based or distance measures when used with ancestral reconstruction methods (matrix correlation applied to ancestor-descendant pairs would yield similarity values between them; distance measures would be more appropriate). Other analytical approaches such as CPCA can be applied in single-pair fashion when the questions justify it or when the number of taxa is very low (two or three), but hierarchical group and ancestral reconstruction methods would be more powerful for larger data sets.

In the single pairs approach, phylogenetic structure is provided by grouping comparisons among taxa according to their taxonomic or phylogenetic level. In Fig. 15.2a, taxa A, B, and C are in clade (e.g., genus) 1, and D, E, and F are in clade (e.g., genus) 2. A simple categorization would group all comparisons between species of the same genus (shown by the solid arrows) together and all comparisons between different genera (dashed arrows) together. One could subdivide the within-genus category to distinguish between-sister species comparisons (A–B and E–F) from between-sister clade comparisons (equivalent to second-level sister groups; C–A, C–B, D–E, and D–F). In my study of *Phyllotis* (Steppan 1997a) I referred to a population-level phylogeny for the species group, and the deeper-level comparisons were the progressively more cumbersome levels of sister species, second-sister clades (sister species once removed), and third-sister clades.

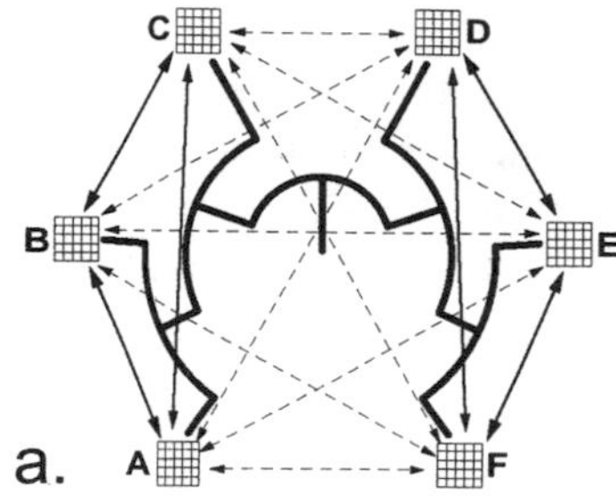


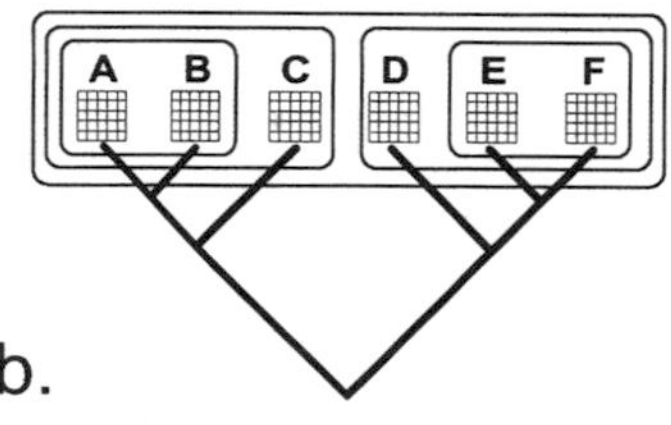


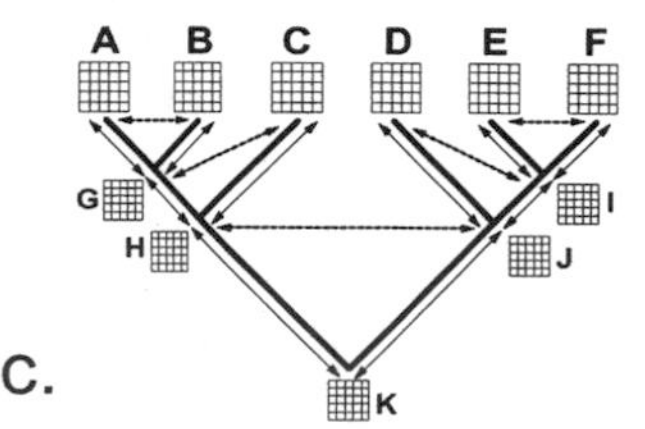


Figure 15.2 Three different approaches to the phylogenetic structuring of comparisons among matrices are illustrated. The phylogeny is the same in all three, with the first split leading to two major clades; clade 1 contains species A, B, and C while clade 2 contains species D, E, and F. A and B are sister species as are E and F. (a) Single-pair approach consists of all possible pairwise comparisons. Comparisons within each major clade are indicated by solid arrows, comparisons between species of different major clades are indicated by dashed arrows. There are six within-major-clade and nine among-major-clade comparisons for these six species, $N(N-1)/2$ overall. (b) Hierarchical group analysis involves including all members of a clade within each analysis. The relevant taxa are enclosed in the Venn diagram sets illustrated. Analyses are repeated for all clades. There are $N - 1$ clades for $N$ taxa in a rooted phylogeny; in this example, five clades are analyzed. (c) Illustration of the two primary methods using an ancestral reconstruction approach with phylogenetically independent comparisons. Matrices must be estimated for the hypothetical ancestors G–K from the observed descendant taxa A–F if not observed directly. The dashed arrows indicate comparisons (contrasts) using the phylogenetic independent contrasts method (PIC). In addition to contrasts between extant taxa (A–B, E–F), there are contrasts between ancestors and extant taxa (G–C, I–D) and between estimated ancestors (H–J). All contrasts are between sister taxa. The solid arrows indicate comparisons between ancestors and descendants using the minimum evolution approach (ME). ME allows more explicit partitioning of evolutionary change on the phylogeny but requires greater reliance on the ancestral reconstructions to do so since a greater proportion of comparisons involve ancestors than the PIC method. There are $N - 1$ contrasts and $2(N - 1)$ ME comparisons.

An alternative grouping criterion is by formal taxon level, such as species, subgenus, genus, and so on (Marroig and Cheverud 2001). This criterion has the advantages of grouping more comparisons within fewer categorical levels, and it does not require an explicit phylogenetic hypothesis. For most groups, fully resolved phylogenies do not exist, and one must rely on the less resolved systematic classifications. One disadvantage of using formal taxonomy is that not all sister-species pairs in a phylogeny are equally old or divergent. For example, if one sampled one clade (say a genus) much better than its sister group, the sister-species pairs within the better-sampled genus are likely to be closely related, but if the poorly sampled sister clade includes only two species, say from different subgenera or genera, then that species pair could be equivalent in divergence time to third- or fourth-sister species groups in the other genus. Equivalent levels of divergence would not then be grouped for analysis. One solution to this problem is to use the branching information (topology) from the phylogeny and categorize the level of grouping by formal taxonomy. That was the approach used in Steppan (1997a) and Ackermann and Cheverud (2000). A more sophisticated approach is to incorporate branch-length information, as discussed below. Branch lengths are distinct from branching sequence and comparative biologists have been much slower to incorporate branch-length information because only with a very good fossil record or the recent accumulation of molecular data can we have good estimates of them.

A variant comparison strategy was used by Podolsky et al. (1997) wherein only a subset of all possible comparisons was made. For each higher-level comparison, only the most similar pairs between members of sister clades (or UPGMA clusters) were compared. They used this approach to reduce computation time for maximum likelihood analyses, rather than to avoid statistical problems, and it is not the preferred method.

## Hierarchical Group Analysis

Hierarchical group analysis may be the most appropriate method for analyses that allow multiple groups but where ancestral state reconstruction is problematic. At present, CPCA (common principal components analysis) is becoming the most popular technique for data with these properties and will be used as the example. In hierarchical group analysis, a CPC analysis is conducted on all members of a clade for each clade in turn (Fig. 15.2b). Thus, A and B would be analyzed together, as would be E and F. Then A, B, and C would be analyzed together as would be D, E, and F. Finally, all taxa would be analyzed at once (Steppan 1997a; Ackermann and Cheverud 2000; Badyaev and Hill 2000). One goal is to identify the branches in the phylogeny at which specific components are no longer shared by all members of a clade. Another goal is the inverse, to identify clades where the hypothesis of common structure can be accepted.

Several questions remain unanswered about the behavior of CPCA with respect to phylogenetic analyses. How does it respond to increasing numbers of taxa? Resampling suggests that rejection of shared structure becomes more likely with more taxa, perhaps because of single outlier matrices (Steppan 1997a). How is CPCA affected by outlier or divergent matrices in an analysis (Steppan et al. 2002)? A further problem is that of repeated comparisons, although the nature of the test (acceptance or rejection of homogeneity of covariance structure) makes this statistical violation only slightly problematic.

Although hierarchical group analysis is more appropriate statistically than single-pair comparisons because it makes fewer repeated comparisons, the latter does provide a complementary perspective. For example in *Phyllotis*, CPCA indicated no shared structure at any level in the PCA hierarchy except between closely related populations, suggesting major divergence in covariance structure; but single-pair matrix correlations found that the matrices were still very similar ($r^2 > 0.90$; Steppan 1997a). Ackermann and Cheverud (2000) and Pigliucci and Kolodynska (2002) found a similar situation in *Saguinus* tamarin monkeys and *Arabidopsis*, respectively. With large sample sizes, CPCA may be very powerful at detecting small deviations in covariance structure and thus reject hypotheses of shared structure, while single-pair comparisons can estimate the magnitude of differences (i.e., significance level is not a measure of magnitude).

## Ancestral Reconstruction

The term "ancestral reconstruction" groups together a variety of comparative methods that use reconstructed ancestral values. The most common types of methods are phylogenetic independent contrasts (PIC; Felsenstein 1985), minimum evolution (ME; Huey and Bennett 1987), and more recently phylogenetic generalized least squares (PGLS; Rohlf 2001). Of these, only minimum evolution has been applied to multivariate data (Steppan 1997b) to my knowledge, although Rohlf (2001) suggests that the multivariate extension of these methods that reconstructs variances and covariances could be computationally straightforward. Provided that ancestors can be measured or estimated accurately, this approach is the most valid statistically and uses all of the available phylogenetic information. Comparisons are made either between sister species, be they two extant taxa or ancestors (PIC), or between ancestor-descendant pairs (ME). These methods were specifically developed to estimate evolutionary correlations (coevolution) between some environmental factor or proximate phenotypic trait (the independent variable) and an adaptive response to it (the dependent variable; e.g., Huey and Bennett 1987). Their strength is that they partition the comparisons such that all are phylogenetically independent of each other. Additionally, the ME method, essentially a variant of character mapping, can be used to partition multivariate evolution to specific branches on the phylogeny. The ME method polarizes ancestor-descendant comparisons while PIC and PGLS do not.

Nearly all the matrix-comparison methods can be used with the ancestral reconstruction approach. I used the distance metric disparity to examine rates of covariance evolution in *Phyllotis* (Steppan 1997b). When branch lengths were scaled by matrix disparity, I found that branches leading to populations and subspecies clades were significantly longer than deeper branches leading to species and more inclusive clades. This pattern contrasted sharply with those reconstructed from Euclidean distance in means or parsimony estimates of DNA sequence evolution (Steppan 1997b). This result suggests that the tempo of covariance evolution or the shape of the adaptive landscape is qualitatively different for covariance structure than for the gross phenotype or selectively neutral DNA substitutions (most substitutions were synonymous). Further extensions of this approach could be employed, such as testing for constant rates by the relative rates test or likelihood ratio tests under a "covariance clock," analogous to a molecular clock, using maximum likelihood. It should be noted that the sample sizes for variable characters may be much smaller for multivariate data (i.e., number of covariance elements) than for molecular data (nucleotide sites), and thus rate tests are likely to lack power.

## Phylogenetic Correlation

Three methods that use different correlational strategies are grouped under phylogenetic correlation. Cheverud and Dow (1985) and Cheverud et al. (1985) introduced phylogenetic autocorrelation (PA) as a phylogenetic extension of spatial autocorrelation methods. The rationale behind all autocorrelation methods (phylogenetic, spatial, serial) is that observations in close proximity will be relatively similar because they share many processes or histories that are not of direct interest to the investigator. Autocorrelation (actually autoregression) removes the effect of proximity from the overall pattern and identifies the scale over which the underlying (but unknown) processes operate. After removal of proximity effects, the remaining pattern is the focus of the researcher, most often adaptation in comparative studies. The special function of PA is to estimate the proportion of variance across taxa in a trait that is explained by phylogenetic relatedness. The phylogenetic component has been called phylogenetic inertia. PA uses a connection matrix that summarizes phylogenetic relationships among taxa. The connection matrix can be determined from a variety of phylogenetic descriptors, ranging from integers representing taxonomic category (e.g., species, genus, family), to number of nodes separating taxa on a tree, to methods that incorporate branch-length information and the height of the common ancestors above the root (Rohlf 2001). Residuals from the autoregression procedure can be used as new characters after the effects of phylogeny have been removed. Again, modifications to the implementation would be needed to use PA with multivariate data.

An alternative to PA, the phylogenetic eigenvector regression (PVR), regresses trait values against the principal coordinate eigenvectors of a pairwise phylogenetic distance matrix (Diniz-Filho et al. 1998). The performance of this method has not been widely explored, nor has its application to multivariate data.

In one of the few applications of phylogenetic correlation to multivariate data, Marroig and Cheverud (2001) calculated all pairwise matrix correlations for a group of New World monkeys and then built a taxon-by-taxon covariance similarity matrix from them. This covariance similarity matrix was compared with genetic and ecological (dietary) distance matrices to determine that diet was more important than phylogeny in explaining similarity in covariance structure.

## Branch Length Information (Rate of Evolution)

Incorporating information on branch lengths allows analyses that could not be conducted without a phylogenetic approach. Figure 15.3 illustrates the potential importance of branch-length information. The two phylograms presented have identical topologies but very different branch lengths that are drawn proportional to time. A common assumption in studies of quantitative trait evolution is that the expected variance in a trait increases proportionally to time since divergence

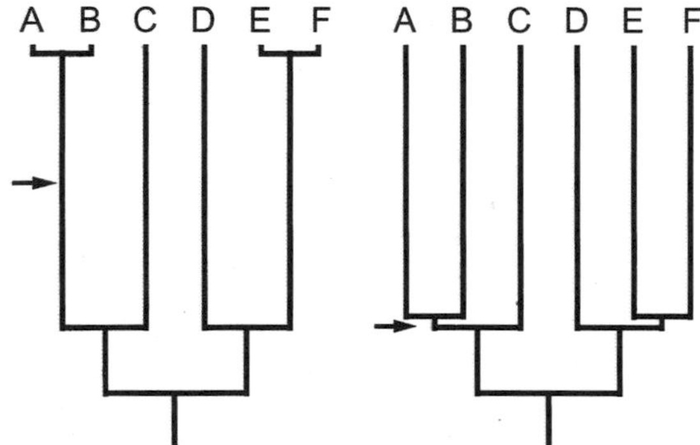


Figure 15.3 Contrast between two phylogenies with identical topologies but different branch-length distributions. Branch lengths could represent time, amount of molecular evolution, or other characteristics. In this example, the branches are proportional to time, for example, DNA sequence data evolving under a molecular clock. On the left-hand tree, sister species pairs A–B and E–F are very closely related with a long period of shared history (i.e., long branch), while on the right-hand tree, the pairs are only slightly more closely related to each other than either member is to the pair's sister taxa, C or D (i.e., short branch, as indicated by arrow). If expected divergence in covariance structure is in any way correlated with time, that is, any common model of evolution except punctuated equilibrium, then we would expect A and B to be very similar given the left tree but not so given the right. Failure to incorporate branch-length information could lead to faulty conclusions about the rate of covariance evolution or about the patterns of similarity as predicted by models being tested.

(Brownian motion model). Therefore, on the left tree, A and B are very closely related (i.e., have diverged relatively recently), and we would expect them to have very similar covariance matrices. In contrast, on the right tree, A and B shared very little unique history, and we would not expect them to be much more similar to each other than either is to C. Strong trait divergence between A and B would be remarkable on the left tree but unremarkable on the right. Likewise, if A and B were very similar on the right tree but the other four taxa were divergent, this observation would be contrary to expectations and would suggest some mechanism for this conservation, possibly canalization or stabilizing selection. The exception to these expectations would be the case in which one assumes a punctuated model of evolution, in which all change occurs at speciation and no subsequent anagenetic evolution takes place along lineages, making branch lengths irrelevant.

Branch-length information either can be added to or is already an important part of all the phylogenetic approaches discussed above. For example, when the single pairs approach is used, matrix similarity values can be plotted against or regressed on a continuous character describing relatedness (e.g., height of each pair's common ancestor above the root) rather than their taxonomic category (e.g., within genus). PICs are often calculated with contrasts scaled by the square root of the branch lengths assuming a Brownian motion model of evolution.

As outlined above, in addition to the use of independently derived branch lengths (usually from DNA sequence data) to inform comparative multivariate analyses, branch lengths can be calculated from the multivariate data themselves. "Covariance lengths" represent the amount of evolution in covariance structure and when scaled by an independent estimate of branch lengths, they represent evolutionary rates. High rates of covariance evolution appeared associated with population-level processes (sampling error, drift, or morphological divergence of populations/phylogenetic species) rather than with the evolution of putative reproductive barriers between biological species or divergence of clades in *Phyllotis* (Steppan 1997b). This result suggests that variation in covariation structure is becoming "saturated," perhaps because stabilizing selection on developmental pathways limits the divergence of covariance structures much beyond some fluctuation produced by drift. It also suggests the possibility of exploring the mechanisms underlying covariance evolution by means of saturation curves analogous to those plotted for nucleotide sequence data. In saturation curves, covariance distance (e.g., disparity) would be plotted against genetic distance or time. Deviations from a linear (for corrected DNA sequence distances) or quadratic (for time, the expectation under Brownian motion) relationship would give evidence for mechanisms limiting variation. Lineage-specific changes in relative branch lengths (in which all members of a clade share particularly long or short branches as detected by a relative-rates test) could be evidence for accelerated or retarded rates of the evolution of covariance structure. Single anomalously long branches (relative to time) might suggest a reorganization of the developmental or genetic program.

## What To Do?

The comparative analysis of multivariate data presents the researcher with a wide variety of less than optimal choices for methods of analysis. There are two sets of choices to be made. The first is how to compare matrices and how to measure differences in patterns of covariation. The choice of analytical method will depend in part on the aspect of covariance evolution of interest, and each method has strengths and weaknesses. There does not appear to be a clearly superior method for all circumstances, but maximum likelihood, model-based ordination techniques, and distance methods show the most promise for future improvement. The second is how to incorporate phylogenetic information. None of the comparative methods developed explicitly for phylogenetic data is optimized for multivariate data. Again, the choice of phylogenetic approach will vary with the question of interest and all may have their place in a study. Ancestral reconstruction and hierarchical group approaches are perhaps the most powerful and flexible for use with multivariate data. With the surge of molecular phylogenies in the literature, it will become increasingly valuable to include branch length information in the analyses to address both increasingly sophisticated and old questions with greater precision.

*Acknowledgments* I thank Anne Thistle, Katherine Preston, and Massimo Pigliucci for comments improving the quality of the chapter. Financial support was provided NSF grant DEB-0108422.

## *Literature Cited*

Ackermann, R. R., and J. M. Cheverud. 2000. Phenotypic covariance structure in tamarins (genus *Saguinus*): a comparison of variation patterns using matrix correlation and common principal component analysis. Am. J. Phys. Anthropol. 111:489–501.

Badyaev, A. V., and G. E. Hill. 2000. The evolution of sexual dimorphism in the house finch. I. Population divergence in morphological covariance structure. Evolution 54:1784–1794.

Begin, M., and D. A. Roff. 2001. An analysis of G matrix variation in two closely related cricket species, *Gryllus firmus* and *G. pennsylvanicus*. J. Evol. Biol. 14:1–13.

Berg, R. L. 1960. The ecological significance of correlation pleiades. Evolution 14:171–180.

Brodie, E. D., III. 1993. Homogeneity of the genetic variance-covariance matrix for anti-predator traits in two natural populations of the garter snake *Thamnophis ordinoides*. Evolution 47:844–854.

Brooks, D. R., and D. A. McLennan. 1991. Phylogeny, Ecology, and Behavior. University of Chicago Press, Chicago.

Cane, W. P. 1993. The ontogeny of post-cranial integration in the common tern, *Sterna hirundo*. Evolution 47:1138–1151.

Chernoff, B., and P. M. Magwene. 1999. Afterword. Pp. 319–348 in E. C. Olson and R. L. Miller, Morphological Integration. University of Chicago Press, Chicago.

Cheverud, J. M. 1982. Phenotypic, genetic, and environmental morphological integration in the cranium. Evolution 36:499–516.

Cheverud, J. M., and M. M. Dow. 1985. An autocorrelation analysis of genetic variation due to lineal fission in social groups of *Rhesus* macaques. Am. J. Phys. Anthropol. 67:113–121.

Cheverud, J. M., M. M. Dow, and W. Leutenegger. 1985. The quantitative assessment of phylogenetic constraints in comparative analyses: sexual dimorphism in body weight among primates. Evolution 39:1335–1351.

Cheverud, J. M., G. P. Wagner, and M. M. Dow. 1989. Methods for the comparative analysis of variation patterns. Syst. Zool. 38:201–213.

Coddington, J. A. 1988. Cladistic tests of adaptational hypotheses. Cladistics 4:3–22.

Cunningham, C. W., K. E. Omland, and T. H. Oakley. 1998. Reconstructing ancestral character states: a critical reappraisal. Trends Ecol. Evol. 13:361–366.

Dietz, E. J. 1983. Permutation tests for association between two distance matrices. Syst. Zool. 32:21–26.

Diniz-Filho, J. A. F., C. E. R. De Sant'ana, and L. M. Bini. 1998. An eigenvector method for estimating phylogenetic inertia. Evolution 52:1247–1262.

Felsenstein, J. 1985. Phylogenies and the comparative method. Am. Nat. 125:1–15.

Flury, B. 1988. Common Principal Components and Related Multivariate Models. John Wiley, New York.

Flury, B. K. 1987a. A hierarchy of relationships between covariance matrices. Pp. 31–43 in A. K. Gupta (ed.), Advances in Multivariate Statistical Analysis. Reidel, Dordrecht, The Netherlands.

Flury, B. K. 1987b. Two generalizations of the common principal component model. Biometrika 74:59–69.

Garland, T., and S. C. Adolph. 1994. Why not to do two-species comparative-studies: limitations on inferring adaptation. Physiol. Zool. 67:797–828.

Harvey, P. H., and M. D. Pagel. 1991. The Comparative Method in Evolutionary Biology. Oxford Series in Ecology and Evolution. Oxford University Press, New York.

Houle, D., J. Mezey, and P. Galpern. 2002. Interpretation of the results of partial principal components analysis. Evolution 56:433–440.

Huelsenbeck, J. P., B. Rannala, and J. P. Masly. 2000. Accommodating phylogenetic uncertainty in evolutionary studies. Science 288:2349–2350.

Huey, R. B., and A. F. Bennett. 1987. Phylogenetic studies of co-adaptation: preferred temperatures versus optimal performance temperatures in lizards. Evolution 41:1098–1115.

Kirkpatrick, M., D. Lofsvold, and M. Bulmer. 1990. Analysis of the inheritance, selection and evolution of growth trajectories. Genetics 124:979–993.

Klingenberg, C. P., and M. Zimmermann. 1992. Static, ontogenetic, and evolutionary allometry: a multivariate comparison in nine species of water striders. Am. Nat. 140:601–620.

Klingenberg, C. P., B. E. Neuenschwander, and B. D. Flury. 1996. Ontogeny and individual variation: analysis of patterned covariance matrices with common principal components. Syst. Biol. 45:135–150.

Magwene, P. M. 2001. New tools for studying integration and modularity. Evolution 55:1734–1745.

Marroig, G., and J. M. Cheverud. 2001. A comparison of phenotypic variation and covariation patterns and the role of phylogeny, ecology, and ontogeny during cranial evolution of New World monkeys. Evolution 55:2576–2600.

Olson, E. C., and R. L. Miller. 1958. Morphological Integration. University of Chicago Press, Chicago.

Phillips, P. C., and S. J. Arnold. 1999. Hierarchical comparison of genetic variance-covariance matrices. I. Using the Flury hierarchy. Evolution 53:1506–1515.

Pigliucci, M., and A. Kolodynska. 2002. Phenotypic plasticity and integration in response to flooded conditions in natural accessions of *Arabidopsis thaliana* (L.) Heynh (Brassicaceae). Ann. Bot. 90:199–207.

Podolsky, R. H., R. G. Shaw, and F. H. Shaw. 1997. Population structure of morphological traits in *Clarkia dudleyana*. II. Constancy of within-population genetic variance. Evolution 51:1785–1796.

Rice, W. R. 1989. Analyzing tables of statistical tests. Evolution 43:223–225.

Roff, D. A. 2000. The evolution of the G matrix: selection or drift? Heredity 84:135–142.

Roff, D. A. 2002. Comparing G matrices: a MANOVA approach. Evolution 56:1286–1291.

Roff, D. A., T. A. Mousseau, and D. J. Howard. 1999. Variation in genetic architecture of calling song among populations of *Allonemobius socius*, *A. fasciatus*, and a hybrid population: drift or selection? Evolution 53:216–224.

Rohlf, F. J. 1998. On applications of geometric morphometrics to studies of ontogeny and phylogeny. Syst. Biol. 47:147–158.

Rohlf, F. J. 2001. Comparative methods for the analysis of continuous variables: geometric interpretations. Evolution 55:2143–2160.

Schluter, D., T. Price, A. O. Mooers, and D. Ludwig. 1997. Likelihood of ancestor states in adaptive radiation. Evolution 51:1699–1711.

Service, P. M. 2000. The genetic structure of female life history in *D. melanogaster*: comparisons among populations. Genet. Res. 75:153–166.

Shaw, R. G. 1987. Maximum-likelihood approaches applied to quantitative genetics of natural populations. Evolution 41:812–826.

Shaw, R. G. 1991. The comparison of quantitative genetic parameters between populations. Evolution 45:143–151.

Steppan, S. J. 1997a. Phylogenetic analysis of phenotypic covariance structure. I. Contrasting results from matrix correlation and common principal component analyses. Evolution 51:571–586.

Steppan, S. J. 1997b. Phylogenetic analysis of phenotypic covariance structure. II. Reconstructing matrix evolution. Evolution 51:587–594.

Steppan, S. J., P. C. Phillips, and D. Houle. 2002. Comparative quantitative genetics: evolution of the G matrix. Trends Ecol. Evol. 17:320–327.

Swofford, D. L., G. J. Olsen, P. J. Waddell, and D. M. Hillis. 1996. Phylogenetic inference. Pp. 407–514 in D. M. Hillis, C. Moritz, and B. K. Mable (eds.), Molecular Systematics. Sinauer Associates, Sunderland, MA.

Voss, R. S., and L. F. Marcus. 1992. Morphological evolution in muroid rodents. II. Craniometric factor divergence in seven Neotropical genera, with experimental results from *Zygodontomys*. Evolution 46:1918–1934.

Voss, R. S., L. F. Marcus, and P. P. Escalante. 1990. Morphological evolution in muroid rodents. I. Conservative patterns of craniometric covariance and their ontogenetic basis in the Neotropical genus *Zygodontomys*. Evolution 44:1568–1587.

Wagner, G. P. 1984. On the eigenvalue distribution of genetic and phenotypic dispersion matrices: evidence for a nonrandom organization of quantitative character variation. J. Math. Biol. 21:77–96.

Willis, J. H., J. A. Coyne, and M. Kirkpatrick. 1991. Can one predict the evolution of quantitative characters without genetics? Evolution 45:441–444.

Zelditch, M. 1987. Evaluating developmental models of integration in the laboratory rat using confirmatory factor analysis. Syst. Zool. 36:368–380.

Zelditch, M. L., D. O. Straney, D. L. Swiderski, and A. C. Carmichael. 1990. Variation in developmental constraints in *Sigmodon*. Evolution 44:1738–1747.

Zelditch, M. L., W. L. Fink, D. L. Swiderski, and B. L. Lundrigan. 1998. On applications of geometric morphometrics to studies of ontogeny and phylogeny: a reply to Rohlf. Syst. Biol. 47:159–167.

# 16

# The Evolution of Genetic Architecture

DEREK ROFF

## What Is Genetic Architecture?

What one means by "genetic architecture" depends to some extent upon one's perspective. At one level we might view it as the DNA coding, while at a more abstract level we can view it as the statistical structure arising from interactions within the DNA code. It is the latter sense that I adopt here, regarding the DNA as the "bricks" from which the genotype and phenotype are constructed. This statistical approach, sometimes known as biometrical genetics, has a long history stretching back to before the discovery of DNA or chromosomes or even Mendelian genetics (e.g., Galton's biometrical analysis of quantitative trait variation preceded the rediscovery of Mendel's work). The unification of biometrical genetics with Mendelian genetics to create the modern field of quantitative genetics was one of the great intellectual achievements of the past century.

The statistical analysis of genetic architecture begins with the assessment of phenotypic variation and then, on the basis of principles of Mendelian genetics, uses pedigree information to divide this variation into environmental and genetic components. With suitable sets of pedigrees, the latter can be further divided into additive, dominance, and epistatic subcomponents. Additive effects have been the primary concern of breeders and, at least initially, of evolutionary biologists. This focus arises because the response to selection, which was (and is) the central interest of animal and plant breeders, can be described by the breeder's equation, $R = h^2S$, where $R$ is the response to selection, $S$ is the selection differential, and $h^2$ is the heritability of the trait. Using the Mendelian perspective, the heritability is equated to the ratio of the additive genetic variance of the trait to the total phenotypic variance. Thus, estimating the additive genetic variance became the principal goal in breeding experiments. Nonadditive effects were considered a nuisance to be ignored whenever possible. The goal of the breeder is relatively short-term evolutionary change of generally single, or at most a few traits. In contrast, the aim of the evolutionary biologist is to understand evolutionary change at a range of temporal scales from that of a single generation to time spans encompassing geological epochs. Further, it is recognized that selection acts upon many traits simultaneously and hence interactions at the genetic level could potentially be of great significance. It is this different scope of research that has brought an interest in the evolutionary change in genetic architecture, including both addititive and nonadditive effects on the phenotype.

In this chapter I shall consider five factors that current data suggest are important in influencing the evolution of genetic architecture: mutation, selection, drift, inbreeding, and antagonistic pleiotropy. These are not necessarily the only factors, but they illustrate the potential complexity of evolutionary change. The message that I wish to convey is that this complexity makes it virtually impossible to understand the evolution of genetic architecture simply through a statistical analysis of genetic variation per se. In addition, we must bring to bear other information such as the relationship of the traits to known selective factors. After describing the five main factors mentioned above, I illustrate this more inclusive approach with three examples.

## The Fundamental Equation of Evolutionary Quantitative Genetics

The importance of additive genetic variance has already been noted; it is the central element in the equation describing the response to selection, be it artificial or natural. The breeder's equation is simplistic in that it focuses only upon a single trait. For many artificial selection regimes this is appropriate, but in the natural world selection acts upon several traits simultaneously, requiring us to extend the equation to cover the case of multivariate selection. The multivariate extension of the breeder's equation is $\Delta\overline{\mathbf{Z}} = \mathbf{GP}^{-1}\mathbf{S}$, where $\Delta\overline{\mathbf{Z}}$ is the vector of change in trait means, $\mathbf{G}$ is the matrix of additive genetic variances and covariances, $\mathbf{P}$ is the matrix of phenotypic variances and covariances, and $\mathbf{S}$ is a vector of selection differentials. Two important points can be made about this equation. First, the **G**-matrix is clearly an important component, and understanding how it can change under the three forces of drift, selection, and mutation is fundamental to understanding both the evolution of trait means and the evolution of genetic architecture. Second, the phenotypic variance-covariance matrix is composed of the environmental, additive and nonadditive genetic variances and covariances. Thus, nonadditive effects do play a direct role in evolutionary change, even within the perspective of the breeder's equation and its multivariate extension. If nonadditive genetic variance simply stayed nonadditive, then it would be an uninteresting component of the phenotypic variance. But, as discussed below, nonadditive genetic variance can be converted into additive, with consequent effects on the rates of evolutionary change.

## Mutation, Selection, and Drift

At the simplest level, mutation creates genetic variation whereas selection and drift erode it. Considering a single trait subject to stabilizing selection, mutation, and drift, the equilibrium additive genetic variance in the absence of dominance or epistasis is predicted by the stochastic house of cards (SHC) model to be (Burger et al. 1989)

$$\sigma_A^2 = \frac{4n\mu\sigma_S^2 N_e \alpha^2}{N_e\alpha^2 + \sigma_S^2}$$

where $\sigma_S^2 = \gamma + \sigma_E^2$, $\sigma_E^2$ is the environmental variance, $\gamma$ is the parameter of the Gaussian stabilizing selection function, $\mu$ is the mutation rate per locus assuming two alleles per locus, $\alpha^2$ is the variance of the mutational effect, and $N_e$ is the effective population size. The product $n\mu$ is the per trait gametic mutation rate, designated $\mu_g$. The average amount of new genetic variance introduced per zygote per generation by mutation, $\sigma_m^2$, is equal to $2\mu_g\alpha^2 = 2n\mu\alpha^2$. There is considerable variation in parameter estimates from real data, and depending upon the plausibility of the particular combination chosen, the standing heritability can be smaller than typically found in either life-history (approx. 0.25) or morphological (approx. 0.4) traits (Roff 2002b; Fig. 16.1).

As shown by the SHC model, at equilibrium the additive genetic variance is roughly proportional to the effective population size, as is also the variance of the additive genetic variance (Burger and Lande 1994). Over time, the effect of drift is to cause **G**-matrices to diverge but remain proportional to each other

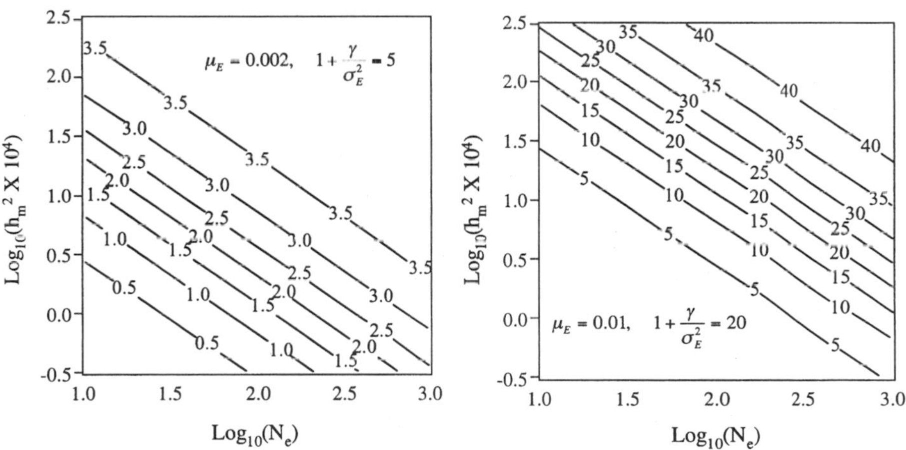


Figure 16.1 Isoclines of heritabilities (×100) maintained at mutation-selection balance as a function of effective population size ($N_e$) and mutational heritability ($h^2_m$). Other parameters: $\sigma_E^2$ is the environmental variance, $\gamma$ is the parameter of the Gaussian stabilizing selection function, $\mu_g$ is the per trait gametic mutation rate. The left panel shows the results for a low set of parameter values and the right panel the results for a high set of values. (For details of the estimation of parameter values see Roff 1997, p. 360).

(Lande 1979; Lofsvold 1988). At first sight this behavior would appear to provide us with a means of distinguishing evolution due to drift from that due to selection, assuming that selection acts differentially on different characters. Where significant differences between **G**-matrices in natural populations (but typically reared in the laboratory) have been found, the differences can be accounted for primarily as proportional changes between the matrices (Roff 2000). Using only the argument developed above, this suggests that much of the evolutionary change in **G**-matrices is due to drift rather than selection. There are two reasons why this conclusion is suspect: first, selection may produce proportional changes, and second, drift can produce nonproportional changes.

## Selection and Proportional Changes in the G-Matrix

The basic argument to be developed in this section is that proportional changes are expected under selection if the traits under study are all highly correlated with each other. For simplicity of exposition I shall focus upon just two traits, say X and Y. Genetic correlation between these two traits is caused by loci that affect both traits. Let the number of loci unique to X ("x" loci) be $n_x$ and the number unique to Y ("y" loci) be $n_y$ and the number that are in common ("c" loci) be $n_c$. For simplicity, absorb the necessary "2" for the diploid condition into the $n$'s. Linkage disequilibrium is ignored and each locus treated as independent (i.e., no epistasis). The trait genotypic means are then

$$\mu_X = n_x\mu_x + n_c\mu_c$$

$$\mu_Y = n_y\mu_y + n_c\mu_c$$

where $\mu_k$ is the mean at the "$k$th" type of locus ($x$, $y$, $c$ and X, Y) and the trait genetic variances are

$$\sigma_X^2 = n_x\sigma_x^2 + n_c\sigma_c^2$$

$$\sigma_Y^2 = n_y\sigma_y^2 + n_c\sigma_c^2$$

where $\sigma_k^2$ is the variance at the "$k$th" type of locus ($x$, $y$, $c$ and X, Y). The only loci that contribute directly to the covariance are those that are in common. The covariance, $\sigma_{XY}$, is thus

$$\begin{aligned}
\sigma_{XY} &= E\{(x-\mu_x)(y-\mu_y)\} \\
&= E\{([x+c]-[n_x\mu_x+n_c\mu_c])([y+c]-[n_y\mu_y+n_c\mu_c])\} \\
&= E\{(x-n_x\mu)(y-n_y\mu_y)\} + E\{(x-n_x\mu_x)(c-n_c\mu_c)\} \\
&\qquad + E\{(y-n_y\mu_y)(c-n_c\mu_c)\} + E\{(c-n_c\mu_c)^2\} \\
&= 0+0+0+E\{(c-n_c\mu_c)^2\} \\
&= n_c\sigma_c^2
\end{aligned}$$

The **G**-matrix is therefore

$$\begin{bmatrix} n_x\sigma_x^2 + n_c\sigma_c^2 & n_c\sigma_c^2 \\ n_c\sigma_c^2 & n_y\sigma_y^2 + n_c\sigma_c^2 \end{bmatrix}$$

Notice that all terms include $n_c\sigma_c^2$; thus genetic variances and covariances will tend to be proportional to each other to the extent that they share loci. For example, suppose that there are two alleles per locus having the values 0 or 1, the latter with frequency $p$. The **G**-matrix is then

$$\begin{bmatrix} n_x p_x q_x + n_c p_c q_c & n_c p_c q_c \\ n_c p_c q_c & n_y p_y q_y + n_c p_c q_c \end{bmatrix}$$

where $p_k$ is the frequency at the "$k$th" type of locus and $q = 1 - p$. The genetic correlation between traits depends upon the number of shared loci and their allelic frequency. The allelic frequencies are clearly not independent but a function of the selection acting on X and/or Y. Consider the extreme case in which selection acts strongly on trait X but very weakly on trait Y. The allelic frequency of the shared loci will be more strongly influenced by the selection on trait X than by the much weaker selection on trait Y. Approximately, therefore, $p_x = p_c$, giving the **G**-matrix

$$\begin{bmatrix} p_x q_x(n_x + n_c) & p_x q_x n_c \\ p_x q_x n_c & p_y q_y n_y + p_x q_x n_c \end{bmatrix}$$

Most of the tests of variation in **G**-matrices have used morphological traits, a category of traits for which the genetic correlation is typically very high (Roff 1996). An example of the distribution of the genetic correlation as a function of the number of shared loci ($n_c$) and $p_x$ (keeping other parameters constant at $n_x = n_y = 10$ and $p_y = 0.5$) is shown in Fig 16.2. Over most of the parameter space there is little change in the genetic correlation, a situation that results from the approximately proportional change in variances and covariances.

The above model is very simplistic, but it captures the essential point that if the suite of traits under study are tightly coupled, then their evolution will tend to produce proportional changes in the **G**-matrix. For a simulation model showing a similar pattern of change under selection see Reeve (2000). To distinguish the proportional change due to drift from that due to selection, the best suite of traits to use is one in which the traits are uncorrelated!

Whereas selection can result in proportional changes in the **G**-matrix, drift can itself produce nonproportional changes in **G**, potentially creating the impression that selection has occurred. This is particularly likely when populations pass through bottlenecks, a case discussed in the next section.

## Genetic Drift and Nonproportional Changes in the G-Matrix

Consider a very large population that passes through a single generation in which the population size is reduced to $N$ individuals. We focus on a single locus with two alleles $A_1$ and $A_2$ at frequencies $p$ and $q(= 1 - p)$, respectively. Due to drift,

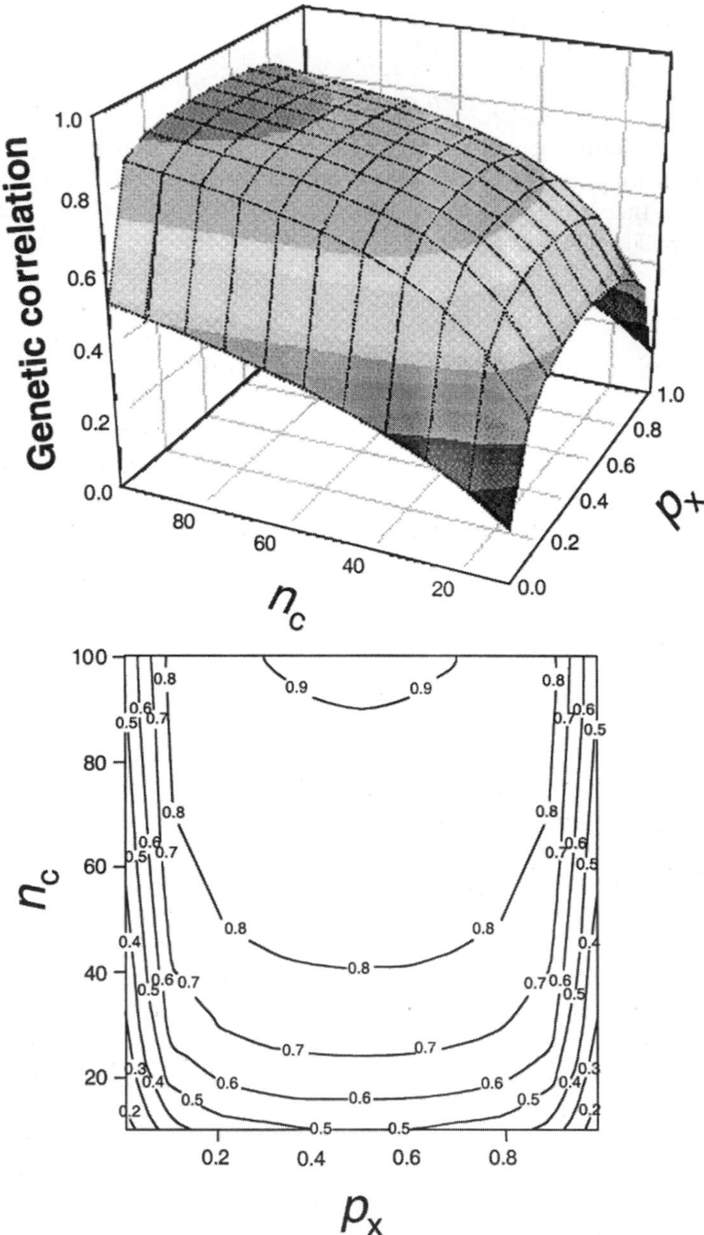


Figure 16.2 The genetic correlation for the two-trait diallelic model described in the text as a function of the number of loci in common ($n_c$) and the allele frequency at loci coding for trait X ($p_x$). The **G**-matrix is

$$\begin{pmatrix} p_x q_x (n_x + n_c) & p_x q_x n_c \\ p_x q_x n_c & p_y q_y n_y + p_x q_x n_c \end{pmatrix}.$$

$n_x = n_y = 10$ and $p_y = 0.5$. The top figure shows the 3D representation, the bottom figure the contour surface.

the allelic frequency will move away from its initial position; depending on its initial position, a change in $p$ in any particular population can generate an increase or decrease in additive genetic variance, with a consequent change in evolutionary response. In the initial population the additive and dominance variances are, respectively, $\sigma_A^2 = 2pq[(1 + d(q - p)]^2$ and $\sigma_D^2 = (2pqd)^2$. The probability that the additive genetic variance will increase varies with the degree of dominance ($d$) and the allele frequency ($p$). Over a wide range of allele frequencies, the probability that the additive genetic variance increases can be extremely high. This conclusion is also obtained for larger bottlenecks (Willis and Orr 1993). Thus, populations passing through a single bottleneck may experience large increases in heritability due to the shift in allelic frequency, but only if $d > 0$. The degree of dominance variance certainly varies among traits, with morphological traits showing much lower levels relative to the additive variance than life-history traits (Crnokrak and Roff 1995). Consequently, for life-history traits at least, passage through a population bottleneck can disproportionally alter the components of the **G**-matrix.

The above analysis ignored epistatic effects, which require by definition at least two loci. Similar increases in additive genetic variance can occur when epistasis is present. An epistatic model involving two loci with additive ($\sigma_A^2$) and additive-by-additive ($\sigma_{AA}^2$) variance but no dominance has been analyzed by Goodnight (2000). For a single-generation bottleneck, Goodnight showed that $\sigma_A^2 \approx \sigma_A^2 + (3\sigma_{AA}^2 - \sigma_A^2)F$, where $F$ is the inbreeding coefficient. This result states that whenever the epistatic variance is greater than one-third of the additive genetic variance, the additive genetic variance will be increased after the bottleneck. A change in allele frequency due to the bottleneck effect changes not only the additive genetic variance, but also the mean breeding values and the underlying average effects. As a consequence, the evolutionary trajectory is altered.

Empirical investigations (Frankham 1980; Lints and Bourgois 1982; Bryant et al. 1986; Brakefield and Saccheri 1994; Wang et al. 1998; Cheverud et al. 1999; Meffert 2000) have verified that the additive genetic variance can indeed increase following a bottleneck. While these findings are restricted to experimental situations, a detailed analysis of dominance and epistatic variance among natural populations of the pitcher-plant mosquito, coupled with knowledge of its northward colonization history, suggests that drift has been followed by conversion to additive genetic variance (Bradshaw and Holzapfel 2000).

## The Role of Inbreeding in Reducing Genetic Variation

Inbreeding has several effects on the evolution of genetic architecture. First, inbreeding increases homozygosity, and hence reduces the additive genetic variance. While, on average, the **G**-matrix of an inbred line will deviate from its ancestral outbred line in a proportional manner, any particular line can deviate quite widely and be far from proportional (Phillips et al. 2001). Second, inbreeding can generate inbreeding depression, defined as a decline in the fitness value of a trait as a result of inbreeding. Inbreeding depression can potentially alter the selection gradient and hence the evolutionary trajectory of a trait, which can in turn feed back onto the components of the **G**-matrix.

Inbreeding depression will not occur, on average, if there is only additivity. When there is dominance, the expected trait value declines linearly with the degree of inbreeding. When there is epistasis but no dominance, there will be no inbreeding depression, and when both epistasis and dominance are present the relationship between inbreeding depression and $F$ may be linear or more likely quadratic (Crow and Kimura 1970, pp. 77–85).

Empirical data from nondomestic species shows overwhelming evidence for significant inbreeding depression in gymnosperms, angiosperms, and animals (see Tables 8.7, 8.8, and 8.9 and Fig. 8.17 in Roff 1997). Falconer (1989) suggested that the severity of inbreeding depression for a trait should increase with the trait's effect on fitness. There are three possible reasons why life-history traits such as survival or fecundity might express greater inbreeding depression than morphological traits such as adult body size or bristle number: (1) there is simply no dominance variance in morphological traits; (2) morphological traits possess dominance variance but the dominance is not directional (i.e., the genes are not predominantly dominant in one direction such as larger size); or (3) there is directional dominance associated with morphological traits but their coefficient of dominance variance ($\sqrt{\text{dominance/trait mean}}$) is lower than it is for life-history traits (DeRose and Roff 1999). Because mutations affecting fitness-related traits (e.g., life-history traits) are typically deleterious and recessive, fitness-related traits are expected to have directional dominance (Lynch and Walsh 1998, p. 270). Traits weakly related to fitness or under stabilizing selection for an intermediate optimum (e.g., morphological traits) may show only slight directional dominance since mutations that cause a change in trait mean in either direction will be selectively equivalent (Lynch and Walsh 1998, p. 270). Thus, there is indirect support for the hypothesis that inbreeding depression will be greater in life-history traits than in morphological traits because the latter will exhibit little or no directional dominance. In an extensive survey of published data, DeRose and Roff (1999) showed that life-history traits are characterized by greater inbreeding depression than morphological traits. To put this into perspective, using average values of the slope of the line relating trait mean to $F$, at an inbreeding level of $F = 0.25$ (brother-sister mating) life-history traits showed a median reduction in trait value due to inbreeding of 11.8%, whereas morphological traits showed a reduction of 2.2%. The reduction in morphological trait values is relatively minor, but the one in life-history traits represents a large decrement in fitness.

The above estimates of decrement in fitness due to inbreeding included estimates for both laboratory and wild conditions. Estimates of inbreeding depression in plants indicate that the degree of depression is increased under field or stressful conditions (see references in Roff 1997). Similarly, inbreeding depression in animals exposed to natural conditions was both significant and higher than those measured under captive conditions (Crnokrak and Roff 1999). Further, in a review of inbreeding in the wild, Keller and Waller (2002) concluded that it occurred with sufficient frequency to both affect fitness and increase the probability of extinction.

In summary, inbreeding is a common phenomenon in nature and can lead to disproportionate changes in the **G**-matrix. Part of this effect is a consequence of drift while part is a consequence of differential effects due to inbreeding depression. The impact on those traits potentially subject to inbreeding depression is dependent on the environment in which the inbreeding occurs. Consequently, while it is apparent that the **G**-matrix will be altered by inbreeding, the actual changes are going to depend upon the inbreeding history, the set of traits included, and the environment.

## The Potential Importance of Antagonistic Pleiotropy

In general, directional selection and stabilizing selection erode genetic variation. Genetic drift and inbreeding also erode additive genetic variance, although, as noted above, drift can also be "creative" in converting nonadditive effects into additive. Of the factors discussed so far, this leaves only mutation as a factor generating new additive variation. Whether mutation is sufficient to replenish genetic variation is still an open question. There are, however, other factors that also impact genetic variation; at least for life-history traits, probably one of the most important of these is antagonistic pleiotropy.

The suggestion that antagonistic pleiotropy may be an important factor preserving genetic variation was made by Hazel (1943) and later by Falconer (1960). However, it was only relatively recently that the mathematical justification for the conjecture has been explored (Rose 1982, 1985; Curtsinger et al. 1994). These studies showed that stable equilibria are most frequent when deleterious alleles are partially recessive (beneficial reversal) and least likely when deleterious alleles are partially dominant (deleterious reversal).

That equilibria are most likely with dominance of beneficial alleles raises two questions: (1) what level of dominance variance is expected, and (2) do we observe this level? Curtsinger et al. (1994) addressed the first question by Monte Carlo simulation, computing the equilibrium genetic variances for those cases in which a stable polymorphism was obtained. These simulations led Curtsinger et al. (1994, p. 221) to conclude, "if antagonism of fitness components often plays a role in maintaining polymorphisms, then the dominance variance for fitness components should, on average, be about half as large as the additive genetic variance for those same fitness components." They concluded, primarily from the data reported in Mousseau and Roff (1987), that dominance variance typically comprises a very small fraction of the total variance, and hence that antagonistic pleiotropy is unlikely to play a major role in the maintenance of genetic variation. There are two reasons why this conclusion is not merited. First, inbreeding depression experiments suggest that dominance variance due to partially recessive alleles is in fact very common for life-history traits, but possibly not for morphological ones (Crnokrak and Roff 1995). The data summarized by Mousseau and Roff (1987) consisted almost entirely of morphological traits, and hence cannot be used as an adequate test of the antagonistic pleiotropy hypothesis. Second, a detailed review of the literature on life-history traits showed that direct estimates of the ratio of dominance to additive genetic variances are in accord with the prediction made by Curtsinger et al.: of the 20 estimates obtained from the literature, 70% (14) have a ratio of dominance to additive variance greater than 0.5, and in 65% (13) of cases $\sigma_D^2 \geq \sigma_A^2$ (see Table 9.14 in Roff 1997).

## Empirical Studies of the Evolution of Additive Genetic Variance

The previous sections suggest that while drift and selection are potentially important factors in the evolution of genetic architecture, it may be impossible to separate their effects simply through statistical analysis, because there is no unambiguous effect attributable to each phenomenon. Here I consider three situations in which knowledge of the biology of the organism can aid in separating these factors. These examples further illustrate that both processes can be important.

### Evolution of Pesticide Resistance in the Oblique-Banded Leafroller

Consider the situation of a population at genetic equilibrium where there is a negative genetic correlation between two traits X and Y. For example, physioogical constraints might cause a negative tradeoff between two life-history traits such as development time (X) and immune defense (Y). Suppose that initially the environment is relatively benign and immune defense is maintained only at a very low level. Using the simple model developed earlier, the **G**-matrix can be written as

$$\begin{bmatrix} n_x p_x q_x + n_c p_c q_c & -n_c p_c q_c \\ -n_c p_c q_c & n_y p_y q_y + n_c p_c q_c \end{bmatrix}$$

Because of the negative tradeoff, selection would maintain the Y loci, including those also affecting trait X, at a relatively low frequency, yielding a low additive genetic variance in trait Y. Now suppose there is a change in the environment such that selection favors a greatly increased allocation to immune defense. Alleles that favor defense will increase in frequency, and hence the additive genetic variance of this trait will also increase. At the same time, the increase in frequency of the shared loci will increase the additive genetic variance of the correlated trait X. This process can be readily visualized if we assume only a single common locus that is initially fixed giving a **G**-matrix of

$$\begin{bmatrix} n_x p_x q_x & 0 \\ 0 & n_y p_y q_y \end{bmatrix}$$

In the face of a challenge requiring increased immune defense, a mutation arises that, by chance, increases immune defense but negatively affects trait X. The **G**-matrix is now

$$\begin{bmatrix} n_x p_x q_x + \sigma_c^2 & -\sigma_c^2 \\ -\sigma_c^2 & n_y p_y q_y + \sigma_c^2 \end{bmatrix}$$

where $\sigma_c^2 = 2p_c q_c(1 + d[q_c - p_c])^2$. The second component in the parentheses is due to the dominance deviation, $d$. The influx of the mutation increases both additive genetic variances. Suppose that selection is sufficiently strong that it eventually drives the mutant allele to fixation. As the allele increases in frequency, the variance contributed by this locus at first increases ($p_c q_c$ increases), but then decreases until at fixation it is zero. The actual turning point of the curve depends on the degree of dominance (Carrière and Roff 1995), but in all cases those components of the **G**-matrix pleiotropically linked to the trait under selection will first increase and then decrease as the alleles approach fixation. This process does not require the creation of new alleles by mutation, simply the increase in frequency of alleles previously at very low values.

The foregoing scenario potentially occurs during the evolution of resistance to an applied pesticide, herbicide, or rodenticide. Such a situation was investigated by Carrière and Roff (1995) using the oblique-banded leafroller (*Choristoneura rosaceana*), a major lepidopteran pest of apple orchards in North America. The application of various types of pesticides led to an increase in pesticide resistance of the larvae in orchards in the area of Oka in Quebec, Canada (Fig. 16.3). Populations from orchards not subject to spraying were highly susceptible to all three commonly used pesticides (Carrière et al. 1995). Test exposure of full-sib families to these three pesticides demonstrated that resistance was heritable and that resistance alleles were in low frequency in populations from untreated orchards (Carrière and Roff 1995). There were highly significant negative correlations between 16-day larval weight and resistance, between pupal mass and resistance, and a positive correlation between development time and resistance, all indicators of tradeoffs between resistance and life-history components (Carrière et al. 1994). Further evidence for tradeoffs is the loss of resistance within five generations in a laboratory colony (Smirle et al. 1998). The mechanism underlying resistance, and probably the tradeoffs, is the increased production of esterases (Carrière et al. 1996; Smirle et al. 1998). The biochemical details of detoxification remain to be resolved (Smirle et al. 1998), while the effect on life-history traits is probably a simple consequence of the metabolic costs of producing the esterases.

To test the predictions outlined above, Carrière and Roff (1995) estimated the additive genetic variances for the two life-history traits larval weight and diapause incidence. Both traits were genetically correlated with pesticide resistance, and hence were predicted to show changes in additive genetic variance as the resistance alleles increased in frequency in the populations. We were not able to measure the frequencies of the resistance alleles directly and so used average resistance to the three insecticides as a surrogate index. As predicted, the additive genetic variances of both life-history traits showed a significant increase with average resistance (Carrière and Roff 1995; Fig. 16.3). With larval weight there was also a significant quadratic term ($P = 0.034$, one-tailed test), while the quadratic was marginal for diapause ($P = 0.062$). The significant quadratic terms are in accord with the prediction that at high allele frequencies the additive genetic variance would decline.

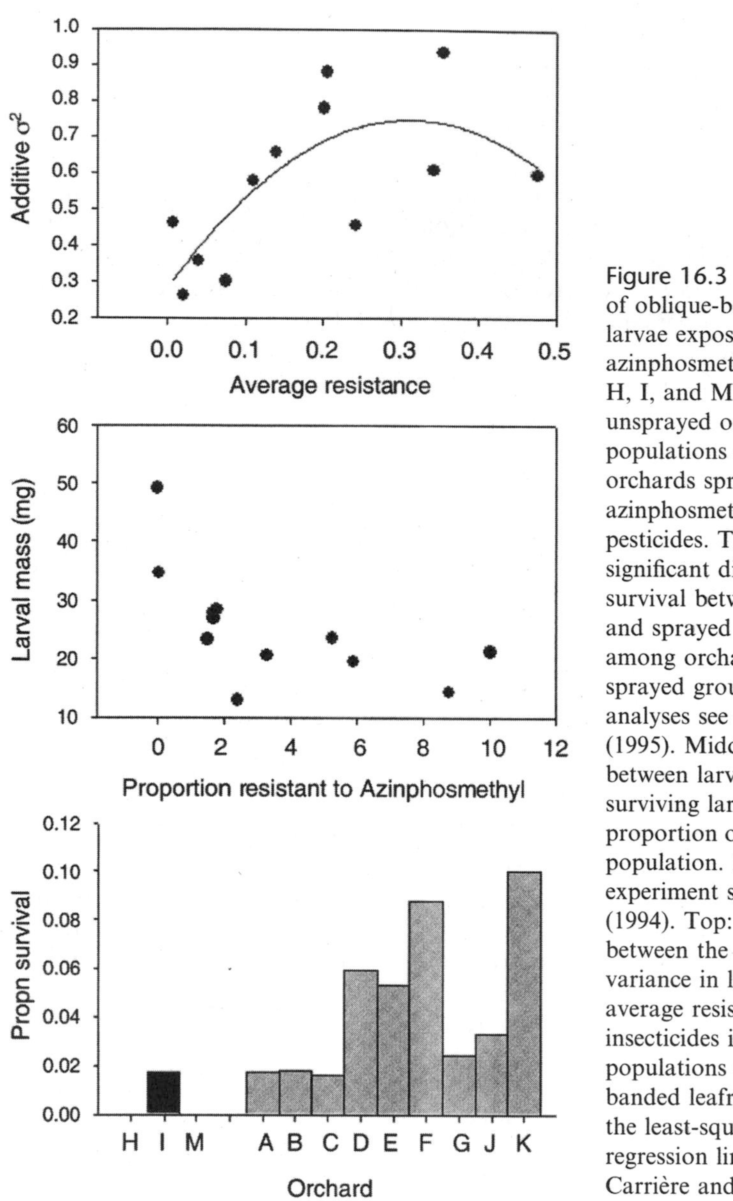


Figure 16.3 Bottom: Survival of oblique-banded leafroller larvae exposed to the pesticide azinphosmethyl. Populations H, I, and M came from unsprayed orchards. All other populations (A–K) came from orchards sprayed with azinphosmethyl and other pesticides. There was a significant difference in survival between unsprayed and sprayed orchards and also among orchards within the sprayed group. For statistical analyses see Carrière and Roff (1995). Middle: Relationship between larval weight of surviving larvae and proportion of survivors in each population. For details of the experiment see Carrière et al. (1994). Top: Relationship between the additive genetic variance in larval weight and average resistance to insecticides in the 12 populations of the oblique-banded leafroller. Line shows the least-squares quadratic regression line. Redrawn from Carrière and Roff (1995).

## Evolution of Calling Song in Two Cricket Species

The components of the calling song in Orthoptera typically show continuous variation, and genetic analyses suggest that this variation is polygenic. Despite the apparent general importance of calling song in mate recognition, empirical observations on mate selection in the two cricket species *Allonemobius fasciatus* and *A. socius* suggest that divergence in song has proceeded largely by genetic drift. If this is the case, we predict proportionality between species in the song components of the **G**-matrix. To eliminate the possibility that selection has caused proportional changes, we need to (1) demonstrate using other types of data that drift is more plausible than selection, and (2) measure song components with low pairwise trait correlations, thereby removing the effect of correlated changes due to shared loci (see above).

*Allonemobius fasciatus* and *A. socius* are closely related, occupy the same type of habitat, and have the same phenology, but are geographically separated over much of their ranges. However, in several places they come together along a mosaic hybrid zone in which hybrids are formed less often than expected under random mating (Howard and Waring 1991). There are highly significant quantitative differences between the species in the components of the calling song (Mousseau and Howard 1998). However, studies of female phonotaxis suggest that conspecific sperm precedence is the cause of present-day isolation rather than variation in male calling song (Doherty and Howard 1996). This result then meets our first condition by providing evidence, independent of the structure of the **G**-matrix, that genetic differences between *A. fasciatus* and *A. socius* in calling song are the result of genetic drift rather than sexual selection.

To assess genetic variation in calling song, crickets were collected from six populations within a 50 km radius of Camden, New Jersey, USA. In this region, *A. fasciatus* and *A. socius* meet and form a mosaic hybrid zone. Two populations were composed of *A. socius* genotypes, three were composed of *A. fasciatus* genotypes, and a sixth population contained a mix of *A. socius, A. fasciatus*, and hybrid genotypes. Genetic variance and covariances for song components were estimated using a full-sib breeding design (Roff et al. 1999).

The calling song of *Allonemobius* males is made up of a series of chirps, each consisting of a number of pulses. Seven calling-song parameters were estimated: mean number of pulses per chirp (M), mean chirp period (C), mean carrier frequency (= dominant frequency, F), mean pulse period (PP), mean pulse rate (PR), and mean pulse duration (PD). To increase the power of the statistical comparisons among the matrices, we eliminated three traits (PD, PR, PP) that had very low levels of additive genetic variance. Traits under strong selection are likely to be those that show little genetic variance. However, including them in the analysis would make the matrices more similar simply because of low statistical power, thus increasing the probability of failing to reject the null hypothesis that the genetic architecture of the songs had not changed.

In the first step of the analysis we compared the matrices among the populations within each species, using the T method, which is a variant of matrix-disparity analysis. In no case were there significant differences among populations in either the **P**- or **G**-matrices. Therefore, in subsequent analyses we combined the populations within each species. Trait heritabilities for both species were large, but the genetic correlations were mostly very low (Fig. 16.4). This satisfies the second condition outlined above that genetic drift is best detected when traits are not strongly correlated. In contrast to the within-species estimates, both the heritabilities and genetic correlations of hybrids were large (Fig. 16.4), indicating considerable interspecific genetic variation. Note that some of this variation could be a result of dominance variance, which inflates full-sib estimates of additive genetic variance.

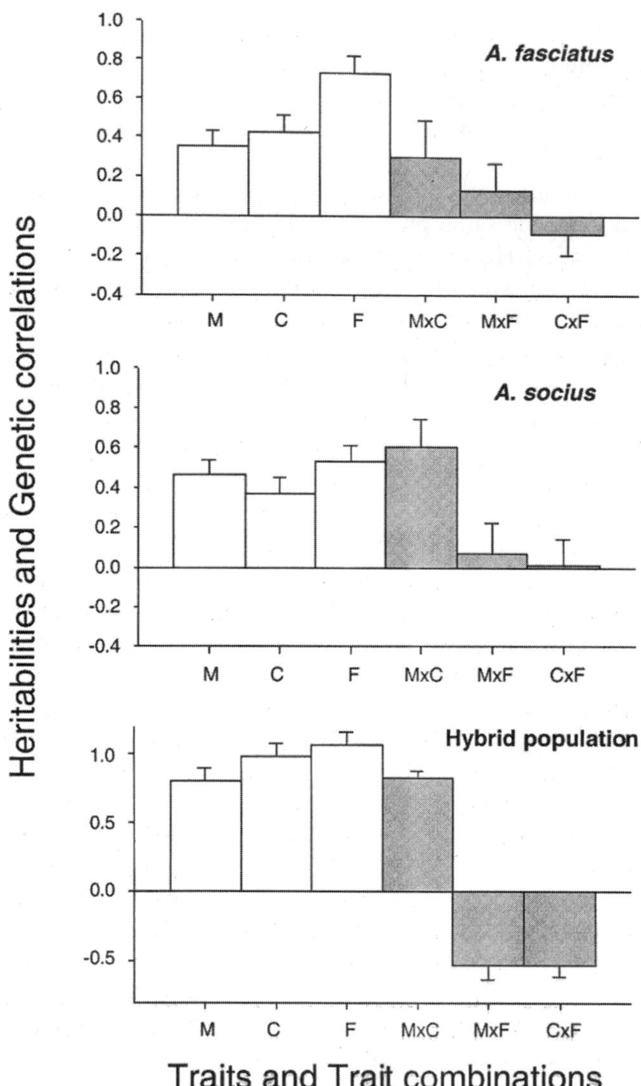


Figure 16.4 Song component heritabilities (white) and genetic correlations (gray) estimated from full-sib pedigrees in two *Allonemobius* species and a population that shows hybridization between the two. M = mean number of pulses per chirp, C = mean chirp period, F = mean carrier frequency. Data from Roff et al. (1999).

Pairwise plots of the additive genetic variances and covariances suggest that the **G**-matrices of the two species differ by a constant of proportionality (Fig. 16.5). A randomization test of the reduced major axis regression indicated an intercept not significantly different from zero (estimate = 0.02, $P = 0.6832$), but a slope significantly different from unity (estimate = 0.60, $P = 0.0072$). These results are consistent with a proportional difference between the two matrices. Comparison

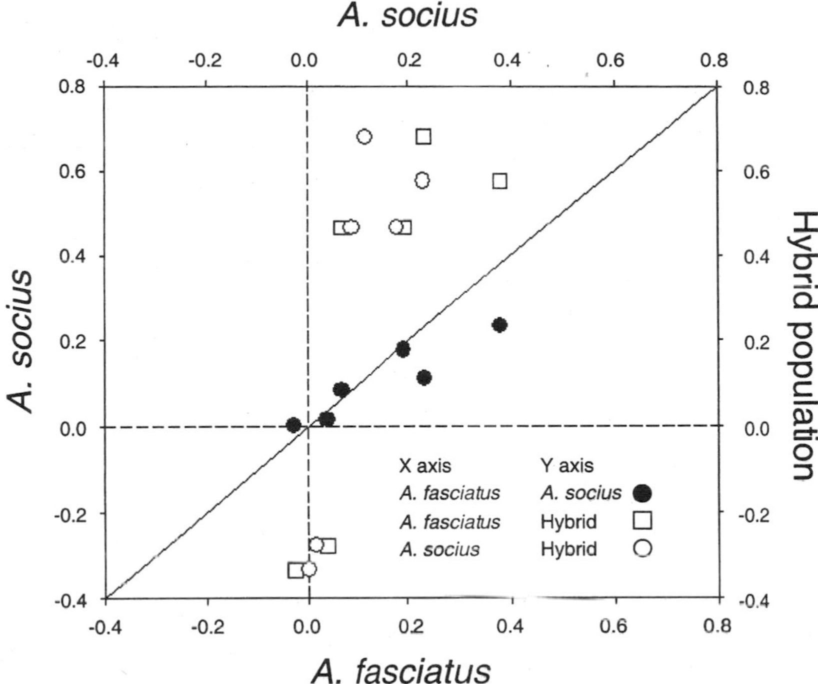


Figure 16.5 Bivariate plots of genetic variances (left 3) and covariances (right 3) of song components in two *Allonemobius* species and a population that shows hybridization between the two. Data from Roff et al. (1999).

between *A. fasciatus* and the hybrid population also showed no significant difference from zero in the intercept, but a significant difference from unity in slope. By contrast, both intercept and slope were significantly different in the comparison of *A. socius* and the hybrid population. In this case the deviation of the two matrices is a consequence of the mixing of two very different genomes.

Taken together, the observation of lack of selectivity by females and the proportionality between the **G**-matrices are both consistent with the hypothesis that the **G**-matrices have diverged primarily under the influence of genetic drift rather than selection. In the next example I present a situation where selection appears to be the primary agent of divergence.

### The Evolution of Morphological Variation in *Gammarus minus*

Consider a single population that becomes divided into two allopatric ones. Over time, the gene frequencies within the two populations will diverge due to drift, the rate of drift being a function of population size and its fluctuations. Consequently, increasing divergence is predicted to result from increased time of separation, decreased population size, and possibly increased frequency of population bottlenecks. Selection also may act on the two populations. We would not expect selection to act with equal intensity on all traits in the matrix, and hence the divergence of the matrices will show fluctuations in the matrix elements, some showing major changes and others changes due to drift alone. In principle, this would generate nonproportional changes in the matrices. However, as described above, drift could also generate nonproportional changes. Thus, detecting the footprint of selection by an analysis of the patterns of variation among matrices without reference to anything else is probably not possible.

In the previous analysis we used the lack of selectivity in female choice to argue a priori for drift. Similarly, if there is some geographic or habit characteristic that covaries with a trait, then this can be taken as indirect evidence for natural selection acting on that trait (Endler 1986, p. 56). The same principle can be applied to the **G**-matrix. In most cases we need to differentiate the effects of both history (i.e., drift) and natural selection. To illustrate this, suppose we have the following situation, schematically illustrated in Fig 16.6, and directly relevant to the case of the amphipod *Gammarus minus*. We have an ancestral population living in a stream, and streams within two separate drainage basins are colonized by migrants from this population. Later, migrants from each of these two separate areas colonize caves within each drainage basin. Assume that there is little or no further migration between the above-ground streams and the caves. There are now two forces acting on morphology and its genetic architecture. First, there is the historical element coming from the fact that (1) all populations are derived from a single ancestral population, and (2) within each drainage basin the populations from the caves are derived from the separate above-ground stream populations. The second force is that of natural selection, which will act similarly on the two cave populations favoring a reduction in organs of sight and an increase in tactile organs (Culver et al. 1995). If only the historical element of drift is important, then the two populations in each drainage basin will show a greater affinity than the corresponding populations between the basins. On the other hand, natural selection will tend to make populations in similar habitats converge to common values. What we must thus do is to disentangle these two factors, or at least be able to assess their relative importance.

*Gammarus minus* is a common amphipod species of karst areas (areas of calcareous rock in which sinkholes and caves are formed by solution, not erosion) throughout the central and eastern United States. Fong (1989) and Jernigan et al. (1994) compared the **G**-matrices of four populations in West Virginia: (1) a population from Benedict's Cave; (2) a population from Davis Spring, which resurges from Benedict's Cave; (3) a population from Organ Cave; and (4) a population from Organ Spring, which resurges from Organ Cave. The two caves and their associated springs are in two separate drainage basins. Fong (1989) assumed that the drainage basins represented replicate areas and that the cave populations were derived from the associated stream populations. In the present analysis I use the following eight traits: head width, number of ommatidia in the compound eye, eye length, eye width, length of the peduncle of the first antennae, length of the peduncle of the second antennae, length of the flagellum of the first antennae, and length of the flagellum of the second antennae.

Individuals from the two populations do not differ much in overall size, as indexed by head width, but do differ significantly in eye and antennal compo-

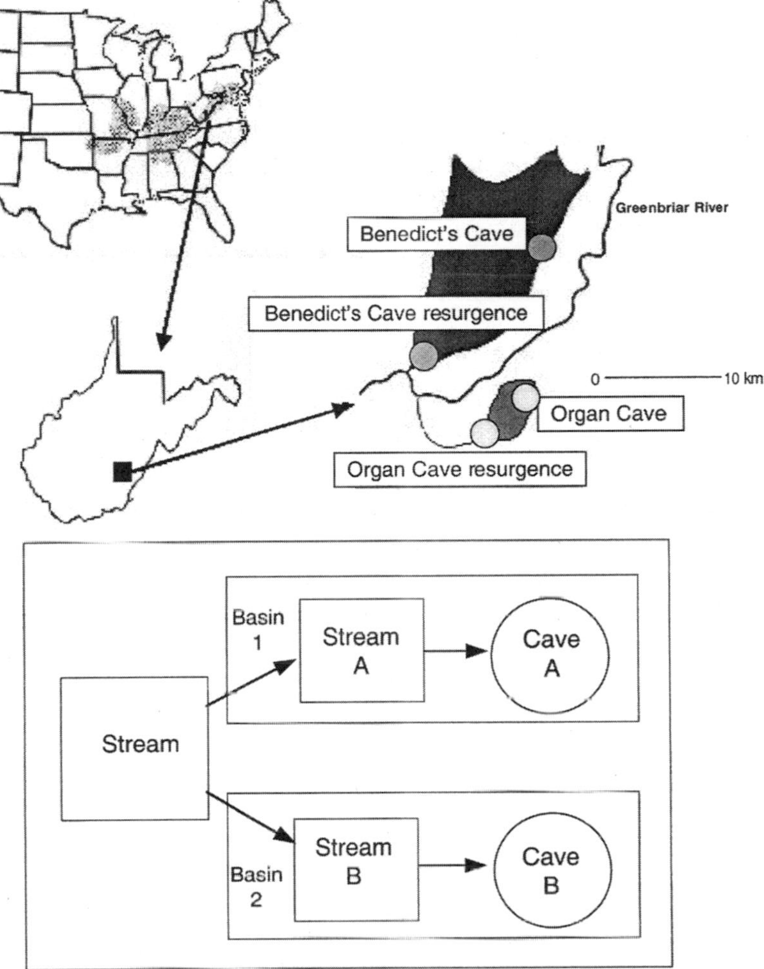


Figure 16.6 Bottom: Schematic illustration of a scenario in which both drift (history) and selection may play a role in the evolution of genetic architecture of a species. If history (drift) is the driving factor, the variation between **G**-matrices within basins will be less than that between basins. If selection is more important, then the variation between **G** matrices within habitats (stream, cave) will be less than that between habitats. Top: Map showing the sampling sites for the four populations of *G. minus* discussed in the text. Adapted from Culver et al. (1995).

nents: individuals from cave populations have a reduced number of ommatidia, smaller eyes, and larger antennal components than those from the spring populations (Fong 1989). These differences are consistent with the hypothesis of adaptive evolutionary change in the cave populations, selection favoring an increase in tactile sensory organs, and a loss of ocular sensory organs.

Fong (1989) estimated the genetic correlations using a full-sib breeding design and compared populations pairwise using a Mantel test. Although the levels of significance obtained from this test may be suspect (reviewed in Roff 1997, pp. 107–109; also see Chapter 15 by Steppan, this volume), the pattern of pairwise correlations between population matrices is nevertheless informative. Fong (1989) found a significant correlation between the two cave populations and between the two spring populations, but not between populations in the same drainage basin. This result suggests that patterns of genetic correlations are more strongly related to habitat (cave or spring) than to drift (common ancestry).

Differences among **G**-matrices can arise simply as a function of scale effects if the groups being compared differ markedly in size. Therefore, I transformed all variables to a log scale, which eliminated the large differences in phenotypic variances. I compared the **G**-matrices using the "jackknife followed by MANOVA" method described in Roff (2002a). The two-way MANOVA shows a highly significant effect of habitat and drainage basin but no significant interaction (Table 16.1). The probability associated with habitat is considerably lower than that associated with basin, indicating that the former is statistically far more important (but see Chapter 7 in this volume for a cautionary note on interpreting $p$-values as a measure of strength of a statistical effect).

To determine which trait (co)variances contribute most to the variation, we can examine the results of univariate tests, keeping in mind that individual significance levels are inflated because of multiple tests. The univariate tests are used to detect a pattern of variation rather than isolating specific (co)variances (which are unlikely to be significant because of the necessary Bonferroni correction). A clear pattern does emerge for the differences associated with habitat, with covariances involving eye length or antennal components generally being "significant" (i.e., $P < 0.05$ without Bonferroni correction: Roff 2002a). In contrast there are only three "significant" ($P < 0.05$) univariate tests for the effect of drainage basin, all involving eye length. This difference between habitat and drainage basin indicates that much greater differences arise between habitats than between drainage basins. These results support Fong's conclusion that habitat is more important than drift (history) in *G. minus* in molding genetic variation.

Table 16.1 A two-way MANOVA examining the influence of habitat (cave, spring) and drainage basin on the **G**-matrices of four populations of the amphipod *Gammarus minus*.

| | Wilkes λ | Approx. $F$ | df | $P$ |
|---|---|---|---|---|
| Habitat | 0.583 | 3.770 | 36,158 | $<0.000$ |
| Basin | 0.671 | 2.154 | 36,158 | 0.001 |
| Habitat × basin | 0.766 | 1.341 | 36,158 | 0.114 |

## Conclusions

Theoretical research is increasingly showing that the statistical description of quantitative genetic variation cannot be viewed as a static phenomenon: that is, the **G**-matrix does not remain constant and nonadditive sources of genetic variation are themselves changing in response to selection and drift. The challenge is to incorporate these new theoretical insights into a model relevant to natural populations. Nonadditive effects should not be viewed as "nuisance" effects, but as potentially major factors in the evolution of genetic architecture. They contribute directly to evolutionary response via the **P**-matrix and by the conversion of nonadditive to additive variance when populations pass through bottlenecks. Additionally, nonadditive variance is the source of inbreeding depression that must necessarily alter the vector of selection differentials, **S**. Nonadditive effects are also critical for the phenomenon of antagonistic pleiotropy. We have strong empirical evidence that the levels of dominance variance in life-history traits is sufficient for antagonistic pleiotropy to be a potent force in preserving additive genetic variance.

In the simplest view of quantitative genetic models, drift causes proportional changes in the **G**-matrix whereas selection alters it in unpredictable ways. However, suites of traits that are highly intercorrelated will show proportional changes in the face of directional selection. Further, population bottlenecks can cause drift effects that convert nonadditive to additive genetic variance. Consequently, there is no single unambiguous signal of drift versus selection in variation among **G**-matrices. Indeed, it is highly likely that both effects are operating at varying magnitudes depending on circumstances. In this chapter I have presented three studies that illustrate how these two effects can be disentangled. In all cases the analysis relies upon supplementary biological information. A narrow focus upon the statistical properties of measures of genetic architecture is unlikely to be very revealing, but can be a powerful tool when combined with other avenues of investigation.

### *Literature Cited*

Bradshaw, W. E., and C. M. Holzapfel. 2000. The evolution of genetic architectures and the divergence of natural populations. Pp. 245–263 in J. B. Wolf, E. D. I. Brodie, and M. J. Wade, eds. Epistasis and the Evolutionary Process. Oxford University Press, Oxford.

Brakefield, P. M., and I. J. Saccheri. 1994. Guidelines in conservation genetics and the use of the population cage experiments with butterflies to investigate the effects of genetic drift and inbreeding. Pp. 165–179 in V. Loeschcke, J. Tomiuk, and S. K. Jain, eds. Conservation Genetics. Birkäuser, Basel.

Bryant, E. H., S. A. McCommas, and L. M. Combs, 1986. The effect of an experimental bottleneck upon quantitative genetic variation in the housefly. Genetics 114:1191–1212.

Burger, R., and R. Lande. 1994. On the distribution of the mean and variance of a quantitative trait under mutation-selection-drift balance. Genetics 138:901–912.

Burger, R., G. P. Wagner, and F. Stettinger. 1989. How much heritable variation can be maintained in finite populations by mutation-selection balance? Evolution 43:1748–1766.

Carriere, Y., and D. A. Roff. 1995. Change in genetic architecture resulting from the evolution of insecticide resistance: a theoretical and empirical analysis. Heredity 75:618–629.

Carriere, Y., J.-P. Deland, D. A. Roff, and C. Vincent. 1994. Life-history costs associated with the evolution of insecticide resistance. Proceedings of the Royal Society of London B 258:35–40.

Carriere, Y., D. A. Roff, and J.-P. Deland. 1995. The joint evolution of diapause and insecticide resistance: a test of an optimality model. Ecology 76:35–40.

Carriere, Y., J.-P. Deland, and D. A. Roff, 1996. Obliquebanded leafroller (Lepidoptera: Tortricidae) resistance to insecticides: among-orchard variation and cross-resistance. Journal of Economic Entomology 89:577–582.

Cheverud, J. M., T. T. Vaughn, L. S. Pletscher, K. King-Ellison, J. Bailiff, E. Adams, C. Erickson, and A. Bonislawski. 1999. Epistasis and the evolution of additive genetic variance in populations that pass through a bottleneck. Evolution 53:1009–1018.

Crnokrak, P., and D. A. Roff. 1995. Dominance variance: associations with selection and fitness. Heredity 75:530–540.

Crnokrak, P., and D. A. Roff. 1999. Inbreeding depression in the wild. Heredity 83:260–270.

Crow, J. F., and M. Kimura. 1970. An Introduction to Population Genetics Theory. Harper & Row, New York.

Culver, D. C., T. C. Kane, and D. W. Fong. 1995. Adaptation and Natural Selection in Caves. Harvard University Press, Cambridge, MA.

Curtsinger, J. W., P. M. Service, and T. Prout. 1994. Antagonistic pleiotropy reversal of dominance and genetic polymorphism. American Naturalist 144:210–28.

DeRose, M. A., and D. A. Roff. 1999. A comparison of inbreeding depression in life-history and morphological traits in animals. Evolution 53:1288–1292.

Doherty, J. A., and D. J. Howard. 1996. Lack of preference for conspecific calling songs in female crickets. Animal Behaviour 51:981–990.

Endler, J. A. 1986. Natural Selection in the Wild. Princeton University Press, Princeton, NJ.

Falconer, D. S. 1960. Selection of mice for growth on high and low planes of nutrition. Genetical Research 1:91–113.

Falconer, D. S. 1989. Introduction to Quantitative Genetics. Longmans, New York.

Fong, D. W. 1989. Morphological evolution of the amphipod *Gammarus minus* in caves: quantitative genetic analysis. American Midland Naturalist 121:361–378.

Frankham, R. 1980. The founder effect and response to artificial selection in Drosophila. Pp. 87–90 in A. Robertson, ed. Proceedings of the International Symposium on Selection Experiments on Laboratory and Domestic Animals. Commonwealth Agricultural Bureau, London.

Goodnight, C. J. 2000. Modeling gene interaction in structured populations. Pp. 129–145 in J. B. Wolf, E. D. I. Brodie, and M. J. Wade, eds. Epistasis and the Evolutionary Process. Oxford University Press, Oxford.

Hazel, L. N. 1943. The genetic basis for constructing selection indices. Genetics 28:476–490.

Howard, D. J., and G. L. Waring. 1991. Topographic diversity, zone width, and the strength of reproductive isolation in a zone of overlap and hybridization. Evolution 45:1120–1135.

Jernigan, R. W., D. C. Culver, and D. W. Fong. 1994. The dual role of selection and evolutionary history as reflected in genetic correlations. Evolution 48:587–596.

Keller, L., and D. M. Waller. 2002. Inbreeding effects in wild populations. Trends in Ecology and Evolution 17:230–241.

Lande, R. 1979. Quantitative genetic analysis of multivariate evolution applied to brain:-body size allometry. Evolution 33:402–416.

Lints, F. A., and M. Bourgois. 1982. A test of the genetic revolution hypothesis of speciation. Pp. 423–436 in S. Lakovaara, ed. Advances in Genetics, Development and Evolution of *Drosophila*. Plenum Press, New York.

Lofsvold, D. 1988. Quantitative genetics of morphological differentiation in *Peromyscus*. II. Analysis of selection and drift. Evolution 42:54–67.

Lynch, M., and B. Walsh. 1998. Genetics and Analysis of Quantitative Traits. Sinauer Associates, Sunderland, MA.

Meffert, L. M. 2000. The evolutionary potential of morphology and mating behavior: the role of epistasis in bottlenecked populations. Pp. 177–196 in J. B. Wolf, E. D. I. Brodie, and M. J. Wade, eds. Epistasis and the Evolutionary Process. Oxford University Press, Oxford.

Mousseau, T. A., and D. J. Howard. 1998. Genetic variation in cricket calling song across a hybrid zone between two sibling species. Evolution 52:1104–1110.

Mousseau, T. A., and D. A. Roff. 1987. Natural selection and the heritability of fitness components. Heredity 59:181–198.

Phillips, P. C., M. C. Whitlock, and K. Fowler. 2001. Inbreeding changes the shape of the genetic covariance matrix in *Drosophila melanogaster*. Genetics 158:1137–1145.

Reeve, J. P. 2000. Predicting long-term response to selection. Genetical Research 75:83–94.

Roff, D. A. 1996. The evolution of genetic correlations: an analysis of patterns. Evolution 50:1392–1403.

Roff, D. A. 1997. Evolutionary Quantitative Genetics. Chapman & Hall, New York.

Roff, D. A. 2000. The evolution of the G matrix: selection or drift? Heredity 84:135–142.

Roff, D. A. 2002a. Comparing **G** matrices: a MANOVA method. Evolution 56:1286–1291.

Roff, D. A. 2002b. Life History Evolution. Sinauer Associates, Sunderland, MA.

Roff, D. A., T. A. Mousseau, and D. J. Howard. 1999. Variation in genetic architecture of calling song among populations of *Allonemobius socius*, *A. fasciatus* and a hybrid population: drift or selection? Evolution 53:216–224.

Rose, M. R. 1982. Antagonistic pleitropy, dominance, and genetic variation. Heredity 48:63–78.

Rose, M. R. 1985. Life history evolution with antagonistic pleiotropy and overlapping generations. Theoretical Population Biology 28:342–358.

Smirle, M. J., C. Vincent, C. L. Zurowski, and B. Rancourt. 1998. Azinphosmethyl resistance in the obliquebanded leafroller, *Choristoneura rosaceana*: reversion in the absence of selection and relationship to detoxication enzyme activity. Pesticide Biochemistry and Physiology 61:183–189.

Wang, J., A. Caballero, P. D. Keightley, and W. G. Hill. 1998. Bottleneck effect on genetic variance: a theoretical investigation of the role of dominance. Genetics 150:435–447.

Willis, J. H., and H. A. Orr. 1993. Increased heritable variation following population bottlenecks: the role of dominance. Evolution 47:949–957.

# 17

# Multivariate Phenotypic Evolution in Developmental Hyperspace

JASON B. WOLF
CERISSE E. ALLEN
W. ANTHONY FRANKINO

Complex phenotypes are constructed during development by interactions among an intricate array of factors including genes, gene products, developmental units, and environmental influences. Although patterns of genetic variation and covariation play a clear role in determining patterns of phenotypic evolution (Lande 1979), it can be difficult to understand how various patterns of developmental interactions affect the evolutionary process, due to the complexity of developmental systems. However, recent advances in theory (e.g., Cowley and Atchley 1992; Atchley et al. 1994; Rice 1998, 2002; Gilchrist and Nijhout 2001; Wolf et al. 2001), and an explosion of interest in evolution and development, make it clear that a consideration of developmental interactions provides major insights into character evolution, including the evolution of phenomena such as canalization (Rice 1998), phenotypic integration (Cheverud 1996; Rice 2000), and genetic variance-covariance structure (Atchley 1984; Atchley and Hall 1991; Atchley et al. 1994; Atchley and Zhu 1997). These insights might allow one to predict (Rice 2002) or reconstruct (Lande 1979; Schluter 1996) evolutionary trajectories, and to identify the proximate basis of differentiation within or among species (e.g., Stern 1998; Kopp et al. 2000; Sucena and Stern 2000).

The "phenotype landscape" (see Rice 2000) has emerged as a useful tool for understanding the way that developmental processes impact evolutionary processes, through the effects of development on phenotypic variation (Rice 2002). The surface of a phenotype landscape (see Fig. 17.1) defines the phenotype associated with a particular combination of underlying factors (e.g., single genes, multigene complexes, developmental modules). The topographical features of the landscape are determined by the developmental system that governs the interactions between the underlying factors. The number of underlying factors contributing to phenotypic variation defines the number of dimensions of space in which the landscape exists.

In theory there is no limit to the number of underlying factors that can influence the expression of a particular trait and thus landscapes can exist in very high-dimensional space (i.e., hyperspace). The terminology we use to describe the topography of a three-dimensional landscape can be applied to hyperdimensional landscapes, but the intuitive interpretation of terms like "slope" or "curvature" become increasingly abstract as the dimensionality increases. However, this does not alter the usefulness of the approach, though it does mean that one should be cautious when interpreting the topography of hyperdimensional landscapes since descriptors like hilly or rugged, which have intuitive meanings in three-dimensional space, may be misleading when applied to higher-dimensional space (see Gavrilets 1997; Gavrilets and Gravner 1997).

The landscape provides a concise summary of the pattern of developmental interactions and the resulting relationship between variation in underlying factors and the phenotype. If there are multiple traits, each is expressed as a separate landscape, influenced by a common set of underlying factors. Pleiotropy is captured by the topographical relationships among such landscapes (Rice 2000; Wolf et al. 2001). The distribution of phenotypes in a population is reflected by the location of the population within the overall phenotype landscape, and movement of a population within the phenotype landscape can be used to describe evolution, providing an intuitive understanding of various evolutionary processes (much like the adaptive landscape: Wright 1932, 1977).

The location of each individual in the phenotypic hyperspace is determined by the value of its underlying factors. Because a population appears as a distribution of individuals plotted on the surface of the landscape, the population "experiences" only the region of the landscape covered by that distribution at any given time. By examining the geometry of the landscape in the local region occupied by a population, we can determine how underlying factors contribute to patterns of trait (co)variation. The slope of the landscape in this region measures the phenotypic variance associated with variation in the underlying factors; more steeply sloped regions correspond to areas of high phenotypic variance and less steep regions correspond to lower variance (Rice 2000). The additive genetic variance (and the covariance when more than one trait is considered) corresponds directly to the slope of the landscape when the underlying factors are genetic. When a landscape is curved so that the slope is not constant, the resulting variance components are weighted averages of the topography of the landscape in the region occupied by the population. Therefore, the properties of the phenotype distribution and the additive genetic (co)variances are determined by the location of the population on the landscape (Wolf et al. 2001).

Figure 17.1 illustrates the relationship between the landscape and properties of the phenotype distribution. Two populations that have equal distributions (bivariate normal) but different means for the underlying developmental factors are plotted on the surface. Population A, shown by the shaded circle, lies in a region of the landscape that is essentially a rising plane. Since individuals lying away from the mean are just as likely to be uphill as downhill on the landscape,

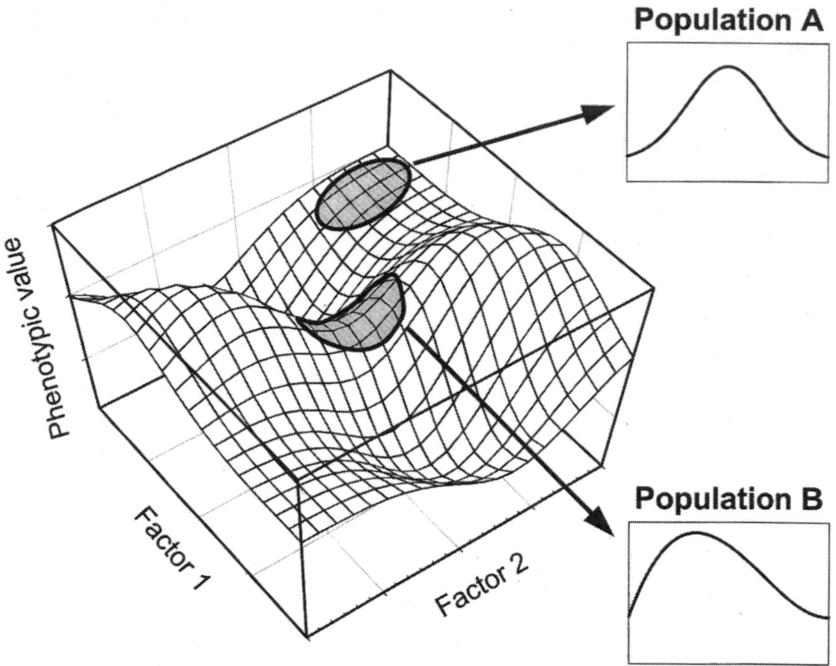


Figure 17.1 An example of a phenotype landscape. The expected phenotypic value of an individual is a function of the value of two underlying factors (1 and 2) that interact during trait development. Populations may lie anywhere on this surface. Two populations (shaded regions) are shown, each having a different mean value for factors 1 and 2. The phenotype distribution of each population is defined by the mean and variance of factors 1 and 2; phenotypes in population A are normally distributed but in population B the phenotype distribution is skewed.

the phenotypes are normally distributed in this population. However, population B lies in a rugged region, located between two hills. As a result, its phenotype distribution is skewed toward positive values since individuals away from the mean are more likely to be uphill than downhill on the landscape. Despite the fact that the two populations have the same distribution of underlying factors (with only the means differing) and share the same developmental system, their phenotype distributions are very different. Thus, we see that, depending on where the population is on the landscape, the distribution of phenotypes need not be normal, even when the set of underlying factors is normally distributed in the population. The landscape in Fig. 17.1 also shows that, when the topography of the landscape is rugged due to the complexity of developmental interactions, we need to know the value of both underlying factors in order to predict the phenotype of any individual or to predict the influence of either factor on phenotype expression.

To predict evolution of a population on the landscape we would need to know the relationship between the phenotype and fitness. The fitness function (the fitness landscape when there are multiple traits) can be used to assign fitness values to each location on the phenotype landscape, creating a new landscape, which defines the relationship between the underlying factors and fitness. This composite surface (the adaptive landscape with respect to underlying factors) can then be used to predict movement of a population on the surface (see Rice 2002 for a general model).

In this chapter we explore the metaphor of the phenotype landscape and the insights it provides into questions about trait evolution, such as evolution of canalization and phenotypic integration. Canalization evolves when the developmental system is altered such that variation in underlying factors leads to less variation in the phenotype; canalization buffers the phenotype from genetic or environmental variation (Wagner et al. 1997; Debat and David 2001), depending on the nature of the underlying factors. Phenotypic integration evolves when traits are coinherited, that is, when they evolve to become developmentally integrated (Wagner 1996), or when they are influenced in the same manner by underlying factors during development (Rice 2000). Our goal is to examine these issues using the phenotype landscape approach to illustrate how the conceptual model can be applied to data from empirical systems, and to suggest future analyses that will elucidate the roles of developmental interactions in phenotypic evolution. Although it is likely that the phenotype is determined by a multitude of underlying factors, meaning that development is often more complex than our models suggest, we use a simple two-factor surface to illustrate most points in the chapter for two reasons. First, landscapes with more than two factors are difficult to visualize since their surfaces exists in a hypercube, and thus do not lend themselves to the intuitive understanding that two-factor landscapes provide. Second, the basic concepts illustrated for two-factor systems apply equally well to any higher-dimensional surfaces (Rice 2002).

In our presentation we focus on genetically based underlying factors such as developmental modules influenced by genetic effects. Developmental modules are the discrete subunits of organization that build phenotypes (Raff 1996; Wagner 1996). They include distinct entities like limb buds or imaginal discs, but can also include more dynamic entities such as morphogenetic fields. Modules exist within a hierarchy of organization, where each module is built from a set of interacting modules at lower levels of organization and usually contributes to modules at higher levels of organization (Wagner and Altenberg 1996; Magwene 2001). At the bottom of this hierarchy lies the genome, and at the top exists the gross phenotype, which itself can be viewed as a hierarchy of phenotypic units. We begin by examining the simplest landscapes, those in which only additive interactions among modules occur. We then explore more complex, nonadditive surfaces to examine the influence of developmental interactions in evolution.

## Evolution on Additive Phenotype Landscapes

The assumption of additivity is the most widely used paradigm in evolutionary genetics (Falconer and MacKay 1996), and so we begin by exploring how additivity between factors affects evolution with the phenotype landscape. Under the additive paradigm, the values of underlying factors (modules) simply sum to build the phenotype, and consequently the phenotype landscape is a plane with constant gradient (i.e., the landscape has no curvature). The gradient is analogous to the slope in two dimensions and is described by a vector pointing uphill on the surface in the steepest direction. The gradient determines the contribution of the underlying factors to components of phenotypic variance. Thus, on a flat landscape (zero gradient), all individuals lie at the same elevation on the landscape and consequently have the same phenotype, regardless of their values for the underlying factors. However, on a sloped landscape, differences between individuals at the genetic level (different values for the factors) are translated into differences at the phenotypic level by development, leading to phenotypic variation. Planar landscapes can be flat or sloped (e.g., a rising plane). However, the gradient is, by definition, constant across the surface of a planar landscape. Therefore, given constant variation in the underlying developmental factors, the amount and components of phenotypic variation are the same at every location on a plane. Thus, the relatively simple geometry of planar landscapes also restricts the possibilities for the independent evolution of the phenotype mean and variance (Rice 2000; Wolf et al. 2001).

The most important evolutionary consequence of planar landscapes is that, regardless of where the population moves on the landscape, the relationship between underlying factors and the phenotype does not evolve (unless the shape of the landscape itself can evolve, which simply implies an additional dimension to the landscape). Because of this, changes in the mean values of factors like developmental modules cannot affect phenotypic variance on a planar landscape. As a result, if the phenotype landscape is planar, movement on that landscape involves only two possible outcomes: (1) Populations move uphill or downhill on the landscape, so that the mean phenotype increases or decreases. The means of developmental factors can change as this movement occurs, but the developmental contribution to (co)variance remains constant because the gradient is constant at all locations. The rate of phenotypic evolution is determined by the gradient of the landscape; a steeply sloped landscape means that the population mean can change rapidly as the population climbs or descends the phenotype plane (Rice 2000). (2) Populations move across the landscape but remain at the same elevation. This can be visualized as a population sliding along a phenotype isocline across a landscape. Thus, the mean phenotype remains constant while the means of the developmental factors change. This can occur due to a drift process, where changes in underlying factors balance in such a way that the changes are essentially neutral with respect to the phenotype. Again, the developmental contribution to (co)variance is constant, regardless of location on the landscape.

For genetically based factors that interact additively, neither the mapping of developmental variance to phenotypic variance nor the genetic architecture depends on the location of the population on the phenotype landscape. As a result, genetic canalization cannot evolve since the developmental system does not allow the decoupling of underlying variation and phenotypic variation. Likewise, if multiple traits all show additive landscapes, then integration cannot evolve since the developmental relationship and the resulting covariance between traits cannot evolve. These processes require a landscape with curvature, allowing the gradient of the landscape experienced by a population to change as a population moves on the landscape.

Although strictly additive interactions between developmental factors may be relatively rare given the myriad of ways factors might interact during development (Rice 2000), additive interactions among factors may be found in the development of holometabolous insects. Most external adult structures (e.g., eyes, wings, antennae) of holometabolous insects develop from semiautonomous modules, the imaginal discs, and interactions among these discs influence the growth rate and final size of the adult traits. From early in larval ontogeny, the sequestered fields of epidermis that comprise the discs grow slowly and somewhat independently from the rest of the insect. Later in larval ontogeny, the cell populations in the discs grow exponentially and differentiate, eventually developing into incipient adult structures. Experiments indicate that additive communication among discs probably affects disc ontogeny and, ultimately, the size of the adult structures. For example, removal of the hindwing discs of the Buckeye Butterfly, *Precis coenia*, increases the final size of other morphological structures developing in the same vicinity (Nijhout and Emlen 1998). If both hindwing discs are removed, the change in size of the forewings is nearly twice that resulting from the excision of just a single disc. These results suggest that discs may compete for a limited pool of resource or morphogen, and that this competition affects disc growth in a nearly additive way; the effect of disc removal is proportional to the rate at which the disc acquires or processes the factor, which is presumably correlated with disc size. Moreover, because disc removal affects the growth of only those discs in close proximity to the manipulated disc, disc-disc interactions may integrate the ontogeny of local "neighborhoods" of developing traits. These disc-disc interactions appear to be conserved across holometabolous insects, as reflected in negative correlations among the size of morphological traits that develop in proximity to one another (Emlen 1996, 2001; Emlen and Nijhout 2000).

## Evolution on Nonadditive Phenotype Landscapes

The additive paradigm is predominant because, by assuming additivity, one can develop simplified models where many higher-order terms vanish and the long-term dynamics of a system remain predictable. Among these models are those that form the foundations of quantitative genetic theory, which generally assumes that nonadditive (i.e., epistatic) effects are absent (Roff 1997). It is important to note that the assumption of additivity is often based more on computational necessity than on biological reality. However, this assumption does not necessarily reflect biology, since many developmental systems can lead to nonadditive interactions between genes, cells, tissues, or modules. Thus, incorporating nonadditive interactions into models may improve the "biological reality" of theory. Perhaps more important than the increased biological reality is the fact that evolutionary processes that lead to evolution of variances and covariances, which are required for canalization and integration to occur, cannot occur unless nonadditive interactions are present (Rice 2000). Thus, a consideration of nonadditive developmental interactions should yield insight into how quantitative genetic parameters change during phenotypic evolution (Wolf et al. 2001).

The most important implication of nonadditivity of the phenotype landscape for evolution is that the phenotype distribution (characterized by variances and covariances) can evolve as the population moves on the landscape. In the following subsections we examine the impact of nonadditive interactions between developmental factors on evolution of the phenotypic mean, the evolution of variance and canalization, and the evolution of covariance and integration. We end by considering the contribution of developmental processes to population differentiation and propose ways to apply the phenotype landscape approach to empirical questions. Throughout, we contrast evolution on additive and nonadditive phenotype landscapes, paying particular attention to how the scale of population variation relative to that in the topography of the landscape affects evolutionary outcomes over the shorter and longer term.

### Evolution of the Phenotypic Mean

When the phenotype landscape is curved, movement on the landscape can be associated with changes in the variances and covariances that characterize the phenotype distribution (Fig. 17.1). Evolutionary changes in (co)variance structure can occur because the curvature and average gradient are not constant across the landscape. These changes in (co)variance result from developmental interactions and can accompany evolution of the mean phenotype, as populations move uphill or downhill on a landscape. Alternatively, (co)variance can evolve independently of the mean phenotype, for example as a population moves along a contour of equal phenotype to a part of the landscape where average slope and curvature differ (Fig. 17.2). Models of phenotypic evolution generally deemphasize evolutionary change in genetic variance components (e.g., Lande 1979); however, it is clear that if variances are to remain constant, the population must exist in a very limited parameter space where the phenotype landscape occupied by a population is invariant.

Even a very rugged landscape can appear relatively invariant if the scale of the topography is either much larger or smaller than the scale of variation in the underlying factors. The importance of scale of variation on the phenotype landscape is analogous to the problem of environmental grain in ecology (Levins 1968): populations with little variation in developmental factors relative to the scale of ruggedness in the landscape experience only the local, perhaps nearly additive (i.e., planar), portion of the landscape. In other words, due to the coarse grain of the phenotype landscape, the population does not "see" the ruggedness since the local region of the landscape is additive, and the nonlinearity lies outside the range of variation present in the population (at a given point in time). As a result, populations evolve in this local region of the landscape in a fashion that is described by an additive model, even though the developmental system that produces the phenotype is not really additive. Hence, nonadditivity of the phenotype landscape does not necessarily render the additive model useless. Nevertheless, as the population explores more of the topography of the landscape over greater evolutionary time, the additive model fails. This may explain why additive models work, at least to some degree, when describing short-term dynamics but often fail over a greater evolutionary scale. The shape of the phenotype landscape deter-

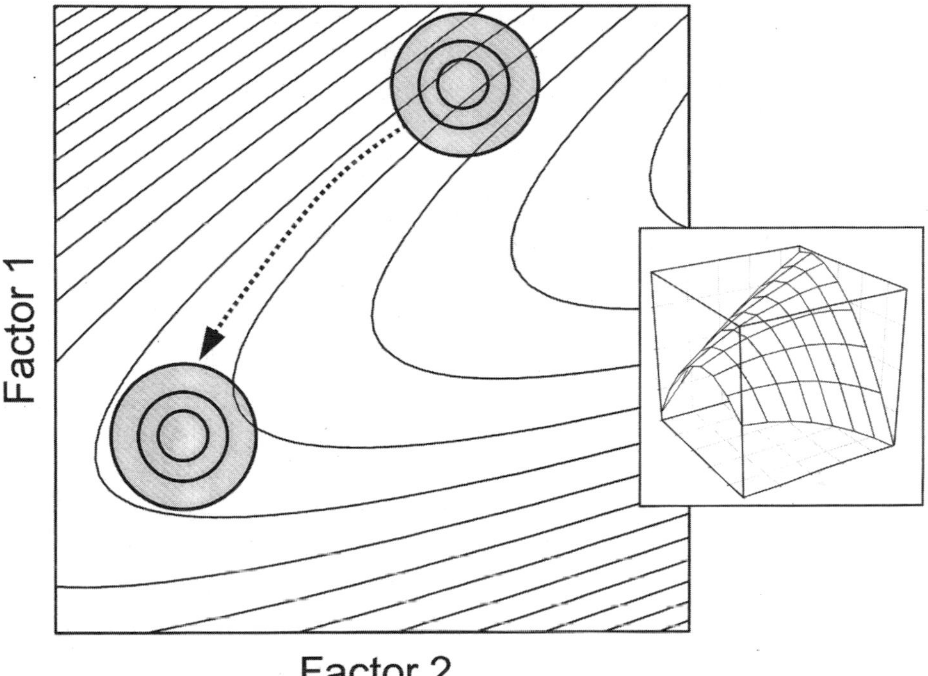


Figure 17.2 Variance can evolve independently of the mean phenotype. A population (shaded circle) starts on a steep region of the landscape, at upper right. Canalization evolves as the population, under stabilizing selection, slides along a phenotype isocline to the relatively flat area along the ridge where phenotypic variance is reduced. Note that the population starts in a location where it covers four isoclines and moves to a location where it essentially lies between a single pair of isoclines. In three dimensions, this landscape is a rising ridge (inset).

mines long-term evolutionary patterns, independent of biases imposed by our limited view of the local domain occupied by a population.

Nonadditivity also may be relatively unimportant when the ruggedness of the landscape occurs at a scale that is much smaller than the scale of variation. In this case, the population covers enough of the landscape to average over all of the variation introduced by the ruggedness. Hence, nonadditivity adds only noise to the system in this case and therefore does not significantly alter the evolutionary dynamics (Fisher 1930). In a quantitative genetic framework, this nonadditivity (i.e., the epistatic or dominance variance) is considered evolutionarily unimportant because it measures deviation from the additive model and does not contribute to heritability. Together, these considerations of scale suggest that nonadditive developmental interactions remain unimportant only if populations continue to experience such equally rugged regions of the landscape through evolutionary time.

Artificial selection experiments are often used to analyze phenotypic evolution and to infer genetic variance components based on the response of a population over many generations of selection. The "realized heritabilities" (or realized genetic correlations) estimated in this way do not provide information about the shape of the landscape because realized parameters are averages, taken from the region of the phenotype landscape over which the populations traveled as the mean evolved. While stochastic variation in the response to selection is expected each generation, some of the presumably "stochastic" variation observed might have important evolutionary implications—if the variation is actually due to nonlinearity of the true phenotype landscape. Since the parameter estimates produced by this approach provide no information about the shape of the phenotype landscape, it is unclear how useful they are because they describe only the path already traveled by the population—they do not allow one to predict future evolutionary trajectories. For example, imagine a population evolving on a phenotype landscape that is a ridge (Fig. 17.3) (a topography that would be due to a combination of epistasis and dominance in the developmental system: Rice 2000; Gilchrist and Nijhout 2001). If the population is selected in the direction of the ridge and starts well downhill (the first position), it can evolve for

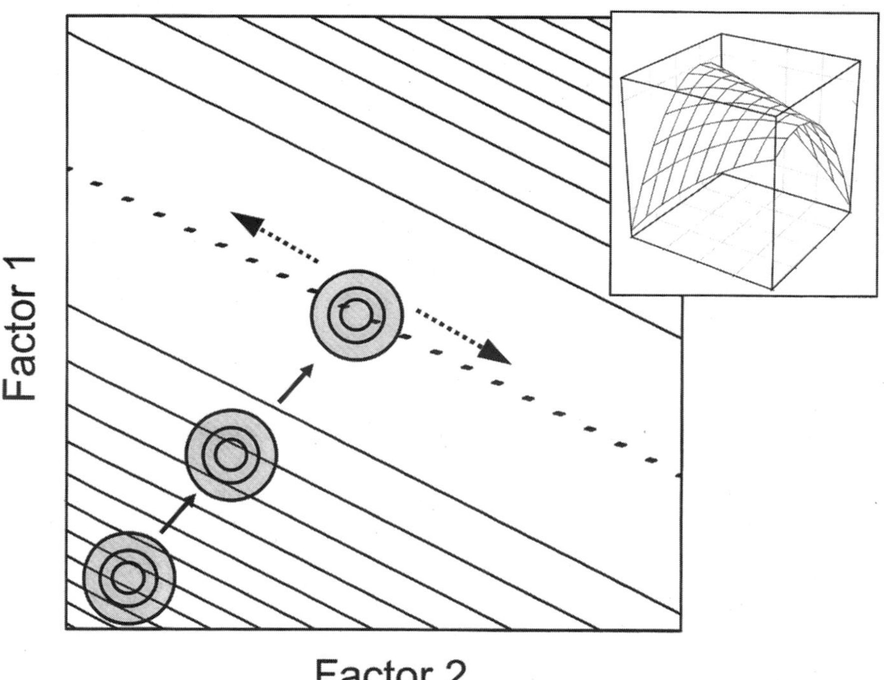


Figure 17.3 Evolution on a landscape that is a rising ridge. The population (shaded circles) starts at the lower left corner, in a region that has a steep uphill gradient. Directional selection favoring larger trait values pushes the population up the hill, but as it climbs the hill the gradient of the landscape decreases. The population comes to rest on the ridge, where there is little variation produced by development. The population is now free to drift along the ridge since all locations on the ridge have the same phenotypic value. Inset: a three-dimensional view of this surface.

many generations in that direction (going from the first to the second position over several generations). In this case, the realized heritability would describe the average gradient of the hill climbing toward the ridge. However, that parameter is not informative as the population nears the ridge where the landscape flattens out, nor does it identify the fact that the population will reach a point beyond which the mean cannot increase.

Despite the fact that the same loci (and perhaps the same distributions of genotypic variation) contribute to phenotypic variation in different regions of the landscape, changes in variance components occur as the phenotypic mean evolves. These changes are accompanied by an apparent change in genetic and developmental architectures, despite the fact that the same developmental system maps genetic variation to phenotypic variation in different regions of the landscape. This suggests two different views of genetic or developmental architecture: the local view, where the architecture describes the pattern of genetic or developmental effects that contribute to phenotypic variation *within* a population; and the global view, where the architecture describes the entire range of variation possible. The degree to which curvature of the phenotype landscape influences the evolution of real populations remains a largely unanswered empirical question. However, the landscape view suggests that, when the ruggedness of the landscape is of the proper scale, the mapping from underlying factors to the phenotype will evolve as a population explores different regions of the landscape.

## Canalization and Evolution of Genetic Variance

Stabilizing selection for a single optimal phenotype favors a canalized developmental system, one that reduces phenotypic variation when a population's mean reaches the optimum. Away from the optimum, selection favors a developmental system that maps genetic variation onto phenotypic variation, allowing the population to climb the hill to a fitness peak. Thus, selection will push populations to various parts of a phenotype landscape with topography that increases population mean fitness (Rice 2000, 2002), and the direction of that movement depends on the initial location of the population on the landscape. Figure 17.2 shows an example where a population starts in a region of a landscape that is steeply sloping (indicated by the fact that the population distribution overlaps four phenotype isoclines). If the population lies at a fitness peak, selection for canalization (i.e., low phenotypic variance) pushes that population along a phenotype isocline to a region of the landscape that is very flat (indicated by the fact that the population now overlaps only a single isocline). Since the mean of the population moves from the first location to the second location along a phenotypic isocline, the mean phenotype remains constant while the contribution of the underlying factors to phenotypic variation is diminished (i.e., canalization is maximized).

Canalized phenotypes can evolve via factors that buffer developmental pathways against underlying genetic variation. One striking example of such a buffering system is the protein family that includes heat-shock protein 90 (Hsp90), which acts to stabilize proteins such as transcription factors that are important constituents of developmental pathways. When Hsp90 function is disrupted, by exposing developing organisms either to drugs or to extreme environments, genetic variation is revealed that produces an abnormal range of phenotypes (Rutherford and Lindquist 1998; Queitsch et al. 2002). The action of a system like Hsp90 could allow populations to move to a region of the landscape where phenotypic variation is reduced and is decoupled from underlying allelic and developmental variation. Such regulatory mechanisms that promote canalization might mask developmental divergence among populations. For example, two populations that share an ancestral, canalized phenotype might begin to accumulate subtle differences in values of developmental factors as the populations drift along phenotype isoclines in canalized regions of the landscape (like the ridge in Fig. 17.3). Under normal conditions, the buffering system would mask underlying variation, but extreme conditions that disrupt the buffering system could reveal underlying developmental differences between populations or closely related species. Hybridization among closely related *Drosophila* species provides strong evidence for hidden genetic variation underlying canalized phenotypes (True and Haag 2001). The shared, canalized thoracic bristle pattern exhibited by both *Drosophila melanogaster* and *D. simulans* is disrupted in the $F_1$ hybrid, and the disruption is attributed to epistatic interactions between a number of autosomal loci and a single X-linked factor (Takano-Shimizu 2000). Disruptions of the buffering system might move populations from canalized (less steep) regions of the landscape to more steeply sloped regions where underlying variation maps onto the phenotype, allowing selection to increase or decrease the mean phenotype (e.g., Queitsch et al. 2002).

Whereas stabilizing selection can lead to canalization, other types of selection might favor increased phenotypic variance. Various forms of diversifying selection (such as disruptive selection) may push a population to a region of greater variance (i.e., a region of greater slope). Directional selection can also act to decanalize the phenotype, by pushing a population either uphill or downhill on the phenotype landscape (depending on the direction of selection) while also pushing the population toward a more steeply sloped region of the landscape (Rice 2000).

## Integration and Evolution of Genetic Covariance

Integration occurs when underlying factors lead to variation in more than one trait and covariation between traits (Rice 2000). Thus, in order to understand the evolution of integration and genetic covariance, we can construct landscapes for multiple traits and examine the topographical relationships between these landscapes. In the simplest case there are two factors, each influencing two traits (in more complex multifactor space some factors might influence only one of the traits). The degree of integration between two traits is determined by the extent to which the underlying factors influence the two traits in a similar way, leading to a phenotypic correlation between the traits.

In order to illustrate the relationship between landscapes and the genetic covariance, we begin with an additive example where the two factors have purely additive effects on two traits. Both phenotype landscapes are therefore planar and the same two factors define the dimensions of the landscapes. If the two planes slope in the same direction, individuals with larger values for the first trait will necessarily have larger values for the second trait and the two traits will be correlated (Fig. 17.4a). However, if the planes are perpendicular to each other, moving uphill on one plane moves along an isocline of the second plane, and in this case, individuals with larger values for trait A may have any value for trait B (Fig. 17.4b). Thus, the degree of correlation between two traits is determined by the similarity of the gradients of their phenotype landscape. This relationship holds true even if the two landscapes are not planar (Rice 2000).

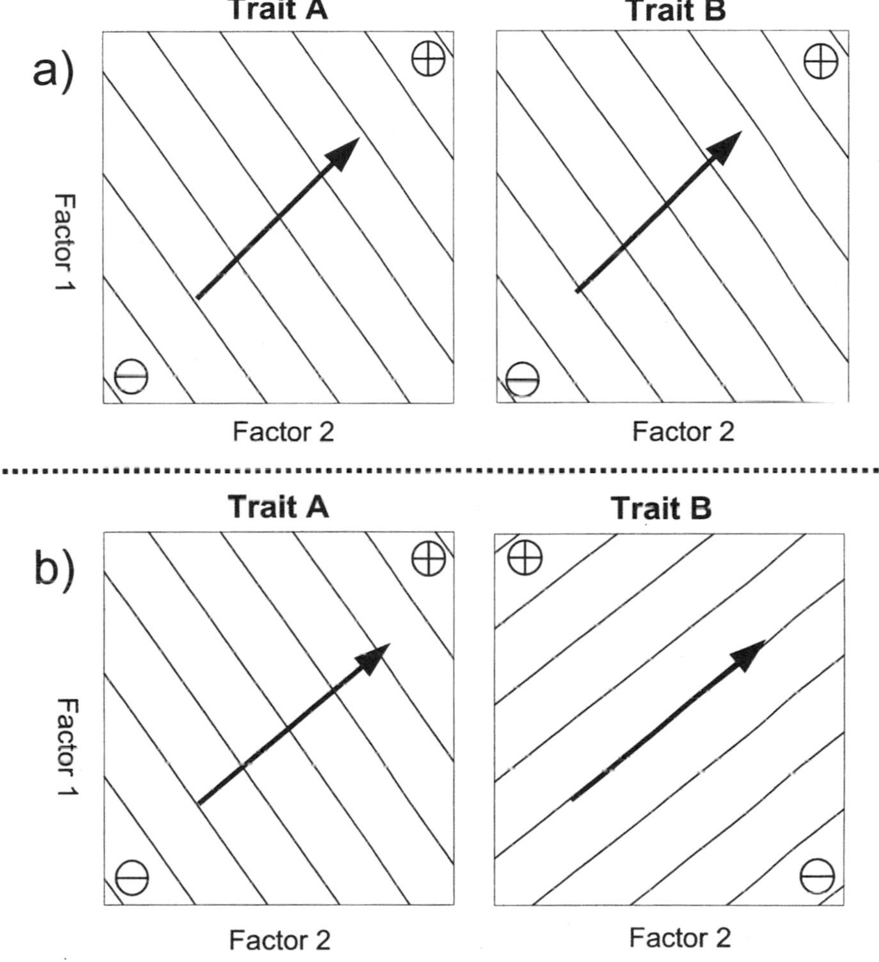


Figure 17.4 Additive phenotype landscapes for two traits. The same two factors influence both traits. The high point on each landscape is marked with a + sign. (a) The landscapes for traits 1 and 2 are similarly sloped, leading to a positive covariance between the traits. Under directional selection for larger trait values, the two traits evolve in the same direction (arrows). (b) The slope of the landscapes for traits 1 and 2 are perpendicular, leading to a zero covariance between traits (see Wolf et al. 2001). If trait 1 responds to directional selection by moving uphill, trait 2 will slide along an isocline and will not evolve (arrows).

For integration to evolve, only one of the traits needs to have a nonadditive landscape. Therefore, we can illustrate the evolution of integration with a simple example: one trait has an additive landscape and the other has a curved landscape. If there are just two underlying factors, the correlation between a pair of traits cannot evolve unless the mean of at least one trait changes. However, it is important to note that, although we use an example where trait means change, selection for integration in higher-dimensional systems may favor a system of trait development that allows trait means to remain at some optimum while genetic correlations change. Figure 17.5 shows two traits, where the value of trait A is an additive function of the underlying factors while the other trait shows a nonadditive, peaked function. The population moves from a location of zero correlation

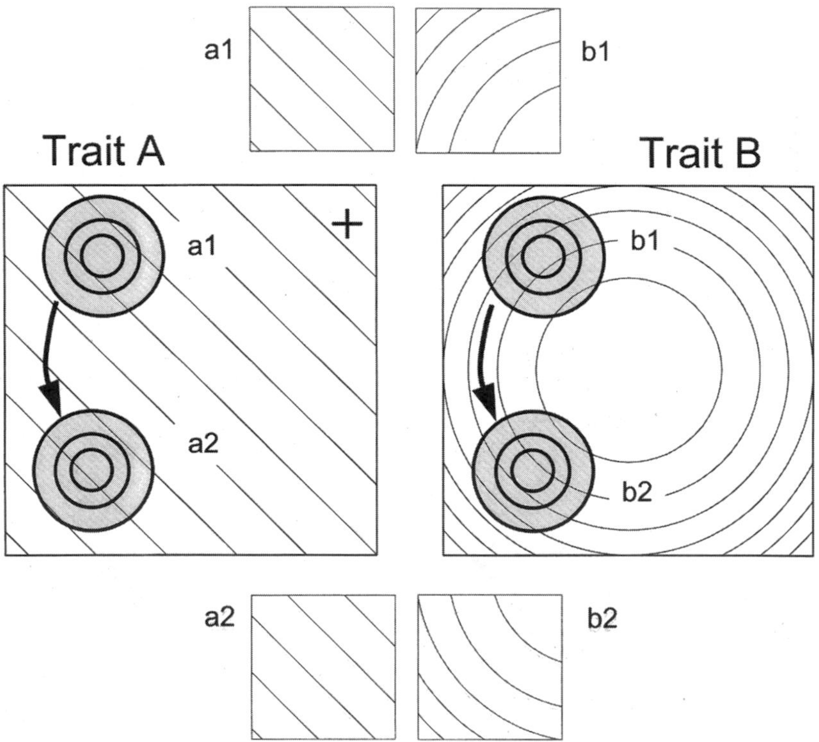


Figure 17.5 Phenotype landscapes for a pair of traits influenced by the same set of factors. Trait A is an additive function of these factors, and trait B is a nonadditive function of the same factors. The two large figures represent the landscapes, and the shaded circles represent a population evolving on the surface. The small surfaces (top and bottom) show an enlarged picture of each landscape at two locations occupied by the population. At the initial location (a1 and b1), the gradients of the two landscapes are perpendicular and there is zero covariance between the traits. As the population moves to the lower left (toward location a2 and b2), trait A evolves downhill on the landscape while trait 2 slides along a phenotype isocline. As the population moves, the values of the underlying factors evolve, and the covariance between the traits changes. At the ending location (a2 and b2 on the respective landscapes), there is positive covariance between traits A and B.

(indicated by the a1 and b1 labels), where the gradients of the local regions of the landscapes occupied by the population are roughly perpendicular, to a region of large positive correlation (indicated by a2 and b2). This is more obvious when the local regions of the landscape are isolated and magnified, as in the insets of Fig. 17.5. Note that the local regions of the landscape occupied by population B in both the starting and the ending location appear nearly planar in magnified view. This demonstrates that a nonadditive landscape may appear nearly additive at a local scale. Note also that the value of trait B remains constant because the population moves along an isocline, while the mean of trait A evolves as the population moves downhill. This type of phenotype landscape, where the pattern of genetic covariance can evolve due to changes in pleiotropic relations, has been demonstrated by Cheverud (2001) in an analysis of QTL effects on the mouse mandible and has been termed "differential epistasis."

The results shown in Fig. 17.5 imply that the genetic correlation between two traits can be very labile, evolving as the populations move through phenotypic hyperspace. Although conserved genetic covariance patterns may constrain evolutionary trajectories (e.g., Schluter 1996), a number of studies demonstrate that strong directional (e.g., Leamy and Atchley 1984; Roff et al. 2002) or antagonistic selection (e.g., Sheridan and Barker 1974; Weber 1990) can disrupt covariance structure over relatively short time periods (but see Chapter 11 in this volume). Current theory does not allow us to predict when changes in covariance structure will occur, or whether covariance will increase, decrease, or remain constant under selection (Turelli 1988). However, viewing evolution on a phenotype landscape shows that properties of the underlying developmental system can have important effects on genetic covariance.

One particularly well-understood system, where there is considerable information available on both trait development and evolution, is the expression of color patterns on butterfly wings. A number of studies have revealed the presence of nonadditive developmental landscapes. Experimental evolution resulting from selection on these traits has also demonstrated considerable evolutionary lability of genetic covariance structure. Many studies focus on one kind of taxonomically widespread color pattern, sets of concentric colored rings typically called "eyespots" (reviewed by Brunetti et al. 2001). Individual eyespots appear to be separate traits influenced by a number of identifiable underlying factors. At the most basic level, these interactions are nonadditive because the color exhibited in different regions of the eyespot is determined by the level of morphogen to which individual wing cells are exposed, and the nonlinear threshold response of the cells to the morphogen (reviewed in Brunetti et al. 2001; Beldade and Brakefield 2002). Within a wing, groups of eyespots are integrated, showing coordinated expression, with positive correlations among groups of eyespots within specific hypothesized regulatory regions (reviewed in McMillan et al. 2002). These correlations, however, are not barriers to the response to selection of at least some eyespot characters in even the short term. The high genetic correlation among eyespots in size, for example, can be broken by disruptive selection on different size combinations among eyespots (Beldade and Brakefield 2002). The developmental basis of variation in eyespot expression has been well characterized (e.g., Nijhout 1980; Carroll et al. 1994; Weatherbee et al. 1999; Beldade and Brakefield 2002; reviewed in McMillan et al. 2002; Nijhout 1991; Brunetti et al. 2001) and so it is likely that phenotype landscapes could be constructed to predict the patterns of change found in experimental evolution experiments. The proximate basis of evolutionary change is known or strongly suggested in some cases (e.g., Monteiro et al. 1994, 1997; Brakefield et al. 1996, Beldade et al. 2002), providing a particularly powerful test of predictions.

Evolutionary lability of genetic covariance structure underlying integrated traits in natural populations has been elegantly demonstrated in the pitcher-plant mosquito *Wyeomyia smithii* (reviewed in Bradshaw and Holzapfel 2000). Populations of *W. smithii* exist along a very large south-north gradient from the Gulf of Mexico (30°N) to Saskatchewan (54°N). In the northern regions (>40°N), populations appear to have been established by a series of founder events after the last glaciation. Across the range, selection favors different optimal photoperiods for diapause induction, increasing with latitude. Within populations, faster-developing individuals are selected to have shorter critical photoperiods since they can complete one more generation per year than slower-developing genotypes. This creates a situation where selection favors integration between critical photoperiod and development time (a positive covariance is indeed observed and is large) and shorter development times within all populations, but divergent critical photoperiods along the latitudinal gradient. From an additive viewpoint the large genetic covariance within populations would appear as a constraint on the evolution of increasing critical photoperiod as populations moved north. Although genetic correlations less than unity do not absolutely preclude the independent evolution of critical photoperiod and development time, the fact that critical photoperiod has diverged 10 standard deviations along the gradient while development has changed less than one standard deviation, coupled with the relatively short evolutionary history, suggests that the correlation itself does not represent a constraint (Bradshaw and Holzapfel 2000). This sort of evolutionary change matches the scenario depicted in Fig. 17.5. A population can move to a region where the traits are not correlated, and critical photoperiod can evolve directionally while development time remains on a phenotype isocline.

Cases where the developmental basis of genetic correlations allows for rapid evolution of the correlations cannot easily be analyzed within the classic quantitative genetic framework, because this framework generally assumes constancy or proportionality of these parameters over evolutionary time (Lande 1979). On the other hand, the understanding we achieve using the phenotype landscape approach can be more sophisticated since it is not constrained by simplifying assumptions (Rice 2002). Thus, it can be used to gain insight into a number of evolutionary processes that are not accessible to classic, variance-partitioning quantitative genetics.

## Differentiation of Populations

One of the most important processes in evolution is the differentiation of populations. Although population differentiation is the foundation for speciation, differentiation also plays an important role within species. Evolutionary divergence can lead to various distributions of phenotypic and genetic variation across the range of a species, to the extent that outbreeding depression can occur when differentiated populations later cross (e.g., Edmands 1999; Gharrett et al. 1999; Fenster and Galloway 2000). Such population and species differences provide snapshots of various regions of the phenotype landscape, and enhance our understanding of evolutionary processes by allowing for a larger-scale understanding of the genetic and developmental basis of phenotypic variation (Johnson 2001).

Populations can become differentiated with respect to the mean values of their phenotypes, or with respect to the values of the underlying factors. Although populations can also differentiate with respect to the shape of the phenotype landscape, this simply implies that there are additional dimensions through which populations can move, which alter the apparent shape of a landscape as viewed in reduced dimensional space. For example, if evolution at one locus alters the shape of the landscape describing the phenotype as a function of alleles present at two other loci, this third locus is simply an additional dimension. The third locus is epistatic with the two loci and the epistatic interaction appears to change the shape of the landscape in the reduced dimensional (i.e., two-locus) system.

As discussed previously, changes in underlying factors are not always accompanied by changes in the mean phenotype. Figure 17.6a shows an example of population differentiation, where two lineages evolve independently from a single ancestral population. The populations experience the same directional selection pressure for larger phenotypic values, but by random chance take different routes to reach the same phenotypic value at the point where we now find them. These two populations appear phenotypically identical, with the same phenotypic mean and variance. However, the genetic basis for the phenotype is completely different. Crossing these two populations would produce a genetically intermediate population that would lie in a valley of lower phenotypic value, and thus, of lower fitness (since there is directional selection for larger values) (Fig. 17.6b). In this case, outbreeding depression results because hybridization breaks up the developmentally integrated (i.e., coadapted) combinations of factors.

One important result of artificial selection analyses is that replicate populations exposed to uniform selection (of equivalent intensity and direction) often reach the same phenotypic endpoint via very different genetic or developmental changes. For example, replicate lines of mice selected for increased tail length (Rutledge et al. 1974) differed with respect to the underlying mechanism, but not the degree, of tail elongation: one replicate increased the number of tail vertebrae, while the other increased the size of individual vertebrae without increasing vertebral number. This response is similar to that shown in Fig. 17.6a. In response to directional selection, the two replicate populations can move independently to regions of the phenotype landscape characterized by identical tail lengths but different combinations of underlying factors that control the growth or condensation of vertebral elements that make up the tail. In this case, the direct response to selection (tail elongation) is the same in the two populations, but the correlated response to selection (change in the size or number of vertebrae in response to selection on tail length) differs between the populations.

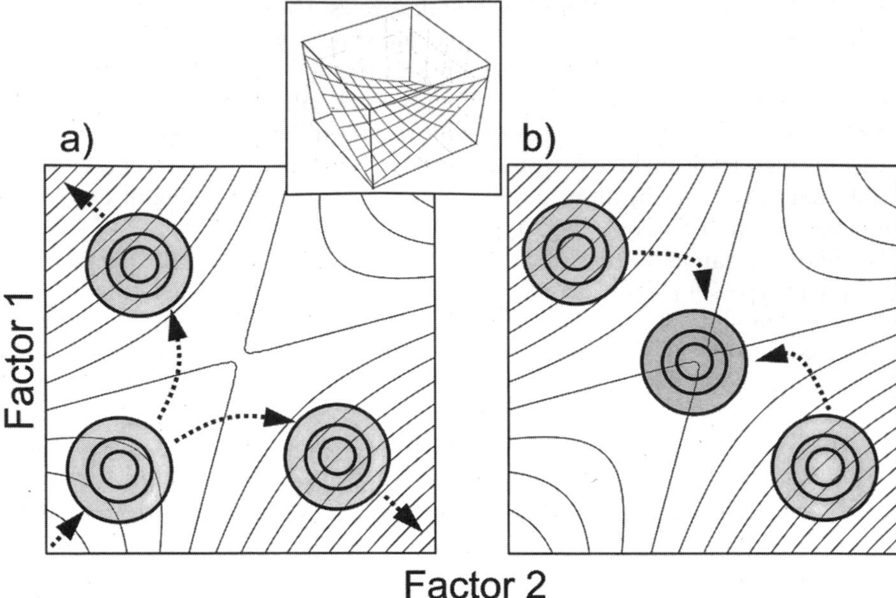


Figure 17.6 Population differentiation in a saddle-shaped phenotype landscape. (a) A population starts at the lower left corner (the dark circle) and directional selection for larger trait values pushes it uphill (illustrated by the dashed arrows). As the population reaches the saddle area it can move uphill on either side of the saddle since the landscape is symmetrical. The two alternatives are shown in the upper left and lower right regions. (b) Two populations that took different routes up the saddle are differentiated with respect to underlying factors but have the same phenotypic mean since they lie at the same elevation on the landscape. Crossing these two populations (illustrated by the dashed arrows) produces a hybrid population that lies at the saddle point and, because it has a lower phenotypic value, it has low fitness. Inset: a three-dimensional view of this surface.

Additional work on murine skull characters (e.g., Atchley et al. 1982, 1990) elegantly demonstrates that selection on whole-body traits such as weight gain or fat content produces predictable direct responses to selection, but unpredictable responses of correlated skull and mandible characters. Variation in the correlated response among replicate populations can be traced to divergent developmental mechanisms among replicates, such as differences in the onset, offset, and duration of weight gain and changes in timing of interactions between skeletal elements.

Although the developmental basis for many quantitative traits is still poorly understood, unpredictability of the correlated response is a common outcome of artificial selection experiments (Bohren et al. 1966; Gromko et al. 1991). The correlated response may bear little resemblance to the response predicted from genetic correlation estimates in the base population, and can differ markedly in both sign and magnitude between replicate populations exposed to the same intensity and direction of selection (Palmer and Dingle 1986; Gromko et al. 1991; Deng et al. 1999). Differences in the correlated response to selection alter patterns of phenotypic correlation among characters (Atchley et al. 1990) and may alter the long-term trajectories of populations by exposing new kinds of variation to selection. From the viewpoint of traditional quantitative genetics, such results highlight the limited usefulness of parameters such as the genetic correlation for making predictions about response to selection, and underline the need for models with enhanced predictive power.

## Empirical Analysis

The phenotype landscape approach has been used primarily as a tool to develop theory—to describe the potential for developmental interactions to impact evolutionary processes through the evolution of canalization (Rice 1998, 2000) and through impacts on genetic variance and covariance that determine the evolutionary trajectory of populations (Wolf et al. 2001). Whether these advances in theory lead to critical insights or a better ability to predict evolutionary outcomes depends on the application of phenotype landscape models to empirical questions. Critically, researchers must be able to characterize landscapes in real systems (see Rice 2002) if the phenotype landscape approach is to be useful. Understanding developmental landscapes requires the ability to predict the resulting phenotype from a given combination of underlying genotypes. However, because a real population may occupy only a small region of space on the landscape, it may be difficult to characterize the shape of the landscape by examining only the region currently occupied by a population.

There are two basic ways to overcome this limitation, and both have been previously suggested for dealing with similar limitations in analyses of selection surfaces (Phillips and Arnold 1989; Arnold et al. 2001). The first is to maximize the distribution of individuals across the landscape, sampling in a way that avoids oversampling any one part of the distribution of underlying factors. This method contrasts with the usual approach of random sampling from a population, which biases the estimate of surface shape to reflect the space around the mean, where most of the population lies. The region around the mean is generally not the region of the landscape that is of most interest to us for evolutionary analyses.

For example, consider a population that lies at the top of a peak on the landscape in a region where there are many surrounding peaks (Fig. 17.7), which may be common if multiple combinations of underlying factors can produce a large phenotypic value. Since most of the population lies near the top of the peak, the estimated surface will appear simply to have a positive curvature since the region around the mean will determine the apparent shape. However, choosing individuals from across the distribution and weighting all classes equally would provide a less biased estimate of the shape of the surface. In this hypothetical example, one might then detect the negative curvature at the edge of the range.

Although it yields a better picture of the landscape, this approach is still inadequate for some evolutionary analyses because the range of available values for the underlying factors and the phenotypes limits our view of the landscape. This limitation may be overcome using the second of these approaches, where the

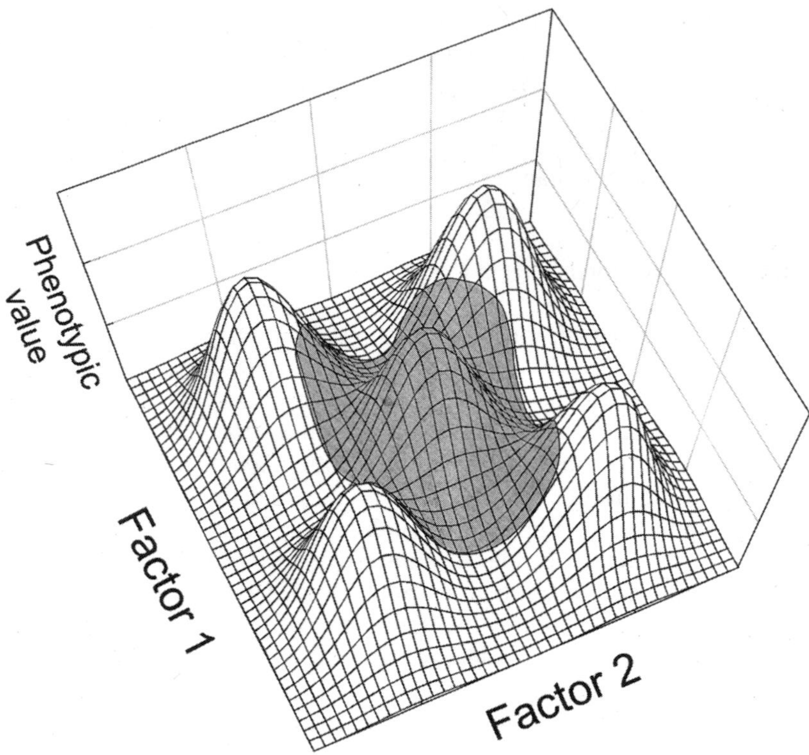


Figure 17.7 A rugged phenotype landscape with many peaks. A population is shown as the shaded region on the landscape. Because the population mean lies on the central peak, random sampling will underestimate the curvature of the landscape.

range of values for the underlying factors is expanded by experimental manipulation. In this experimental approach, one may either modify the values of the underlying factors to values outside of their naturally occurring ranges or modify values of some factors to produce combinations of factors that do not occur naturally. One such approach, called phenotypic engineering, has been successfully used in selection studies, giving researchers a broader picture of the fitness landscape (Sinervo and Huey 1990; Ketterson et al. 1992; Sinervo et al. 1992). There are a variety of techniques that can be brought to bear on this approach, such as direct manipulations of development by removing glands or applying exogenous hormones, using mutants (see Chapter 11 in this volume) or transgenics, and culturing cells or organs in different physiological cocktails. Returning to Fig. 17.7, we would expect the phenotypic engineering approach to reveal the true ruggedness of the landscape, if we were able to create individuals in the regions beyond that currently occupied by the population.

We can gain a number of significant insights from a more complete understanding of the phenotype landscape. For example, under directional selection, our population might become trapped on a phenotype peak or, when considering higher-dimensional spaces, in a local maximum that is not connected to other high-fitness regions. When this occurs, a population may be unable to evolve to a higher peak because moving to the higher peak by directional selection would require the population to evolve through a phenotype valley associated with lower fitness. These phenotype peaks, created by nonadditive interactions in the genetic and/or developmental system, may not be identified using traditional quantitative genetic techniques, but could have a major impact on character evolution. The phenotype landscape approach could provide insights by identifying these topographic features that affect evolutionary trajectories.

The engineering approach may also provide much-needed insight into the evolution of canalization, if certain nonadditive topographies reveal situations where genetic (i.e., additive, dominant, and epistatic) and environmental variances are labile characteristics of a population, easily evolving to new values. The engineering approach might provide important insights into multivariate evolution by revealing the mechanistic basis for genetic correlations, allowing us to understand how integration evolves. It is also likely that, by understanding the developmental basis of the genetic correlation, we would be able to understand to what degree it represents a constraint or is, again, a labile characteristic of a population, molded by functional relationships between traits to achieve patterns of integration.

Finally, the landscape itself can replace quantitative genetic parameters when modeling evolutionary dynamics. It is a particularly powerful method because it requires few or no simplifying assumptions about the shape of the landscape used to analyze trait evolution. Rice (2002) presents an eloquent model in which the geometry of the hyperdimensional landscape can be used to explore how evolution alters any moment of a phenotype distribution, from the mean and variance to more complex moments such as skewness and kurtosis. The model interfaces well with the sorts of data that are derived from developmental studies, such as changes in trait values as a function of factors such as morphogen concentrations. Developmental modules of all sorts are often useful underlying factors since their contribution to the gross phenotype is often obvious, and experimental analyses of epigenetic interactions between modules may be feasible. Rice also suggests that underlying factors such as QTL or gene expression patterns estimated from microarrays could be used to construct the landscape, providing a way to integrate theoretical studies with cutting-edge developmental biology. The ability to predict evolutionary changes in all aspects of the multivariate phenotype distribution using theory that integrates the complex developmental systems that build traits is likely to provide great insights into the evolution of complex traits, and may emerge as a unifying theory of trait evolution (see Rice 2002).

*Acknowledgments* Thanks to S. Rice for sharing unpublished material and to W. Bradshaw for providing details for the example from his experimental work. W.A.F. was supported by a National Science Foundation (NSF) Postdoctoral Fellowship in Biological Informatics. C.E.A. was supported by a grant from the NSF (IBN-0092873) to D. J. Emlen.

*Literature Cited*

Arnold, S. J., M. E. Pfender, and A. G. Jones. 2001. The adaptive landscape as a bridge between micro- and macroevolution. Genetica 112:9–32.

Atchley, W. R. 1984. Ontogeny, timing of development, and genetic variance-covariance structure. American Naturalist 123:519–540.

Atchley, W. R., and B. K. Hall. 1991. A model for development and evolution of complex morphological structures. Biological Reviews 66:101–157.

Atchley, W. R., and J. Zhu. 1997. Developmental quantitative genetics, conditional epigenetic variability and growth in mice. Genetics 147:765–776.

Atchley, W. R., D. E. Rutledge, and D. E. Cowley. 1982. A multivariate statistical analysis of direct and correlated response to selection in the rat. Evolution 36:677–698.

Atchley, W. R., D. E. Cowley, E. J. Eisen, H. Prasetyo, and D. Hawkins-Brown. 1990. Correlated response in the developmental choreographies of the mouse mandible to selection for body composition. Evolution 44:669–688.

Atchley, W. R., S. Xu, and C. Vogl. 1994. Developmental quantitative genetic models of evolutionary change. Developmental Genetics 15:92–103.

Beldade, P., and P. M. Brakefield. 2002. The genetics and evo-devo of butterfly wing patterns. Nature Reviews Genetics 3:442–452.

Beldade, P., P. M. Brakefield, and A. D. Long. 2002. Contribution of *Distal-less* to quantitative variation in butterfly eyespots. Nature 415:315–318.

Bohren, B. B., W. G. Hill, and A. Robertson. 1966. Some observations on asymmetrical correlated responses to selection. Genetical Research 7:44–57.

Bradshaw, W. E., and C. M. Holzapfel. 2000. Epistasis and the evolution of genetic architectures in natural populations. Pp. 245–263 in J. B. Wolf, E. D. Brodie III, and M. J. Wade, eds. Epistasis and the Evolutionary Process. Oxford University Press, Oxford.

Brakefield, P. M., J. Gates, D. Keys, F. Kesbeke, P. J. Wijngaarden, A. Monteiro, V. French, and S. B. Carroll. 1996. Development, plasticity and evolution of butterfly eyespot patterns. Nature 384:236–242.

Brakefield, P. M., F. Kesbeke, and P. B. Koch. 1998. The regulation of phenotypic plasticity of eyespots in the butterfly *Bicyclus anynana.* American Naturalist 152:853–860.

Brunetti, C. R., J. E. Selegue, A. Monteiro, V. French, P. M. Brakefield, and S. B. Carroll. 2001. The generation and diversification of butterfly eyespot color patterns. Current Biology 11:1578–1585.

Carroll, S. B., J. Gates, D. N. Keys, S. W. Paddock, G. E. F. Panganiban, J. E. Selegue, and J. A. Williams. 1994. Pattern formation and eyespot determination in butterfly wings. Science 265:109–114.

Cheverud, J. M. 1996. Developmental integration and the evolution of pleiotropy. American Zoologist 36:44–50.

Cheverud, J. M. 2001. The genetic architecture of pleiotropic relations and differential epistasis. Pp. 411–433 in G. P. Wagner, ed. The Character Concept in Evolutionary Biology. Academic Press, San Diego.

Cowley, D. E., and W. R. Atchley. 1992. Quantitative genetic models for development, epigenetic selection, and phenotypic evolution. Evolution 46:495–518.

Debat, V., and P. David. 2001. Mapping phenotypes: canalization, plasticity, and developmental stability. Trends in Ecology and Evolution 16:555–561.

Deng, H.-W., V. Haynatzka, K. Spitze, and G. Hayhatzki. 1999. The determination of genetic covariances and prediction of evolutionary trajectories based on a genetic correlation matrix. Evolution 53:1592–1599.

Edmands S. 1999. Heterosis and outbreeding depression in interpopulation crosses spanning a wide range of divergence. Evolution 53:1757–1768.

Emlen, D. J. 1996. Artificial selection on horn length-body size allometry in the horned beetle *Onthophagus acuminatus* (Coleopter: Scarabaeidae). Evolution 50:1219–1230.

Emlen, D. J. 2001. Costs and the diversification of exaggerated animal structures. Science 291:1534–1536.

Emlen, D. J., and H. F. Nijhout. 2000. The development and evolution of exaggerated morphologies in insects. Annual Review of Entomology 45:661–708.

Falconer, D. D., and T. F. C. Mackay. 1996. Introduction to Quantitative Genetics, 4th edn. Longman, Harlow, UK.

Fenster, C. B., and L. F. Galloway. 2000. Inbreeding and outbreeding depression in natural populations of *Chamaecrista fasciculata* (Fabaceae). Conservation Biology 14:1406–1412

Fisher, R. A. 1930. The Genetical Theory of Natural Selection. Oxford University Press, London.

Gavrilets, S. 1997. Evolution and speciation on holey adaptive landscapes. Trends in Ecology and Evolution 12:307–312.

Gavrilets, S., and J. Gravner. 1997. Percolation on the fitness hypercube and the evolution of reproductive isolation. Journal of Theoretical Biology 184:51–64

Gharrett, A. J., W. W. Smoker, R. R. Reisenbichler, and S. G. Taylor. 1999. Outbreeding depression in hybrids of even- and odd-broodline pink salmon (*Oncorhynchus gorbuscha* Walb.). Aquaculture 173:117–129.

Gilchrist, M. A., and H. F. Nijhout. 2001. Nonlinear developmental processes as sources of dominance. Genetics 159:423–432.

Gromko, M. H., A. Briot, S. C. Jensen, and H. H. Fukui. 1991. Selection on copulation duration in *Drosophila melanogaster*: predictability of direct response versus unpredictability of correlated response. Evolution 45:69–81.

Johnson, N. A. 2001. Towards a new synthesis: population genetics and evolutionary developmental biology. Genetica 112:45–58.

Ketterson, E. D., V. Nolan, M. J. Cawthorn, P. Parker, and C. Ziegenfus. 1992. Phenotypic engineering: using hormones to explore the mechanistic and functional bases of phenotypic variation in nature. Ibis 138:70–86.

Kopp, A., I. Duncan, and S. B. Carroll. 2000. Genetic control and evolution of sexually dimorphic characters in *Drosophila*. Nature 408:553–559.

Lande, R. 1979. Quantitative genetic analysis of multivariate evolution, applied to brain:-body size allometry. Evolution 33:402–416.

Leamy, L., and W. R. Atchley. 1984. Static and evolutionary allometry of osteometric traits in selected lines of rats. Evolution 38:47–54.

Levins, R. 1968. Evolution in Changing Environments: Some Theoretical Explorations. Princeton University Press, Princeton, NJ.

Magwene, P. M. 2001. New tools for studying integration and modularity. Evolution 555:1734–1745

McMillan, W. O., A. Monteiro, and D. D. Kapan. 2002. Development and evolution on the wing. Trends in Ecology and Evolution 17:125–133.

Monteiro, A. F., P. M. Brakefield, and V. French. 1994. The evolutionary genetics and developmental basis of wing pattern variation in the butterfly *Bicyclus anynana*. Evolution 48:1147–1157.

Monteiro, A. F., P. M. Brakefield, and V. French. 1997. The genetics of development of an eyespot pattern in the butterfly *Bicyclus anynana*: response to selection for eyespot shape. Genetics 146:287–294.

Nijhout, H. F. 1980. Pattern formation on lepidopteran wings: determination of an eyespot. Developmental Biology 80:267–274.

Nijhout, H. F. 1991. The Development and Evolution of Butterfly Wing Patterns. Smithsonian Institution Press, Washington, DC.

Nijhout, H. F., and D. J. Emlen. 1998. Competition among body parts in the development and evolution of insect morphology. Proceedings of the National Academy of Sciences USA 95:3685–3689.

Palmer, J. O., and H. Dingle. 1986. Direct and correlated responses to selection among life-history traits in milkweed bugs (*Oncopelus fasciatus*). Evolution 40:767–777.

Phillips, P. C., and S. J. Arnold. 1989. Visualizing multivariate selection. Evolution 43:1209–1222.

Queitsch, C., T. A. Sangster, and S. Lindquist. 2002. Hsp90 as a capacitor of phenotypic variation. Nature 417:618–624.

Raff, R. 1996. The Shape of Life: Genes, Development, and the Evolution of Animal Form. University of Chicago Press, Chicago.

Rice, S. H. 1998. The evolution of canalization and the breaking of von Baer's laws: modeling the evolution of development with epistasis. Evolution 52:647–656.

Rice, S. H. 2000. The evolution of developmental interactions: epistasis, canalization, and integration. Pp. 82–98 in J. B. Wolf, E. D. Brodie III, and M. J. Wade, eds. Epistasis and the Evolutionary Process. Oxford University Press, Oxford.

Rice, S. H. 2002. A general population genetic theory for the evolution of developmental interactions. Proceedings of the National Academy of Sciences USA. 99:15518–15523.

Roff, D. A. 1997. Evolutionary Quantitative Genetics. Chapman & Hall, New York.

Roff, D. A., S. Mostowy, and D. J. Fairbairn. 2002. The evolution of trade-offs: testing predictions on response to selection and environmental variation. Evolution 56:84–95.

Rutherford, S. L., and S. Lindquist. 1998. Hsp90 as a capacitor for morphological evolution. Nature 396:336–342.

Rutledge, J. J., E. J. Eisen, and J. E. Legates. 1974. Correlated response in skeletal traits and replicate variation in selected lines of mice. Theoretical and Applied Genetics 46:26–31.

Schluter, D. 1996. Adaptive radiation along genetic lines of least resistance. Evolution 50:1766–1774.

Sheridan, A. K., and J. S. F. Barker. 1974. Two-trait selection and the genetic correlation. II. Changes in the genetic correlation during two-trait selection. Australian Journal of Biological Science 27:89–101.

Sinervo, B., and R. B. Huey. 1990. Allometric engineering: an experimental test of the causes of interpopulational differences in locomotor performance. Science 248:1106–1109.

Sinervo, B., P. Doughty, R. B. Huey, and K. Zamudio. 1992. Allometric engineering: a causal analysis of natural selection on offspring size. Science 258:1927–1930.

Stern, D. L. 1998. A role of *Ultrabithorax* in morphological differences between *Drosophila* species. Nature 396:463–466.

Sucena, E., and D. L. Stern. 2000. Divergence of larval morphology between *Drosophila schellia* and its sibling species due to *cis*-regulatory evolution of ovo/shaven-baby. Proceedings of the National Academy of Sciences USA 97:4530–4534.

Takano-Shimizu, T. 2000. Genetic screens for factors involved in the notum bristle loss of interspecific hybrids between *Drosophila melanogaster* and *D. simulans*. Genetics 156:269–282.

True, J. R., and E. S. Haag. 2001. Developmental system drift and flexibility in evolutionary trajectories. Evolution and Development 3:109–119.

Turelli, M. 1988. Phenotypic evolution, constant covariances, and the maintenance of additive variance. Evolution 42:1342–1347.

Wagner, G. P. 1996. Homologues, natural kinds, and the evolution of modularity. American Zoologist 36:36–43.

Wagner, G. P., and L. Altenberg. 1996. Complex adaptations and the evolution of evolvability. Evolution 50:967–976

Wagner, G. P., G. Booth, and H. Bagheri-Chaichian. 1997. A population genetic theory of canalization. Evolution 51:329–347.

Weatherbee, S. D., H. F. Nijhout, L. W. Grunert, G. Halder, R. Galant, J. Selegue, and S. Carroll. 1999. *Ultrabithorax* function in butterfly wings and the evolution of insect wing patterns. Current Biology 9:109–115.

Weber, K. E. 1990. Selection on wing allometry in *Drosophila melanogaster*. Genetics 126:975–989.

Wolf, J. B., W. A. Frankino, A. F. Agrawal, E. D. Brodie III, and A. J. Moore. 2001. Developmental interactions and the constituents of quantitative variation. Evolution 55:232–245.

Wright, S. 1932. The roles of mutation, inbreeding, crossbreeding and selection in evolution. Proceedings of the 6th International Congress of Genetics 1:356–366.

Wright, S. 1977. Evolution and the Genetics of Populations, Vol. 3: Experimental Results and Evolutionary Deductions. University of Chicago Press, Chicago.

# 18

# The Relativism of Constraints on Phenotypic Evolution

KURT SCHWENK
GÜNTER P. WAGNER

Nought may endure but Mutability.
*Percy Bysshe Shelley*

## Mutability versus Stability and the Constraint Problem

The evolution of complex phenotypes entails a central paradox. On the one hand, phenotypes must be mutable and therefore responsive to the constantly changing demands of the environment over evolutionary time. This is necessary for organisms to remain in harmony with their surroundings, a state of adaptation. On the other hand, phenotypes must be stable so that the complex dynamics of their developmental and functional systems are not disrupted. This is necessary for organisms to maintain functionality through evolutionary time, regardless of environment. It is this fundamental tension—between mutability and stability—that much of current evolutionary theory seeks to explain.

Historically, evolutionary biology has been the study of mutability. This preoccupation with change is logical, given that a necessary first step for evolutionary theory was to establish the fact of evolution and its mechanistic underpinnings. These foci were clearly evident in the neo-Darwinian synthesis of the 1940s and 50s, and they remain central to most evolutionary studies today. Our recognition that evolution is a universal attribute of living systems and our emphasis on the mechanisms of mutability, however, have led to the unacknowledged supposition that change is as inevitable as it is desirable. It is therefore the examples of phenotypic persistence that have tended to perplex us most. Certainly the synthesis did not deal effectively with stasis (Eldredge and Gould 1972; Gould and Eldredge 1977; Wake et al. 1983; Gould 2002), typically falling back on the unlikely notion of long-term environmental constancy to explain it (e.g., Simpson 1953; see Wagner and Schwenk 2000).

Evolutionary biology's emphasis on mutability is also logically related to its attention on natural selection as the principal agent of phenotypic change. Since Darwin (1859), organismal evolution has been tied directly to demands of the biotic and abiotic environment. Because this environment is constantly changing, organisms must adapt to keep pace. Mutability is therefore regarded as a necessary and inevitable phenotypic response to external, environmental factors mediated by natural selection.

The triad of mutability, natural selection, and environment has contributed to an "externalist" (or "functionalist") approach to phenotypic evolution that continues to dominate thinking in evolutionary biology. Nonetheless, there is also a long, if somewhat subversive, history of an "internalist" (or "structuralist") perspective (e.g., Woodger 1929, 1945; de Beer 1930, 1938; Whyte 1965; Raup 1966; Seilacher 1972, 1973; Gould 1977; 2002; Riedl 1977; Ho and Saunders 1979, 1984; Gould and Lewontin 1979; Dullemeijer 1980; Alberch 1982; Wake 1982, 1991; Raff and Kaufman 1983; Maynard Smith et al. 1985; Roth and Wake 1985; Wagner 1986, 1988; Wake and Larson 1987; Wake and Roth 1989; Smith 1992; Amundson 1994, 1996; Raff 1996; Wagner and Altenberg 1996; Webster and Goodwin 1996; Arthur 1997; Hall 1998; Schlichting and Pigliucci 1998; Galis 1999; Wagner and Schwenk 2000, to name a few contributions). To a large extent, the internalist perspective on evolution grew out of nineteenth- and early twentieth-century studies of embryology. The strictures of early development and the apparent conservation of ontogenetic patterning seemed to suggest a role for intrinsic, organismal attributes in determining evolutionary outcomes. Some extreme statements of this view go so far as to recast selection as a secondary player in the evolution of phenotypes, with internally generated "rules of form" assuming the greater role (e.g., Ho and Saunders 1984; Webster and Goodwin 1996). More commonly, however, the importance ascribed to development stems from its potential role in mitigating, opposing, biasing, or otherwise modulating the effects of environmental selection. As such, the internalist perspective emphasizes the role of internal (organismal) factors in determining the efficacy of environmental selection acting on the phenotype. Most often this effect is seen as negative in the sense that the evolutionary response of the phenotype to environmental selection is reduced (as compared to some expected value) or eliminated altogether. In other words, intrinsic, organismal attributes are thought to promote phenotypic stability by attenuating the effects of environmental selection, the engine of mutability.

The internalist perspective has become increasingly conspicuous with the recent growth of evolutionary developmental biology as a separate field, largely owing to the introduction of new molecular genetic techniques (e.g., Hall 1998; Stern 2000; Carroll et al. 2001). Furthermore, the internalist approach to phenotypic evolution was to some extent formalized with the publication of Gould and Lewontin's seminal 1979 paper in which they argued for increased attention to "constraints" on adaptive evolution. Constraints, although vaguely defined, were attributed to intrinsic qualities of the organism associated with the rules of development or constructional principles. Thus, the notion of constraint accords well with the internalist view, particularly in the sense that constraint was portrayed by Gould and Lewontin (1979), and especially by many subsequent workers, as a "force" acting in opposition to selection (for a critique of the applicability of the concept of force in evolutionary biology, see Pigliucci, Chapter 11, this volume). This has led to a "dichotomous approach" in which constraint is conceptually divorced from natural selection and pitted against it in a kind of evolutionary battle for dominance over the phenotype (Schwenk 1995; Schwenk and Wagner 2003). Although this may sound overstated, much of the constraint literature over the last twenty-five years has explicitly sought to explain evolutionary outcomes as *either* the result of selection *or* of constraint.

The conceptually crisp distinction implied by an "either/or" approach to phenotypic evolution was engendered, in part, by the historical and conceptual links of internalism to embryology. In the context of development, it seems possible to separate the generation of phenotypic variation from the action of selection on that variation once produced. Obviously, phenotypic evolution can proceed only if variation exists in a population upon which selection can operate. If a developmental system fails to generate variant morphs in the first place, the population will be devoid of variation and no (adaptive) phenotypic evolution can occur—no matter what the strength of the environmental selection. Thus, in the dichotomous approach, constraints are manifested as limitations or biases in the generation of phenotypic variation imposed by the developmental system, an intrinsic attribute of the organism (e.g., Maynard Smith et al. 1985). This is an apparently neat solution to the problem of integrating internalist and externalist perspectives and resolving the tension between mutability and stability: the former is attributable to selection (an external factor), the latter to the failure of development to generate variation (an internal factor).

Unfortunately, this dichotomous solution, advocated implicitly or explicitly in much of the constraint literature, is based on a false premise and is really no solution at all (see also Pigliucci, Chapter 11, this volume). In attempting to distinguish between selection and constraint, it requires that selection on the phenotype begins only upon completion of development. Obviously this is not the case; selection operates on the phenotype seamlessly throughout its ontogeny. It is therefore not possible to separate mechanistically the generation of phenotypic variation during development from the action of selection because the very failure of a developmental system to produce variant phenotypes might itself be due to selection (Hall 1996; Schlichting and Pigliucci 1998; Schwenk and Wagner 2003). As such, random variation arises during development but the variants are not viable, possibly because of disruption to subsequent developmental stages. In other words, the embryonic variants are selected against. The phenotypic stability ensuing from this process, if iterated over generations, would seem to have no special provenance in development. Rather, the cause of stability would be the same as the cause of mutability—natural selection. In this case, however, it is selection acting at an early life stage during the generative process rather than after its completion.

If the role of development in limiting variation simply devolves to another case of selection, it would seem that the internalist perspective has little to contribute to our understanding of phenotypic evolution. There is, however, the case of canalization. As traditionally understood, canalization refers to developmental "buffering" that promotes consistency in a trait's phenotype despite background variation in genotype and epigenetic milieu (Waddington 1957). In this case, the failure of development to produce variation is related directly to the internal dynamics of the system and not to viability selection. It would therefore seem to present a valid case of "constraint" acting to stymie environmental selection by limiting the introduction of variation into the population. However, canalization introduces yet another paradox within this context because it is itself thought to arise historically from the action of stabilizing selection (Waddington 1957; Clarke and McKenzie 1987; A. Wagner 1996; Wagner et al. 1997). Once again, the same factor promoting mutability (selection) is revealed as the ultimate source of stability.

Finally, there is the case of pleiotropy and other kinds of genetic correlation. Such genetic correlations can cause correlated phenotypic responses to selection on a given trait that have negative fitness consequences for the organism (e.g., Galis 1999; Galis and Metz 2001). This might serve as a constraint on phenotypic evolution. However, genetic correlations can also promote adaptive phenotypic responses to selection (Wagner 1988). This suggests that genetic correlations, in and of themselves, cannot be regarded as constraints. Furthermore, many genetic correlations are built and maintained by selection, and changes in the selection regime can break or modify them (e.g., Cheverud 1984, 1988, 2001; Beldade et al. 2002). Once again, it appears difficult to separate constraint from selection at a mechanistic level. The elusive notion of constraint seems to be in danger of slipping away altogether.

We are led, naturally, to the purpose of this chapter, which is to explore the conundrum of constraint and its role in our theories of phenotypic evolution. It is our view that a workable constraint concept is not only possible, but highly desirable. In devising a useful conception of constraint, we shall need to address two important aspects of the problem introduced above. First, the relationship of constraint to selection is ambiguous and the widely accepted dichotomous approach leads to self-contradiction. A reconciliation of constraint and selection is therefore necessary. Second, many developmental mechanisms, widely accepted as constraints, are either direct manifestations of selection, were built by selection historically, or in some contexts contribute directly to adaptation. It appears, therefore, that no process or mechanism can be universally identified as constraint. We are left with the uncomfortable recognition that evolutionary constraint might sometimes result from selection, the very thing said to be limited by constraint, and that some "constraints" might contribute to adaptive evolution, the province of natural selection. Under what conditions, therefore, can constraint be recognized? To answer this question we clarify some surrounding concepts and issues, and introduce an explicit framework in which to formulate a constraint concept. In particular, we emphasize the simple notion of a "focal life stage" and illustrate its important role in resolving the ambiguities of constraint. Our goal is a concept of constraint that is internally consistent, biologically defensible, and genuinely useful in crafting explanations for the evolution of complex phenotypes.

## Establishing a Framework for Constraint

### Constraint as a Relativistic Concept

We suspect that one of the roots of the constraint problem is in the simple fact that it is a standard English word that has been called upon to play a technical role. Scientific usage of the word is therefore burdened by our commonplace understanding of its meaning—a meaning that is far too broad to be of much use in scientific discourse. Most people understand intuitively the basic meaning of constraint in the context of evolutionary studies, but few apply the word with the rigor necessary to avoid its semantic pitfalls. Words newly minted for scientific purposes circumvent the problem of historical burden (e.g., "synapomorphy"), but the long, if speckled, use of constraint in the evolutionary literature suggests that to coin a new word now would only be a disservice. As David Wake (1994) has warned, however, technical words require technical meanings and the failure to provide them leads to conceptual ambiguity. What does science require, after all, if not the precise use of language?

Perhaps the single greatest roadblock to arriving at a technical meaning for constraint has been a general failure to recognize its relativism (Schwenk and Wagner 2003). In other words, one cannot say absolutely that constraint is one thing or another, but rather must specify it in relation to other factors. As such, constraint is meaningfully applied only within a local context (e.g., Stearns 1986; Gould 1989; Fusco 2001; Schwenk and Wagner 2003). The necessary context for constraint is rarely, if ever, made explicit in the literature that seeks to apply it and it is mostly this inadequacy that has led to the apparent paradoxes introduced earlier.

A critical point is to recognize that changes in the factors delineated below are likely to lead to different meanings of constraint. For example, different null models for constraint can lead to diametrically opposed conceptions of the responsible mechanism. The invocation of constraint in a particular case is justified by a local context, but outside that context it may not apply. We believe that recognition of constraint's relativism is necessary to operationalize the concept. The advantages that relativism introduces are rigor and flexibility. Rigor follows from the need to specify explicitly a context for constraint, case by case, in order to identify the way in which it is being applied. It is this rigor that permits constraint to be used as a technical term. Flexibility is possible because, using the guidelines suggested here, one can adapt the application of constraint to any particular case by changing the specifications. For example, a molecular evolutionist might specify a different null model as compared to someone interested in phenotypic evolution, as we are here, leading to a different characterization of constraint (see Schwenk and Wagner 2003).

### Ancillary Concept: Internal versus External Selection

A key to resolving the externalist/internalist conflict in the context of constraint, and to reconciling constraint and selection, lies in recognizing that natural selection is resolvable into "external" and "internal" components (Whyte 1965; Riedl

1977; Cheverud 1984, 1988; Arthur 1997; Schlichting and Pigliucci 1998; Wagner and Schwenk 2000; Fusco 2001). External selection refers to the traditional, Darwinian notion of selection stemming from the pressures of the biotic and abiotic environment surrounding an organism, that is, the habitat in which it lives. The fitness consequences of character variation manifested among individuals in a population are determined within this environmental arena. If the environment changes, either through time or because the organism moves to another place, so too does the selective regime and therefore the relative fitness of each variant. As such, given two individuals that vary in some trait, one might be relatively more fit in one environment, but relatively less fit in another. The environmental dependence of external selection identifies it as the force that drives adaptive evolution in the traditional sense of maintaining the "match" between an organism's phenotype and its environment.

Internal selection results from the fitness consequences of trait variation as determined within the internal arena of organismal function. In this case, selection pressures derive from the internal dynamics of a functioning organism. When system-level functioning (e.g., the prey-capture system or the developmental system) requires the coordinated action of two or more characters in a chain or web of interaction, a phenotypic change in a single character can affect the performance of the system as a whole (a character complex). If the interactions are critical and functional tolerances between characters are tight, even a slight phenotypic change in one character can cause complete system failure leading to death of the organism. Thus, highly integrated systems exert strong stabilizing selection on their constituent characters. The crucial point is that this selection is internal in the sense that it does not depend on external factors of the particular environment inhabited. It is an intrinsic attribute of the organism, traveling with it wherever it goes. As such, given two individuals that vary in some internally selected trait, a change in environment is very unlikely to cause a change in their rank order according to relative fitness—in contrast to the situation described above for external selection. This is because the relative performance (and therefore relative fitness) of the two individuals is not determined by the character's interaction with the external environment, but by its interaction with other characters of a system within the internal milieu. These internal interactions remain more or less constant across a wide range of environments because the contribution of the system, as a whole, to organismal survival is unchanging (e.g., it is necessary to capture prey in any environment, even if the particular prey species vary; or it is necessary to produce viable offspring, no matter where the mother is living). Thus, the stabilizing internal selection exerted by these intrinsic, functional interactions also persists across environments and through time. Its potential importance in the context of constraint is obvious. Arthur (1997) and Wagner and Schwenk (2000) discuss internal selection in great detail and provide formal definitions.

Obviously, internal selection is not irrelevant to adaptation; the maintenance of system functionality is critical to an organism's survival within the environments it encounters and therefore to its state of adaptation. However, the two components of selection vary in their relative dependence on environmental factors, with external selection environment-dependent and internal selection environment-independent (within a broad but finite range of environments; see Wagner and Schwenk 2000). Thus, each component of selection plays complementary roles in character evolution: maintaining the "fit" between an organism and its habitat on the one hand (external selection) and maintaining the intrinsic coordination of interacting characters within integrated systems on the other (internal selection). The former conforms to our traditional notion of adaptation. At any point in time a given character is subject to both internal and external components of selection. The realized strength and direction of selection on the character is the sum of these effects integrated over an individual's life.

An analogy for the potential significance of internal selection is provided by the example of sexual selection. Although sexual selection contributes in an obvious way to fitness, it can often be in direct conflict with the adaptive state of an organism as determined by its (nonsexual) environment. It is this potential for conflict that led Darwin (1871) and most subsequent workers to distinguish sexual selection from natural selection proper. For example, Endler (1980, 1983) showed that bright coloration in male guppies makes them more attractive to females, but also more visible to predators. As such, a sexually selected trait is maladaptive in most environments. Likewise, recent work has demonstrated that the brightly ornamented male fowl preferred by females are immunologically compromised (e.g., Zuk and Johnsen 1998; Verhulst et al. 1999), again illustrating the potential for conflict between sexual and natural selection in shaping the phenotype.

As for sexual selection, the importance of distinguishing internal and external components of selection is most apparent when they are in conflict. For example, one can imagine a scenario in which environmental conditions favor modification of a character in a particular way. A mutant evincing such a modification would have enhanced performance in this environment and so would be better adapted. Taken in isolation, such a mutant would have higher fitness compared to unmodified forms. External selection would therefore cause the character to change. However, the same character might participate in a tightly integrated functional system. Modifying the character in the direction favored by external selection pressures would cause reduced performance of this system as a whole. Taken in isolation, such a mutant would have lower fitness compared to unmodified forms. Internal selection would therefore enforce stability of the character. Whether or not selection *in toto* acts to modify the character in this case will depend on the balance between the fitness gain associated with the external component of selection and the fitness loss associated with the internal component of selection. Given a complexly integrated and critically important system, however, the fitness cost associated with system failure, or even reduced system performance, is likely to exceed any gain associated with improvement of a single character within a particular environment.

An empirical example of this scenario was developed at length in Wagner and Schwenk (2000), Schwenk (2000), and Schwenk and Wagner (2001), who applied it to the evolution of lizard feeding and chemosensory systems. It was shown that the remarkable, long-term phenotypic stability of the prey-capture system in one clade of lizards could be accounted for by an internal selection model. A limited test of the model was possible using phylogenetic character analysis (Schwenk and Wagner 2001).

The strength of internal selection is proportional to the degree of functional integration within a system. If the functional dependencies among characters are critical and the tolerances[1] tight, stabilizing selection on a particular character will be strong. Functional dependencies or interactions occur most obviously at two time scales: instantaneous (or virtually so), as implied by "function" or "behavior" (e.g., the execution of a prey-capture sequence); and sequentially, over a longer time period, as in the dependencies of developmental sequences and epigenetic interactions. These might be thought of as "horizontal" and "vertical" dependencies, respectively (Raff 1996). Thus, the notion of internal selection has special relevance in the realms of function and development.

In the hypothetical scenario presented above, adaptive evolution by external selection would be thwarted by internal selection, an intrinsic attribute of the organism. This would seem to accord well with the internalist perspective, as well as our intuitive notion of constraint. However, this example does nothing but expose selection, once again, as the culprit behind stability. In a purely dichotomous view, in which constraint is pitted against selection, there is therefore no evidence of constraint (Schwenk 1995; Schlichting and Pigliucci 1998). In a relativistic view, however, this may be a valid case of constraint. In order to clarify this situation we need additional context.

## Some Important Attributes of Constraint

In establishing the boundaries within which constraint can be understood, it is necessary to state some of our initial assumptions about its nature. Our long experience in grappling with constraint suggests that these assumptions are necessary for a viable concept, but we readily acknowledge that other formulations are possible. In the following sections we are guided by the outline developed in Schwenk and Wagner (2003).

First, constraints are intrinsic attributes of living organisms and therefore not variable among the environments in which the organisms can live. This assumption has two important consequences. It suggests that constraints have an evolutionary history consonant with the organisms within which they are manifested, meaning that, like organisms, they arise, persist, and disappear. Accordingly, they have a phylogeny and will be represented taxonomically within monophyletic and paraphyletic groups (the latter situation occurs when the constraint is lost in one or more subclades within a larger clade) (Wagner and Schwenk 2000; Schwenk and Wagner 2001). Also, if constraint is an intrinsic organismal attribute, environmental selective pressures cannot act as constraints. Conflicting demands of the environment, for example, are sometimes labeled "ecological" or "selective constraints." However, these conflicts arise as a transient interaction between the organism and a local environment, and therefore do not have historical reality as an evolving entity. In any case, competing selection pressures are resolvable into a single vector of given strength and direction that we would call the external, adaptive component of selection. This vector is likely to change frequently, both spatially and temporally—it does not travel with the organism. If unusual environmental stability results in long-term phenotypic stability, it is attributable to standard notions of stabilizing selection. It is neither necessary nor desirable to conflate this with constraint (Schwenk 1995).

Second, constraints are revealed over deep time—geological rather than ecological time scales. If putative constraints do not persist for long enough, they will not be manifested in the phenotype and would not be identified as constraints in the first place. Pragmatically, this suggests that nonhistorical approaches can provide only limited insight into constraint. Ongoing populational processes identified as potential constraints (e.g., Arnold 1992) must be inferred to have operated historically over very long periods of time. The lack of genetic variation, for example, often cited as a putative constraint, is unlikely to persist long enough in the face of mutation to contribute importantly to phenotypic stability over deep time (see Chapter 11 in this volume). It therefore follows that a mechanistic hypothesis of constraint (e.g., pleiotropic effects or internal selection) can be tested only by reference to a historical pattern predicted as its consequence (McKitrick 1993; Schwenk 1995; Schwenk and Wagner 2001). Predicted patterns depend on the null model of evolution assumed (see below) and on the interests of the investigator. Most often constraint is thought to be revealed as phenotypic stability or stasis in the characters of interest, but other predicted patterns include unoccupied areas of a theoretical morphospace, parallelism, and suboptimal design.

Finally, hypotheses of constraint apply to characters and not to whole organisms (Schwenk 1995). The effects of constraint are therefore evident in patterns of character evolution rather than in some measure of evolutionary "success." Critically, this allows us to decouple the negative effects of evolutionary limitations (constraints) on characters from the consequences of these limitations on the organism, which may be positive, negative or neutral (Wagner 1988; Roth and Wake 1989; Amundson 1994; Schwenk 1995; *contra* Gould 2002). Thus, constraints are always negative in the sense that they limit character evolution, but they may at the same time contribute positively to evolvability in response to natural selection (Wagner 1988), or in biasing the direction of phenotypic change (e.g., Wake 1991; Jacobs et al. 1995).

## Specifying the Local Context

Armed with background information and some basic assumptions, we can proceed to aspects of the local context that require specification. If, as we believe, constraint is a relativistic concept, then the explicit specification of the local context is essential both in defining constraint and in judging its valid application to a particular case.

(1) Given that constraint hypotheses apply to characters and not to whole organisms, it is necessary to specify the character(s) putatively constrained and in what manner. For example, Wagner and Schwenk (2000) listed thirteen characters of the lizard feeding system that they argued were phenotypically stabilized by internal selection (see also Schwenk and Wagner 2001). Wake (1991) described aspects of salamander carpal morphology putatively biased in phenotypic evolution by the dynamics of the developmental system. It is worth noting that, in both examples, the organisms and clades evincing putative constraints are diverse and

evolutionarily "successful." It is only the focal characters that are constrained in terms of the phenotypic space they occupy.

(2) The constraint hypothesis must be made in reference to a specified phylogenetic node. The constraint will apply to the entire specified clade, or to a paraphyletic group within the clade. The latter case would obtain when one or more descendant lineages modifies, loses, or otherwise "breaks" the constraint. Importantly, specification of a phylogenetic node implies the existence of a sister or related clade lacking the putative constraint. This is essential if we are to test constraint hypotheses against historical patterns of character evolution. Furthermore, if the object of constraint theory is to account for different patterns of phenotypic variation among clades, it follows that the constraint hypothesis is only relevant at the specified phylogenetic level and not to nested subclades. In other words, the hypothesis is applicable only to the level at which we can compare putatively constrained and unconstrained clades. If, for example, we argue that limited variation in bird vocalization reflects evolutionary constraint (e.g., Podos 1997), the hypothesis might have good explanatory value in understanding the evolution of bird vocalization as compared to vocalization in other tetrapods, but the same hypothesis does not inform our understanding of vocal evolution in one genus of songbird as compared to another. In the latter case, the putative constraint is a shared, ancestral trait that cannot account for observed differences between genera (analogously, a synapomorphy at one node of a phylogeny is a symplesiomorphy for nested nodes, and is therefore not informative about relationships among nested taxa).

(3) Given that constraints limit the evolution of characters over long periods of time, it is necessary to have a benchmark against which to measure the effects of the putative constraint, that is, a null model of evolution (Gould 1989; Antonovics and van Tienderen 1991; Schlichting and Pigliucci 1998). A null model represents the pattern of character evolution expected in the absence of constraint. The choice of a null model is to some extent arbitrary and will depend, minimally, on the characters of interest. As noted above, molecular versus phenotypic characters, for example, require different null models of evolution, leading to radically different conceptions of constraint (Schwenk and Wagner 2003). The null model implicit in most studies of phenotypic evolution is adaptive evolution by natural selection (Antonovics and van Tienderen 1991; but see discussion of this null model by Pigliucci in Chapter 11, this volume). Under this model, our expected pattern of character evolution is adaptive change in characters through time. If the focal characters are subject to constraint, we would therefore expect their pattern of evolution to deviate from the null model in a predicted way (e.g., Schwenk 1995; Schwenk and Wagner 2001). For character evolution we usually expect constraints to be manifested as long-term phenotypic stability (stasis) unaccounted for by environmental stability. As already discussed, adaptive evolution in the sense of phenotype-environment matching is driven by external selection, therefore we can refine our null model as "adaptive evolution by external selection."

(4) Finally, and perhaps most critically, it is necessary to designate a focal life stage. This is the life stage for which character evolution is putatively constrained. In most studies this is implicitly the adult stage, but this is rarely acknowledged,

leading to many of the problems discussed here. By designating a focal life stage, constraint can be decoupled from a particular mechanism as well as a given ontogenetic stage (e.g., development). This is possible because the focal life stage (in conjunction with the factors discussed above) provides a fixed point of reference for assessing putative mechanisms of constraint. By fixing this point, we interrupt the temporal continuity of ontogeny and establish a "before" and an "after" so that the consequences of earlier-acting processes can be assessed for their impact on the focal stage. Our statements about phenotypic evolution can then be directed specifically to the focal life stage. This stage essentially "roots" ontogeny to provide a sequence of cause and effect in much the same way that rooting an unrooted phylogeny establishes a sequence of branching (Schwenk and Wagner 2003). If, for example, the evolution of adult characters is our interest (as is typical), then external selection acting to modify a character at this stage will obviously be limited by the amount of variation present in the adult population. It is likewise apparent that a variety of processes might have acted earlier in ontogeny to limit or bias the variation present at the adult stage. From the point of view of external selection acting during the focal adult stage, the precise nature of these processes is unimportant (with one exception; see next section). All will diminish the efficacy of external selection acting at the focal stage and therefore will constrain character adaptation at that stage. External selection at the focal stage might also be mitigated by simultaneously opposing internal selection on the same character(s) (see discussion above).

## A Definition of Evolutionary Constraint

Given the assumptions and specifications presented above, we are now able to propose a definition of evolutionary constraint (Schwenk and Wagner 2003):

> Evolutionary constraint is a mechanism or process that limits the evolutionary response of a character or set of characters to external selection acting during a focal life stage.

Given that a historical approach is necessary to identify the effect of constraint, "limits" are assessed by comparing patterns of character evolution in putatively constrained and unconstrained sister clades. Character evolution in the unconstrained clade should reflect the pattern of the null model (e.g., adaptive change), whereas constrained characters should conform to predictions suggested by the constraint (e.g., stasis, bias, or unoccupied theoretical morphospace) (Schwenk and Wagner 2001).

Inferring the past action of selection is problematic and may be questioned on philosophical grounds (e.g., Leroi et al. 1994; Reeve and Sherman 1992). However, an inference of past external selection is reasonable if the historical pattern of character evolution is consistent with adaptive change. It is strengthened if there is evidence of similar, ongoing selection (e.g., Wagner and Schwenk 2000). There is no avoiding the fact that present circumstances must be used to infer past constraint. Although a currently operational constraint might limit future phenotypic evolution, only a historical perspective allows us reasonably to test whether the constraint is persistent enough to affect phenotypic evolution.

Because of its relativism, the proposed definition resolves the paradoxes introduced at the beginning of this chapter. Constraint is manifested by its effect on adaptive external selection acting during a specified life stage. Therefore, any processes that weaken or modulate the effect of this selection are candidates for constraint, including selection itself, in the form of internal selection. The definition thus salvages the traditional notion of developmental constraint propounded by Maynard Smith et al. (1985), because it allows that internal selection acting during development might limit the availability of phenotypic variation available to external selection acting at a later (focal) stage. Without the notion of internal selection and the relativism of the focal life stage approach, Maynard Smith et al.'s (1985) own dichotomous view renders their definition self-contradictory.[2]

External selection is the one exception to the plurality of mechanisms encompassed by our definition. Although external viability selection acting at earlier life stages would decrease phenotypic variation available at the focal stage, and therefore limit the potential for adaptive phenotypic evolution at that stage, this limitation is not the result of an intrinsic, organismal attribute. It does not reside within the organism and thus is not a historical, evolving entity. We see no conceptual clarification emerging from the conflation of constraint and external selection in a technical sense.

## An Example of the Focal Life Stage Approach to Constraint

### The Evolution of Teleost Fins

Ray-finned teleost fishes and sarcopterygians (including tetrapods) both have endoskeletal elements in their paired fin/limb skeleton. However, these elements differ fundamentally in how they develop. In sarcopterygians, the endoskeleton develops from individualized cell condensations forming distinct cell populations. These chondrify and eventually fuse or ossify individually. The skeletal elements generally follow a branching rule, starting with a single element articulating with the girdle (Shubin and Alberch 1986). Within sarcopterygians, this pattern of development has permitted a wide array of skeletal forms conforming to the adaptive needs of the various taxa. In contrast, the radials of teleost fins develop in a nearly invariant manner (Peters 1981, 1983; Potthoff et al. 1984; Watson 1987; Mabee and Trendler 1996; Grandel and Schulte-Merker 1998). The radials secondarily individuate from a cartilaginous plate that dedifferentiates into a consistent pattern of four elements arranged in an anteroposterior series. Although teleosts are the most speciose of living vertebrates, and arguably the most diverse, this fundamental pattern of four anteroposterior radials is conserved in nearly all 24,000 species. We might surmise the action of constraint in maintaining this character state.

A central difference between teleost and sarcopterygian life cycles is that teleosts tend to hatch at much smaller body sizes with relatively undeveloped skeletons that are neither fully differentiated nor mechanically competent. Despite the lack of a mature endoskeleton, larval fish soon require functional pectoral fins. Structural support for the developing fin rays is provided at this early stage by the cartilaginous plate. As growth and differentiation proceed, the radial elements of the mature fin skeleton must therefore develop from this preceding larval condition rather than arising de novo as separate condensations, as in sarcopterygians.

In considering ultimate causes, small hatchling size in teleosts and its associated skeletal immaturity is probably related to life-history selection favoring the production of many propagules (owing to low offspring survivorship). Although initially planktonic, larval fish soon require mobility to survive: they hold station in moving water, migrate offshore from estuarine habitats, or migrate vertically in the water column. The origin of the cartilaginous plate can be considered a larval adaptation to the need for mobility at this immature life stage. In the manner discussed, the dynamic interactions occurring during development of the cartilaginous plate, as well as those occurring during fin function, impose functional dependencies among constituent characters. These dependencies are consistent in nearly all habitats encountered by teleost fish (e.g., fresh water, marine, estuarine, pelagic, with predators, without predators) and therefore reflect internal, stabilizing selection on the larval fin skeleton.

A clear picture of constraint emerges when we specify the adult phenotype as our focal life stage. In adult teleosts, it is the remarkable conservation of four anteroposterior radials, in contrast to the diversity of patterns evident in sarcopterygians, that bears scrutiny. The diversity of sarcopterygians reflects our expectations under a null model of adaptive evolution. Given the vast phenotypic and ecological diversity of adult teleosts, and the commensurate diversity of their environments, we can reasonably infer that the fin skeleton has been subject to external selection tending to favor change in the number and arrangement of radials. For example, the stonefish fin approaches a condition *functionally* similar to a salamander arm, but that anatomically remains a secondary modification of the four anteroposterior radials. From the perspective of the adult stage, stability of the fin skeleton is attributable to a constraint arising from the developmental "bottleneck" imposed by the larval cartilaginous plate and its consequences for the development of the adult skeleton. This unique larval phenotype determines the initial condition for the development of the adult endoskeleton. The form of the plate, and the mode of individuation of radials from the plate, impose limits on the number and arrangement of radials that can be formed. Thus, functional integration of the larval life stage, together with the developmental pathway connecting larval and adult stages, leads to a constraint on phenotypic evolution of the adult life stage. In contrast to traditional, dichotomous views (e.g., Maynard Smith et al. 1985), however, the mechanistic basis of this developmental constraint is an interaction between (internal) selection and development that limits the amount of phenotypic variation available in the limb skeleton at the adult stage. The rationale for invoking selection as part of the constraint is provided by the focal life stage perspective and recognition of an internal component of natural selection.

It is worth mentioning that there are a few exceptions to the rule of four radials in teleost fins. These are mostly found among the anglerfishes, Order Lophiiformes (A. Ward, pers. comm.). Many anglerfishes are benthic with highly modified pectoral fins used as props, or in the case of batfishes (Ogcocephalidae) as limbs to propel the fish across the substrate (Nelson 1994; A. Ward, unpublished data). The novel functions of these fins might account for disruption of the putative constraints on their skeletons. It would be interesting to investigate the developmental basis for these deviations from the "typical" phenotype. These exceptions highlight the important fact that constraints, while persistent, are not immutable. The rules governing even highly integrated systems can be overcome under certain circumstances (Wagner and Schwenk 2000; Schwenk 2001).

An interesting question arises when we shift our focal life stage to the larva. Does conservation of the cartilaginous plate at this stage reflect constraint? In general, highly integrated systems are expected to be self-stabilizing owing to the long-term effects of internal selection (Wagner and Schwenk 2000). This stability has been described as a kind of evolutionary homeostasis (Wake et al. 1983; Roth and Wake 1985). The phenotypic stability of characters within such systems would deviate from our null model expectation of adaptive change. Nonetheless, in a relativistic view the mere presence of a particular mechanism, such as internal selection, is not justification for the invocation of constraint. The constraint only "exists" relative to a force tending to change it—external selection in our model. If we used a null model of random change caused by mutation, then the self-stabilization of internal selection could be seen as constraint. Over the long term, it is likely that most integrated systems will be subject to opposing external selection, in which case constraint would be justified in our model. However, in the case of the cartilaginous plate we have no evidence for such external selection at the larval stage, hence there is no need to invoke constraint.

## Conclusions

In this chapter we have taken a dialectical approach suggesting that insight into the processes of phenotypic evolution are to be found in the resolution of opposing themes—mutability versus stability, externalism versus internalism, and selection versus constraint. These dichotomies seem to reflect a real evolutionary tension between the forces of change and the forces of stability. In focusing historically on the former, evolutionary theory has tended to neglect the latter. Internalism and constraint theory are responses to this neglect. They emphasize the role of intrinsic organismal attributes in resisting or otherwise modulating the external forces of change.

As heuristically useful as dichotomies are in structuring our thoughts and arguments, they can sometimes lead us astray if interpreted too literally. Such is the case with selection and constraint. Since natural selection is viewed as the primary engine of phenotypic change, it was perhaps logical that constraint would be construed as an opposing engine of stability. In this chapter we have sought to show that this dichotomy cannot be sustained in a historical context. Constraints are forged in the fires of selection and to require their separation leads inevitably to paradox. Our approach, therefore, has been to develop a conceptual framework that reconciles selection and constraint by pointing to the interaction between selective forces and developmental factors in causing constraints. We hope that such an approach will facilitate a fruitful application of constraint to problems of phenotypic evolution.

We see constraint as a central concept in resolving the tension between mutability and stability. Although Shelley may ultimately be right—nothing endures in life but the inevitability of change—constraints might at least account for some of the pauses along the way.

*Acknowledgments* We thank Massimo Pigliucci and Katherine Preston for inviting us to contribute to this volume, and for their thoughtful comments on the chapter. We are also grateful to Kentwood Wells for discussion of sexual selection, and to Eric Schultz for discussion of larval fish biology. Andrea Ward's information on batfish locomotion was presented at the Northeast Regional Meeting of the Division of Vertebrate Morphology, Society for Comparative and Integrative Biology, September 2002, at Harvard University.

### Notes

1 We use "tolerance" here in the engineering sense, referring to "an allowable amount of variation in the dimensions of a machine or part" (*The New Shorter Oxford English Dictionary*, 1993, L. Brown, ed., Clarendon Press, Oxford, p. 3330). In other words, the functional interaction between two parts or elements of a system might depend on a very tight "fit" between them. This is most easily visualized for discrete anatomical parts, such as upper and lower incisor teeth in a rodent: if the teeth don't occlude very precisely, they are entirely nonfunctional, i.e., their cutting/shearing function depends on their exact alignment while closing against one another (for the same reason, if the blades of a pair of scissors are loose, the scissors will not cut). Mutations disrupting this alignment will have severe consequences for the feeding system as a whole, implying there will be strong internal selection against them. Other functional interactions might have larger tolerances, implying weaker internal selection and a broader range of permissible phenotypes (see Schwenk 2001). It is interesting to consider what role compensatory plasticity might play in this context (K. Preston, pers. comm.). Plasticity might mitigate the strength of internal selection implied by a tight functional tolerance. In the rodent scenario, for example, a precise fit between upper and lower incisors is maintained by self-sharpening through wear between the ever-growing teeth. Furthermore, tooth alignment can be actively manipulated by changing lower jaw position. Thus, plasticity might both broaden the range of permissible phenotypes and act ontogenetically to "fine tune" the functionality of a system.

2 Maynard Smith et al.'s (1985) definition of developmental constraint is widely accepted in the literature and was briefly noted in this chapter's introduction. It states that constraints are "biases on the production of variant phenotypes or limitations on phenotypic variability caused by the structure, character, composition, or dynamics of the developmental system" (p. 265). Although Maynard Smith et al.'s (1985) view of constraint was dichotomous (constraint is independent of selection: see Schwenk and Wagner 2003), their definition fails to exclude selection as the cause of developmental bias or limitation.

*Literature Cited*


Alberch, P. 1982. Developmental constraints in evolutionary processes. Pp. 313–332 in J. T. Bonner, ed. Evolution and Development. Springer-Verlag, Berlin.

Amundson, R. 1994. Two concepts of constraint: adaptationism and the challenge from developmental biology. Phil. Sci. 61:556–578.

Amundson, R. 1996. Historical development of the concept of adaptation. Pp. 11–53 in M. R. Rose and G. V. Lauder, eds. Adaptation. Academic Press, San Diego.

Antonovics, J., and van Tienderen, P. H. 1991. Ontoecogenophyloconstraints? The chaos of constraint terminology. Trends Ecol. Evol. 6:166–168.

Arnold, S. J. 1992. Constraints on phenotypic evolution. Am. Nat. 140 (Suppl.):S85–S107.

Arthur, W. 1997. The Origin of Animal Body Plans. Cambridge University Press, Cambridge.

Beldade, P., K. Koops, and P. M. Brakefield. 2002. Developmental constraints versus flexibility in morphological evolution. Nature 416:844–847.

Carroll, S. B., J. K. Grenier, and S. D. Weatherbee. 2001. From DNA to Diversity: Molecular Genetics and the Evolution of Animal Design. Blackwell Science, Malden, MA.

Cheverud, J. M. 1984. Quantitative genetics and developmental constraints on evolution by selection. J. Theor. Biol. 110:155–171.

Cheverud, J. M. 1988. The evolution of genetic correlation and developmental constraints. Pp. 94–101 in G. de Jong, ed. Population Genetics and Evolution. Springer-Verlag, Berlin.

Cheverud, J. M. 2001. The genetic architecture of pleiotropic relations and differential epistasis. Pp. 411–433 in G. P. Wagner, ed. The Character Concept in Evolutionary Biology. Academic Press, San Diego.

Clarke, G. M., and J. A. McKenzie. 1987. Developmental stability of insecticide resistant phenotypes in blowfly: a result of canalizing natural selection. Nature 325:345–346.

Darwin, C. 1859. On the Origin of Species by Means of Natural Selection. John Murray, London (1964 facsimile reprint of the first edition, Harvard University Press, Cambridge, MA).

Darwin, C. 1871. The Descent of Man, and Selection in Relation to Sex. John Murray, London (1989 reprint, New York University Press, New York).

de Beer, G. R. 1930. Embryology and Evolution. Oxford University Press, Oxford.

de Beer, G. R. 1938. Embryology and evolution. Pp. 57–78 in G. R. de Beer, ed. Evolution: Essays on Aspects of Evolutionary Biology. Oxford University Press, Oxford.

Dullemeijer, P. 1980. Functional morphology and evolutionary biology. Acta Biotheor. 29:151–250.

Eldredge, N., and S. J. Gould. 1972. Punctuated equilibria: an alternative to phyletic gradualism. Pp. 82–115 in T. J. M. Schopf, ed. Models in Paleobiology. Freeman, Cooper, San Francisco.

Endler, J. A. 1980. Natural selection on color patterns in *Poecilia reticulata*. Evolution 34:76–91.

Endler, J. A. 1983. Natural and sexual selection on color patterns in poeciliid fishes. Environ. Biol. Fishes 9:173–190.

Fusco, G. 2001. How many processes are responsible for phenotypic evolution? Evol. Dev. 3:279–286.

Galis, F. 1999. Why do almost all mammals have seven cervical vertebrae? Developmental constraints, Hox genes, and cancer. J. Exp. Zool. (Mol. Dev. Evol.) 285:19–26.

Galis, F., and J. A. J. Metz. 2001. Testing the vulnerability of the phylotypic stage: on modularity and evolutionary conservatism. J. Exp. Zool. (Mol. Dev. Evol.) 291:195–204.

Gould, S. J. 1977. Ontogeny and Phylogeny. Harvard University Press, Cambridge, MA.

Gould, S. J. 1989. A developmental constraint in *Cerion*, with comments on the definition and interpretation of constraint in evolution. Evolution 43:516–539.

Gould, S. J. 2002. The Structure of Evolutionary Theory. Harvard University Press, Cambridge, MA.

Gould, S. J., and N. Eldredge. 1977. Punctuated equilibria: the tempo and mode of evolution reconsidered. Paleobiology 3:115–151.

Gould, S. J., and R. C. Lewontin. 1979. The spandrels of San Marco and the Panglossian paradigm: a critique of the adaptationist programme. Proc. R. Soc. London B 205:581–598.

Grandel, H., and S. Schulte-Merker. 1998. The development of the paired fins of the zebrafish (*Danio rerio*). Mech. Dev. 79:99–120.

Hall, B. K. 1996. *Baupläne*, phylotypic stages, and constraint: why there are so few types of animals. Evol. Biol. 29:215–261.

Hall, B. K. 1998. Evolutionary Developmental Biology, 2nd ed. Chapman & Hall, London.

Ho, M.-W., and P. T. Saunders. 1979. Beyond neo-Darwinism: an epigenetic approach to evolution. J. Theor. Biol. 78:573–591.

Ho, M.-W., and P. T. Saunders. 1984. Beyond Neo-Darwinism: An Introduction to the New Evolutionary Paradigm. Academic Press, London.

Jacobs, S. C., A. Larson, and J. M. Cheverud. 1995. Phylogenetic relationships and orthogenetic evolution of coat color among tamarins (genus *Saguinus*). Syst. Biol. 44:515–532.

Leroi, A. M., M. R. Rose, and G. V. Lauder. 1994. What does the comparative method reveal about adaptation? Am. Nat. 143:381–402.

Mabee, P. M., and T. A. Trendler. 1996. Development of the cranium and paired fins in *Betta splendens* (Teleostei: Percomorpha): intraspecific variation and interspecific comparisons. J. Morphol. 227:249–287.

Maynard Smith, J., R. Burian, S. Kauffman, P. Alberch, J. Campbell, B. Goodwin, R. Lande, D. Raup, and L. Wolpert. 1985. Developmental constraints and evolution. Quart. Rev. Biol. 60:265–287.

McKitrick, M. C. 1993. Phylogenetic constraint in evolutionary biology: has it any explanatory power? Annu. Rev. Ecol. Syst. 24:307–330.

Nelson, J. S. 1994. Fishes of the World, 3rd ed. John Wiley, New York.

Peters, K. M. 1981. Reproductive biology and developmental osteology of the Florida blenny, *Chasmodes saburrae* (Perciformes: Blenniidae). Northeast Gulf Sci. 4:79–98.

Peters, K. M. 1983. Larval and early juvenile development of the frillfin Goby, *Bathygobius soporator* (Perciformes: Gobiidae). Northeast Gulf Sci. 6:137–153.

Podos, J. 1997. A performance constraint on the evolution of trilled vocalizations in a songbird family (Passeriformes: Emberizidae). Evolution 51:537–551.

Potthoff, T., S. Kelley, M. Moe, and F. Young. 1984. Description of porkfish larvae (*Anisotremus virginicus*, Haemulidae) and their osteological development. Bull. Mar. Sci. 34:21–59.

Raff, R. A. 1996. The Shape of Life. University of Chicago Press, Chicago.

Raff, R. A., and T. C. Kaufman. 1983. Embryos, Genes, and Evolution. Macmillan, New York.

Raup, D. M. 1966. Geometric analysis of shell coiling: general problems. J. Paleontol. 40:1178–1190.

Reeve, H. K., and Sherman, P. W. 1992. Adaptation and the goals of evolutionary research. Quart. Rev. Biol. 68:1–32.

Riedl, R. 1977. A systems-analytical approach to macro-evolutionary phenomena. Quart. Rev. Biol. 52:351–370.

Roth, G., and D. B. Wake. 1985. Trends in the functional morphology and sensorimotor control of feeding behavior in salamanders: an example of the role of internal dynamics in evolution. Acta Biotheor. 34:175–192.

Roth, G., and D. B. Wake. 1989. Conservatism and innovation in the evolution of feeding in vertebrates. Pp. 7–21 in D. B. Wake and G. Roth, eds. Complex Organismal Functions: Integration and Evolution in Vertebrates. John Wiley, Chichester, UK.

Schlichting, C. D., and M. Pigliucci. 1998. Phenotypic Evolution: A Reaction Norm Perspective. Sinauer, Sunderland, MA.

Schwenk, K. 1995. A utilitarian approach to evolutionary constraint. Zoology 98:251–262.

Schwenk, K. 2000. Feeding in lepidosaurs. Pp. 175–291 in K. Schwenk, ed. Feeding: Form, Function and Evolution in Tetrapod Vertebrates. Academic Press, San Diego.

Schwenk, K. 2001. Functional units and their evolution. Pp. 165–198 in G. P. Wagner, ed. The Character Concept in Evolutionary Biology. Academic Press, San Diego.

Schwenk, K., and G. P. Wagner. 2001. Function and the evolution of phenotypic stability: connecting pattern to process. Am. Zool. 41:552–563.

Schwenk, K., and G. P. Wagner. 2003. Constraint. Pp. 52–61 in B. K. Hall and W. M. Olson, eds. Keywords and Concepts in Evolutionary Developmental Biology. Harvard University Press, Cambridge, MA.

Seilacher, A. 1972. Divaricate patterns in pelecypod shells. Lethaia 5:325–343.

Seilacher, A. 1973. Fabricational noise in adaptive morphology. Syst. Zool. 22:451–465.

Shubin, N. H., and P. Alberch. 1986. A morphogenetic approach to the origin and basic organization of the tetrapod limb. Evol. Biol. 20:319–387.

Simpson, G. G. 1953. The Major Features of Evolution. Simon & Schuster, New York.

Smith, K. C. 1992. Neo-rationalism versus neo-Darwinism: integrating development and evolution. Biol. Phil. 7:431–451.

Stearns, S. C. 1986. Natural selection and fitness, adaptation and constraint. Pp. 23–44 in D. M. Raup and D. Jablonski, eds. Patterns and Processes in the History of Life. Springer-Verlag, Berlin.

Stern, D. L. 2000. Evolutionary developmental biology and the problem of variation. Evolution 54:1079–1091.

Verhulst, S., S. J. Dieleman, and H. K. Parmentier. 1999. A tradeoff between immunocompetence and sexual ornamentation in domestic fowl. Proc. Natl. Acad. Sci. USA 96:4478–4481.

Waddington, C. H. 1957. The Strategy of the Genes. Macmillan, New York.

Wagner, A. 1996. Does evolutionary plasticity evolve? Evolution 50:1008–1023.

Wagner, G. P. 1986. The systems approach: an interface between development and population genetic aspects of evolution. Pp. 149–165 in D. M. Raup and D. Jablonski, eds. Patterns and Processes in the History of Life. Springer-Verlag, Berlin.

Wagner, G. P. 1988. The significance of developmental constraints for phenotypic evolution by natural selection. Pp. 222–229 in G. de Jong, ed. Population Genetics and Evolution. Springer-Verlag, Berlin.

Wagner, G. P., and L. Altenberg. 1996. Complex adaptations and the evolution of evolvability. Evolution 50:967–976.

Wagner, G. P., and K. Schwenk. 2000. Evolutionarily stable configurations: functional integration and the evolution of phenotypic stability. Evol. Biol. 31:155–217.

Wagner, G. P., G. Booth and H. Bagheri-Chaichian. 1997. A population genetic theory of canalization. Evolution 51:329–347.

Wake, D. B. 1982. Functional and developmental constraints and opportunities in the evolution of feeding systems in urodeles. Pp. 51–66 in D. Mossakowski and G. Roth, eds. Environmental Adaptation and Evolution. G. Fischer, Stuttgart.

Wake, D. B. 1991. Homoplasy: the result of natural selection, or evidence of design limitations? Am. Nat. 138:543–567.

Wake, D. B. 1994. Comparative terminology. Science 265:268–269.

Wake, D. B., and A. Larson. 1987. Multidimensional analysis of an evolving lineage. Science 238:42–48.

Wake, D. B., and G. Roth. 1989. The linkage between ontogeny and phylogeny in the evolution of complex systems. Pp. 361–377 in D. B. Wake and G. Roth, eds. Complex Organismal Functions: Integration and Evolution in Vertebrates. John Wiley, New York.

Wake, D. B., G. Roth, and M. H. Wake. 1983. On the problem of stasis in organismal evolution. J. Theor. Biol. 101:211–224.

Watson, W. 1987. Larval development of the endemic Hawaiian blenniid, *Enchelyurus brunneolus* (Pisces: Blenniidae: Omobranchini). Bull. Mar. Sci. 41:856–888.

Webster, G., and B. Goodwin 1996. Form and Transformation: Generative and Relational Principles in Biology. Cambridge University Press, Cambridge.

Whyte, L. L. 1965. Internal Factors in Evolution. George Braziller, New York.

Woodger, J. H. 1929. Biological Principles: A Critical Study. Kegan Paul, Trench, Trubner, London.

Woodger, J. H. 1945. On biological transformations. Pp. 95–120 in W. E. Le Gros Clark and P. B. Medawar, eds. Essays on Growth and Form Presented to D'Arcy Wentworth Thompson. Oxford University Press, Oxford.

Zuk, M., and T. S. Johnsen. 1998. Seasonal changes in the relationship between ornamentation and immune response in red jungle fowl. Proc. R. Soc. London B 265:1631–1635.

# 19

# The Developmental Systems Perspective

## *Organism–Environment Systems as Units of Development and Evolution*

PAUL E. GRIFFITHS
RUSSELL D. GRAY

Undergraduate textbooks typically define evolution as change in gene frequency.[1] This reflects the conventional view of natural selection and the conventional view of heredity. Natural selection occurs because individuals vary, some of these variations are linked to differences in fitness, and some of those variants are heritable (Lewontin 1970). Because variants that are not heritable cannot play a role in natural selection, and because the mechanism of inheritance is presumed to be genetic, evolution is defined as change in gene frequencies. In the 1960s and 1970s this gene-centered vision of inheritance was extended to yield a gene-centered view of selection (Williams 1966; Dawkins 1976). According to gene-selectionism, the fact that individual genes are integrated into larger units, from genetic modules to entire phenotypes, is merely a special case of the fact that the fitness of any evolutionary unit is a function of the environment in which it happens to find itself (Sterelny and Kitcher 1988). The study of development and the study of how phenotypes are integrated are thus doubly divorced from the study of evolution. First, all causal factors in the development of the phenotype other than genes are excluded as potential sources of evolutionary change. Second, the study of development and phenotypic integration cease to be possible sources of theoretical insight into the evolutionary process. The phenotypes to which an individual gene contributes and the developmental processes by which it makes that contribution are simply environmental factors like rainfall or predator density. The distribution of each individual allele across these environmental parameters determines the fitness of that allele. As far as theoretical population genetics is concerned, developmental biology and the nature of complex phenotypes are part of ecology and can be adequately represented by the varying fitness values of competing alleles. Insights into how organisms develop and how their phenotypes decompose into meaningful units do not yield any general insight into evolutionary dynamics, although, naturally, to understand the particular selective pressures on particular gene lineages it will be necessary to study their "ecology."

The rapid advance of molecular developmental biology and the emergence of the new field of evolutionary developmental biology (EDB) has done a great deal to counter the atomistic approach to organisms and their evolution represented by gene-selectionism. It is now widely accepted that a meaningful decomposition of the organism—or its genome—into parts that can be considered to have their own evolutionary history must reflect an understanding of the developmental biology of the organism. This has led to renewed attention to the concept of homology and extensive research on the newer concept of developmental modularity (Hall 1992, 1994; Raff 1996; Arthur 1997; Wagner 2001). However, most evolutionary developmental biologists still accept that the developmental systems they study emerge from combinations of genes and genes alone. Jason Robert, Brian Hall, and Wendy Olson have noted that "EDB ... continues to show a tendency toward reductionism and gene-centrism; developmental mechanisms are ultimately genetic ... and there is no such thing as epigenetic inheritance" (Robert et al. 2000, p. 959). In contrast, developmental systems theory (DST)[2] questions both elements of the gene-centered perspective, integrating an emphasis on the relevance of development to evolution with an emphasis on the evolutionary potential of extragenetic inheritance. The result is an account of evolution in which the fundamental unit that undergoes natural selection is neither the individual gene nor the phenotype, but the life cycle generated through the interaction of a developing organism with its environment. In our usage, the "developmental system" is the whole matrix of resources that interacts to reconstruct that life cycle.

## Inheritance

An organism inherits more than its nuclear DNA. A viable egg cell must contain a variety of membranes, both for its own viability as a cell and to act as templates for the assembly of proteins synthesized from the DNA into new membrane. A eukaryote cell must contain a number of organelles, such as mitochondria, with their own distinctive DNA. But the full variety of the contents of the cell is only now being uncovered. For normal gene transcription to occur, DNA must be accompanied by the elements of the chromatin marking system. For normal differentiation of the embryo, initial cytoplasmic chemical gradients must be set up within the cell. The essential role of still further parts of the package, such as microtubule organizing centers, is becoming apparent. But unpacking the inherited resources in the cell is not the end of unpacking inheritance. In multicellular organisms the parental generation typically contributes extracellular resources. An ant in a brood cell is exposed to a variety of chemical influences that lead it to develop as a worker, a queen, or a soldier. A termite inherits a population of gut endosymbionts by coprophagy. In viviparous organisms the environment of the womb provides not only nutrition but a range of stimulation essential for the normal development of the nervous system (for examples, see Gottlieb 1992, 1997,

2001). This stimulation continues after birth. The effects of severe deprivation of conspecific stimulation in infant primates, including humans, have been well documented (Harlow and Harlow 1962; Money 1992). Nor are these effects confined to animals. Many eucalypt species have seeds that cannot germinate until they have been scorched by a bushfire. For eucalypts to increase the frequency of bushfires to the point where this system works reliably, local populations of trees must create forests scattered with resinous litter and hung with bark ribbons. These are carried aloft by the updraft as blazing torches and spread the fire to new areas (Mount 1964). Even after the resources created by the population as a whole are added in, a range of other factors must be present before the sum of the available resources adds up to a viable package. Development frequently requires gravity or sunlight or, for a hermit crab, a supply of discarded shells. These factors are unaffected by the activities of past generations of the species that rely on them. Nevertheless, the organism must position itself so that these factors interact with it and play their usual role in development. While the evolving lineage cannot make these resources, it can still make them part of its developmental system.

It is uncontroversial to describe all these resources as playing a role in development. But it is highly controversial to say that these same resources are "inherited." With the exception of genes, and more recently the chromatin marking system, their roles are not supposed to extend to the intergenerational processes of evolution. Nongenetic factors, it is generally supposed, do not have the capacity for replication through many generations, and lack the potential to produce the kind of variation upon which natural selection can act: "The special status of genetic factors is deserved for one reason only: genetic factors replicate themselves, blemishes and all, but non-genetic factors do not" (Dawkins 1982, p. 99). Or, more bluntly: "Differences due to nature are likely to be inherited whereas those due to nurture are not; evolutionary changes are changes in nature, not nurture" (Maynard Smith 2000). The continued popularity of this argument is puzzling. Many nongenetic resources are reliably passed on across the generations. Variations in these resources can be passed on, causing changes in the life cycle of the next generation (Jablonka and Lamb 1995; Avital and Jablonka 2001). The concept of inheritance is used to explain the stability of biological form from one generation to the next. In line with this theoretical role, DST applies the concept of inheritance to any resource that is reliably present in successive generations, and is part of the explanation of why each generation resembles the last. This seems to us a *principled* definition of inheritance. It allows us to assess the evolutionary potential of various forms of inheritance empirically, rather than immediately excluding everything but genes and a few fashionable extras.

## Natural Selection

Armed with a thoroughly epigenetic view of development and an expanded view of inheritance, let us now turn to the concept of natural selection. In principle, there seems no reason why this concept should not be decoupled from gene-centered theories of development and evolution. After all, Darwin developed the theory of natural selection prior to the mechanisms of inheritance being discovered. The three requirements for natural selection (variation, fitness differences, heritability) are agnostic about the details of inheritance. In Daniel Lehrman's classic phrase, "Nature selects for outcomes" (Lehrman 1970, p. 28) and the developmental routes by which differences are produced do not matter as long as the differences reliably reoccur.

Consider the following two cases: Newcomb et al. (1997) found that a single nucleotide change in blowflies can change the amino acid at an active site of an enzyme (carboxylesterase). This change produced a qualitatively different enzyme (organophosphorous hydrolase), which conferred resistance against certain insecticides. This case fulfills the three requirements for natural selection. There are phenotypic differences in insecticide resistance, these differences are likely to produce differences in fitness, and these differences are heritable. Moran and Baumann (1994) discuss a similar, fascinating example of evolution in action. Certain aphid species reliably pass on their endosymbiotic *Buchnera* bacteria from the maternal symbiont mass to either the eggs or developing embryo. The bacteria enable their aphid hosts to utilize what would otherwise be nutritionally unsuitable host plants. Aphids that have been treated with antibiotics to eliminate the bacteria are stunted in growth, reproductively sterile, and die prematurely. A lineage that inherits bacteria is clearly at an advantage over one that does not. Once again there is variation (lineages with either different *Buchnera* bacteria or without *Buchnera*), these differences confer differences in fitness, and they are heritable. All biologists would recognize the first case as an example of natural selection in action, but they would probably balk at categorizing the aphid/bacteria system in the same way. Yet why should these cases be treated differently when both meet the three criteria for natural selection?

An obvious response would be to claim that if there is selection in this case, then it can be reduced to selection of genetic differences. Aphids with genes for passing on their endosymbionts have evolved by outcompeting aphids with genes for not passing on endosymbionts. However, it is possible to have differential reproduction of the aphid/bacteria system without any genetic difference between the two lineages involved. An aphid lineage that loses its bacteria will produce offspring without bacteria. These offspring remain genetically identical to the lineages with which they compete, but have a lower expected reproductive output. A naturally occurring instance of this sort of selectively relevant nongenetic variation is found in the North American fire ant *Solenopsis invicta* (Keller and Ross 1993). Colonies containing large, monogynous queens and colonies containing small, polygynous queens were shown to have no significant genetic differences. Differences between queens are induced by the type of colony in which they have been raised, as shown by cross-fostering experiments. Exposure of eggs from either type of colony to the pheromonal "culture" of a polygynous colony produces small queens who found polygynous colonies, leading to more small queens, and so forth. Exposure of eggs from either type of colony to the pheromonal "culture" of a monogynous colony produces large queens who found monogynous colonies, leading to more large queens, and so forth. What appears to happen here is that a "mutation" in a nongenetic element of the developmental matrix can induce a new self-replicating variant of the system that may differ in fitness from the original.

The moral that proponents of DST draw from the comparison of these cases is that the power of selective explanations need not be limited to genetic changes. The range of phenomena that can be given selective explanation should be expanded to include differences dependent upon chromatin marking systems (Jablonka and Lamb 1995), prions (Lansbury 1997; Lindquist 1997), dietary cues in maternal milk, cultural traditions, and ecological inheritance (Gray 1992; Laland et al. 2001). Selection for differences in one of these heritable developmental resources is likely to have consequences for other aspects of the developmental system. Whitehead (1998) has argued that cultural selection has led to genetic changes in this way. He observed that in species of whales with matrilineal social systems, mitochondrial DNA diversity is ten times lower than in those with nonmatrilineal social systems. He suggested that differences in maternally transmitted cultural traits, such as vocalizations and feeding methods, have conferred a sufficient advantage to lead to the spread of some maternal lineages, and thus their mtDNA. The mtDNA that exists today remains because it hitchhiked along with the cultural traits that were selected for.

At this point, orthodox gene-centered biologists might concede that natural selection can indeed be generalized to cover cases of expanded inheritance. Having made this concession, they might then attempt to minimize its significance. We now turn to discuss some well-known strategies for marginalizing the role of expanded inheritance in evolution and argue that they are unsuccessful.

## Do Only Genes Contain Developmental Information?

Genes are widely believed to contain a program that guides development and to contain information about the evolved traits of the organism. Perhaps the best-known aspect of DST is the rejection of this claim in Susan Oyama's book *The Ontogeny of Information* (Oyama 1985/2000). The obvious way to explicate information talk in biology is via information theory. In the mathematical theory of information as a quantity (Shannon and Weaver 1949) and its semantic relatives (Dretske 1981), a signal sender conveys information to a receiver when the state of the receiver is correlated with the state of the sender. The conditions under which this correlation exists constitute a "channel" between sender and receiver. Changes in the channel affect which state of the receiver corresponds to which state of the sender. The information conveyed by a particular state of the receiver is as much a function of the channel, the context, as it is of the sender. In the case of development, the genes are normally taken to be the source, the life cycle of the organism is the signal, and the channel conditions are all the other developmental resources needed for the life cycle to unfold. But it is a fundamental feature of information theory that the role of source and channel condition can be reversed. A source/channel distinction is imposed on a causal system by an observer. The source is merely the channel condition whose current state the signal is being used to investigate. If all other resources are held constant, a life cycle can give us information about the genes, but if the genes are held constant, a life cycle can give us information about whichever other resource we decided to let vary. This fact is exploited whenever a biologist uses a clonal population to measure the effects of some aspect of the environment. Thus, so far as information theory and its relatives are concerned, every resource whose state affects development is a source of developmental information (Johnston 1987; Gray 1992, 2001; Griffiths and Gray 1994; Griffiths 2001).

A common response to the fact that genes and other physical causes are equally good sources of developmental information has been to look for a more demanding notion of information that allows the traditional distinction to be drawn. Several biologists and philosophers have suggested that "teleosemantic" information can play this role. Teleosemantics is a proposal originating in the philosophy of language to find a place in the material world for "meaning" in the sense that human thoughts and utterances have meaning. The teleosemantic approach reduces meaning to teleology and then reduces teleology to natural selection in the usual manner: the purpose of a biological entity is the outcome for which it is an adaptation (Pittendrigh 1958). John Maynard Smith has offered one such teleosemantic account of biological information (Maynard Smith 2000). He compares natural selection to computer programming using the "genetic algorithm" technique. The genetic algorithm programmer randomly varies the code of a computer program and selects variants for their performance. In the same way, natural selection randomly varies the genes of organisms and selects those organisms for their fitness. Just as the purpose of the final computer program is to perform the task for which it was selected, the biological purpose of successful genes is to produce the developmental outcomes by virtue of which they were selected. This biological purpose constitutes the teleosemantic meaning of the gene. For example, the defective hemoglobin gene in some human populations that has been selected because it confers resistance to malaria, carries teleosemantic information about malaria resistance.

Unfortunately, teleosemantic information is fundamentally unsuited to the aim of avoiding parity between genes and other developmental causes. Extragenetic inheritance systems of the various kinds discussed above are designed by natural selection to cause developmental outcomes in offsprings. So all forms of extragenetic inheritance transmit teleosemantic information about development. The most fully developed teleosemantic account of developmental information is the "extended replicator theory" (Sterelny et al. 1996; Sterelny 2000), which recognizes from the outset that teleosemantic information exists in both genetic replicators and in at least some extragenetic replicators. Griffiths and Gray (1997) have argued that teleosemantic information is carried by all the material traces that play a role in inheritance in the extended sense defined above.

In conclusion, while many concepts of information can be applied to the role of genes in development, it appears unlikely that any of these captures the intuition that genes supply information and other developmental causes do not. The various senses in which genes "code for" phenotypic traits, "program" development, or contain developmental "information" can be equally well applied to other factors required for development. This is not to say, of course, that there is no difference between the actual role of genes in development and the roles of membrane templates or host imprinting. Genes play a unique role in templating for proteins and a distinct role as nodes in the causal networks regulating cell metabolism. The point is that these empirical differences between the role of DNA and that of other inherited developmental factors do not imply the metaphysical distinction between "form" and "matter" that is often inferred from them (Griffiths and Knight 1998). The concept of information does not supply the missing link between these empirical differences and the conclusion that only genetic change is of evolutionary significance.

## Extragenetic Inheritance Systems Have Limited Evolutionary Potential

Maynard Smith and Szathmáry (1995) have introduced a distinction between "limited" and "unlimited" systems of heredity. They argue that it distinguishes genes and languages from all other forms of heredity. Most nongenetic inheritance systems, they argue, can only mutate between a limited number of states. In contrast, they note that the genome and language both have recursive, hierarchical structures, and hence an indefinite number of possible heritable states. This unlimited range of combinatorial possibilities enables microevolutionary change and cumulative selection to take place. These points are all perfectly legitimate, but from a developmental systems perspective the significance of unlimited inheritance should not be oversold for three reasons.

First, the unlimited nature of an inheritance system is a property of the developmental system as a whole, not only of the resource in which we find the recursive structure. The vast coding potential of genes, language, and perhaps pheromones is created by the way in which combinations of these factors "mean something" to the rest of the developmental system. Asking if a system is limited or unlimited holds the current developmental system fixed, and asks what can be achieved by ringing the changes on one of the existing developmental resources. But the lesson of the major evolutionary transitions—the introduction of whole new levels of biological order such as multicellularity—is that evolution can change developmental systems so as to massively expand the possible significance of existing developmental resources. A base-pair substitution in a multicellular organism has potentials that it lacked in a unicellular ancestor. If it occurs in a regulatory gene, it could mean a new body plan. The role of systems of "limited heredity" in these evolutionary transitions is considerable, as Maynard-Smith and Szathmáry themselves have made clear.

Second, from a selectionist viewpoint the combinatorial richness of an inheritance system must be measured in terms of the number of different phenotypic effects, not just the number of combinations of components. If the rest of the developmental system were such that the indefinitely many base-pair combinations of DNA collapsed into only a few developmental outcomes, then for all its combinatorial structure DNA would not be an unlimited heredity system. It is not hard to imagine cellular machinery with this result, as the existing genetic code is substantially redundant in just this way: several codons produce the same amino acid. Hence "unlimitedness" is a property of the developmental system as a whole, not of one of its components.

A third and final reason not to place too much emphasis on the limited/unlimited distinction is that it treats genetic and extragenetic inheritance as if they acted separately. This is manifestly not the case. Adding one form of inheritance to another causes a *multiplication* of evolutionary possibilities, not just an *addition* to them. Extragenetic inheritance expands the set of possible heritable combinations, rather than merely offering a supplement to it. Moreover, it makes accessible possibilities that would not be accessible to genetic inheritance acting alone. One of the distorting effects of gene-centrism is that it forces biologists who are interested in the evolutionary potential of extragenetic inheritance to focus on the rare cases in which extragenetic inheritance is relatively decoupled from genetic inheritance (as in the case of *Solenopsis invicta* described above). This is because when genetic and extragenetic inheritance act in conjunction with one another, the extragenetic element is inevitably treated as a mere agent or assistant of the genetic element. But it is evident that the real importance of extragenetic inheritance lies in the contribution it makes to the multifaceted system which, as a whole, generates the heritable variation on which evolution acts.

## Epigenetic Potential and Epigenetic Processes

As noted above, current work in evolutionary developmental biology (EDB) retains the traditional idea that the developmental system of an organism emerges from its genes, with the environment acting only as some kind of background or enabling cause. In a recent comparison of the ideas of EDB and DST, Jason Robert, Brian Hall, and Wendy Olson use a distinction between the actual units of inheritance with epigenetic potential and the epigenetic processes to which these contribute in an attempt to explain why EDB remains more focused on the genetic material (Robert et al. 2001). They suggest that at least one "hard" version of EDB "identifies the gene (defined as the actual genetic material) as the sole unit of inheritance" (p. 960). The difference between EDB and DST, they suggest, is that EDB regards development as the expression of *epigenetic potentials* of genes in an environment, whereas DST regards development itself—*epigenetic processes*—as if they were passed on from one generation to the next: "Developmental systems theorists . . . define inheritance as the reliable reconstruction of interactive causal networks" (p. 961). Robert et al. also suggest that, whereas DST sees aspects of the environment being passed on from one generation to the next, EDB sees the control of development in the next generation by genes from the previous generation that act via those aspects of the environment—a sort of "extended genotype" (p. 961).

As one might expect in an article entitled "Bridging the gap between developmental systems theory and evolutionary developmental biology," Robert et al. are concerned to show that the two views of inheritance just described are often only two ways to describe the very same biological phenomena. We do not think that DST and EDB are best compared in this way. DST is not an alternative way of conceptualizing the same phenomenon, but a challenge to different aspects of conventional, gene-centered thought. DST accepts the important theoretical advances of EDB but calls for other changes to theory that have not to date been part of EDB. Robert et al. are on the right track, we think, when they compare DST to Scott Gilbert's call for EDB to embrace the role of the environment in development and forge an "ecological developmental biology" (Gilbert 2001, and see note 2).

Understandably, Robert et al.'s attempt to treat DST as an alternative approach to the focal questions of EDB leads them to misinterpret DST. The alternative picture of inheritance they identify is an interesting one, but it is not the view outlined in canonical presentations of DST (e.g., Oyama et al. 2001). The units of inheritance in DST are entities with epigenetic potential and not epigenetic processes, just like those of EDB. The units of inheritance are developmental resources that reliably reoccur in each generation and interact with the other resources to reproduce the life cycle. The difference is that DST identifies more things with epigenetic potential than does (hard) EDB. Some of these were briefly described above: membrane templates, chemical gradients in the egg, microtubule organizing centers, enbosymbionts, hosts and habitats on which organisms are imprinted or with which they are passively biogeographically associated, the environment of the hive in insects with castes, cultural traditions, and constructed features of a niche such as the acidity of the soil in a pine forest or the periodicity of fire in a eucalypt forest. There are important distinctions between these developmental resources. Some, but not all, are the immediate causal consequence of the expression of maternal genes. Some, but not all, are actively reproduced, either by the parents of the developing organism or by the wider population. The property that all these resources share, and in terms of which we defined inheritance above, is that they are reliably present in each generation and causally necessary for the production of the life cycle of the evolutionary lineage. It is an important empirical question whether and to what extent the inheritance of *variation* in each of these resources has the potential to drive evolutionary change (for a overview of this question, see Sterelny 2001). A broad, principled definition of inheritance leaves this empirical question open for investigation, rather than prejudging it. Eva Jablonka and Marion Lamb make the same comment in their recent reply to Robert et al.: "It seems to us that refusing to call the transmission of non-DNA variations inheritance precludes a discussion of the evolutionary effects of the consequences of such transmission" (Jablonka and Lamb 2002, p. 291)

## Adaptation and Niche Construction

The broadest form of extragenetic inheritance is the effect of niche construction on future generations. The idea of niche construction finds its ultimate origin in three seminal papers in which Richard Lewontin criticized the metaphors that have traditionally been used to represent the process of adaptation by natural selection (Lewontin 1982, 1983a, 1983b). The metaphorical conception that Lewontin criticized is the so-called lock-and-key model of adaptation. Adaptations are solutions (keys) to the problems posed by the environment (locks). Organisms are said to be adapted to their ways of life because they were made to fit those ways of life. In place of the traditional metaphor of adaptation as "fit," Lewontin suggested a metaphor of construction. Organisms and their ecological niches are co-constructing and co-defining. Organisms both physically shape their environments and determine which factors in the external environment are relevant to their evolution, thus assembling such factors into what we describe as their niche. Organisms are adapted to their ways of life because organisms and their way of life were made for (and by) each other. Lewontin also revised the popular metaphor of a "fitness landscape." In this image, populations occupy a rugged landscape with many fitness peaks and evolve by always trying to walk uphill. But because organisms construct their niches, the landscape is actually much like the surface of a trampoline. As organisms climb the hills they change the shape of the landscape. Lewontin's metaphor of construction is not merely a new way to describe the same evolutionary process. It is the public face of a substantially revised model of the actual process of natural selection, redefining the causal relationships that ecology and evolutionary biology must seek to model.

The most detailed attempt to develop the new metaphor of construction is that of F. J. Odling-Smee and his collaborators (for a brief overview, see Laland et al. 2001). The current prominence of the term "niche construction" is due to this group. The first two columns in Table 19.1 give the traditional model of adaptation as "fit" and the model of adaptation as construction as these two models are described by Lewontin. In the conventional picture, change in organisms over time is a function of the state of the organism and its environment at each previous instant. The environment acts on the existing state of organisms by selecting from the pool of variation those individuals best fitted to the environment. The environment itself changes over time too, but as the bottom equation shows, these changes are not a function of what organisms are doing at each previous instant. In Lewontin's alternative picture, shown in the center column of Table 19.1, organisms and their environments play reciprocal roles in each other's change. Change in the environment over time is a function of the state at each previous instant of both the environment and the organisms evolving in that environment.

The right-hand column of Table 19.1 shows Odling-Smee's model of evolution as the co-construction of organism and environment (Odling-Smee 1988). Odling-Smee's "general coevolutionary model" differs from Lewontin's in two ways. First, Odling-Smee hoped to generate a common framework in which to represent both development and evolution. This explains why the terms $E_{\text{pop}}$ and $O_{\text{pop}}$ occur in the equations in Table 19.1. Evolution is a process in which *populations* and their environments co-construct one another over time. If the terms were $E_{\text{i}}$ and $O_{\text{i}}$, then in Odling-Smee's notation the equations would describe the co-construction of an individual organism and its developmental environment as the organism's life cycle unfolds. By introducing these indices Odling-Smee is making explicit what was already implicit in the explanation of Lewontin's equations given in the last paragraph: the term $O$ in those equations refers to populations of organisms, not to some individual organism. Earlier versions of DST (e.g, Oyama 1985/2000; Gray 1992) and some of Lewontin's writings are sympathetic to this idea that there is a significant parallelism between the way populations of organisms and their environments reciprocally influence one another and the way

Table 19.1 Three pictures of the dynamical equations for evolution.

| Traditional neo-Darwinism | Lewontin's constructionism | Odling-Smee's general coevolution |
|---|---|---|
| $dO/dt = f(O, E)$ | $dO/dt = f(O, E)$ | $dO_{pop}dt = f(O_{pop}, E_{pop})$ |
| $dE/dt = g(E)$ | $dE/dt = g(O, E)$ | $dE_{pop}/dt = g(O_{pop}, E_{pop})$ |
| | | $d(O_{pop}, E_{pop})/dt = h(O_{pop}, E)$ |

$E$ = Environment, $O$ = organism, $E_{pop}$ = organism-referent environment of a population, $O_{pop}$ = population of organisms. These variables are related by functions $f$, $g$, $h$. See text for explanation.

in which individual organisms and their developmental environments do so. But this is not the place to give this idea the attention it deserves.

The second way in which Odling-Smee's treatment differs from Lewontin's is that he is concerned not to represent the organism–environment system as a closed system, as the equations in the center column would seem to imply. Although the eucalypt–bushfire relationship, for example, is one of mutual construction, the change in this system over time is externally driven by the progressive drying of the Australian continental climate. Organisms feel the impact of changes in the environment in its traditional sense of their total biotic and abiotic surroundings; but they experience these impacts via the environment as it appears in relation to them, and thus different lineages experience "the same changes" quite differently. Odling-Smee tries to respect this situation by assigning separate roles to the environment of a particular lineage of organisms and what he calls the "universal physical environment." The former, organism-referent description of the environment is the source of evolutionary pressures on that organism, and the organism is the source of niche-constructing forces on that environment. The latter, the universal physical environment, is a source of exogenous change in the organism's environment (see Brandon 1990 for a similar treatment of the concept of environment).

The developmental systems model of evolution (Gray 1992; Griffiths and Gray 1994, 1997) can be clarified and improved by the insights of Odling-Smee and his collaborators. In particular, the insight that exogenous factors can affect the availability of developmental resources has not been sufficiently stressed in previous presentations. There remains, however, one major difference between DST and work on niche construction up to and including the present time. Niche construction is still a fundamentally dichotomous account of evolution (and, indeed, of development). There are two systems of heredity—genetic inheritance and environmental inheritance. There are, correspondingly, two causal processes in evolution—natural selection of the organism by the niche and construction of the niche by the organism. The niche-construction model could be modified to take account of recent work on narrow epigenetic inheritance, with a category like "intracellular inheritance" taking the place of genetic inheritance. This, however, would seem merely to substitute one rigid boundary for another. A central theme of the DST research tradition has been that distinctions between classes of developmental resource should be fluid and justified by particular research interests, rather than built into the basic framework of biological thought. Fundamentally, the unit of both development and evolution is the developmental system, the entire matrix of interactants involved in a life cycle. The developmental system is not two things, but one, albeit one that can be divided up in many ways for different theoretical purposes. Hence we would interpret niche-construction models "tactically," as a method for rendering tractable some aspects of evolution. We would not interpret them "strategically" as a fundamental representation of the nature of the evolutionary process.

The DST model of evolution can be represented in such a way as to make it directly comparable with the models in Table 19.1. We can aptly represent the developmental system with the symbol *Œ*. We retain Odling-Smee's insight that evolutionary change in organism–environment systems is often exogenously driven by using *E* to represent the universal physical (external) environment. We end up with the equation:

$$d\text{Œ}_{\text{pop}}/dt = f(\text{Œ}_{\text{pop}}, E)$$

Evolution is change in the nature of populations of developmental systems. This change is driven both endogenously, by the modification by each generation of developmental systems of the resources inherited by future generations, and exogenously, by modifications of these resources by factors outside the developmental system.

## Fitness and Adaptation

This representation of developmental systems evolution allows us to answer a persistent objection to DST. Since we claim that there is no distinction between organism and environment, where do evolutionary pressures on the developmental system come from? What causes adaptation?[3] To give a clear answer we must go back to the definition of the developmental system given in Griffiths and Gray (1994). The developmental system of an individual organism contains all the unique events that are responsible for individual differences, deformities, and so forth. Just as a traditional model of evolution abstracts away from the unique features of individual phenotypes, developmental systems theory must abstract away from these features in order to tackle evolutionary questions. In evolutionary terms, the developmental system contains all those features that reliably recur in each generation and that help to reconstruct the normal life cycle of the evolving lineage. Of course, many species have more than one normal life cycle, either because there are different types of organism in a single evolving population, each reproducing its own differences (polymorphism), or because there are variations in the developmental matrix from one generation to the next (facultative development). For example, there are tall and short human families and heights also vary from one generation to another due to nutrition. These features are handled in the same way as in characterizations of "the" phenotype of an evolving lineage (Griffiths and Gray 1997). The resultant description of the idealized developmental system of a particular lineage at some stage in its evolution is highly self-contained. Because the focus is on how the complete life cycle is achieved, everything needed for that life cycle is assumed to be present. So everything that impinges on the process is an element of the system itself. It is this that creates the impression that all change in the system must be endogenously driven and creates the apparent puzzle about the source of selection pressures.

The puzzle is only apparent, because to think about evolution we need to switch from describing the developmental system characteristic of an evolving lineage at a time to describing an evolving population of individual developmental systems. We need to look at the causes of variation, as well as how the characteristics of the lineage are reliably reconstructed. Hence we need to look at the causes of idiosyncratic development in particular individuals. These causes lie "outside" the description we have constructed of the typical developmental system of the lineage. A population of individual developmental systems will exhibit variation and differential reproduction for a number of reasons. Parental life cycles may fail to generate the full system of resources required to reconstruct the life cycle. Resources generated by the activities of an entire population (such as bushfires in eucalypt forest) may also be scarce, or patchily distributed, so that some individuals lack an important element of their developmental system. Finally, persistent resources—those developmental factors whose abundance is independent of the activities of the lineage—may be scarce or patchy and so some individuals may be unable to reestablish the relationship to these resources that is part of their life cycle. The external environment ($E$) can impinge on developmental systems by any of these routes. But this does not mean that we can go back to thinking of evolution as a response to the demands of the external environment. The effect of changes in the external environment on the evolution of a lineage can be understood only when those changes are described in terms of how they change the organism-referent environment ($E_{pop}$). "Changes" in parameters of the external environment which are developmentally equivalent are not changes from the point of view of the evolving system. People in different regions of Britain experience substantially different quantities of dissolved limestone in their drinking water, but this is generally of no ecological significance. Conversely, apparently trivial changes may seem momentous when described in terms of a particular developmental system. Far smaller changes in the concentration of lead from one region to another would have momentous consequences. This is, of course, the point already made by Lewontin, Odling-Smee and collaborators, and Brandon (1990).

So far we have concentrated on how failures of development can lead to evolutionarily significant variation. But positive innovations are possible as well. An individual difference in the system of developmental resources may allow some individuals to cope better when both are deprived of some developmental resource because of exogenous change. Alternatively, an individual difference may simply alter the life cycle in such a way that it gives rise to a greater number of descendants. The source of novelty can be a mutation in any of the developmental resources—parentally generated, population-generated, or independently persistent. To make this discussion more concrete, imagine a typical population of hermit crabs. A key component of the developmental system in this lineage is a succession of discarded shells of other species. A dearth of shells would be an exogenous cause of selective pressure on the lineage. Variants with a beneficial set of behaviors or a beneficial habitat association that allowed them to continue to reliably reestablish their relationships to shells would be favored by selection. Shells will typically be an independently persistent resource, and the case in which an independently persistent developmental resource acts as a limiting resource has obvious resonance with traditional ideas of selection of the organism by an independent environment. But, to fictionalize the example slightly, suppose the crab life cycle includes disturbing the soil in such a way as to expose a greater supply of discarded shells. That would make shells a population-generated resource, but they might still act as a limiting resource. Or suppose a lineage evolves behaviors that allows crabs to bequeath shells to their offspring when they themselves seek a larger home. Shells would then be parentally generated, but exogenous change in the availability of shells might still leave some offspring without them, just as a shortage of a trace element in the parental diet may lead to a birth defect in a viviparous species.

One factor that really is missing from this picture is the idea that the external (universal physical) environment poses definite problems that lineages must seek to solve. Instead, the lineage helps to define what the problems are. A dearth of shells is a feature of the ecological environment of a hermit crab and a problem for the hermit crab, but it is completely invisible to a blue-swimmer crab. The number of discarded shells per square meter is a feature of the external environment of both species, but it is only a feature of the ecological environment of one of them. So it is true that the developmental systems treatment of evolution does not incorporate Darwin's original, intuitive idea of fitness as a measure of the match between an organism and an independent environment (e.g., Darwin 1859/1964, p. 472). But this is a feature that the developmental systems treatment shares with conventional neo-Darwinism. Adaptation is no longer defined intuitively, as the sort of organism–environment relationship that a natural theologian would see as a sign of God's beneficent plan. Darwin set out to explain the fact that the biological world is full of adaptations in this sense, but as so often happens in science, the phenomenon to be explained became redefined in the process of explaining it. In modern usage, an adaptation is whatever results from natural selection, even when what results is intuitively perverse and "inefficient."

We hope it is now clear how DST can explain adaptation, in the modern sense of that term. Change over time in the developmental system of a lineage is driven by the differing capacity of variant developmental systems to reconstruct themselves, or, in a word, differential fitness. What is fitness? In contemporary evolutionary theory, fitness is a measure of the capacity of a unit of evolution to reproduce itself (Mills and Beatty 1979). Fitness differences are caused by physical and behavioral differences between the individuals in the population. So fitness can be translated on a case-by-case basis into a detailed causal explanation of evolutionary success. Fitness in general, however, does not correspond to any single physical property (Rosenberg 1978). The only general account of fitness describes its role as a parameter in population dynamic equations. It is clear that this orthodox account of fitness applies equally well to the developmental systems theory. There is no puzzle about how developmental systems that incorporate the whole range of resources that reconstruct the life cycle could come to vary in their success in reconstructing themselves and be selected on that basis.

## Individuals, Lineages, and the Units of Evolution

A coherent theory of evolution requires an accurate conception of its fundamental units. According to DST, an evolutionary individual is one cycle of a complete developmental process—a life cycle. We have shown that natural selection can act on populations of developmental systems and give rise to adaptation, but in doing so we have assumed that developmental systems are the sort of things that can be counted, that they have clear boundaries, and that they do not overlap so much that they cannot be distinguished from one another. We now turn to justifying this assumption. Developmental systems include much that is outside the traditional phenotype. This raises the question of where one developmental system and one life cycle ends and the next begins There is an enormous amount of cyclical structure in most biological lineages. As well as the life cycles associated with traditional physiological individuals, there are "repeated assemblies" (Caporael 1995) within a single individual, such as cells or morphological parts like the leaves of a tree. There are also repeated assemblies of whole individual organisms, such as lichens or ant–acacia symbioses. In previous publications we have tried to identify what makes a repeated assembly a developmental system in its own right, as opposed to a part of such a system or an aggregate of several different systems (see especially Griffiths and Gray 1997). While we still see some merit in our previous suggestions, we have learned a great deal from the work of David Sloan Wilson and Elliott Sober on trait-group selection, and also from Kim Sterelny's work on higher-level selection (Wilson and Sober 1994; Sterelny 1996; Wilson 1997; Sober and Wilson 1998).

Wilson and Sober's work centers on the idea of a trait group—a set of organisms relative to which some adaptation is, in economic terms, a public good. The beavers that share a lodge form a trait group with respect to dam-building adaptations because it is not possible for one beaver to increase its fitness by dam-building without increasing the fitness of its lodge mates. Wilson and Sober argue that trait groups are units of evolution. That is to say, it makes sense to assign fitnesses to trait groups and to track the evolution of adaptations due to the differential reproduction of their associated replicators. But the emergence of a new level of evolutionary individuality seems to require more than this. The emergence of communal living in the beaver lineage, or of a symbiotic association between ant lineages and acacia lineages, does not mark the same sort of fundamental transition as occurred with the origin of the eukaryote cell, the emergence of multicellularity, or the evolution of eusociality in social insects. Not every trait group is a "superorganism." There are a number of features that seem to mark the difference between mere trait groups and superorganisms, such as the functional differentiation of parts and the dependence of parts on the whole for their viability. Sterelny and Griffiths have argued that the fundamental feature of a superorganism is that many traits of the component organisms are selected with respect to the very same trait group (Sterelny and Griffiths 1999, pp. 172–177; see also Wilson 1997). The ants in a nest and the cells in a human body have a shared fate not just with respect to one part of their activities, but with respect to all of them. A liver cell does not have some adaptations with respect to the whole body and others with respect to the liver alone. This is because the only way the liver cell can reproduce itself is via the success of the whole organism. Similarly, the only way an ant can contribute to its own reproduction is via the success of the nest as a whole. This broad congruence of interests results from evolved features that suppress competition between the component parts of the superorganism. The best known of these is the segregation of the germ line. However, it is easy to overstate the importance of this particular mechanism. Plants typically do not have germ-line segregation, so it cannot be a prerequisite for complex multicellular life. Leo Buss has explored some of the very different mechanisms that are used to bind the interests of cell lineages together in plants and fungi (Buss 1987). Again, in bee nests, the queen marks her eggs with a pheromone that inhibits workers from eating them. Eggs laid by workers are eaten by other workers, so the only realistic way for workers to bring about the reconstruction of their life cycle is via the larger, colony life cycle (Ratnieks and Visscher 1989). The worker bee is reduced to a part of a larger cycle as effectively as a metazoan cell is reduced to a part by segregation of the germ line.

We suggest, then, that a repeated assembly is a developmental system in its own right, as opposed to a part of such a system or an aggregate of several different systems when specific adaptations exist, presumably due to trait-group selection, which suppress competition between the separate components of the assembly. This account of the evolution of individuality can actually explain why the distinction between a colony of organisms or a symbiotic association and an individual organism is not a sharp one. The mechanisms that bind the trait group together can be more or less effective. They may also keep the evolutionary interests of the same group aligned across a wider or narrower range of traits. The metazoan organism and the unicellular eukaryotic cell are clearly individuals. Jellyfish, lichens, eusocial insect colonies, and the ant–acacia symbiosis are progressively less clearly individuals. Each has a life cycle and a developmental system that feeds into its development. But in most of these latter cases, it is possible to describe evolutionary pressures with respect to which the smaller life cycles nested within the larger cycle do not form a trait group. The more forced and implausible these scenarios, the less theoretical role there is for a description in which these cycles are treated as independent and not as parts.

## Building Bridges: The Complementary Programs of DST and EDB

At the end of a well-known paper on adaptation, Lewontin (1978) notes that adaptive evolution requires quasi-independence: selection must be able to act on a trait without causing deleterious changes in other aspects of the organism. If all the features of an organism were so closely developmentally integrated that quasi-independent variation did not exist, then "organisms as we know them could not exist because adaptive evolution would have been impossible" (Lewontin 1978, p.169). The requirement for quasi-independence means that we must add a caveat to the slogan that "Nature selects for outcomes" and not for how those outcomes are produced (Lehrman 1970, p. 28). The reliable reoccurrence of an advantageous variant is not enough. The developmental process that produces the variation must be quasi-independent (modular) if it is to be the basis for cumulative selection. There has been considerable recent interest in the extent to which the organization of development really is modular. Günther Wagner and Lee Altenberg (1996) suggest that directional selection might act on developmental systems to reduce pleiotropic effects between characters with different functions, thereby enhancing the modularity and evolvability of these developmental systems. They speculate that there should be evolutionary trends toward increased modularity. Brandon (1999) goes so far as to suggest that developmental modules at each level in the evolutionary hierarchy are the units of selection at that level. The study of developmental modularity is still in its infancy and the extent of that modularity far from resolved. However, it is clear that understanding the extent and nature of developmental modularity is essential to understanding how evolution actually unfolds—and has unfolded—in practice.

The modularity issue is at the heart of the new discipline of evolutionary developmental biology (EDB), so it is striking that DST has had little to say about the issue of modularity, or the older, related issue of the extent of developmental constraints (Maynard Smith et al. 1985; for an exception to this remark, see Oyama 1992). What, one might ask, is *developmental* about developmental systems theory? The answer is that the term "developmental" in this context stands in contrast to "innate." To think developmentally is to focus on the many factors that must be present for a fertilized egg to give rise to a normal life cycle and on the process—development—in which those factors interact. Whereas EDB stems from the impact of the molecular revolution on developmental biology and comparative morphology, DST grew out of the continued efforts of developmental psychobiologists to resist the idea that evolved features of mind and behavior are outside the sphere of developmental psychology because they are programmed in the genes and so do not *have* a developmental psychology. Most developmental systems theorists trace their intellectual ancestry to Daniel S. Lehrman's 1953 "Critique of Konrad Lorenz's theory of instinctive behavior" and to the work in comparative psychology reflected in that critique (Gottlieb 2001; Johnston 2001). Research in the developmental systems tradition has thus had a strong emphasis on demonstrating the contingency of developmental outcomes on extragenetic factors and, latterly, demonstrating the evolutionary potential of those factors. From this perspective, to understand their development is to understand how each stage of the developing organism interacts with its environment to give rise to the next. DST, with its emphasis on the life cycle in place of the adult phenotype and on the role of extragenetic inheritance, is an attempt to provide a general conception of development and evolution in the spirit of this research tradition of developmental psychobiology.

In stark contrast, much of the criticism directed by developmental morphologists at conventional neo-Darwinism has emphasized, not the contingency of development, but its fixity. The literature on "developmental constraints" suggests that some aspects of phenotypes are too strongly integrated to be altered by natural selection acting on the genes, let alone by natural selection manipulating nongenetic developmental parameters! In recent years, however, EDB has been able to move beyond treating development as a set of constraints on selection, in part because of conceptual advances and in part because technical advances have made it possible to elucidate developmental "constraints" mechanistically and not merely describe them phenomenologically on the basis of gaps in the comparative data

In our view, these two critiques of earlier evolutionary thought are essentially complementary. DST does not provide a theory of phenotypic integration and modular evolution, but rather stands in need of one, and EDB is beginning to supply such a theory. Conversely, nothing in the fundamental ideas of EDB excludes applying that theory to a wider conception of the developmental system, not as emerging from interactions between genes, but as emerging from interactions between the whole matrix of resources that are required for development. In fact, the need to extend the research agenda of EDB in this manner has been recognized in recent calls for greater attention to the ecological context of development and the developmental basis of phenotypic plasticity (Sultan 1992; Schlichting and Pigliucci 1998; Gilbert 2001; this volume). Research under the banner of "phenotypic plasticity" has emphasized the fact that the environment plays an informative, not merely a supporting, role in development. For understandable reasons, however, the emphasis of research on phenotypic plasticity has been on *adaptive* plasticity. Organisms have evolved to use the environment as a source of information for the deployment of facultative adaptations. DST can enrich this perspective by emphasizing that the dependence of development on the ecological context is a fundamental feature, not a last flourish of adapted complexity. EDB recognizes that the nature of development explains the nature of evolution as much as evolution explains the nature of development. In the same way, the fact that what changes over evolutionary time are organism–environment systems is an essential part of any adequate theory of adaptive evolution. "Ecological developmental biology" is not merely a framework for studying adaptive plasticity, it is a fundamentally better, more inclusive way of approaching the evolution of development and the implications of development for evolutionary theory.

The ubiquity of environmental effects on development can be nicely illustrated using neuroscientist Terence Deacon's concept of "addiction to the environment" (Deacon 1997). During primate evolution, an abundance of dietary vitamin C caused the loss of the normal mammalian pathway for ascorbic acid synthesis (Jukes and King 1975). Because this vital developmental resource could be inherited passively, rather than via the genes previously involved in its synthesis, the primate lineage became dependent on ("addicted to") this form of extragenetic inheritance. In the same way, hermit crabs are "addicted" to discarded shells and almost all large organisms are "addicted" to the earth's gravity. In fact, evolved lineages are "addicted" to innumerable aspects of the environment with which they have coevolved, although most of these aspects are reproduced so reliably that this does not give rise to significant variation, and so is overlooked. Nevertheless, any account of how organisms develop that neglects these factors, and the evolutionary processes that led to their incorporation in development, is seriously incomplete.

## Conclusion

DST yields a representation of evolution that is quite capable of accommodating the traditional themes of natural selection and also the new results that are emerging from evolutionary developmental biology. But it adds something unique—a framework for thinking about development and evolution without the distorting dichotomization of biological processes into gene and nongene and the vestiges of the "black-boxing" of developmental processes in the modern synthesis, such as the asymmetric use of the concept of information. Phenomena that are marginalized in current gene-centric conceptions, such as extragenetic inheritance, niche construction, and phenotypic plasticity, are placed center stage.

*Acknowledgments* Some sections of this chapter are adapted from Griffiths, P. E., and Gray, R. D., Darwinism and developmental systems, in S. Oyama, P. E. Griffiths, and R. D. Gray (eds.), *Cycles of Contingency: Developmental Systems and Evolution* (pp. 195–218). Cambridge, MA.: MIT Press, 2001. Material from that chapter is reproduced by permission of the MIT Press.

*Notes*

1 Here is a random selection: "Evolution 1. Process by which organisms come to differ from generation to generation. 2. Change in the gene pool of a population from generation to generation." (Arms and Camp 1987, p. 1121). "Evolution is the result of accumulated changes in the composition of the gene pool." (Curtis and Barnes 1989, p. 989).

2 DST is an attempt to sum up the ideas of a research tradition in developmental psychobiology that goes back at least to Daniel Lehrman's work in the 1950s (Gottlieb 2001; Johnston 2001). Robert, Hall, and Olson (2001) set out to compare DST with the more gene-centered ideas of contemporary evolutionary developmental biology (EDB), but they miss the fact that DST has its roots in developmental psychobiology. They note that "EDB has yet to draw extensively from behavior/psychology" (p. 958) and cite the work of Gilbert Gottlieb as an exception. But Gottlieb, of course, is one of the seminal figures in DST and his "developmental-psychobiological systems view" is quite unlike most work in EDB because it stresses the role in development of a highly structured, species-specific environment and of extragenetic inheritance. As developmental psychologists David Bjorklund and Anthony Pellegrini state in their recent book on evolutionary developmental psychology, "In this book we adopt a specific model, the developmental systems approach (Gottlieb 2000; Oyama 2000a) ... strengthening considerably, we believe, evolutionary psychologists' arguments that genes are not necessarily destiny" (Bjorklund and Pellegrini 2002). It is encouraging to see that the tradition represented by Gottlieb and Oyama has received increased attention in the last few years. It has been embodied in a textbook by George Michel and Celia Moore (1995) and popularized by authors like David Moore (2001; see also Bateson and Martin 1999). Much of our own work has been designed to demonstrate the relevance of DST concepts outside their original home in the study of behavioral development. DST may still be "virtually unknown among biologists" (Robert et al. 2001, p. 954), if this means biologists working on the evolutionary developmental biology of morphological structures, but as calls for an "ecological developmental biology" suggest, those biologists could learn much from the tradition of developmental psychobiology.

3 We are not aware of any published version of this criticism, but it was first suggested to Griffiths in conversation by Lindley Darden in 1994 and has also been raised by Alexander Rosenberg (pers. comm.).

*Literature Cited*

Arms, K., and Camp, P. S. 1987. Biology, 3rd ed. New York: CBS College Publishing.

Arthur, W. 1997. The Origin of Animal Body Plans: A Study in Evolutionary Developmental Biology. Cambridge: Cambridge University Press.

Avital, E., and Jablonka, E. 2001. Animal Traditions: Behavioural Inheritance in Evolution. Cambridge: Cambridge University Press.

Bateson, P. P. G., and Martin, P. 1999. Design for a Life: How Behaviour and Personality Develop. London: Jonathan Cape.

Bjorklund, D. F., and Pellegrini, A. D. 2002. The Origins of Human Nature: Evolutionary Developmental Psychology. Washington, DC: American Psychological Association.

Brandon, R. 1990. Adaptation and Environment. Princeton: Princeton University Press.

Brandon, R. 1999. The units of selection revisited: the modules of selection. Biology and Philosophy 14:167–180.

Caporael, L. R. 1995. Sociality: Coordinating Bodies, Minds and Groups. Hostname: princeton.edu Directory: pub/harnad/Psycoloquy/1995.volume.6 File: psycoloquy.95.6.01.group-selection.1.caporael.

Curtis, H. and Barnes, N. S. 1989. Biology, 5th ed. New York: Worth Publishers.

Darwin, C. 1859/1964. On the Origin of Species: A Facsimile of the First Edition. Cambridge, MA: Harvard University Press.

Dawkins, R. 1976. The Selfish Gene. Oxford: Oxford University Press.

Dawkins, R. 1982. The Extended Phenotype. Oxford: Freeman.

Deacon, T. W. 1997. The Symbolic Species: The Coevolution of Language and the Brain. New York: W. W. Norton.

Dretske, F. 1981. Knowledge and the Flow of Information. Oxford: Blackwell.

Gilbert, G. 2000. Environmental and behavioral influences on gene activity. Current Directions in Psychological Science 9:93–102.

Gilbert, S. F. 2001. Ecological developmental biology: developmental biology meets the real world. Developmental Biology 233:1–12.

Gottlieb, G. 1992. Individual Development and Evolution. Oxford: Oxford University Press.

Gottlieb, G. 1997. Synthesizing Nature-Nurture: Prenatal Roots of Instinctive Behavior. Hillsdale, NJ: Lawrence Erlbaum.

Gottlieb, G. 2001. A developmental psychobiological systems view: early formulation and current status. In S. Oyama, P. E. Griffiths, and R. D. Gray (eds.), Cycles of Contengency: Developmental Systems and Evolution (pp. 41–54). Cambridge, MA.: MIT Press.

Gray, R. D. 1992. Death of the gene: developmental systems strike back. In P. E. Griffiths (ed.), Trees of Life. Dordrecht: Kluwer.

Gray, R. D. 2001. Selfish genes or developmental systems? Evolution without interactors and replicators? In R. Singh, K. Krimbas, D. Paul, and J. Beatty (eds.), Thinking about Evolution: Historical, Philosophical and Political Perspectives. Festschrift for Richard Lewontin (pp. 184–207). Cambridge: Cambridge University Press.

Griffiths, P. E. 2001. Genetic information: a metaphor in search of a theory. Philosophy of Science 68:394–412.

Griffiths, P. E., and Gray, R. D. 1994. Developmental systems and evolutionary explanation. Journal of Philosophy 91:277–304.

Griffiths, P. E., and Gray, R. D. 1997. Replicator II: judgement day. Biology and Philosophy 12:471–492.

Griffiths, P. E., and Knight, R. D. 1998. What is the developmentalist challenge? Philosophy of Science 65:253–258.

Hall, B. K. 1992. Evolutionary Developmental Biology. New York: Chapman & Hall.

Hall, B. K. (ed.). 1994. Homology: The Hierarchical Basis of Comparative Biology. San Diego: Academic Press.

Harlow, H. F., and Harlow, M. K. 1962. Social deprivation in monkeys. Scientific American 207:136–146.

Jablonka, E., and Lamb, M. J. 1995. Epigenetic Inheritance and Evolution: The Lamarckian Dimension. Oxford: Oxford University Press.

Jablonka, E., and Lamb, M. 2002. Creating bridges or rifts? Developmental systems theory and evolutionary developmental biology. Bioessays 24:290–291.

Johnston, T. D. 1987. The persistence of dichotomies in the study of behavioural development. Developmental Review 7:149–182.

Johnston, T. D. 2001. Towards a systems view of development: an appraisal of Lehrman's critique of Lorenz. In S. Oyama, P. E. Griffiths, and R. D. Gray (eds.), Cycles of Contingency: Developmental Systems and Evolution (pp. 15–23). Cambridge, MA: MIT Press.

Jukes, T. H., and King, J. L. 1975. Evolutionary loss of ascorbic acid synthesizing ability. Journal of Human Evolution 4:85–88.

Keller, L., and Ross, K. G. 1993. Phenotypic plasticity and "cultural transmission" of alternative social organisations in the fire ant *Solenopsis invicta*. Behavioural Ecology and Sociobiology 33:121–129.

Laland, K. N., Odling-Smee, F. J., and Feldman, M. W. 2001. Niche construction, ecological inheritance, and cycles of contingency in evolution. In S. Oyama, P. E. Griffiths, and R. D. Gray (eds.), Cycles of Contingency: Developmental Systems and Evolution (pp. 117–126). Cambridge, MA: MIT Press.

Lansbury, P. 1997. Yeast prions: inheritance by seeded protein polymerisations? Current Biology 7:R617.

Lehrman, D. S. 1953. Critique of Konrad Lorenz's theory of instinctive behavior. Quarterly Review of Biology 28:337–363.

Lehrman, D. S. 1970. Semantic and conceptual issues in the nature-nurture problem. In D. S. Lehrman (ed.), Development and Evolution of Behaviour (pp. 17–52). San Francisco: W. H. Freeman.

Lewontin, R. 1970. The units of selection. Annual Review of Ecology and Systematics 1:1–14.

Lewontin, R. C. 1978. Adaptation. Scientific American 239:157–169.

Lewontin, R. C. 1982. Organism and environment. In H. Plotkin (ed.), Learning, Development, Culture (pp. 151–170). New York: John Wiley.

Lewontin, R. C. 1983a. Gene, organism and environment. In D.S. Bendall (ed.), Evolution: From Molecules to Men (pp. 273–285). Cambridge: Cambridge University Press.

Lewontin, R. C. 1983b. The organism as the subject and object of evolution. Scientia 118:65–82.

Lindquist, S. 1997. Mad cows meet psi-chotic yeast: the expansion of the prion hypothesis. Cell 89:495.

Maynard Smith, J. 2000. The concept of information in biology. Philosophy of Science 67:177–194.

Maynard Smith, J., and Szathmáry, E. 1995. The Major Transitions in Evolution. Oxford: W. H. Freeman.

Maynard Smith, J., Burian, R., Kauffman, S., Alberch, P., Campbell, J., Goodwin, B., Lande, R., Raup, D., and Wolpert, L. 1985. Developmental constraints and evolution. Quarterly Review of Biology 60:265–287.

Michel, G. F., and Moore, C. L. 1995. Developmental Psychobiology: An Interdisciplinary Science. Cambridge, MA: MIT Press.

Mills, S., and Beatty, J. 1979. The propensity interpretation of fitness. Philosophy of Science 46:263–286.

Money, J. 1992. The Kaspar Hauser Syndrome of "Psychosocial Dwarfism": Deficient Statural, Intellectual, and Social Growth Induced by Child Abuse. Buffalo, NY: Prometheus Books.

Moore, D. S. 2001. The Dependent Gene: The Fallacy of "Nature versus Nurture." New York: W. H Freeman/Times Books.

Moran, N., and Baumann, P. 1994. Phylogenetics of cytoplasmically inherited microorganisms of arthropods. Trends in Ecology and Evolution 9:15–20.

Mount, A. B. 1964. The interdependence of the eucalpyts and forest fires in southern Australia. Australian Forestry 28:166–172.

Newcomb, R. D., Campbell, P. M., Ollis, D. L., Cheah, E., Russell, R. J., and Oakeshott, J. G. 1997. A single amino acid substitution converts a carboxylesterase to an organophosphorous hydrolase and confers insecticide resistance on a blowfly. Proceedings of the National Academy of Sciences USA 94:7464–7468.

Odling-Smee, F. J. 1988. Niche-constructing phenotypes. In H. C. Plotkin (ed.), The Role of Behavior in Evolution (pp. 73–132). Cambridge, MA: MIT Press.

Oyama, S. 1985/2000. The Ontogeny of Information, 2nd revised ed. Durham, NC: Duke University Press.

Oyama, S. 1992. Ontogeny and phylogeny: a case of metarecapitulation? In P. E. Griffiths (ed.), Trees of Life: Essays in Philosophy of Biology (pp. 211–240). Dordrecht: Kluwer.

Oyama, S., Griffiths, P. E., and Gray, R. D. 2001. Introduction: What is developmental systems theory? In S. Oyama, P. E. Griffiths, and R. D. Gray (eds.), Cycle of Contingency: Developmental Systems and Evolution. Cambridge, MA: MIT Press.

Pittendrigh, C. S. 1958. Adaptation, natural selection and behavior. In A. Roe and G. Simpson (eds.), Behavior and Evolution (pp. 390–416). New York: Academic Press.

Raff, R. 1996. The Shape of Life: Genes, Development and the Evolution of Animal Form. Chicago: University of Chicago Press.

Ratnieks, F. L. W., and Visscher, P. K. 1989. Worker policing in honeybees. Nature 342:796–797.

Robert, J. S., Hall, B. K., and Olson, W. M. 2001. Bridging the gap between developmental systems theory and evolutionary developmental biology. BioEssays 23:954–962.

Rosenberg, A. 1978. The supervenience of biological concepts. Philosophy of Science 45:368–386.

Schlichting, C. D., and Pigliucci, M. 1998. Phenotypic Evolution: A Reaction Norm Perspective. Sunderland, MA: Sinauer.

Shannon, C. E., and Weaver, W. 1949. The Mathematical Theory of Communication. Urbana, IL: University of Illinois Press.

Sober, E., and Wilson, D. S. 1998. Unto Others: The Evolution and Psychology of Unselfish Behavior. Cambridge, MA: Harvard University Press.

Sterelny, K. 1996. Explanatory pluralism in evolutionary biology. Biology and Philosophy 11:193–214.

Sterelny, K. 2000. Development, evolution, and adaptation. Philosophy of Science 67(Suppl.):S369–S387.

Sterelny, K. 2001. Niche construction, developmental systems and the extended replicator. In S. Oyama, P. E. Griffiths, and R. D. Gray (eds.), Cycles of Contingency: Developmental Systems and Evolution (pp. 333–349). Cambridge, MA: MIT Press.

Sterelny, K., and Griffiths, P. E. 1999. Sex and Death: An Introduction to the Philosophy of Biology. Chicago: University of Chicago Press.

Sterelny, K., and Kitcher, P. 1988. The return of the gene. Journal of Philosophy 85:339–361.

Sterelny, K., Dickison, M., and Smith, K. 1996. The extended replicator. Biology and Philosophy 11:377–403.

Sultan, S. 1992. Phenotypic plasticity and the neo-Darwinian legacy. Evolutionary Trends in Plants 6:61–70.

Wagner, G. P. (ed.). 2001. The Character Concept in Evolutionary Biology. San Diego: Academic Press.

Wagner, G. P., and Altenberg, L. 1996. Complex adaptations and the evolution of evolvability. Evolution 50:967–976.

Whitehead, H. 1998. Cultural selection and genetic diversity in matrilineal whales. Science 282:1708–1711.

Williams, G. C. 1966. Adaptation and Natural Selection. Princeton, NJ: Princeton University Press.

Wilson, D. S. 1997. Biological communities as functionally organized units. Ecology 78:2018–2024.

Wilson, D. S., and Sober, E. 1994. A critical review of philosophical work on the units of selection problem. Philosophy of Science 61:534–555.

# Conclusion

The breadth of subjects covered in this volume should not be surprising, given that phenotypic complexity is a ubiquitous feature of living systems. To study evolution in complex phenotypes is, in some ways, simply to study organismal processes broadly construed. Yet the converse is only sometimes true: research into even the most complex and dynamic organismal processes need not concern itself with functional integration and its evolution. Despite this asymmetry at the level of principle, however, in practice recent biological research from across the spectrum has contributed fundamentally to dramatic growth in integration research. We have tried to reflect that diversity of influences in the chapters collected here.

The field of integration studies is not young (see Introduction and Chapter 1), but it appears to be in a period of rejuvenation marked by extremely fluid thinking about a variety of central questions. Accordingly, this volume delivers no single message; instead, we have tried to provide a representative sample of some of the best new work on phenotypic complexity and evolution. Still, among the variety of subfields, approaches, and opinions included here, a few major themes resound repeatedly throughout. One common thread traces the consequences of interactions among levels of biological organization, from genes to clades. A second, and related, theme is the dependence of the phenomena of integration on their contexts—including genetic, phenotypic, and phylogenetic contexts. Finally, nearly all the contributors to this volume present new ways of thinking about specific problems in their own fields by making new connections with other fields. We consider each of these themes in turn.

## Levels of Biological Organization

Much of the complexity in biological systems arises from their tendency to be organized into levels. Phylogenetic relationships among lineages show this structure very clearly as a series of nested hierarchies; and various biological processes that contribute to phenotypic complexity also operate at distinct levels of biological organization (plasticity, convergence, speciation). The array of vantage points offered by authors in different fields demonstrates that phenotypic complexity is itself multilayered. For example, the phenotypic consequences of interactions among genes (Murren and Kover) and cells (Klingenberg) are addressed in Part III on genetics and molecular biology; relationships among traits and modules play an important role in analyses of sexual selection (Badyaev), predator-induced defenses (Relyea), and floral evolution (Armbruster et al.); and patterns of trait covariation among lineages are addressed in Part IV on macroevolution and in several other chapters (e.g., Merilä and Björklund, Hansen and Houle).

Within many of the chapters, the relationship between processes operating across levels becomes an important motif. For example, interactions among levels of biological organization give rise to various phenomena such as the apparent tension between constraint and adaptation, the interplay of plasticity and integration, and the problem of stasis. The growing use of molecular phylogenics in evolutionary biology brings the hierarchy from genes to clades full circle, by providing essential tools from the molecular level for use in comparative biology and certain macroevolutionary studies. The connection between genotypic and phenotypic variation is taken up by a number of authors, and in this context, the traditional distinction between the genotype and the phenotype becomes blurred. Genetic architecture and patterns of epistasis give the genotype something akin to its own phenotype. This issue arises in Chapter 6 by Hansen and Houle, in which they distinguish between the quantity and the quality of genetic variation as the raw material for selection. Similarly, it is clear that an important characteristic of the phenotype is the way in which traits are related, not only functionally and morphologically, but also in terms of their genetic and developmental underpinnings. This kind of phenotypic substructuring along developmental genetic lines (i.e., modularity) is explored in several chapters.

## Context Dependence

It follows from the complex and hierarchical structure of biological systems that the expression of many biological phenomena (as well as our interpretation of their significance) depends on the context in which they occur. An example brought out in many chapters involves the status of phenotypic integration as adaptation or constraint. Briefly, trait integration (via genetic or developmental mechanisms) may be adaptive when it enhances organismal function or phenotypic stability within a certain environmental, ontogenetic, or phenotypic context, but it may also constrain the response to selection under a different set of conditions. Chapter 18 by Schwenk and Wagner and Chapter 5 by Merilä and

Björklund highlight the relativism of adaptation and constraint, whereas Pigliucci argues (in Chapter 11) that from some points of view, the distinction between them collapses altogether.

One practical consequence of recognizing that integration depends on the context under consideration is the need to characterize this context. Often phylogenetic distance stands in for phenotypic context, since more closely related groups generally have more characteristics in common, and patterns of trait covariation within and between taxa or populations can be compared (see Chapter 2 by Armbruster et al. and Chapter 14 by Ackermann and Cheverud). There are less hierarchically stratified contexts as well, including spatial position within an organism (Zelditch and Moscarella; Preston and Ackerly) and the functional relationships among modules (e.g., Badyaev). Quantifying these contexts, especially if they are to be comparable across taxa, remains an outstanding challenge for integration studies.

## New Connections

In his final technical work, Stephen Jay Gould (2002, p. 31) credits the current *Zeitgeist* with promoting many of the major biological advances of the last decade. A crucial element of our prevailing scientific spirit is a willingness to grapple with problems of complexity and multidirectional interactions and, as described above, the chapters in this volume certainly exemplify such a spirit. Some biologists may resist Gould's diagnosis of the current conventional wisdom, based on the thought that the real spirit of twentieth-century biological theory centers on our dramatic advances in understanding the micro-level processes that shape the form of living things—a spirit that might be summarized in a quick and crude way by reference to the aptly named "central dogma" of genetics. While these two conceptions of our field may seem to stand in tension, in fact they rest on a single, deeper thought. It is exactly because our understanding of the micro-level causal processes is so much more precise, that we are now able to see what is explained directly by them, and what is not. The path is therefore now open toward a more detailed picture, which apportions to each level of biological organization its share of causal responsibility for the phenomena of life revealed to empirical study. Thus, another crucial facet of the *Zeitgeist* identified by Gould is a general willingness to forge connections among different disciplines from the biological sciences, within an integrated approach to understanding complex systems. This attitude also runs throughout the book, and in many ways is emblematic of the entire field of integration studies.

As we emphasized above, phenotypic integration and its evolution may be studied at many levels of biological organization. Thus, some of the cross-field connections that have developed result from the simple fact that scientists from many different subfields have converged on the quest to understand phenotypic complexity. In other cases, researchers have incorporated tools or methods from disciplines more peripheral to integration studies. Examples from this volume range widely, from the use of model systems, mutagenesis, and other techniques derived from molecular genetics, through statistics, analytical tools, and computer simulations, to developmental biology, comparative phylogenetics, and philosophy of science. A consequence of all these new and fruitful relationships among different lines of research is that we are now redrawing the boundaries that have traditionally defined separate subfields. In some ways, this process is similar to the reorganization of modular structure within a lineage in response to selection. It is heartening to see that the field of integration studies is thriving in the environment created by the current *Zeitgeist*.

Book chapters in a volume such as this typically strive to cover a topic or line of research by presenting a broad, synthetic vision of its current state, and to provide positive suggestions for new work. Such a strategy almost automatically builds in a certain amount of hopeful anticipation about the possibilities for development in the field. Even discounting for that effect, however, the degree of optimism expressed by the authors included here is striking. Moreover, this optimism not only extends to future research within their subfields, but also encompasses innovative use of tools drawn from other areas of biology. In our view, this high level of enthusiasm reflects genuine and important underlying realities about the state of evolutionary biology. In recent decades, a number of new techniques and concepts have coalesced in a way that holds promise for significant advancement along multiple paths. Studies of integration are particularly well positioned to benefit from such an interconnected approach, and it was with this hope in mind that we worked to bring together a variety of different, emerging perspectives on the integrated phenotype in a single volume. In the event, it has been extremely rewarding for us to learn from these different viewpoints, and we are genuinely grateful to all the authors for their contributions.

*Literature Cited*

Gould, S. J. 2002. The Structure of Evolutionary Theory. Belknap Press, Cambridge, MA.

# Index